METHODS IN MOLECULAR BIOLOGY

Series Editor
John M. Walker
School of Life and Medical Sciences
University of Hertfordshire
Hatfield, Hertfordshire, AL10 9AB, UK

For further volumes:
http://www.springer.com/series/7651

Protein-Protein Interactions

Methods and Applications

Second Edition

Edited by

Cheryl L. Meyerkord

Department of Pharmacology and Emory Chemical Biology Discovery Center,
Emory University School of Medicine, Atlanta, GA, USA

Haian Fu

Department of Pharmacology, Department of Hematology & Medical
Oncology, and Emory Chemical Biology Discovery Center, Emory University
School of Medicine, Atlanta, GA, USA

☀ Humana Press

Editors
Cheryl L. Meyerkord
Department of Pharmacology
  and Emory Chemical Biology
  Discovery Center, Emory University
  School of Medicine
Atlanta, GA, USA

Haian Fu
Department of Pharmacology
  Department of Hematology & Medical
  Oncology, and Emory Chemical Biology
  Discovery Center, Emory University
  School of Medicine
Atlanta, GA, USA

ISSN 1064-3745   ISSN 1940-6029   (electronic)
Methods in Molecular Biology
ISBN 978-1-4939-2424-0   ISBN 978-1-4939-2425-7   (eBook)
DOI 10.1007/978-1-4939-2425-7

Library of Congress Control Number: 2015933374

Springer New York Heidelberg Dordrecht London

Printed on acid-free paper

Humana Press is a brand of Springer
Springer Science+Business Media LLC New York is part of Springer Science+Business Media (www.springer.com)

Preface

From regulation of DNA replication, RNA transcription, and protein translation to posttranslational modifications and signal transduction, protein-protein interactions are critical for most biological functions. In recent years, our understanding of the protein-protein interaction landscape has significantly increased due to advances in genomics, proteomics, and functional and network biology. It is also recognized that dysregulated protein-protein interactions or aggregations form the molecular basis for a large number of human diseases ranging from cancer to neurodegenerative disorders. Thus, protein-protein interactions may represent a promising class of targets for therapeutic intervention. The emerging prominence of therapeutic discovery has ignited a rising interest in the analysis of protein-protein interactions, the development of various assays for monitoring protein-protein interactions, and the identification of small molecule protein-protein interaction modulators. Targeting protein-protein interactions can be challenging and demands innovative technological platforms. Such techniques promise to accelerate future advances in our understanding of protein-protein interactions, will help to better define the role of these interactions in complex biological systems, and will aid in the identification of small molecule modulators for targeting protein-protein interactions.

The growing interest in interrogating protein-protein interactions for mechanistic studies and potential therapeutic discovery, coupled with new technological advances, prompted us to update and expand the scope of *Protein-Protein Interactions: Methods and Applications*. In this new edition, we have retained the core technological platforms that are classical cornerstones used to study protein-protein interactions and added new chapters for cutting-edge technologies that reflect recent scientific advances and the emerging focus on therapeutic discovery with miniaturized protein-protein interaction detection platforms. The second edition of *Protein-Protein Interactions: Methods and Applications* covers a wide range of protein-protein interaction detection topics that are separated into five parts. The book begins with overview chapters that describe the fundamental principles of protein-protein interactions, available protein-protein interaction databases, methods for quantitative and computational analysis of protein-protein interactions, and targeting protein-protein interactions for small molecule modulator identification and drug discovery. Part II details methodologies involving label-free biosensors and techniques, which allow researchers to analyze protein-protein interactions without the requirement for traditional radioisotope or fluorophore labeling of test proteins. Part III includes tag- and antibody-based methods that offer the benefit of specific and sensitive detection of a target protein and its interacting partners. Part IV is a compilation of cell-based biomolecular interaction reporter assays, which provide a physiologically relevant context and in some cases allow for the analysis of protein-protein interactions in live cells. The final section consists of case studies describing protein-protein interaction analysis platforms that are applicable in a miniaturized format for high-throughput screening. This section provides examples of methods that have been effectively used to analyze the modulation of protein-protein interactions by various small molecules as well as examples of biophysical methods that can be used for identification of fragment-based inhibitors of protein-protein interactions.

Our hope is that this book will serve as a valuable resource for all readers, from students to experienced scientists. In the chapters of this book, experienced researchers describe in detail the underlying theory and practical application of widely used methods for monitoring protein-protein interactions. In keeping with the series tradition, the Notes section of the chapters contains valuable explanations to sensitive procedures and potential pitfalls, with tips on how to avoid them. This book is expected to empower readers in their quest to elucidate the mechanisms of protein-protein interactions and the role of these interactions in diverse biological processes, and to target protein-protein interactions for therapeutic discovery.

We are grateful to the contributors of each chapter for their effort, commitment, and willingness to share their knowledge and valuable experience with the scientific community. We sincerely thank the series editor, John Walker, for all of his guidance. Together, we present an extensively updated and expanded version of this valuable resource to all those who study protein-protein interactions.

Atlanta, GA, USA   *Cheryl L. Meyerkord*
Atlanta, GA, USA   *Haian Fu*

Contents

Contributors

FABRICE AGOU • *Département de Biologie Cellulaire et Infection, Institut Pasteur, Unité de Signalisation et Pathogenèse, Paris, France*

SYED R. ALI • *Department of Pharmacology and Toxicology, Pharmacology and Toxicology Graduate Program, The University of Texas Medical Branch, Galveston, TX, USA*

MICHELLE R. ARKIN • *Small Molecule Discovery Center, Department of Pharmaceutical Chemistry, University of California San Francisco, San Francisco, CA, USA; UCSF, San Francisco, CA, USA*

YAN BAI • *Department of Pharmacology, Emory University School of Medicine, Atlanta, GA, USA*

MORIAH R. BECK • *Chemistry Department, Wichita State University, Wichita, KS, USA*

MATTHEW A. BENNETT • *Department of Biochemistry, Emory University School of Medicine, Atlanta, GA, USA*

DENZIL BERNARD • *Comprehensive Cancer Center and Departments of Internal Medicine, Pharmacology and Medicinal Chemistry, University of Michigan, Ann Arbor, MI, USA*

KATHERINE BETKE • *Department of Pharmacology, Vanderbilt University Medical Center, Nashville, TN, USA*

JOE B. BLUMER • *Department of Cell and Molecular Pharmacology and Experimental Therapeutics, Medical University of South Carolina, Charleston, SC, USA*

NICOLE E. BROWN • *Department of Pharmacology, Emory University School of Medicine, Atlanta, GA, USA*

SHARON L. CAMPBELL • *Department of Biochemistry and Biophysics, Lineberger Cancer Center, University of North Carolina, Chapel Hill, NC, USA*

LEWIS C. CANTLEY • *Division of Signal Transduction, Department of Medicine, Beth Israel Deaconess Medical Center, Harvard Medical School, Boston, MA, USA; Weill Cornell Cancer Center, Weill Cornell Medical College, New York, NY, USA*

MICHAEL S. CHIMENTI • *Department of Pharmaceutical Chemistry, University of California San Francisco, San Francisco, CA, USA*

YUHONG DU • *Department of Pharmacology, Emory Chemical Biology Discover Center, Emory University School of Medicine, Atlanta, GA, USA*

JOHN F. ECCLESTON • *Division of Physical Chemistry, MRC National Institute for Medical Research, London, UK*

TOBIAS EHRENBERGER • *Department of Biological, Massachusetts Institute of Technology, Cambridge, MA, USA*

ZIAD M. ELETR • *Department of Biochemistry, Emory University School of Medicine, Atlanta, GA, USA*

ERNESTO FREIRE • *Department of Biology, Johns Hopkins University, Baltimore, MD, USA*

DAVID C. FRY • *Roche Research Center, Nutley, NJ, USA*

HAIAN FU • *Departments of Pharmacology, Hematology and Medical Oncology, Emory University School of Medicine, Atlanta, GA, USA; Emory Chemical Biology Discovery Center, Emory University, Atlanta, GA, USA*

NORMA J. GREENFIELD • *Associate Professor (retired), Department of Neuroscience and Cell Biology, Robert Wood Johnson Medical School, Rutgers University, Piscataway, NJ, USA*

KUN-LIANG GUAN • *Department of Pharmacology and Moores Cancer Center, University of California San Diego, La Jolla, CA, USA*

JILLIAN R. GUNTHER • *Department of Chemistry, University of Illinois at Urbana-Champaign, Urbana, IL, USA; Division of Radiation Oncology, University of Texas MD Anderson Cancer Center, Houston, TX, USA*

RANDY A. HALL • *Department of Pharmacology, Emory University School of Medicine, Atlanta, GA, USA*

HEIDI HAMM • *Department of Pharmacology, Vanderbilt University Medical Center, Nashville, TN, USA*

GABRIELLA T. HELLER • *Department of Chemistry, Pomona College, Claremont, CA, USA*

JOHN R. HEPLER • *Department of Pharmacology, Emory University School of Medicine, Atlanta, GA, USA*

ANTHONY A. HIGH • *St. Jude Proteomics Facility, St. Jude Children's Research Hospital, Memphis, TN, USA*

YUN HUA • *Department of Pharmaceutical Sciences, School of Pharmacy, University of Pittsburgh, Pittsburg, PA, USA*

LARS JUHL JENSEN • *Faculty of Health Sciences, Novo Nordisk Foundation Centre for Protein Research, University of Copenhagen, Copenhagen, Denmark*

MALKIAT S. JOHAL • *Department of Chemistry, Pomona College, Claremont, CA, USA*

PAUL A. JOHNSTON • *Department of Pharmaceutical Sciences, School of Pharmacy, University of Pittsburgh Cancer Institute, Pittsburg, PA, USA*

RICHARD A. KAHN • *Department of Biochemistry, Emory University School of Medicine, Atlanta, GA, USA*

JOHN A. KATZENELLENBOGEN • *Department of Chemistry, University of Illinois at Urbana-Champaign, Urbana, IL, USA*

KANISHA KAVDIA • *St. Jude Proteomics Facility, St. Jude Children's Research Hospital, Memphis, TN, USA*

MARK J.S. KELLY • *Department of Pharmaceutical Chemistry, University of California San Francisco, San Francisco, CA, USA*

KIRAN KODALI • *St. Jude Proteomics Facility, St. Jude Children's Research Hospital, Memphis, TN, USA*

SRIRAM KUMARASWAMY • *ForteBio Inc.—A Division of Pall Life Sciences, Menlo Park, CA, USA*

FERNANDA LAEZZA • *Department of Pharmacology and Toxicology, Center for Addiction Research, Mitchell Center for Neurodegenerative Diseases, Center for Biomedical Engineering, The University of Texas Medical Branch, Galveston, TX, USA*

ROB LAVIGNE • *Laboratory of Gene Technology, Katholieke Uniersiteit Leuven, Leuven, Belgium*

STEPHANIE A. LEAVITT • *Gilead Sciences, Inc., Foster City, CA, USA*

IRMA LEMMENS • *Department of Medical Protein Research, VIB, Ghent, Belgium; Department of Biochemistry, Faculty of Medicine and Health Sciences, Ghent University, Ghent, Belgium*

MIN LI • *The Solomon H. Snyder Department of Neuroscience, Johns Hopkins Ion Channel Center and High Throughput Biology Center, School of Medicine, Johns Hopkins University, Baltimore, MD, USA; GSK, King of Prussia, PA, USA*

ROBERT LIDDINGTON • *Sanford-Burnham Medical Research Institute, La Jolla, CA, USA*

SAM LIEVENS • *Department of Medical Protein Research, VIB, Ghent, Belgium; Department of Biochemistry, Faculty of Medicine and Health Sciences, Ghent University, Ghent, Belgium*

LIU LIU • *Comprehensive Cancer Center and Departments of Internal Medicine, Pharmacology and Medicinal Chemistry, University of Michigan, Ann Arbor, MI, USA*

ADAM I. MARCUS • *Department of Hematology and Medical Oncology, Winship Cancer Institute of Emory University, Atlanta, GA, USA*

ALEXA L. MATTHEYSES • *Department of Cell Biology, Emory University School of Medicine, Atlanta, GA, USA*

ALISON R. MERCER-SMITH • *Department of Chemistry, Pomona College, Claremont, CA, USA*

SAMY O. MEROUEH • *Department of Biochemistry and Molecular Biology, Center for Computational Biology and Bioinformatics, Indiana University School of Medicine, Indianapolis, IN, USA; Department of Chemistry and Chemical Biology, Indiana University Purdue University Indianapolis, Indianapolis, IN, USA*

CHERYL L. MEYERKORD • *Department of Pharmacology, Emory University School of Medicine, Atlanta, GA, USA; Emory Chemical Biology Discovery Center, Emory University, Atlanta, GA, USA*

STEPHEN W. MICHNICK • *Département de Biochimie, Université de Montréal, Montréal, QC, Canada*

ASHUTOSH MISHRA • *St. Jude Proteomics Facility, St. Jude Children's Research Hospital, Memphis, TN, USA*

TOSHIYUKI MIYASHITA • *Department of Molecular Genetics, Kitasato University School of Medicine, Sagamihara, Japan*

JEREMY MOGRIDGE • *Department of Laboratory Medicine and Pathology, University of Toronto, Toronto, ON, Canada*

TERRY W. MOORE • *Department of Chemistry, University of Illinois at Urbana-Champaign, Urbana, IL, USA; Department of Medicinal Chemistry and Pharmacognosy, University of Illinois Hospital and Health Sciences System Cancer Center, Chicago, IL, USA*

ZANETA NIKOLOVSKA-COLESKA • *Department of Pathology, University of Michigan Medical School, Ann Arbor, MI, USA*

VISHWAJEETH R. PAGALA • *St. Jude Proteomics Facility, St. Jude Children's Research Hospital, Memphis, TN, USA*

NELI I. PANOVA-ELEKTRONOVA • *Department of Pharmacology and Toxicology, The University of Texas Medical Branch, Galveston, TX, USA*

JUNMIN PENG • *St. Jude Proteomics Facility, Department of Structural Biology, Department of Neurodevelopmental Biology, St. Jude Children's Research Hospital, Memphis, TN, USA*

SAMUEL J. PFAFF • *Small Molecule Discovery Center, Department of Pharmaceutical Chemistry, University of California San Francisco, San Francisco, CA, USA*

CAU D. PHAM • *Department of Pharmacology, Emory University School of Medicine, Atlanta, GA, USA*

MARY C. PUCKETT • *Department of Pharmacology, Emory University School of Medicine, Atlanta, GA, USA*

RONALD T. RAINES • *Department of Biochemistry, University of Wisconsin-Madison, Madison, WI, USA; Department of Chemistry, University of Wisconsin-Madison, Madison, WI, USA*

INGRID REMY • *Département de Biochimie, Université de Montréal, Montréal, QC, Canada*

KATRIN RITTINGER • *Division of Molecular Structure, MRC National Institute for Medical Research, London, UK*

ALEXANDER S. SHAVKUNOV • *Department of Pharmacology and Toxicology, The University of Texas Medical Branch, Galveston, TX, USA*

JACK F. SHERN • *Department of Biochemistry, Emory University School of Medicine, Atlanta, GA, USA*

CHRISTOPHER J. STROCK • *Cyprotex US, Watertown, MA, USA*

DAMIAN SZKLARCZYK • *Faculty of Health Sciences, Novo Nordisk Foundation Centre for Protein Research, University of Copenhagen, Copenhagen, Denmark*

YOSHINORI TAKAHASHI • *Department of Pediatrics, The Pennsylvania State University College of Medicine, Hershey, PA, USA*

HAIYAN TAN • *St. Jude Proteomics Facility, St. Jude Children's Research Hospital, Memphis, TN, USA*

JAN TAVERNIER • *Department of Medical Protein Research, VIB, Ghent, Belgium; Department of Biochemistry, Faculty of Medicine and Health Sciences, Ghent University, Ghent, Belgium*

IAN A. TAYLOR • *Division of Molecular Structure, MRC National Institute for Medical Research, London, UK*

PETER M. THOMPSON • *Department of Biochemistry and Biophysics, Program in Molecular and Cellular Biophysics, University of North Carolina, Chapel Hill, NC, USA*

RENEE TOBIAS • *Forte Bio—A Division of Pall Life Sciences, Menlo Park, CA, USA*

ADRIAN VELAZQUEZ-CAMPOY • *Institute of Biocomputation and Physics of Complex Systems (BIFI), Joint Unit IQFR-CSIC-BIFI, Department of Biochemistry and Molecular and Cell Biology, Universidad de Zaragoza, Zaragoza, Spain; Fundacion ARAID, Government of Aragon, Zaragoza, Spain*

MICHEL VÉRON • *Département de Biologie Structurale et Chimie, Institut Pasteur, Unité de Biochimie Structurale et Cellulaire, Paris, France*

HARIS G. VIKIS • *Department of Pharmacology and Toxicology, MCW Cancer center, Medical College of Wisconsin, Milwaukee, WI, USA*

JEROEN WAGEMANS • *Laboratory of Gene Technology, Katholieke Uniersiteit Leuven, Leuven, Belgium*

BO WANG • *Center for Computational Biology and Bioinformatics, Indiana University School of Medicine, Indianapolis, IN, USA; Department of Chemistry and Chemical Biology, Indiana University Purdue University Indianapolis, Indianapolis, IN, USA*

SHAOMENG WANG • *Comprehensive Cancer Center and Departments of Internal Medicine, Pharmacology and Medicinal Chemistry, University of Michigan, Ann Arbor, MI, USA*

XUSHENG WANG • *St. Jude Proteomics Facility, St. Jude Children's Research Hospital, Memphis, TN, USA*

ZUSEN WENG • *State Key Laboratory of Molecular Vaccinology and Molecular Diagnostics, National Institute of Diagnostics and Vaccine Development in Infectious Diseases, School of Public Health, Xiamen University, Xiamen, P. R. China*

KEITH D. WILKINSON • *Department of Biochemistry, Emory University School of Medicine, Atlanta, GA, USA*

MENG WU • *High Throughput Screening Facility at Univ. of Iowa (UIHTS), Division of Medicinal and Natural Products Chemistry, Department of Pharmaceutical Sciences and Experimental Therapeutics, College of Pharmacy, Department of Biochemistry, Carver College of Medicine, The University of Iowa, Iowa City, IA, USA*

ZHIPING WU • *Department of Structural Biology, Department of Neurodevelopmental Biology, St. Jude Children's Research Hospital, Memphis, TN, USA*

DAVID XU • *Center for Computational Biology and Bioinformatics, Indiana University School of Medicine, Department of BioHealth Informatics, Indiana University School of Informatics, Indianapolis, IN, USA*

YANJI XU • *St. Jude Proteomics Facility, St. Jude Children's Research Hospital, Memphis, TN, USA*

MICHAEL B. YAFFE • *Department of Biology, Department of Biological Engineering, Koch Institute for Integrative Cancer Biology, Massachusetts Institute of Technology, Cambridge, MA, USA; Department of Surgery, Beth Israel Deaconess Medical Center, Harvard Medical School, Boston, MA, USA*

YUN YOUNG YIM • *Department of Pharmacology, Vanderbilt University Medical Center, Nashville, TN, USA*

QINJIAN ZHAO • *State Key Laboratory of Molecular Vaccinology and Molecular Diagnostics, National Institute of Diagnostics and Vaccine Development in Infectious Diseases, School of Public Health, Xiamen University, Xiamen, P. R. China*

Part I

Overviews

Chapter 1

Structural Basis of Protein-Protein Interactions

Robert C. Liddington

Abstract

Regulated interactions between proteins govern signaling pathways within and between cells. Structural studies on protein complexes formed reversibly and/or transiently illustrate the remarkable diversity of interactions, both in terms of interfacial size and nature. In recent years, "domain–peptide" interactions have gained much greater recognition and may be viewed as both pre-translational and posttranslational-dependent functional switches. Our understanding of the multistep regulation of auto-inhibited multi-domain proteins has also grown. Their activity may be understood as the "combinatorial" output of multiple input signals, including phosphorylation, location, and mechanical force. The prospects for bridging the gap between the new "systems biology" data and the traditional "reductionist" data are also discussed.

Key words Structure, Complex, Energetics, Crystallography, Allostery, Regulation, Force, Domain–peptide

1 Introduction

Proteins live, work, and die in a highly crowded environment and must find their cognate partner(s) in a vast sea of non-partners. But all soluble proteins tend to look rather similar: their surfaces are covered by a mix of hydrophobic and hydrophilic residues, the latter in greater abundance, as well as numerous backbone atoms with unsatisfied H-bonding potential. Recent studies have considered how the remarkable feats of co-localization, recognition, and specificity are achieved [1, 2].

But first I review what we know and do not know about productive protein-protein interactions. The obvious starting point is the weekly updated library of crystal structures contained within the Protein Data Bank (PDB) (http://www.rcsb.org/pdb/), which now number in the tens of thousands. In addition, many searchable databases of protein complexes are available. For example, two well-curated sites reported ~9,000 distinct classes of interfaces involving ~5,000 Pfam [3] domains, as of mid-2013 [4]. These numbers may be impressive, but they still represent only a

Cheryl L. Meyerkord and Haian Fu (eds.), *Protein-Protein Interactions: Methods and Applications*, Methods in Molecular Biology, vol. 1278, DOI 10.1007/978-1-4939-2425-7_1, © Springer Science+Business Media New York 2015

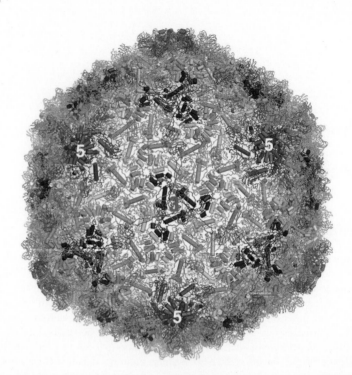

Fig. 1 Atomic structure from single-particle electron microscopy. Structure of the bacteriophage epsilon15 determined directly from cryo-EM images at a resolution of 3–5 Å [92]. There are 14 protein chains (3,122 residues) comprising the asymmetric unit particles (*color-coded*). The other subunits (total of 420 asymmetric units) are generated by the $T = 7$ icosahedral symmetry. View is down one of the threefold axes of the icosahedral particle. Subunits on fivefold axes are *colored red* (labeled *white*)

small subset of the complete human binary protein-protein interactions (the "interactome") which has been estimated at ~500,000 [5, 6].

Crystallography remains the major tool for determining the binary interactome: protein complexes at atomic resolution, with the upper limit defined by rigidity rather than size: the structure of the ribosome may be the crowning achievement for a *relatively* static, asymmetric, structure [7], while the organization of viral capsids captures some of the complexity and much of the beauty of protein-protein interactions. Advances in electron microscopy have recently enabled atomic models to be built directly into EM maps, but only in the case of highly symmetric viruses [8] (Fig. 1). NMR has typically been limited to protein of ≤30 kDa, but it plays an important role in defining interacting surfaces, especially at the membrane (*see* Chapter 16).

Automated crystallization pipelines developed over the past 15 years have greatly increased the database of known structures, extending the coverage of protein "fold-space," as well as pointing to its boundaries [9]. But our knowledge of protein "interaction-space" has not kept pace: crystallizing complexes has not lent itself well to automation, neither has the crystallization of large multidomain proteins that are typical of eukaryotic signaling pathways. The genome-wide studies have given us the first "wiring diagrams" of the vast protein interaction landscape; but by themselves, they tell us little about how a pathway (or cell) works. We still need to understand, at the atomic, energetic, and kinetic levels, the *how*, *where*, and *when* of protein interactions.

So what are the prospects for bridging the gap between the "systems biology" data (comprehensive but lacking mechanism) and the "reductionist" data (atomistic/mechanistic but lacking in quantity)? There are reasons to be optimistic. First, the growth in the number of homologous protein *superfamilies* in the CATH database [10] has nearly plateaued in recent years and is not expected to exceed 3,000 [11], suggesting that protein "fold-space" is finite and of a manageable size. And there are also indications that "interaction-space" may be similarly restricted.

How far can we extrapolate from what we do know, i.e., by building homology models based on known complexes? Some reports have suggested that suitable "docking templates" (derived from diverse organisms) can be found for modeling *all* binary protein complexes in the human interactome [12–15]. However, by the authors' own admission, only ~30 % of these models (rising to ~50 % for homodimers [11]) are of "good" quality, i.e., partly correct, partly wrong, but good enough to warrant experimental verification/refinement. This is not surprising, as earlier studies showed that when sequence identity with the template falls below ~30 %, predictive power becomes very low [16].

And if we are given the structures of two proteins known to interact, but no template, can we predict how they will do so with atomic precision? Well, sometimes yes, sometimes no. While much effort has been expended over the past 10 years into testing and evaluating ab initio docking algorithms [17], there is a problem. In short, proteins are never truly rigid bodies, and they do not operate in a vacuum. In the minimalist case, main chains make minor adjustments (1–2 Å), interfacial residues alter their side-chain torsion angles, and surface-bound water is released or reorganized upon complex formation. These are considered "rigid-body" complexes; and, in fact, modern algorithms and computing power can now predict these reasonably well [18]; *see* Chapter 4. In contrast, ab initio predictions nearly always fail when complex formation triggers large conformational changes in either or both proteins [17, 19, 20]; but these are the complexes that provide the most biological insight into signaling pathways.

And moving beyond the binary interactome, how can we tackle those large, often transient, multiprotein signaling hubs? They do not crystallize, so in order to build atomic models, we need to combine crystallography with other methods. Principal among these complementary methods is electron microscopy, which is actually optimal for larger particles (>500 kDa) and can provide "sub-molecular" (5–15 Å resolution) maps or envelopes [21], into which atomic models can be fitted with high overall precision in the case of rigid-body docking, or when quaternary reorganization of domains occur [22]. Another major technique is Small-Angle X-ray Scattering (SAXS), which is performed in solution and can discriminate (with high precision) between competing models of quaternary organization [23]. Many other complementary techniques are discussed in Parts II, III, and IV of this book.

In this chapter, I begin by classifying and describing the properties of binary complexes, and then review the regulatory mechanisms of binary interactions as well as large multidomain signaling proteins. While the diversity of interactions may be finite, we should continue to expect surprises. For example, over the past 10 years, the profound role of "domain–peptide" interactions in higher eukaryotes (arising from "pre-translational modification") has been established [24]; and giant steps have been made in defining the roles of mechanical force, which can be considered as an additional (and often decisive) allosteric effector in many signaling pathways [25]. Clearly, much remains to be learned.

2   Classification of Binary Protein Complexes

Interacting domain-pairs may be classified as *reversible* and *irreversible* (a very similar classification found in the literature is "transient" and "obligate"). *Irreversible* complexes are typified by large machines such as the proteasome or ribosome. They have high pairwise and/or collective affinities, and remain associated until they are damaged, or cells have no further use for them. By contrast, *reversible* complexes display a spectrum of affinities and lifetimes: (1) Some complexes come together weakly and assemble transiently; and some of these may proceed to form parts of a larger multiprotein complex that localizes to a specific cellular location and is stabilized by multiple interactions (2) Other *reversible* signaling complexes have a high pairwise affinity, but only for a limited time. For example, kinase-mediated phosphorylation on tyrosine may generate a strong bond with a partner; but, sooner or later, a cognate phosphatase will remove the phosphate group, and the complex will disassemble [26]. (3) Some complexes are strong but have built-in timers, such as the G protein complexes; their active lifetime is determined by the rates of binding and hydrolysis of GTP [27]. (4) At the far end of the *reversible* spectrum are antibody–antigen and protease–inhibitor complexes, which may have lifetimes of hours to days; some of these may be classified as *irreversible*, depending on the timescale of the biological process being studied [28].

An independent classification distinguishes "*domain–domain*" and "*domain–peptide*" complexes [29, 30]. In the former case, both partners are pre-folded; in the latter, one component is a pre-folded domain, while the "peptide" is a short linear motif (appended to a different folded domain) that folds only when it binds to its cognate domain. These types of interaction are a vital element of *reversible* signaling in higher eukaryotes (see below). Short motifs may also be observed as flexible, dynamic recognition elements within the context of a larger *irreversible* complex such as the ribosome [31].

There is also an intriguing class of protein that lacks any inherent three-dimensional structure, at least when purified to homogeneity in a test tube, and yet functions as a signaling molecule in cells by acting as a flexible linker, or wrapping around other proteins in order to display its recognition motifs (reviewed in refs. [32–34]).

3 The Architecture and Energetics of Domain–Domain Interfaces

A number of general principles have arisen from the study of known complexes, which have been refined and extended in recent years (reviewed in refs. [35–37]).

3.1 The Standard Patch

Most domain–domain interfaces have a typical layout, which defines a standard "patch" (Fig. 2a). A central solvent-excluded region ("*core*") is surrounded by a partly buried outer ring ("*rim*") that includes water-mediated interactions. The *core*, on average,

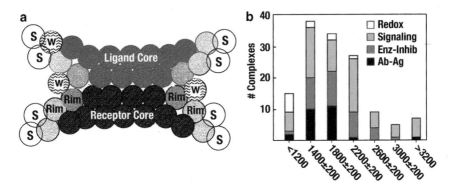

Fig. 2 The standard patch and interface size. (**a**) Slice through an ideal domain–domain interfacial "*patch*" (side view). All the atoms that lose full or partial accessibility to solvent when ligand binds are *interface* atoms. *Dark-shaded* atoms, which were solvent-exposed but become fully buried constitute the *core*, which is surrounded by a partly buried *rim* that forms a water-tight "O-ring." "Hot spot" residues typically comprise a subset of the *core* (see text). (**b**) Distribution of buried surface area ($Å^2$) among a selection of mostly reversible heterodimeric complexes, including transient redox complexes, signaling, enzyme–inhibitor, and antibody–antigen complexes. Interfaces larger than 2,000 Å generally comprise more than one *patch*. Adapted from ref. 35

contributes ~75 % of the buried surface area and the majority of the binding energy. It is typically more hydrophobic than the surface in general but less than the interior of the protein. The *rim* has a similar composition to the protein surface; it contributes ~25 % of the buried surface, and its major role is thought to be the exclusion of water molecules from the core, analogous to the "O-ring" model proposed by Bogan and Thorn [38]. Good shape and charge complementarity is important, such that the packing density is not very different from the interior of the protein. Most interfaces are rather flat, with a typical RMS deviation from planarity of ~2 Å. The interfacial "patch" varies in size, but most are in the range of 1,200–2,000 Å^2 (Fig. 2b). Interfaces larger than this typically comprise two or more patches.

Bringing two hydrophobic surfaces into close apposition in water is generally favorable (the "hydrophobic effect"), although its physical basis is still debated (*see* Chapters 2, 4, and 7). An entropic origin, arising from the release of water molecules into bulk solvent, is often cited; but careful thermodynamic measurements of mutated interfaces point to an enthalpic driving force; moreover, the free energy change ($\Delta\Delta G$) scales linearly with the change in buried surface, ~20 cal/Å^2 at the *rim* and ~45 cal/Å^2 at the center of the *core* [39]. A minimal interface of 1,200 Å^2 would therefore equate to a free energy of ~4 kcal/mol or a K_d ~ 1 μM. For a 2,000 Å^2 interface, this estimate rises to K_d ~ 10 nM. Both values are commensurate with the local concentrations of interacting proteins in vivo.

A priori, the energetics of polar interactions at interfaces are much harder to assess, because complex formation involves the displacement of *bound* water molecules from both sides of the interface (gain in entropy, loss in enthalpy), which must be compensated by a new set of ionic, polar, and van der Waals contacts between the protein partners (gain in enthalpy, loss in entropy), so that the net energetic effect of polar interactions may be positive, negative, or negligible. However, a simplifying feature may be the presence of energetic "hot spots" (see below).

4   Residue Usage at Interfaces

As the number and type of interfaces found in the PDB have increased over the past 20 years, it has been recognized that amino acid (residue) composition is systematically different in different types of complex, e.g., between *irreversible* and *reversible* complexes; and between *homodimeric* and *heterodimeric* complexes [40]. This probably explains many apparently conflicting accounts, especially in the earlier literature, which were based on smaller "mixed" data sets. And to further confound comparative studies, there are at least four distinct ways of describing interfaces, as enumerated below. Note that while this discussion focuses on side chains, it should be borne in mind that about one-third of all interfacial H-bonds are mediated by backbone atoms.

4.1 Residue Count

Perhaps the simplest method of describing an interface is to count the number of residues of a given type, and express them as a percentage of the total. Sometimes, residues counts are weighted by their size (i.e., their $Å^2$ contribution to the buried surface). Recently, Levy and colleagues collated a comprehensive "mixed" set of protein-protein interfaces from human, yeast, and *E. coli* [41]. Figure 3a plots, for human complexes, the relative incidence

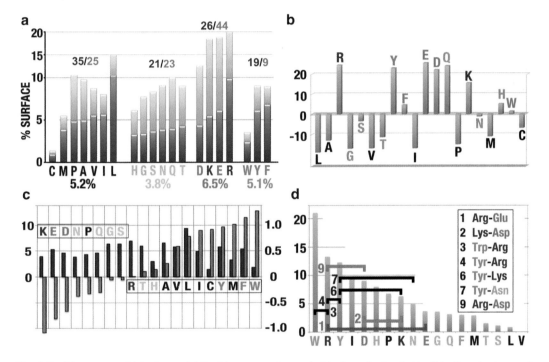

Fig. 3 Residue usage at interfaces. (**a**) Residues are color-coded by type: hydrophobic (*black*); polar (*orange*); charged (positive = *blue*, negative = *red*); or aromatic (*green*), and sorted within their groups by contribution to interface (*blue bars*) and surface (*gray bars* (stacked)). Average interface incidence by type is given as % below residues. Values above bars are total contributions for each type. Plot derived from data on human complexes [41]. (**b**) The difference in residue incidence between *reversible* and *irreversible* heterodimeric complexes. Bars above zero indicate an increase in residue usage at *reversible* interfaces. An enrichment in charged and aromatic residues at the expense of hydrophobic and small polar residues is evident. Plot derived from data in ref. [40]. (**c**) Residues ordered by increasing "Residue Propensity," p (*red bars*; scale on right ordinate), where $p = \ln (f_i/f_s)$ (see main text). *Blue bars* (scale on left ordinate) show the actual incidence (%) of residues at the core. Plot derived from data in ref. [41]. (**d**) Residues ranked in decreasing order of their energetic contribution to hot spots (*vertical bars*), based on a data set of >2,000 alanine mutations [38]. Also indicated, as connecting brackets (*numbers* refer to *boxed inset*), are the most common pairwise interactions reported for a large set of reversible heterodimeric interfaces, ranked by incidence at the *core* [44]. Refer to Table 1 for further information

of each residue, grouped by type in increasing order of their appearance at the interface as well as their overall occurrence on the surface.

Leu is the most prevalent residue at interfaces in general, followed, perhaps surprisingly, by Arg: its special role is described in more detail below. Charged residues are more prevalent than polar residues, and both are more abundant on the surface in general (except for Arg and His). The aromatics have a low abundance on surfaces, but a high (except for Trp) abundance at interfaces. Hydrophobic residues have a generally high incidence at interfaces, and a lower incidence on surfaces. Cys is particularly rare on surfaces and interfaces.

Figure 3b illustrates the difference in residue incidence between *reversible* and *irreversible* heterodimeric complexes, based on data from Ofran and Rost [40]. The differences are modest, but a clear and logical pattern emerges: *reversible* interfaces are enriched in all four charged residues, as well as the polar Gln and the aromatics, especially Tyr, while they are depleted in all hydrophobic and smaller polar residues. A similar comparison for homodimers shows additional changes in the aromatic content (not shown); but reports from different authors vary, presumably because the data sets and criteria for defining reversibility differ [40, 42].

4.2 Interface Propensity

A popular measure of interface composition is the "interface propensity," p, which answers the question, "Given the presence of a particular residue on the protein surface, what is its a priori probability of being found at an interface?" (Fig. 3c). If f_i and f_s are the probabilities of finding a given residue at the interface versus elsewhere on the surface, then $p = \ln(f_i/f_s)$. For example, aromatic residues have high p values: they are poorly represented on surfaces in general; but if you observe a Phe, Tyr, or Trp on the surface, then there is a high a priori probability that it will be involved in a protein-protein interaction. Note that the ratio f_i/f_s defines only the relative probability for a given residue and does not consider its absolute abundance on surfaces. Thus, charged residues have low p values, but are abundant at both surfaces and interfaces. So that while interface *cores* are highly "enriched" in aromatic residues, you are still more likely to find one of Asp, Glu, or Lys in the *core* as one of Phe, Tyr, or Trp.

4.3 Residue Pairs

A third method focuses on the frequency with which specific *pairs* of residues interact *across* interfaces [43]. In a recent study of a diverse set of *reversible* heterodimeric complexes [44], with interfaces ranging from 700 to 8,500 Å^2 and an average size of 2,100 Å^2, the results are quite striking. The authors also divided residue distribution into categories of total interface and *core*, making the results more directly comparable with the approaches above. The ten most abundant interactions are listed in Table 1 and

Table 1
Common pairwise interactions at interfaces

| Rank | | Residue | |
Core	Interface	pair	Interaction type(s)
1	1	Arg–Glu	Mostly salt bridges
2	3	Lys–Asp	Mostly salt bridges
3	9	Trp–Arg	Π–cation (75 %)
4	5	Tyr–Arg	Π–cation (40 %); s/c H-bond 40 %
5	–	Phe–Leu	Hydrophobic s/c packing
6	7	Tyr–Lys	s/c H-bond (55 %); s/c m/c (33 %)
7	6	Tyr–Asn	Π–cation (15 %); s/c H-bond (55 %)
8	–	Trp–Ile	Hydrophobic s/c packing
9	2	Arg–Asp	Mostly salt bridges
–	4	Lys–Glu	Mostly salt bridges

The table lists the top ten most common pairwise interactions in a large collection of reversible heterodimeric complexes [44]. They are ranked according to their incidence at the *core*. Incidence at the interface is also shown. Residues are color-coded as in Fig. 3. *s/c* side chain, *m/c* main chain, *H-bond* hydrogen bond

Fig. 3d. All four charge-charge pairs are present; two of them (Arg–Glu and Lys–Asp) are highly abundant at both the *rim* and *core*, while the other two are abundant at the *rim* but not the *core*. The other six pairs involve aromatic residues: two of them involve purely hydrophobic side-chain packing (Phe–Leu and Trp–Leu), and are restricted to the *core*; the other four are polar interactions, three of which are often mediated by Π–cation bonding (see below).

4.4 Hot Spots and the Special Role of Π–Cation Interactions

A fourth descriptor of interfaces focuses on the energetics of the interaction and relies on the results of alanine scanning mutagenesis. Thus, while many mutations at the interface have a limited effect on binding ($\Delta\Delta G$), certain clusters, called "hot spots" [38], have a large/dominant effect. Hot spots are nearly always buried near the center of the *core* and well shielded from solvent. They pack against hot spots on their cognate domains, forming an intricate cooperative network of interactions. Sequence conservation (between orthologous complexes) is highest at hot spots [45].

Residue preferences at hot spots are a subset of those described in the previous section (Fig. 3d). Trp, Arg, and Tyr are the most common, accounting for more than half of the total. These three residues are versatile in being able to form hydrophobic, aromatic, and polar interactions, all of which are required to bury complementary surfaces and satisfy unmet hydrogen-bonding needs. Moreover, the polar "Π–cation" bond between Arg and either Trp or Tyr is found at more than 50 % of hot spots [46]. The guanidinium group of Arg may lie either coplanar or orthogonal to the aromatic ring. Tyr can also make Π–cation interactions with Arg, but conventional side-chain interactions are more common. By contrast, the most common residue at interfaces, Leu, is rarely found at hot spots (as judged by alanine substitution), while, curiously, Ile is abundant.

When interfaces are viewed this way, it is not surprising, perhaps, to find that algorithms can be trained rather well (using known structures, their experimentally defined hot spots, as well as evolutionary information) to recognize and predict hot spots [45]; and significant success has even been reported based on sequence information alone [47].

5   Complex Formation and Conformational Changes

A very useful starting point for probing the linkage between complex formation and conformational changes is the "benchmark" collection of high-resolution heterodimeric *reversible* complexes last updated in 2010 [48]. The set is restricted to those complexes for which structures of both isolated components are also known. A plot of the RMS change in structure of the component domains upon complex formation *versus* interface size points first to a tight cluster with differences of ~1 ± 0.5 Å; these approximate to "rigid-body" docking, and account for about half of the total set (colored red in Fig. 4a). Most of these interfaces conform to the standard "patch" model described above. Large RMS differences (3–7 Å) are almost exclusively associated with *reversible* signaling complexes, in which conformational changes, including disorder–order transitions and quaternary changes, are commonplace.

A particularly well-studied example is the interactions between the small G-proteins (GTPases) and their effectors, activators, and inhibitors (reviewed in ref. [49]). For example, guanine nucleotide exchange factors ("GEFs"), which activate small G-proteins by recognizing the inactive GDP-bound form and promoting GDP release and GTP binding, show a great variety in their structures and modes of interaction, and several distinct nucleotide exchange mechanisms (Fig. 4b). And while some GEFs are specific for a single G-protein, others are quite promiscuous. Although they all utilize at least part of one or both of the "switch" regions (regions that undergo conformational changes upon hydrolysis of GTP), different GEFs bind to different parts of the G-protein surface and display a large range of interface sizes that extend well beyond the "rigid-body" set.

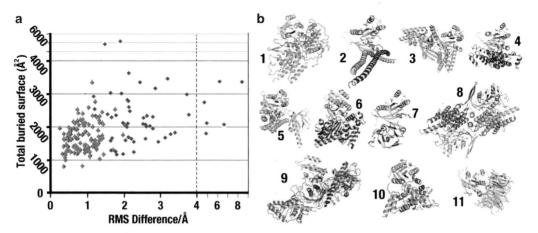

Fig. 4 Conformational changes and complex formation. (**a**) RMS change in protein structure following complex formation, as a function of buried surface area. *Red diamonds* approximate to "rigid-body" docking. *Black diamonds* are typically signaling complexes. See main text for details. Based on data from ref. 48. (**b**) The remarkable variety of interactions made between small G proteins (shown in *gray*, in approximately the same orientation) and their GTP Exchange Factors (GEFs). All GEFs recognize the GDP-bound form by binding to at least one of the switch regions of the G protein (highlighted in *red*). The complexes shown are: *1* Ras:SOS, *2* Sec4p:Sec2p, *3* Cdc42:Dock9, *4* Rab21:Rabex5, *5* Cdc42:Dbs, *6* Rab35:DENN1B, *7* Rab8:MSS4, *8* Rop4: RopGEF8, *9* Ypt1p:TRAPP1, *10* Arf1:Gea2, *11* Ran:RCC1. Adapted from ref. 49, with permission

The integrin family of αβ heterodimeric plasma membrane receptors illustrates a distinct type of protein-protein interaction, one that is mediated in part by a metal ion (Mg^{2+} or Ca^{2+}) that forms a bridge between integrin and ligand (Fig. 5). The α-subunit "I-domain" mediates binding to extracellular matrix and counter-receptors on other cells, via a "metal-ion dependent adhesion site" (MIDAS) [50]. The structure of a complex of the α2 I-domain with a fragment of triple helical collagen shows how a glutamate side chain from the collagen completes the coordination sphere of the Mg^{2+} ion [51] (Figs. 5 and 6).

The total buried surface area—1,200 Å—is near the lower limit of stable interfaces, which is even more surprising given that a substantial part of the binding energy must be used to drive conformational changes (including a 10 Å shift of the C-terminal helix), which are ultimately transduced through the plasma membrane ("outside-in" signaling [52]). Presumably, the strong bond formed between the metal ion and the ligand glutamate represents a high-energy hot spot. This mode of interaction is conserved in the structure of the αL I-domain in complex with ICAM-1 [53], and is likely conserved across the integrin family. The hot spot provides the energy for a general mode of binding, while specificity arises from additional contacts with the surrounding residues. A similar principle, called "dual recognition," rationalizes the structural basis of binding specificity between bacterial endonuclease colicins and a family of immunity proteins [54].

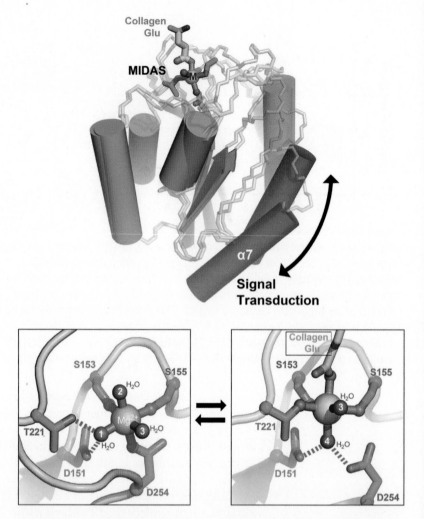

Fig. 5 Conformational changes in the α2 I domain upon binding collagen. *Upper panel* shows the major changes in the I domain, notably a large (10 Å) shift of the C-terminal helix (α7) that promotes the "open" conformation of the integrin head. *Lower panel* compares the MIDAS motif in the absence (*left*) and presence (*right*) of ligand. Ligand binding is dominated by the metal–glutamate bond, which changes the metal coordination (T221 directly coordinates; D254 loses coordination), triggering the tertiary changes seen in the *Upper panel* that underlie signal transduction (from PDB code 1DZI [51])

The β-subunit of integrins also contains an I-domain. Although it lacks detectable sequence homology with the α-I domain, it does contain a DxSxS sequence that is characteristic of the MIDAS motif, and it recognizes proteins containing the linear motif, RGD. Based on the α-I domain observation, the β-I domain fold was correctly predicted [51, 55], including the role of the RGD Asp in mediating binding to the MIDAS metal ion; and even

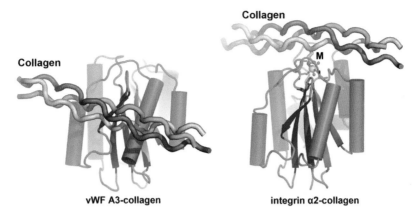

Fig. 6 Homologous domains may bind the same ligand but at different locations. Both domains are members of the same Pfam family (vWF A domain). The five central β-strands superpose with an RMS difference of 0.69 Å for 35 Cα positions. At right, the integrin α2 "I" domain engages triple-helical collagen via the MIDAS motif (labeled "M") at the "top" of the domain (*see* Fig. 5). At *left*, the von Willebrand Factor A3 domain lacks one of the metal-coordinating residues that comprise the MIDAS motif, and does not bind metal (PDB code 4DMU [58]). It also binds collagen, but on the "front" face of the A domain, and no conformational changes are observed

conformational changes within the C-terminal helix linked to signaling [56].

As a caveat, the von Willebrand Factor A1 and A3 domains have detectable (>20 %) sequence identity, as well as high structural similarity with integrin α-I domains. However, they lack one or more of the MIDAS residues required for binding metal, and ligands bind at different surfaces [57, 58] and do not trigger conformational changes (*see* Fig. 6).

6   Protein-Peptide Interactions: A Key Mediator of Complexity in Higher Eukaryotes

As noted above, recent years have witnessed an explosion in the structural characterization and prediction of protein–peptide interactions, and a comprehensive analysis has recently been performed [24]. The "peptides," typically 4–11 residues long, are often referred to as "SLiMs" (*S*hort *Li*near *M*otifs). SLiMs are widely used by higher eukaryotes to perform a variety of targeting and signaling functions [59]. The most common "receiver" domains are PDZ, PTB, SH2, and WW [60]. SLiMs adopt a variety of structures when bound to their receiver domain: an extended structure that lies in a groove (14-3-3 [61]; Rb/E7 [62]); a β-strand that augments a β-sheet (PTB [63] and PDZ [64] domains); α-helices that augment a helical bundle (FAK-paxillin [65]; β-/α-catenin [66]); or even a polyproline helix (SH3 [67]).

SLiMs most often arise by alternative gene-splicing (also by alternative promoter usage and RNA editing), typically as part of a larger *intrinsically disordered region* (IDR) that is encoded by a single (alternative) exon [68]. Several curated websites collect and classify these interactions. The Eukaryotic Linear Motif (ELM) database is a major repository: it currently contains 2,070 validated SLiMs [69], while the "3DID" identified 462 distinct classes of domain–SLiM complexes [4].

The small interfaces (average ~350 Å^2) imply that domain–SLiM complexes are inherently weak (≥ 10 µM) and *reversible*; but SLiMs and their surrounding (IDRs) are also enriched in phosphorylation sites, which make them prime candidates for regulated, dynamic assembly [70]. Moreover, SLiMs are not constrained by a preexisting 3D fold, facilitating rapid evolution [71]. Indeed, a dedicated server provides a collection of more than 700 validated examples of SLiM-based *motif switch mechanisms* that have been categorized into ten different types [72]. Despite their low pairwise binding energy, SLiMs can be highly specific in vivo. Auxiliary mechanisms presumably account for this, including avidity effects, augmentation by (phosphorylated) IDRs, and compartmentalization [73].

The interaction between the NPxY motif of integrin tails with PTB domains provides an example of *motif-switching*. The motif forms a helical turn preceded by a short β-strand that augments a β-sheet. The PTB domain of talin prefers the non-phosphorylated motif, but the "classical" PTB domain of Shc requires tyrosine phosphorylation [74]. The talin–integrin linkage promotes cell adhesion, but subsequent phosphorylation of the NPxY motif leads to a switch in recognition to Shc, resulting in a shift from cell adhesion to cell migration [75].

Translation of alternative mRNAs greatly increases the effective size of the proteome of higher eukaryotes, and facilitates the fine-tuning and diversification of signaling pathways [33]. In humans, it has been estimated that ~86 % of genes are alternatively spliced [68], often with multiple variants selected in a tissue- or developmental-dependent fashion. One example is the γ-isoform of the lipid kinase, PIPKIγ, which has at least five splice variants. One of them has a unique 28-residue C-terminal segment that binds to the talin PTB domain, targeting it to Focal Adhesions (see below) [76], while two recently described variants cause them to localize either to the nucleus or to intracellular vesicles [77].

This phenomenon of "Pre-translational modification" may resolve the conundrum of how a complex organism, such as *Homo sapiens*, can function with a "basis-set" of ~30,000 protein-encoding genes, only ~4 times the size of the genome of the unicellular budding yeast [68]. A recent study of the transcriptomes of six primate species underlines the importance of alternative splicing [78]. Thus, although a good deal of alternative exon usage is species-dependent, and probably arises from neutral drift or noisy splicing, the authors found that a "sizeable minority" of alternative exon usage, which is enriched in SLiMs and IDRs, exhibited strong tissue-dependence that is conserved from human to macaque.

Finally, the linear nature of SLiMs lends themselves to structure-based predictions. Combined structural and sequence information was used to predict models for 46 new domain–SLiM interaction classes [24]. And turning the concept around, a structure-based predictor of novel SLiM targets for PDZ domains was reported [79].

7   Intramolecular Protein-Protein Interactions and Signal Transduction

Signal transduction involves much more than the regulated, sequential association of proteins, one to the next. Multidomain proteins typically adopt a default auto-inhibited globular conformation, stabilized by head–tail interactions. They respond to multiple upstream activating signals (such as phosphorylation, co-localization of binding partners, and mechanical force); and, once a threshold has been reached, a series of increasingly activated states output distinct combinatorial signals [80–82].

Studies on Focal Adhesion Kinase (FAK), a scaffolding and signaling protein, illustrate many of these features. Focal Adhesions [83] are dynamic, multicomponent complexes that assemble around the cytoplasmic face of integrins, controlling the linkage between the Extracellular Matrix (ECM) and the actin cytoskeleton. FAK comprises FERM, kinase, and FAT (Focal Adhesion Targeting) domains, connected by flexible linker regions rich in phosphorylation sites. In the auto-inhibited state, the FERM and kinase domain pack tightly together, inhibiting the functions of the FERM domain [84], and the C-terminal FAT domain inhibits kinase activity. *Step 1*: the FAT domain recognizes paxillin (via a pair of α-helical SLiMs), close to the plasma membrane. *Step 2*: this releases some restraints on the kinase domain, which can now phosphorylate another FAK molecule at position Tyr397, but only if one is nearby (i.e., "coincidence" detection). *Step 3*: A Src-family kinase binds to pTyr397 (via its SH2 domain), and can then further phosphorylate FAK at its kinase domain. *Step 4*: This disrupts the FERM-kinase interaction, fully activating the kinase, and exposing new binding sites on the FERM domain for integrin, phospholipid, and talin that stabilize attachment to the FA. *Step 5*: Subsequent phosphorylation of the FAT domain disrupts paxillin-binding (which is no longer required for FA-targeting), enabling Grb2 binding that provides a hub for Ras/MAPK signaling (reviewed in ref. [85]).

Thus, weak binding of FAK to paxillin suggests that FAK first samples the membrane environment until it finds clusters of paxillin molecules (which occur at nascent FAs), where other FAK molecules are likely to be bound, at least transiently. Auto-phosphorylation *in trans* thus results from "coincidence-detection." Further phosphorylation of FAK then proceeds in an orderly temporal progression, in which new binding surfaces are exposed, new bonds to the membrane are made, and domains switch their function. Interestingly, in neurons, an alternatively spliced variant of FAK is found with a 7-residue insertion adjacent to Tyr 397. This enables FAK to autophosphorylate *in cis*, eliminating the need for coincidence detection and implying a different kind of regulation and function [86].

8   Force and the Single Molecule

Many intracellular signaling/scaffolding molecules switch from a compact auto-inhibited form to an extended, open or "activated" form. Now, if the molecule is anchored at two points—one fixed, the other mobile—then mechanical force applied in the appropriate direction should, in principle, (help to) open up and activate the molecule. This force–function relationship has been demonstrated in spectacular fashion for talin, which unfolds, reversibly, by up to 3,000 Å, in order to maintain its linkage between integrin and actin. Here, the fixed linkage is integrin (bound to the ECM), and the dynamic linkage is the actin cytoskeleton, which undergoes a "retrograde" flow during cell migration [87, 88]. Unfolding of talin exposes multiple binding sites for other cytoskeletal proteins, including vinculin.

Indeed, it is now recognized that mechanical force, either of intracellular (myosin-like motors) or extracellular (hydrodynamic shear) origin, plays a major role in defining or fine-tuning the biological activity of many proteins [89]. Interestingly, there is also recent evidence for force-dependent and force-independent mechanisms among related receptors. For example, integrin $\alpha5\beta1$ requires fibronectin-mediated tension to activate FAK, while collagen-bound integrin $\alpha2\beta1$ activation of FAK is tension-independent [25]. Understanding the structural basis of this selective interplay between mechanical and chemical forces is a fascinating field for future study.

9   Concluding Remarks

The complexity of conformational switches in signal transduction complexes provides perhaps the clearest imperative for further careful biochemical and structural studies at atomic resolution, using a combination of "direct" and hybrid methods. Modern proteomics approaches produce "wiring diagrams" of signaling pathways of whole cells, but these do not by themselves explain how cells work. We must learn the nature of each switch (e.g., phosphorylation, phospholipid-binding, mechanical stress), and precisely how each of the binding functions or catalytic activity of each protein is turned on/off by each switch, both singly and in combination.

But this is only the end of the beginning. Next, we need to observe multiprotein complexes in their *native* environment (*see* Part IV). Recent advances in EM ("Tomography") have enabled direct visualization of protein complexes at molecular resolution (20–60 Å) in flash-frozen (fixed, but unstained/undamaged) thin cell sections [90]. And finally, cells are most definitely not static structures: we must observe single molecules as they go about their business in living cells. For example, during cell migration, multiprotein complexes and higher subsystems (e.g., the cytoskeleton–membrane–matrix linkage) assemble, work, and then disassemble, on a rapid timescale. Ongoing developments that combine light microcopy of living cells (using fluorescently labeled protein) with electron tomography of the same (subsequently frozen) cell (termed "correlative microscopy") provide a glimpse into a future in which we may truly begin to understand how protein-protein interactions drive the organization and dynamics of living cells [91].

References

1. Schreiber G, Keating AE (2011) Protein binding specificity versus promiscuity. Curr Opin Struct Biol 21:50–61

2. Levy ED, De S, Teichmann SA (2012) Cellular crowding imposes global constraints on the chemistry and evolution of proteomes. Proc Natl Acad Sci U S A 109:20461–20466

3. Finn RD, Tate J, Mistry J et al (2008) The Pfam protein families database. Nucleic Acids Res 36:D281–D288

4. Mosca R, Ceol A, Stein A et al (2014) 3did: a catalog of domain-based interactions of known three-dimensional structure. Nucleic Acids Res 42:D374–D379

5. Stumpf MP, Thorne T, de Silva E et al (2008) Estimating the size of the human interactome. Proc Natl Acad Sci U S A 105:6959–6964

6. Venkatesan K, Rual JF, Vazquez A et al (2009) An empirical framework for binary interactome mapping. Nat Methods 6:83–90

7. Voorhees RM, Weixlbaumer A, Loakes D et al (2009) Insights into substrate stabilization from snapshots of the peptidyl transferase center of the intact 70S ribosome. Nat Struct Mol Biol 16:528–533

8. Grigorieff N, Harrison SC (2011) Near-atomic resolution reconstructions of icosahedral viruses from electron cryo-microscopy. Curr Opin Struct Biol 21:265–273

9. Almo SC, Garforth SJ, Hillerich BS et al (2013) Protein production from the structural genomics perspective: achievements and future needs. Curr Opin Struct Biol 23:335–344

10. Cuff AL, Sillitoe I, Lewis T et al (2011) Extending CATH: increasing coverage of the protein structure universe and linking structure with function. Nucleic Acids Res 39: D420–D426

11. Kundrotas PJ, Vakser IA, Janin J (2013) Structural templates for modeling homodimers. Protein Sci 22:1655–1663

12. Kundrotas PJ, Zhu Z, Janin J et al (2012) Templates are available to model nearly all complexes of structurally characterized proteins. Proc Natl Acad Sci U S A 109: 9438–9441

13. Zhang QC, Petrey D, Norel R et al (2010) Protein interface conservation across structure space. Proc Natl Acad Sci U S A 107:10896–10901

14. Gao M, Skolnick J (2010) Structural space of protein-protein interfaces is degenerate, close to complete, and highly connected. Proc Natl Acad Sci U S A 107:22517–22522

15. Zhang QC, Petrey D, Garzon JI et al (2013) PrePPI: a structure-informed database of protein–protein interactions. Nucleic Acids Res 41:D828–D833

16. Aloy P, Ceulemans H, Stark A et al (2003) The relationship between sequence and interaction divergence in proteins. J Mol Biol 332:989–998

17. Janin J (2013) The targets of CAPRI rounds 20–27. Proteins 81:2075–2081

18. Janin J, Rodier F, Chakrabarti P et al (2007) Macromolecular recognition in the Protein Data Bank. Acta Crystallogr D Biol Crystallogr 63:1–8

19. Janin J (2010) The targets of CAPRI Rounds 13–19. Proteins 78:3067–3072

20. Wass MN, David A, Sternberg MJ (2011) Challenges for the prediction of macromolecular interactions. Curr Opin Struct Biol 21:382–390

21. Lander GC, Saibil HR, Nogales E (2012) Go hybrid: EM, crystallography, and beyond. Curr Opin Struct Biol 22:627–635

22. Rouiller I, Xu XP, Amann KJ et al (2008) The structural basis of actin filament branching by the Arp2/3 complex. J Cell Biol 180:887–895

23. Rambo RP, Tainer JA (2013) Super-resolution in solution X-ray scattering and its applications to structural systems biology. Annu Rev Biophys 42:415–441

24. Stein A, Aloy P (2010) Novel peptide-mediated interactions derived from high-resolution 3-dimensional structures. PLoS Comput Biol 6:e1000789

25. Seong J, Tajik A, Sun J et al (2013) Distinct biophysical mechanisms of focal adhesion kinase mechanoactivation by different extracellular matrix proteins. Proc Natl Acad Sci U S A 110:19372–19377

26. Hunter T (2012) Why nature chose phosphate to modify proteins. Philos Trans R Soc Lond B Biol Sci 367:2513–2516

27. Wolfenson H, Lavelin I, Geiger B (2013) Dynamic regulation of the structure and functions of integrin adhesions. Dev Cell 24:447–458

28. Rawlings ND, Tolle DP, Barrett AJ (2004) Evolutionary families of peptidase inhibitors. Biochem J 378:705–716

29. Dice JF (1990) Peptide sequences that target cytosolic proteins for lysosomal proteolysis. Trends Biochem Sci 15:305–309

30. Pawson T, Nash P (2003) Assembly of cell regulatory systems through protein interaction domains. Science 300:445–452

31. Tourigny DS, Fernandez IS, Kelley AC et al (2013) Elongation factor G bound to the ribosome in an intermediate state of translocation. Science 340:1235490

32. Brown CJ, Johnson AK, Dunker AK et al (2011) Evolution and disorder. Curr Opin Struct Biol 21:441–446

33. Dunker AK, Silman I, Uversky VN et al (2008) Function and structure of inherently disordered proteins. Curr Opin Struct Biol 18:756–764

34. Janin J, Sternberg MJ (2013) Protein flexibility, not disorder, is intrinsic to molecular recognition. F1000 Biol Rep 5:2

35. Lo CL, Chothia C, Janin J (1999) The atomic structure of protein-protein recognition sites. J Mol Biol 285:2177–2198

36. Wodak SJ, Janin J (2003) Structural basis of macromolecular recognition. Adv Protein Chem 61:9–73

37. Levy ED, Teichmann S (2013) Structural, evolutionary, and assembly principles of protein oligomerization. Prog Mol Biol Transl Sci 117:25–51

38. Bogan AA, Thorn KS (1998) Anatomy of hot spots in protein interfaces. J Mol Biol 280:1–9

39. Li Y, Huang Y, Swaminathan CP et al (2005) Magnitude of the hydrophobic effect at central versus peripheral sites in protein-protein interfaces. Structure 13:297–307

40. Ofran Y, Rost B (2003) Analysing six types of protein-protein interfaces. J Mol Biol 325:377–387

41. Levy ED (2010) A simple definition of structural regions in proteins and its use in analyzing interface evolution. J Mol Biol 403:660–670

42. Dey S, Pal A, Chakrabarti P et al (2010) The subunit interfaces of weakly associated homodimeric proteins. J Mol Biol 398:146–160

43. Glaser F, Steinberg DM, Vakser IA et al (2001) Residue frequencies and pairing preferences at protein-protein interfaces. Proteins 43:89–102

44. Headd JJ, Ban YE, Brown P et al (2007) Protein–protein interfaces: properties, preferences, and projections. J Proteome Res 6:2576–2586

45. Keskin O, Ma B, Nussinov R (2005) Hot regions in protein–protein interactions: the organization and contribution of structurally conserved hot spot residues. J Mol Biol 345:1281–1294

46. Crowley PB, Golovin A (2005) Cation-Π interactions in protein-protein interfaces. Proteins 59:231–239

47. Chen P, Li J, Wong L et al (2013) Accurate prediction of hot spot residues through

physicochemical characteristics of amino acid sequences. Proteins 81:1351–1362

48. Hwang H, Vreven T, Janin J et al (2010) Protein-protein docking benchmark version 4.0. Proteins 78:3111–3114

49. Cherfils J, Zeghouf M (2013) Regulation of small GTPases by GEFs, GAPs, and GDIs. Physiol Rev 93:269–309

50. Lee JO, Rieu P, Arnaout MA et al (1995) Crystal structure of the A domain from the alpha subunit of integrin CR3 (CD11b/CD18). Cell 80:631–638

51. Emsley J, Knight CG, Farndale RW et al (2000) Structural basis of collagen recognition by integrin α2β1. Cell 101:47–56

52. Hogg N, Harvey J, Cabanas C et al (1993) Control of leukocyte integrin activation. Am Rev Respir Dis 148:S55–S59

53. Shimaoka M, Xiao T, Liu JH et al (2003) Structures of the αL I domain and its complex with ICAM-1 reveal a shape-shifting pathway for integrin regulation. Cell 112:99–111

54. Kuhlmann UC, Pommer AJ, Moore GR et al (2000) Specificity in protein–protein interactions: the structural basis for dual recognition in endonuclease colicin-immunity protein complexes. J Mol Biol 301:1163–1178

55. Xiong JP, Stehle T, Zhang R et al (2002) Crystal structure of the extracellular segment of integrin αVβ3 in complex with an Arg-Gly-Asp ligand. Science 296:151–155

56. Luo BH, Carman CV, Springer TA (2007) Structural basis of integrin regulation and signaling. Annu Rev Immunol 25:619–647

57. Huizinga EG, Tsuji S, Romijn RA et al (2002) Structures of glycoprotein Ibα and its complex with von Willebrand factor A1 domain. Science 297:1176–1179

58. Brondijk TH, Bihan D, Farndale RW et al (2012) Implications for collagen I chain registry from the structure of the collagen von Willebrand factor A3 domain complex. Proc Natl Acad Sci U S A 109:5253–5258

59. Weatheritt RJ, Gibson TJ (2012) Linear motifs: lost in (pre)translation. Trends Biochem Sci 37:333–341

60. Weatheritt RJ, Davey NE, Gibson TJ (2012) Linear motifs confer functional diversity onto splice variants. Nucleic Acids Res 40:7123–7131

61. Yaffe MB, Rittinger K, Volinia S et al (1997) The structural basis for 14-3-3:phosphopeptide binding specificity. Cell 91:961–971

62. Lee JO, Russo AA, Pavletich NP (1998) Structure of the retinoblastoma tumour-suppressor pocket domain bound to a peptide from HPV E7. Nature 391:859–865

63. Eck MJ, Shoelson SE, Harrison SC (1993) Recognition of a high-affinity phosphotyrosyl peptide by the Src homology-2 domain of p56lck. Nature 362:87–91

64. Doyle DA, Lee A, Lewis J et al (1996) Crystal structures of a complexed and peptide-free membrane protein-binding domain: molecular basis of peptide recognition by PDZ. Cell 85:1067–1076

65. Hayashi I, Vuori K, Liddington RC (2002) The focal adhesion targeting (FAT) region of focal adhesion kinase is a four-helix bundle that binds paxillin. Nat Struct Biol 9:101–106

66. Pokutta S, Weis WI (2000) Structure of the dimerization and beta-catenin-binding region of alpha-catenin. Mol Cell 5:533–543

67. Musacchio A, Saraste M, Wilmanns M (1994) High-resolution crystal structures of tyrosine kinase SH3 domains complexed with proline-rich peptides. Nat Struct Biol 1:546–551

68. Nilsen TW, Graveley BR (2010) Expansion of the eukaryotic proteome by alternative splicing. Nature 463:457–463

69. Weatheritt RJ, Jehl P, Dinkel H et al (2012) iELM – a web server to explore short linear motif-mediated interactions. Nucleic Acids Res 40:W364–W369

70. Van Roey K, Gibson TJ, Davey NE (2012) Motif switches: decision-making in cell regulation. Curr Opin Struct Biol 22:378–385

71. Harrison SC (1996) Peptide-surface association: the case of PDZ and PTB domains. Cell 86:341–343

72. Van Roey K, Dinkel H, Weatheritt RJ et al (2013) The switches ELM resource: a compendium of conditional regulatory interaction interfaces. Sci Signal 6:rs7

73. Stein A, Aloy P (2008) Contextual specificity in peptide-mediated protein interactions. PLoS One 3:e2524

74. Garcia-Alvarez B, de Pereda JM, Calderwood DA et al (2003) Structural determinants of integrin recognition by talin. Mol Cell 11:49–58

75. Cowan KJ, Law DA, Phillips DR (2000) Identification of Shc as the primary protein binding to the tyrosine-phosphorylated β3 subunit of αIIbβ3 during outside-in integrin platelet signaling. J Biol Chem 275:36423–36429

76. Di Paolo G, Pellegrini L, Letinic K et al (2002) Recruitment and regulation of phosphatidylinositol phosphate kinase type 1-γ by the FERM domain of talin. Nature 420:85–89

77. Schill NJ, Anderson RA (2009) Two novel phosphatidylinositol-4-phosphate 5-kinase type Igamma splice variants expressed in

human cells display distinctive cellular targeting. Biochem J 422:473–482

78. Reyes A, Anders S, Weatheritt RJ et al (2013) Drift and conservation of differential exon usage across tissues in primate species. Proc Natl Acad Sci U S A 110:15377–15382

79. Hui S, Xing X, Bader GD (2013) Predicting PDZ domain mediated protein interactions from structure. BMC Bioinformatics 14:27

80. Bakolitsa C, Cohen DM, Bankston LA et al (2004) Structural basis for vinculin activation at sites of cell adhesion. Nature 430:583–586

81. Balla T (2005) Inositol-lipid binding motifs: signal integrators through protein-lipid and protein–protein interactions. J Cell Sci 118:2093–2104

82. Carlton JG, Cullen PJ (2005) Coincidence detection in phosphoinositide signaling. Trends Cell Biol 15:540–547

83. Schiller HB, Fassler R (2013) Mechanosensitivity and compositional dynamics of cell-matrix adhesions. EMBO Rep 14:509–519

84. Lietha D, Cai X, Ceccarelli DF et al (2007) Structural basis for the autoinhibition of focal adhesion kinase. Cell 129:1177–1187

85. Arold ST (2011) How focal adhesion kinase achieves regulation by linking ligand binding, localization and action. Curr Opin Struct Biol 21:808–813

86. Toutant M, Costa A, Studler JM et al (2002) Alternative splicing controls the mechanisms of FAK autophosphorylation. Mol Cell Biol 22:7731–7743

87. Margadant F, Chew LL, Hu X et al (2011) Mechanotransduction in vivo by repeated talin stretch-relaxation events depends upon vinculin. PLoS Biol 9:e1001223

88. Hu K, Ji L, Applegate KT et al (2007) Differential transmission of actin motion within focal adhesions. Science 315:111–115

89. Seifert C, Grater F (2013) Protein mechanics: how force regulates molecular function. Biochim Biophys Acta 1830:4762–4768

90. Yahav T, Maimon T, Grossman E et al (2011) Cryo-electron tomography: gaining insight into cellular processes by structural approaches. Curr Opin Struct Biol 21:670–677

91. Zhang P (2013) Correlative cryo-electron tomography and optical microscopy of cells. Curr Opin Struct Biol 23:763–770

92. Baker ML, Hryc CF, Zhang Q et al (2013) Validated near-atomic resolution structure of bacteriophage epsilon15 derived from cryo-EM and modeling. Proc Natl Acad Sci U S A 110:12301–12306

Chapter 2

Quantitative Analysis of Protein-Protein Interactions

Ziad M. Eletr and Keith D. Wilkinson

Abstract

Numerous authors, including contributors to this volume, have described methods to detect protein-protein interactions. Many of these approaches are now accessible to the inexperienced investigator thanks to core facilities and/or affordable instrumentation. This chapter discusses some common design considerations that are necessary to obtain valid measurements, as well as the assumptions and analytical methods that are relevant to the quantitation of these interactions.

Key words Ligand binding, Protein-protein interaction, Fluorescence, Binding equations, Binding equilibria

1 Introduction

In the post-genomic era, the importance of protein-protein interactions is becoming even more apparent [1]. We are coming to recognize that most, if not all, catalytic and regulatory pathways operate as networks, with frequent and extensive input from signaling pathways, feedback, and cross talk. Replication, transcription, translation, signal transduction, protein trafficking, and protein degradation are all accomplished by protein complexes, often temporally assembled and disassembled to accomplish vectoral processes. Often these interactions are driven by interaction of recognized domains in the constituent proteins. We must identify and understand these domain interactions in order to discern the patterns and logic of cellular regulation [2].

1.1 Assumptions

There are several assumptions inherent to any analysis of a simple ligand–receptor interaction (https://tools.lifetechnologies.com/downloads/FP7.pdf).

1. The interactions are assumed to be reversible. In the simplest case, the association reaction is bimolecular while the dissociation reaction in unimolecular.

Cheryl L. Meyerkord and Haian Fu (eds.), *Protein-Protein Interactions: Methods and Applications*, Methods in Molecular Biology, vol. 1278, DOI 10.1007/978-1-4939-2425-7_2, © Springer Science+Business Media New York 2015

2. All receptor molecules are equivalent and independent.

3. The measured response is proportional to the number of occupied receptor sites.

4. The interactions are measured at equilibrium.

5. The components do not undergo any other chemical reactions and exist only in the free or bound states.

Any or all of these assumptions may prove to be unfounded in a more complex case. In fact, it is the deviation from simple behavior that is often the first indication of a more complex binding event and each assumption should be explored to explain deviations from simple behavior. Outlined below are treatments for simple cases. A general method to deriving binding formulas to more complex cases has been derived from statistical thermodynamic principles [3].

1.2 Binding to One Site

The receptor–ligand terminology is useful, even if artificial, in the case of protein-protein interactions. Either protein could be considered the receptor or the ligand.

For the purposes of this chapter we refer to the protein present in fixed and limiting amounts as the receptor and the component that is varied as the ligand.

Thus, for one molecule of L binding to one molecule of R:

$$R_f + L_f \underset{k_2}{\overset{k_1}{\rightleftharpoons}} RL, \tag{1}$$

where R_f is the concentration of free receptor, L_f is the concentration of free ligand, RL is the concentration of the complex, k_1 is the association rate constant, and k_2 is the dissociation rate constant. At equilibrium:

$$\frac{[R_f]\ [L_f]}{[RL]} = \frac{k_2}{k_1} = K_d, \tag{2}$$

where K_d is the dissociation constant.

In most binding titrations, the concentrations of free ligand and receptor are difficult to quantify and one typically measures the fractional saturation, $[RL]/[R_t]$ as function of total ligand $[L_t]$. Determining the dissociation constant K_d can be performed in two ways, depending on the experimental design. If $[L_f]$ can be measured, or if $[L_t] \gg [R_t]$ and we can assume $[L_t] = [L_f]$, a simpler derivation of fractional saturation vs. $[L_t]$ can be applied. In more general cases where $[L_f]$ is not measured yet $[L_t]$ is known, one must solve a quadratic equation to determine the K_d. These derivations are shown below.

Rearranging the fractional saturation term ($[RL]/[R_t]$) by substituting for $[RL]$ in terms of $[R_f]$, $[L_f]$ and K_d (Eq. 2) and applying the conservation of mass assumption $[R_f] = [R_t] - [RL]$ gives:

$$\frac{[RL]}{[R_t]} = \frac{\dfrac{[R_f][L_f]}{K_d}}{[R_f] + [RL]}, \tag{3}$$

which can be simplified to Eq. 4:

$$\frac{[RL]}{[R_t]} = \frac{[L_f]}{K_d + [L_f]}. \tag{4}$$

Thus, a plot of fraction saturation $[RL]/[R_t]$ vs. $[L_f]$ will give the familiar rectangular hyperbola if only one type of binding site is present (Fig. 1a). This equation is valid when one can directly measure $[L_f]$, or when $[L_t] \gg [R_t]$ and it can be assumed that $[L_f] = [L_t]$. This assumption is valid when 5–10 % of the ligand is bound [3]. Alternatively, a plot of fractional saturation vs. $\log[L_t]$ can be used. If free concentrations are actually measured (instead of calculated), we can use the Klotz plot [4], a plot of fractional saturation vs. $\log[L_f]$ (Fig. 1b); or the Scatchard plot, a plot of ligand bound/free vs. ligand free (Fig. 1c).

In the second scenario, when $[L_f]$ is not measured or when $[L_t]$ is not much greater than $[R_t]$, we must apply the conservation of mass assumption to $[L_f]$ as well. Substituting into Eq. 2:

$$K_d = \frac{[R_f][L_f]}{[RL]} = \frac{([R_t] - [RL])([L_t] - [RL])}{[RL]}. \tag{5}$$

Equation 5 can then be rearranged to the quadratic form $ax^2 + bx + c = 0$:

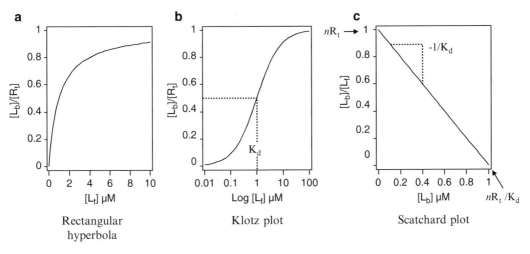

Fig. 1 Plots of simulated data for simple binding. In all cases $n = 1$ and $K_d = 1$ μM. (**a**) Direct plot of fractional saturation vs. free ligand; (**b**) Klotz plot of the same data. Note the log scale; (**c**) Scatchard plot of the same data. The parameters nR_t and nR_t/K_d are estimated from the intercepts

$$0 = [RL]^2 - ([R_t] + [L_t] + K_d) + [R_t][L_t], \qquad (6)$$

where $x = [RL]$, $a = 1$, $b = ([R_t] + [L_t] + K_d)$ and $c = [R_t][L_t]$. One can then fit the fractional saturation vs. $[L_t]$ binding curve to the quadratic equation solution (one root is positive) to determine the unknown K_d using nonlinear least squares regression (*see* Eq. 8). An example of this method which has been used to determine the K_d of a protein-protein interaction monitored by fluorescence anisotropy is given [5].

1.3 Binding to Multiple Sites

It should be noted that if more than one ligand molecule binds to R then the behavior may be more complex. For n multiple binding sites we get:

$$[RL] = [RL_1] + [RL_2] \ldots [RL_n]$$
$$= \frac{[R_t][L_f]}{K_{d1} + [L_f]} + \frac{[R_t][L_f]}{K_{d2} + [L_f]} \cdots \frac{[R_t][L_f]}{K_{dn} + [L_f]}, \qquad (7)$$

where n different sites can be occupied by ligand with the corresponding binding constants.

1.3.1 Identical, Non-interacting Binding Site(s)

If all binding sites are identical and non-interacting (i.e., all bind with the same K_d) then Eq. 7 reduces to:

$$\frac{[L_b]}{n[R_t]} = \frac{[L_f]}{K_d + [L_f]}, \qquad (8)$$

where $n[R_t] = [R_f] + [L_b]$.

Note that this equation is similar to Eq. 4 except for the inclusion of the stoichiometry n. A Klotz plot of fractional saturation vs. $\log[L_f]$ will be sigmoidal and symmetrical about the midpoint. The curve is nearly linear from 0.1 to $10 \times K_d$ and 99 % saturation is achieved when $[L_t]$ is two orders of magnitude above K_d. A complete description of binding and accurate estimation of the plateau values requires that $[L_f]$ vary from two log units below to two log units above K_d. A steeper curve is indicative of positive cooperativity, while a flatter curve could be due to negative cooperativity or the presence of an additional binding site. The stoichiometry is calculated from the plateau value and $[R_t]$ while the K_d is calculated from the midpoint [6], or more accurately using a nonlinear least squares fit to Eq. 8.

If free ligand is not measured then we must use a plot of fractional saturation vs. $\log[L_t]$ and the curve will deviate from sigmoidal by the difference between $\log[L_f]$ and $\log[L_t]$. This condition is often referred to as ligand depletion [7, 8]. It should be recognized, however, that it may not be possible to cover such a large range of concentrations with proteins. At the low end we are often limited by the sensitivity of the technique and at the high end limited solubility or sample amounts that may prevent us from attaining concentrations necessary to reach the plateau.

An alternative way to plot the data is with a Scatchard plot. For the last 40 years this has been the traditional method for the analysis of binding data where $[L_f]$ is measured. The Scatchard plot is described by:

$$\frac{[L_b]}{[L_f]} = \frac{-[L_b]}{K_d} + \frac{n[R_t]}{K_d}. \tag{9}$$

In the simple model, a plot of ligand bound vs. ligand bound/ligand free gives a straight line with the x-intercept = $n[R_t]$, a y-intercept of $n[R_t]/K_d$, and a slope of $-1/K_d$ (Fig. 1c) [6].

Before the advent of computers, estimates of K_d and n were obtained by any of a number of transformations of the relevant equations to give linear plots. These include the double reciprocal plot, and the Scatchard plot. These linearizations are notoriously difficult to fit and generally fraught with problems. *The preferred method of obtaining K_d and n from binding data is direct fitting of the data using a nonlinear least squares fitting algorithm.* Many commercial packages for doing such fits are available today. As discussed in Subheading 1.2, if we do not explicitly measure the concentration of ligand free, an appropriate solution of the binding equation to obtain the dissociation constant requires that we determine and fit the fractional saturation as a function of the concentration of total L added. The solution of the equation for $[RL]/[R_t]$ as a function of $[L_t]$ is a quadratic equation with the following real solution:

$$\frac{[L_b]}{[R_t]} = \frac{([L_t] + n[R_t] + K_d) - \sqrt{(-[L_t] - n[R_t] - K_d)^2 - 4[L_t]n[R_t]}}{2n[R_t]}. \tag{10}$$

1.3.2 Non-identical Binding Sites

While the most common reason for observing multiple non-identical binding sites in a protein-protein interaction is likely to be nonspecific binding (see below), it is always possible that there are two independent and non-interacting sites with different affinities. Either case will manifest itself as a deviation from the expected behavior for a simple binding model. The Scatchard plot is a useful diagnostic tool to point out such deviations (Fig. 2). A Scatchard plot that is concave upward is indicative of nonspecific binding, negative cooperativity, or multiple classes of binding sites. A concave downward plot suggests either positive cooperativity or instability of the ligand. In any case, proper analysis of this behavior requires other information (for instance stoichiometry, stability) and the data are best fitted using nonlinear least squares fitting of the data according to an appropriate model.

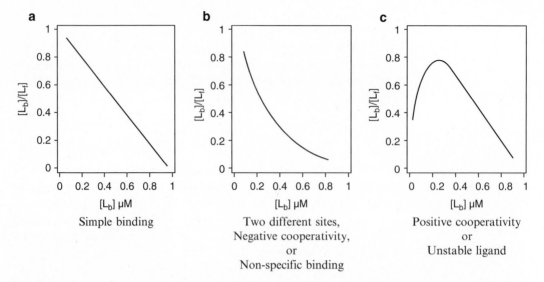

a — Simple binding

b — Two different sites, Negative cooperativity, or Non-specific binding

c — Positive cooperativity or Unstable ligand

(y-axis: $[L_b]/[L_f]$; x-axis: $[L_b]$ μM)

Fig. 2 Effects of complexities on the appearance of the Scatchard plot. (**a**) Represents the expected behavior in the simple case; (**b**) a *concave upward deviation* as shown in this panel could be caused by the presence of two different sites, the presence of negative cooperativity or a significant nonspecific binding component; (**c**) Positive cooperativity or ligand instability would lead to the *curvature* shown in this panel

Most deviations from simple binding are expected to be due to either multiple sites or nonspecific binding which, as discussed below, may be difficult to distinguish. Either case can be fitted with appropriate modifications of the simple binding expressions. Note that a satisfactory analysis of such complicated binding will require measurement of $[L_f]$.

1.4 Cooperativity

Cooperativity is the term used to describe the situation where occupancy of one site changes the affinity for ligand at another site. There have been many treatments of cooperative binding interactions, including analysis by Scatchard and Hill plots, but these are beyond the scope of this discussion. In general, models explaining cooperativity invoke subunit–subunit interactions in oligomeric protein structures and may well be important in cases where multiple proteins are being assembled into a multimeric complex. The reader is referred to any of several other treatments of such binding if complications of this sort are indicated [9–11]. However, it may be simpler to restrict the measurements to conditions where individual subcomplexes are assembled at saturating concentrations before measuring the binding of a subsequent protein.

2 Materials

The only materials relevant to this chapter are a computer and a program to mathematically fit the data. Many commercial and shareware packages capable of nonlinear fitting of equations are available for all platforms, i.e., Prism (GraphPad Software, Inc., San Diego, CA), SigmaPlot (SPSS Science, Chicago, IL), Mathematica (Wolfram Research Inc., Champaign, IL), DynaFit (BioKin, Ltd., Pullman, WA), MATLAB (MathWorks, Natick, MA), and others. There are also published solutions using the popular spreadsheet Microsoft Excel [12, 13]. The choice is largely up to personal preference.

3   Methods

Several chapters in this book describe techniques for determining fractional saturation and/or binding parameters. These basically fall into two categories, direct methods that measure the actual concentration of bound or free ligand, and indirect methods that infer the concentrations from some measured signal. The choice of which technique to use may be limited by the strength of the interactions and the inherent sensitivity of the technique. For instance, NMR may be a poor choice to monitor binding constants tighter than micromolar since one commonly needs mM concentrations of protein to see a signal. Thus, $[R_t]$ may be $\gg K_d$ and we would be restricted to measuring only the stoichiometry under these conditions (see below). Similarly, with an interaction of millimolar affinity it may be difficult to determine the stoichiometry since it may not be possible to attain a concentration of $[R_t] \gg K_d$. See below for a discussion of the relationships between K_d and $[R_t]$.

3.1   Direct Measurement of Free Ligand

Direct methods require that we accurately determine the concentrations of free and bound ligand. Examples of techniques that yield such information include gel filtration, ultracentrifugation, ultrafiltration, or equilibrium dialysis. For binding with slow dissociation rates pull-downs, band shift or electrophoresis techniques *may* be appropriate. If the process of separating the bound and free ligand is fast compared to the rate of dissociation of the complex such methods can yield directly the concentrations of bound and free ligand. If dissociation and separation of bound and free reactants occur on similar time scales such methods are not appropriate for quantitation as the equilibrium will be disturbed by the separation of the reactants. For the same reasons techniques such as cross-linking may overestimate the concentration of RL since the equilibrium will be disturbed by the removal of RL from the equilibrium.

3.2   Indirect Measurements of Bound Ligand

More commonly an indirect measure of saturation is used to monitor binding. These include optical methods such as fluorescence, absorbance, and resonance techniques. These methods all assume that the output signal is directly proportional to the concentration of RL present. For instance, if a fluorescence change is being monitored it is assumed that there are only two states, the bound and the free, and that each has a characteristic value. If S_o is the signal in the absence of binding, S_L the signal in the presence of total ligand concentration L, and S_∞ is the value at saturation, then:

$$\text{fraction saturation} = \frac{(S_L - S_o)}{(S_\infty - S_o)}. \qquad (11)$$

The concentration of free ligand can be calculated by assuming a stoichiometry n and using the expression $[L_f] = [L_t] - n[R_t] = (S_L - S_o)/(S_\infty - S_o)$. Note that if n is incorrect, then the calculated $[L_f]$ will be incorrect also and this will be apparent in the deviation of the data from the theoretical rectangular hyperbola. This is one reason why the determination of n is an important exercise in most binding studies. Alternatively, and preferably, data are fitted using nonlinear least squares methods and n is determined directly from this analysis.

3.3 Competition Methods

Direct methods measure either bound ligand $[RL]$ or free ligand $[L_f]$ as a function of $[L_t]$ and indirect methods usually involve measuring fractional saturation $[RL]/n[R_t]$ as a function of $[L_t]$. However, one of the most useful variations of the binding experiment is the use of competitive binding assays where a single labeled indicator ligand can be bound and subsequently displaced by any of a variety of competitive inhibitors [14–19]. Such experiments are particularly useful if the affinity of a series of inhibitors is to be determined. Methods such as fluorescence polarization or florescence resonance energy transfer are particularly well suited for such measurements. A small amount of the labeled ligand is first bound to the receptor and subsequently displaced by titrating with unlabeled inhibitor. The K_i of the unlabeled inhibitor is then calculated. The labeled ligand does not have to be physiological or bound with a physiological affinity since we are always comparing the K_i of the unlabeled inhibitor. Thus, any adverse effects of labeling the indicator ligand will be unimportant.

The IC_{50} is the concentration of inhibitor necessary to displace half the labeled ligand. If $[R_t] \ll K_d$, IC_{50} is related to K_i, the affinity of the unlabeled ligand by:

$$K_i = \frac{IC_{50}}{1 + {L_t}/{K_d}}, \qquad (12)$$

where $[L_t]$ is the concentration of labeled ligand and K_d is its' dissociation constant. If only relative affinities are to be measured then comparing IC_{50} directly is sufficient. If absolute affinities are desired, then we must also determine the concentration and affinity of the labeled ligand in the assay.

If $[R_t]$ is similar to or greater than K_d and/or K_i, it follows that the concentrations of free ligand and inhibitor are not equal to their respective total concentrations. For this reason, it is simplest to work at conditions where $[R_t] \sim 0.1 \times K_d$ so that less than 10 % of the labeled ligand is bound to the receptor at the start of the experiment.

If higher concentrations of receptor are necessary or if inhibitor binds much tighter than ligand, then one has to fit with a more complex equation [7, 15, 17–19]. The following treatment was first published by Wang in 1995 [19] and is suitable for fitting the data from competitive displacement experiments where absorbance, fluorescence, or fluorescent anisotropy are measured using commercially available fitting programs. Consider, for example, the binding of a fluorescent probe A to a non-fluorescent protein P in the presence or absence of a competitive inhibitor B that prevents binding of A.

Given:

$$K_a = [A_f][P_f]/[PA]$$
$$[A_f] + [PA] = [A_t]$$
$$[P_f] + [PA] + [PB] = [P_t]$$
$$K_b = [B_f][P_f]/[PB]$$
$$[B_f] + [PB] = [B_t],$$

then Eq. 13 describes the fractional saturation:

$$\frac{(S - S_o)}{(S_\infty - S_o)} = \frac{\left\{2\sqrt{(a^2 - 3b)}\cos(\theta/3) - a\right\}}{3K_a + \left\{2\sqrt{(a^2 - 3b)}\cos(\theta/3) - a\right\}}, \qquad (13)$$

$$\text{where}: \theta = \arccos\frac{-2a^3 + 9ab - 27c}{2\sqrt{(a^2 - 3b)^3}},$$

$$a = K_a + K_b + [A_t] + [B_t] - [P_t]$$
$$b = K_b([A_t] - [P_t]) + K_a([B_t] - [P_t]) + K_aK_b$$
$$c = -K_aK_b[P_t].$$

The experiment requires the measurement of the fractional saturation at various concentrations of A_t, B_t, and P_t. Only a small range of measurements are useful: the ones where fractional saturation is >0.05 and <0.95. Fractional saturation of P with the probe A is determined by indirect measurements where it is the fluorescence or the anisotropy of AP which gives rise to the signal. The usual experiment is to measure the full binding curve, i.e., $(S - S_o)/(S_\infty - S_o)$ as a function of P_t. This experiment should then be repeated at three or more concentrations of B_t to calculate K_b. Although this may seem like its only giving you three data points, if the curve is fitted, the actual number of useful data points is equal to the total measurements made where fractional saturation is in a useful range.

3.4   Parameters of Reversible Binding

3.4.1   Stoichiometry

Quantitation of binding often requires accurate estimates of the binding stoichiometry n. Many methods are appropriate for this purpose including cross-linking, pull-downs, and electrophoretic methods (when off rates are slow). If association and dissociation rates are fast these techniques will perturb the equilibrium and give erroneous results. In these cases stoichiometry must be determined from more conventional titrations measuring the equilibrium amounts of RL. To determine stoichiometry an excess of ligand is present and one of the components must be present at concentration well above the K_d in order to assure saturation. Often this is the first experiment that is done as it helps greatly in fitting the data to more complete titrations.

3.4.2   Kinetics

The analysis of binding requires that we conduct the measurements after binding has reached equilibrium or that we measure individually the rate constants involved. The binding constant can then be calculated from the relationship $K_d = k_2/k_1$. From a practical standpoint, assuring that the reaction has reached equilibrium often involves measuring a time course for binding at low ligand concentrations and making all measurements after sufficient time to allow attainment of equilibrium. Several examples of each type of analysis are given in later chapters.

In any case it is instructive to consider the magnitudes of association and dissociation rates. The association rate constants expected for protein-protein interactions are limited by diffusion. If we assume reasonable numbers for the diffusion rate of an average protein, the diffusion limit in aqueous solution is around 10^8–10^9 M^{-1} s^{-1}. There are also additional steric constraints as only a fraction of the collisions occurring at this rate are oriented properly, and it is commonly assumed that the rate limiting association rate (k_1) for two proteins binding to each other is around 10^8 M^{-1} s^{-1}.

It can be shown that the rate of approach to equilibrium is determined by the sum of the association rate and the dissociation rate constants. Further, the concentrations of reagents must be at or near the binding constant for accurate determination of both stoichiometry and affinity in the same experiment (see below). If the dissociation constant (K_d) for such an interaction is moderate (10^{-6} M) then the dissociation rate for such a complex will be $k_2 = k_1 \times K_d = 10^2$ s^{-1}. Thus, binding will be complete in seconds and the half-life of the bound state will be tens of milliseconds. If, however, the binding constant is very tight, as may occur in antibody–antigen interactions, the overall equilibrium may take some time. Consider a binding interaction with a free energy of -16 kcal/mol, an affinity exhibited by many antibodies and other protein-protein interactions [20]. This represents a dissociation constant of 10^{-13} M. Here, binding may take as long as hours and the half-life of the bound state could be as long as 20 h. The latter fact is the reason that tight binding can be detected using techniques like immunoprecipitation and pull-down experiments, but tight binding complicates the determination of accurate binding constants.

3.5   Concentrations of Components to Use

3.5.1   Ligand Concentration

Equation 8 is the equation for the familiar rectangular hyperbola with a horizontal asymptote corresponding to 100 % saturation and half-maximal saturation occurring at $L_f = K_d$. This equation points out that *the concentrations of free ligand present must be similar to the dissociation constant in order to vary the fractional saturation of receptor*, i.e., to measure the strength of binding. The most common form of the experiment then is to titrate a fixed amount of receptor with variable amounts of ligand and to fit the experimental data to the appropriate binding equation to determine the stoichiometry n and the binding constant K_d.

3.5.2   Receptor Concentration

If we consider the concentration of the fixed protein in this binding equation, i.e., $[R_t]$, we can define three limiting conditions: $[R_t] \ll K_d$, $[R_t] \gg K_d$, and $[R_t] \sim K_d$. Figure 3 illustrates the interrelationships between $[K_d]$ and $[R_t]$ in such experiments.

$[R_t] \ll K_d$

Under these conditions saturation is achieved by varying $[L]$ at concentrations from 0.1 to 10 times K_d. Since $[L_t]$ is always much greater than $[RL]$ under these conditions, then $[L_f] \sim [L_t]$. Thus, Eq. 8 can be simplified to give:

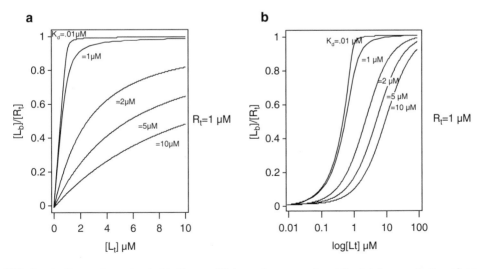

Fig. 3 Binding isotherms for a simple binding equilibrium where $n = 1$ and the total concentration of receptor is 1 μM. (**a**) Direct plot of fractional saturation vs. total ligand added and (**b**) the same data plotted on a log scale. Note that as the K_d approaches $[R_t]$ there is a significant deviation from the rectangular hyperbolic behavior

$$\frac{[L_b]}{n[R_t]} = \frac{[L_t]}{K_d + [L_t]}.$$ (14)

If we only measure the fractional saturation (i.e., the *ratio* $[L_b]/n$ $[R_t]$) as a function of L_t then we cannot calculate $[L_b]$ since $[L_b]$ $= [L_t] - [L_f]$ and we have not measured $[L_f]$. Note that even if we use direct methods and measure free ligand concentration, the calculation of bound ligand is subject to large errors since the bound is the difference between total and free and under these conditions they are about equal [7, 8]. *Thus, under these conditions, we can accurately determine K_d but not n.* Determination of accurate values for n requires that the concentration of R_t be similar to or larger than K_d.

$[R_t] \gg K_d$

If the concentration of R_t is much greater than K_d then Eq. 8 can be rearranged to give:

$$\frac{[L_b]}{[L_f]} = \frac{n[R_t]}{K_d + [L_f]}.$$ (15)

In the first part of the titration curve, when $[L_f]$ is less than K_d (and much less than $n[R_t]$ in this example), the ratio of bound/free ligand is determined solely by the ratio of $n[R_t]/[K_d]$.

If we only measure $[L_t]$ the limiting slope for a plot of saturation vs. $[L_t]$ is $n[R_t]/[K_d]$. For example, if $n[R_t]/[K_d] = 100$, then only about 1 % of the added ligand is free at low ligand concentrations. In order to saturate binding $[L_t]$ must exceed $100 \times K_d$. When $[R_t] \gg K_d$, the saturation curve is really an end-point determination consisting of two lines (first a slope of ~n and then 0 intersecting at $[L_t] = n[R_t]$) with little curvature (Fig. 3). *Under these conditions we can accurately determine n, but not K_d.*

If direct methods to measure free ligand are used, we can, in theory, calculate K_d, but in practical terms the curve will only deviate from its biphasic nature near $L_t = R_t$, and generally there will not be enough data in this region to obtain accurate estimates of K_d.

$[R_t] \sim K_d$

The most useful conditions for determining both K_d and n are when $[R_t] \sim K_d$. The binding curve still resembles a rectangular hyperbola but with small deviations due to the fact that $[L_t] = [L_f] + n[RL]$. Since $[L_f]$ is similar in magnitude to $[L_b]$ each can be measured (or calculated) with good accuracy. *Under these conditions we can determine both K_d and n with a good degree of accuracy from the same experiment.*

4 Notes

Nonspecific Binding: Specificity vs. Affinity
Almost any real-life binding experiment will show some low-affinity binding that is often attributed to "nonspecific binding". If indirect methods are used to monitor binding, one may or may not see this binding step and one must evaluate if the technique being used will reveal nonspecific binding (i.e., does the detection of binding require occupancy of a specific site such as in fluorescence resonance energy transfer techniques). Nonspecific binding usually presents as an additional slope added to the familiar rectangular hyperbola apparent at high ligand concentrations and the temptation is to simply subtract the linear phase from the observed binding to obtain the specific binding profile. The ambiguity as to whether this binding is "specific" (but just low affinity) or whether this is "nonspecific" has, and will, bedevil many studies [7, 16, 21]. Numerous hydrophobic and ionic interactions can lead to nonspecific binding but these may be saturable and show a defined n value when two large proteins are involved. Because the binding may well be saturable the linear subtraction of nonspecific binding may not be appropriate. If we restrict ourselves to consider only two classes of sites, one tight site binding n_1 molecules with affinity K_{d1} and a second weaker site (either due to another specific site or nonspecific binding) binding n_2 molecules with affinity K_{d2}, then we can modify Eq. 7 to give:

$$[L_b] = \frac{n_1[R_t][L_f]}{K_{d1} + [L_f]} + \frac{n_2[R_t][L_f]}{K_{d2} + [L_f]}. \tag{16}$$

Direct fitting of the data to this expression will allow assessment of both classes of sites. If $K_{d2} \gg [L_f]$ then the second term is approximately linear with $[L_f]$ and this is similar to the usual case of nonspecific binding. But if $K_{d2} \sim [L_f]$, then the second term will not be linear. Thus, it is preferable to simply fit the binding as though there are two different but independent binding sites. After the data are analyzed with no assumptions one can question if this interaction occurs at a defined site and in a physiological range of concentrations and is therefore relevant.

Curve Fitting and Adequacy of the Models
Deviations from the simple binding expressions indicate complexity such as multiple sites or cooperativity. However, the simplest model that explains the data is to be preferred. If the data fit a model with two independent binding sites no better than that with one, the one site model should be chosen unless there is independent evidence to suggest two sites. Methods of evaluating the goodness of fitting are beyond the scope of this chapter, but are often available

with available fitting programs and should be evaluated before proposing a more complicated expression [8].

Procedures and Problems

To summarize, the determination of K_d and n for a protein-protein interaction requires that we select a technique appropriate for the binding affinity to be measured. The best concentration of receptor is near the K_d and the concentration of ligand should be varied from two orders of magnitude below to two orders of magnitude above the K_d. The concentration of bound ligand should be determined as a function of the free ligand and the data should be fit to the simplest appropriate model. Generally, n can be determined with a precision of $\pm 20\ \%$ and K_d within a factor of 2.

Several experimental limitations and errors can limit the accuracy and correctness of the observed fits. Common problems (https://tools.lifetechnologies.com/downloads/FP7.pdf) are:

1. Incorrect correction for nonspecific binding or additional loose binding sites. The suggested solution is to fit to Eq. 16.

2. Pooling data from experiments with different receptor concentrations. This will be a problem if the receptor concentrations are near K_d. To avoid this, collect enough data from each titration to do an independent fit and compare the fitted parameters from independent determinations.

3. Presence of a non-binding contaminant in the receptor or labeled ligand. This may be relevant when labeling the ligand damages the protein, when recombinant proteins are used and there is undetected heterogeneity due to misfolded protein.

4. Use of a labeling method for the ligand that alters the binding behavior of that ligand. Use of truncated constructs or incorporation of epitope tags or fluorescent labels may be particularly troublesome. Such problems may be revealed if one compares the apparent affinity from direct experiments using titration with labeled ligand to experiments where unlabeled ligand is used to displace labeled ligand.

5. Inadequate number of data points or range of ligand concentrations. This is avoided by collecting enough data points, especially at high ligand concentrations.

References

1. Auerbach D, Thaminy S, Hottiger MO et al. (2002) The post-genomic era of interactive proteomics: facts and perspectives. Proteomics 2:611–623

2. Pawson T, Raina M, Nash P (2002) Interaction domains: from simple binding events to complex cellular behavior. FEBS Lett 513:2–10

3. Johnson ML, Straume M (2000) Deriving complex ligand-binding formulas. Methods Enzymol 323:155–167

4. Klotz IM (1985) Ligand–receptor interactions: facts and fantasies. Q Rev Biophys 18:227–259

5. Eletr ZM, Huang DT, Duda DM et al. (2005) E2 conjugating enzymes must disengage from

their E1 enzymes before E3-dependent ubiquitin and ubiquitin-like transfer. Nat Struct Mol Biol. 12:933–934

6. Munson PJ, Rodbard D (1983) Number of receptor sites from Scatchard and Klotz graphs: a constructive critique. Science 220:979–981

7. Swillens S (1995) Interpretation of binding curves obtained with high receptor concentrations: practical aid for computer analysis. Mol Pharmacol 47:1197–1203

8. Motulsky HJ, Ransnas LA (1987) Fitting curves to data using nonlinear regression: a practical and nonmathematical review. FASEB J 1:365–374

9. Tuk B, van Oostenbruggen MF (1996) Solving inconsistencies in the analysis of receptor-ligand interactions. Trends Pharmacol Sci 17:403–409

10. Koshland DE Jr (1996) The structural basis of negative cooperativity: receptors and enzymes. Curr Opin Struct Biol 6:757–761

11. Forsen S, Linse S (1995) Cooperativity: over the Hill. Trends Biochem Sci 20:495–497

12. Hedlund PB, von Euler G (1999) EasyBound – a user-friendly approach to nonlinear regression analysis of binding data. Comput Methods Programs Biomed 58:245–249

13. Brown AM (2001) A step-by-step guide to non-linear regression analysis of experimental data using a Microsoft Excel spreadsheet. Comput Methods Programs Biomed 65:191–200

14. Jezewska MJ, Bujalowski W (1996) A general method of analysis of ligand binding to competing macromolecules using the spectroscopic

signal originating from a reference macromolecule. Application to Escherichia coli replicative helicase DnaB protein nucleic acid interactions. Biochemistry 35:2117–2128

15. Schwarz G (2000) A universal thermodynamic approach to analyze biomolecular binding experiments. Biophys Chem 86:119–129

16. van Zoelen EJ (1992) Analysis of receptor binding displacement curves by a nonhomologous ligand, on the basis of an equivalent competition principle. Anal Biochem 200:393–399

17. van Zoelen EJ, Kramer RH, van Moerkerk HT et al (1998) The use of nonhomologous Scatchard analysis in the evaluation of ligand-protein interactions. Trends Pharmacol Sci 19:487–490

18. van Zoelen EJ, Kramer RH, van Reen MM et al (1993) An exact general analysis of ligand binding displacement and saturation curves. Biochemistry 32:6275–6280

19. Wang ZX (1995) An exact mathematical expression for describing competitive binding of two different ligands to a protein molecule. FEBS Lett 360:111–114

20. Brooijmans N, Sharp KA, Kuntz ID (2002) Stability of macromolecular complexes. Proteins 48:645–653

21. Rovati GE, Rodbard D, Munson PJ (1988) DESIGN: computerized optimization of experimental design for estimating Kd and Bmax in ligand binding experiments. I. Homologous and heterologous binding to one or two classes of sites. Anal Biochem 174:636–649

Chapter 3

Protein-Protein Interaction Databases

Damian Szklarczyk and Lars Juhl Jensen

Abstract

Years of meticulous curation of scientific literature and increasingly reliable computational predictions have resulted in creation of vast databases of protein interaction data. Over the years, these repositories have become a basic framework in which experiments are analyzed and new directions of research are explored. Here we present an overview of the most widely used protein-protein interaction databases and the methods they employ to gather, combine, and predict interactions. We also point out the trade-off between comprehensiveness and accuracy and the main pitfall scientists have to be aware before adopting protein interaction databases in any single-gene or genome-wide analysis.

Key words Protein-protein interactions, Functional associations, Protein-protein interaction databases, Pathways, Protein-protein interaction prediction, Biochemical pathways, Selection bias

1 Introduction

The continual cost decrease of high-throughput experiments and the development of computational prediction methods have produced vast numbers of protein-protein interactions (PPIs). This ability to provide fairly comprehensive and reliable sets of PPIs prompted the development of many databases aiming to gather and unify the available data, each with a different focus and different strengths.

PPI databases can be categorized into three broad types: pathway databases like Reactome [1] and KEGG [2] in which expert curators collect consensus knowledge, databases of experimentally verified PPIs like IntAct [3] and BioGRID [4] that collect primary experimental data, and databases like STRING [5] and GeneMANIA [6] that also include computationally predicted interactions and text mining but perform no manual curation. Although databases from the latter category may contain many false positives, each interaction generally has a confidence score associated with it, which allows the user to filter out the most likely errors.

Cheryl L. Meyerkord and Haian Fu (eds.), *Protein-Protein Interactions: Methods and Applications*, Methods in Molecular Biology, vol. 1278, DOI 10.1007/978-1-4939-2425-7_3, © Springer Science+Business Media New York 2015

Some protein interaction databases are focused on a particular organism [7–9], disease [10, 11], or type of interaction, e.g., kinase–substrate interactions [12]. Other databases, especially resources with detailed pathway data [1, 2, 13], in addition to PPIs, include interactions with other macromolecules (mainly RNA and DNA) and small molecules such as drugs and metabolites.

Another differentiating feature of PPI databases is the ability to graphically visualize interaction networks instead of just showing a list of interaction partners of a query protein. This is not only a visual gimmick, as the network view gives the user an overview of the interactions between first (or more) degree neighbors, which in turn allows for quick visual recognition of highly interconnected functional modules of proteins. The significance of network visualization has been widely recognized and most databases provide some kind of network view; if not as a native application working in a browser, then by providing the network as a downloadable file that can easily be imported into visualization tools such as Cytoscape [14] or NAViGaTOR [15]. These tools and some databases also allow users to lay out networks, annotate nodes, and perform various types of network analysis such as clustering and term enrichment analysis (e.g., for Gene Ontology terms [16]). For different PPI visualization methods, *see* Fig. 1.

Because of the versatility and diverse set of features, PPI resources are now commonly used for data analysis, data interpretation, and hypothesis testing. A comprehensive list of more than 300 pathways and interaction databases is available from Pathguide (www.pathguide.org). However, the extent to which any of the published PPI datasets reflect the biological interactome is unknown, and it is thus essential to carefully evaluate the advantages and drawbacks of each interaction data source before using them. Undoubtedly till now, none can capture the full complexity of biological systems with different protein variants, modifications, and spatial and temporal dependencies.

2   The Many Faces of Interactions

Much like the term "function" does not only encompass molecular functions such as enzymatic catalysis, the term "interaction" in addition to direct physical binding covers a variety of indirect links such as complex co-membership, regulatory relationships, and genetic interactions. These are collectively referred to as "functional associations"; however, the terms "interaction" and "functional association" are often used interchangeably in the literature and in databases too.

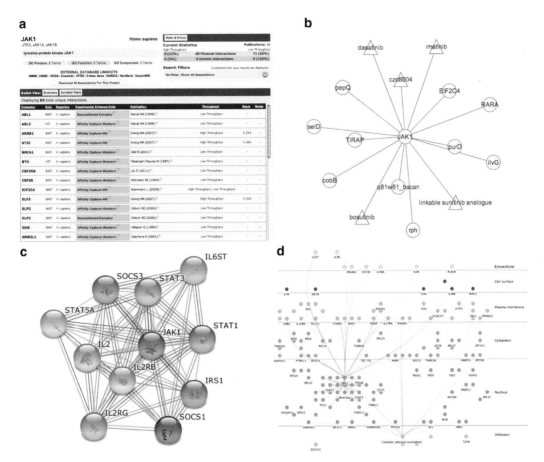

Fig. 1 Default visualizations of JAK1 interactions by different databases. (**a**) BioGRID HTML table showing proteins that physically interact with JAK1 (indicated by the *yellow color* of "experimental evidence"). (**b**) JAK1 interactions shown in the IntAct "graph viewer" with small molecules being depicted as *triangles*. The network view shows only the interactions with JAK1 and not between any other two nodes. (**c**) JAK1 STRING network. Each *different colored line* that connects proteins indicates a separate evidence channel for the particular interaction, such as text mining (*green*), experiments (*magenta*), and databases (*blue*). (**d**) InnateDB's Cytoscape with Cerebral plug-in view uses "cellular component" GO annotation to lay out proteins based on their localization inside the cell with the *horizontal lines* indicating boundaries between different components

This conceptual mixing of interactions and functions is in part due to data limitations and in part because the two concepts are closely related. Many proteins carry out their functions as parts of complexes. Deciphering the exact topology of a complex requires data on direct physical binding (e.g., yeast two-hybrid assays). By contrast, many of the experimental (most notably Tandem Affinity Purification) and computational methods identify highly interconnected clusters of proteins that represent complexes or other functional modules of unknown topology [17].

3   Pathway Databases

Reactome [1] is an open resource of, primarily human, curated pathway data. Contrary to other interaction databases, Reactome focuses on the accuracy of rather than the comprehensiveness of interactomes. Interactions are curated by Ph.D. level curators and reviewed by an expert from the relevant field. Each of the interactions is annotated in depth; this includes the directionality (if applicable) of interaction, type, substrates, stoichiometry of the reaction, its localization, and any known disease associations. In the default view, the interactive interface gives the user an ability to select specific pathway, or its part, using a pathway hierarchy browser, which then highlights its corresponding genes on the interaction map. Each pathway in Reactome has an in-depth description, relevant references, and should reflect current expert consensus. Reactome gives users the ability to conduct pathway enrichment analyses for an uploaded set of genes, which then can be browsed in the hierarchical pathway viewer, with pathways color-coded according to the found enrichment. The database also allows users to upload expression data for given set of genes, which are then mapped on the pathways with genes color-coded according to the user-supplied data.

One of the most widely used pathway map resources is KEGG (Kyoto Encyclopedia of Genes and Genomes) PATHWAY [2]. With over 1,500 different genomes, it is the most comprehensive species-wise. Its pathways span different cellular processes ranging from metabolism to genetic and environmental information processing, and each pathway aims to represent the complete knowledge about all existing reactions within it. Each unified manually curated map is referred to as a reference pathway and is not made with respect to any one specific species; instead each protein node represents a group of genes from different species. To create a species-specific pathway, the genes from each organism are mapped onto the reference pathway via semiautomatically inferred orthology relationships, and the map is color-coded to reflect which parts of the pathway are present in the particular species. KEGG PATHWAY is tightly coupled with other KEGG resources on genes and genomes (KEGG GENES), functional units (KEGG MODULE), genetic and environmental perturbations (KEGG DISEASES), and drugs (KEGG DRUG).

4   Databases of Experimentally Verified PPIs

It is very hard to assess how much knowledge about molecular interactions there is as most of the information on interactions exists only in the form of tables, supplementary data files, or free text in separate research articles, thus in practice rendering it inaccessible or hard to parse reliably by automatic means. Till this day manual curation of research articles is the only reliable way to accurately extract primary interaction data from the literature.

The BioGRID database [4] with its 370,000 unique PPIs is one of the largest repositories of experimentally verified PPIs, curated from 35,000 low- and high-throughput experiments covering 40 species. This includes full coverage of the literature for *Saccharomyces cerevisiae*, *Schizosaccharomyces pombe*, and *Arabidopsis thaliana*, with special focus on conserved networks and pathways. The database makes it easy to visually differentiate physical from genetic interactions as well as from high to low-throughput experiments. Except standard searches by gene name and publication, BioGRID allows user to construct complex Boolean queries with wildcards. BioGRID also allows user to search for keywords, sentences, and authors of abstracts associated with the interactions. Each interaction is annotated with easily discernible annotations such as organism (color-coded), BAIT/HIT directionality, experimental method used, and publication. In addition, all high-throughput interactions are annotated by the confidence score used by the author in the publication, with low-confidence interactions not being imported into the database. BioGRID does not include a native network visualization tool, but it allows the user to export data in a format compatible with Cytoscape [14]. There are also two Cytoscape plug-ins that facilitate import of BioGRID data handling both redundancy and annotations. It is also possible to access BioGRID via REST API, which allows for building web sites and scripts that directly communicate with the BioGRID database.

The DIP database [18], like BioGRID, is a manually curated set of experimentally determined PPIs retrieved from research articles. It does not focus on a particular organism, though more than 30 % of its interactions are from *S. cerevisiae*. The interactions stored in DIP can be viewed using Cytoscape's MiSink plug-in [19], which provides an interactive platform to analyze and visualize DIP data. Additionally, the web interface of DIP provides services for other external data analysis tools that allow the user to assess the reliability of interactions, for example based on expression profiles [20], comparison with interactions of paralogous proteins [20] and protein domain composition of the interacting proteins [21]. DIP also provides a subset of its database that consists of only signal transducing ligand–receptor pairs (DLRP) [22].

IntAct [3], a part of the EMBL-EBI database ecosystem, is another important repository of validated PPIs, both derived from literature and through user submissions. The database extends the list of binary interaction through "spoke" automatic co-complexes expansion, when an accurate list of binary interactions could not be derived from the experiment. The IntAct database allows the user to browse and search PPI data based on various categories including

GO terms [16], taxonomy, ChEBI ontology [23], Reactome pathway associations, mRNA expression, chromosomal location, and protein domains. The results of search queries can be visualized through a basic cytoscape plug-in working in a web browser or opened as a java application on the user's PC. Each experimentally inferred binary interaction stored in the IntAct database has an associated confidence score (MIscore, compliant with PSI-MI standard), which is cumulative and weighted based on available evidence. IntAct also includes interactions of proteins with small molecules associated with the ChEBI dictionary.

The MINT database [24] is a resource of experimentally validated molecular interactions extracted from peer-reviewed publications. In addition to protein-protein interactions, it also covers interactions of proteins with mRNA and genes' promoter regions. The MINT database, like IntAct, provides universal confidence scores, which range from 0 to 1 and are based on the number and reliability of the interaction's evidences. The results of the user query can be viewed in the web browser (MINT Viewer) as a list or as a modifiable network. The MINT Viewer allows user to arrange nodes, expand a network, filter links based on confidence, and show the links between all nodes in the network. The MINT Viewer gives the ability to view additional information about the protein, such as synonyms, proteins domains, and diseases associations. In addition to main database, MINT provides access to sister databases: HomoMINT [25] that covers human proteins extended by interactions from orthologous proteins from different species (the interactions stored in homoMINT are automatically imported from the main MINT database as soon as relevant interactions are uploaded); VirusMINT [26], a collection of human–viral protein interactions integrated with the human protein network; and DOMINO [27], which is a database of protein interactions mediated by a wide range of 200 protein-interaction domains.

5   Databases with Distinct Focus

The Human Protein Reference Database [7] is a resource of literature-mined interactions involving human proteins, small molecules, and nucleic acids. The HPRD, in addition to interactions, also includes information about protein posttranslational modifications, domain structures, localization, expression sites (differentiating between cell lines, normal and disease tissues), and OMIM [28] disease associations. The database features comprehensive browsing capabilities including searching by protein length, weight, name, chromosome locus, molecular class, posttranslational modifications (PTM), cellular component, domain, motif, and expression site. The PTM information contains residue-specific information, such as location, type, and upstream enzyme. The interactions are divided into direct (binary) and complex (when the topology of interaction is unknown) types and each is annotated with experiment type and link to the source publication. Each protein sequence can be visualized in the build-in viewer with the domains, motifs, modifications, and corresponding genomic region highlighted. If available, HPRD can also provide isoform-specific information for all of the above annotations. All the proteins are cross-referenced with NetPath pathway database [29] and Human Proteinpedia with which HPRD is tightly integrated. HPRD also includes PhosphoMotif Finder, built upon a literature-derived set of motifs, which reports found motifs in a submitted protein sequence alongside the annotations such as matched sequence, matched motif, upstream kinase, and link to the original publication from which the motif is derived.

The InnateDB [30] database is a resource for manually curated PPI and pathway data that focuses primarily on interactions associated with innate immunity. The database covers three organisms: *Homo sapiens, Mus musculus, and Bos Taurus.* In addition to 18,000 interactions extracted by the InnateDB curators, the database integrates various other resources including MINT, BioGRID, IntAct, DIP, and BIND [31]. The InnateDB also incorporates interolog transfer generated by in-house pipeline. Further it cross-references various pathway databases including but not limited to KEGG, Reactome, and NetPath. The web interface allows a user to query, apart from gene and protein names, for interactions in specific pathways and for interactions according to various criteria, among others: host system, cell type, tissue type, interaction type, molecule type, and interaction detection method. InnateDB provides data analysis tools for submitted protein/gene lists including gene ontology overrepresentation analysis, transcription factor biding sites overrepresentation analysis, and pathway enrichment analysis. The PPIs can be viewed as HTML or visualized, along with user submitted annotations (i.e., p-values, expression values), and analyzed using versatile tools including two Cytoscape plug-ins, CyOog [32] and Cerebral [33].

The Human Immunodeficiency Virus Type 1 (HIV-1), Human Protein Interaction Database [34] is a resource focused on cataloging all known interactions between HIV-1 and human proteins. It consists of 2,589 unique interactions targeting 1,448 human proteins. All these interactions are referenced and annotated with one of the 42 different types of associations including *binds, inhibits, complexes with, upregulates, cleaves,* and *co-localizes.* Although the primary interface shows only the rudimentary list of interactions along with their type, the user can download a subset or all the interactions as a tab-separated value file, which contains both the references to research articles from which the interaction was curated from and a brief description of each interaction.

Alternatively, the tight coupling of the database with NCBI [35] allows a user to view each HIV-1 or human protein in NCBI gene viewer along with its genome location, intron-exon structure, PubMed references, pathway associations, gene ontology terms, human–HIV and human–human PPIs from other databases along with their references to relevant research articles.

6 Interolog Prediction

Homology is evidence of functional similarity. Orthology and paralogy form the two major types of homologous relationships between genes and it is widely believed that, from the two, orthologs have greater ability to retain the same function [36]. As a consequence, most interolog prediction methods solely rely on ortholog mapping by transferring interaction found between orthologous genes in different species. Interolog predictions depend principally on ortholog mapping therefore the quality of the genome annotation, distance between species and number of paralogs in both species have a major impact on the confidence of these predictions.

As interolog prediction can be a reliable source of interaction information, especially between closely related genomes where orthology mapping could be accurately resolved, some of the databases of manually curated interaction incorporate these predictions as an additional source of interaction. The most obvious case is KEGG PATHWAY where the concept of using reference pathways depends solemnly on ortholog mapping. Reactome on the other hand does not try to predict interactions but instead allows a user to compare and visualize (only on human maps) pathway coverage from 20 selected genomes.

Some of the databases of experimentally validated interactions such as homoMINT and InnateDB also utilize interolog predictions, but I2D (Interologous Interaction Database) [37] stands out as it incorporates and transfers several major repositories of interaction data. The data stored in the I2D database can be categorized into two sets: the combined literature-derived human interaction from DIP, MINT, HPRD, and BIND databases; and computational prediction of interlogs between *Homo Sapiens, Saccharomyces cerevisiae, Caenorhabditis elegans, Drosophila melanogaster,* and *Mus musculus.* The interologs are inferred using a custom built pipeline using a best-hit approach and evaluated based on co-expression, Gene Ontology terms, and domain pairs co-occurrence. The results of the queries can be viewed either as an HTML table, custom graph viewer, or exported to NAViGa-TOR software [15]. In table view, each pair of interactions is annotated with additional supporting information, including domain co-occurrence, protein co-localization, co-mentioning in abstracts, expression correlation, and GO similarity. The java-based custom graph viewer allows for visual modification of the network, including changing the size, spread, opacity, and shape of the nodes. NAViGaTOR software is a comprehensive network analysis and visualization tool developed in the same lab as I2D, but it has the ability to import data of various formats including BioPAX (www.biopax.org) and PSI [38].

7 Predicted PPIs

It has been estimated that human proteins could give rise to as many as 200,000–300,000 direct interactions [39] and the most comprehensive databases to date, HPRD and BioGRID, both list a little more than 30,000 direct unique binary human protein interactions (*see* Table 1). Although the total number of experimentally validated associations (not only physical interactions) in all databases specialized in manual curation of literature is fivefold

Table 1
Comparison of different interaction databases

Database	Total number of interactions (thousands)	Number of interactions in human (thousands)	Number of species	Computational predictions	Built in network view	Unified scoring	IMEx
BioGRID	372	75,9	40	No	No	No	Yes (C)
DIP	68	3,4	485	No	Yes	No	Yes
GeneMANIA	96208	24220,1	7	Yes	Yes	Yes	No
HPRD	37	37,1	1	No	No	No	Yes
I2D	666	152,4	6	Yes (A)	No	No	Yes
InnateDB	104	84,0	3	Yes (A)	Yes	No	Yes
IntAct	2331	44,0	397	No	Yes	Yes	Yes
MINT	95	21,3	434	Yes (A, B)	Yes	Yes	Yes
PIPs	78	78,4	1	Yes	No	Yes	No
STRING	224346	1540,7	1133	Yes	Yes	Yes	No
Reactome	2161	121,9	28	No	Yes	No	No
KEGG (D)	5880	63,3	1509	Yes (B)	Yes	No	No

Binary protein interaction count does not include self-associations. For databases focused on computational predictions, the total number of interactions stored is not a valid predictor of quality of comprehensiveness of the database as most of the interactions are low-scoring and not relevant for noncomputational analysis. *A* only interologs prediction, *B* only homoMINT, *C* observing member. *D* the version of KEGG PATHWAY included in this breakdown is from July 2011, which was the last set available from KEGG under free license

higher, the overall number of existing associations inside the human cell could be at least a magnitude larger. In order to get a more complete picture of any interactome, it will be necessary to augment the existing knowledge with computational prediction.

In the late 1990s, access to the quickly expanding fields of genomics, transcriptomics, and proteomics gave researchers data-sets needed for the development of new computational methods for prediction of functional associations. Some of the developed methods, most notably interolog predictions, can infer interac-tomes for species for which the only existing data is a sequenced genome. These methods also gather more and more focus as it is becoming effectively impossible to acquire experimentally inferred interactomes with the pace new genomes are being sequenced. Although computational predictions could contain many false posi-tives, it is common that each computationally predicted interaction is annotated with a confidence score. Filtering to include only the highest scoring associations can thus yield interaction networks with a very low false-positive rate.

Access to fully sequenced and annotated genomes allows for the identification of protein fusion events. If such an event occurred, in one or more lineages, different parts (domains) of one fused protein could be found as being separate full-length proteins in other species [40]. Therefore, the existence of a fusion event is a strong predictor that, prior to it, the proteins were also functionally associated [41].

Another method of predicting protein-protein interaction from genomic data is based on phylogenetic profiles. This method relies on the assumption that genes that interact with each other or function together tend to be inherited together during speciation events. Therefore, genes that are functionally associated should exhibit similar patterns of absence and presence across many sequenced genomes [42].

If, in addition to the gene sequence data, we have knowledge about location and synteny of genes, we can utilize this as another source for PPI predictions. Both in prokaryotes and in eukaryotes, genes located near each other in the genome often form operons—a cluster of genes which is under a single regulatory signal. The conservation of gene order across several genomes [43] and the concurrent directionality of transcription [44] of genes located within these clusters point to their functional associations.

In order to computationally predict protein interactions, researchers can also utilize transcriptomics data from various RNA-seq or microarray experiments. A single experiment has the ability to cover a considerable part of the genome, which provides researchers an unparalleled look into the regulation of proteins across various species and conditions. Similar to gene neighborhood conservation, the co-expression of two proteins is considered to be an indication of co-regulation and, consequently, functional association [45]. Leveraging, for example, the vast library of over 30,000 different experiments stored in the GEO database [46] researchers can explore co-expression across numerous experimental conditions.

Only 0.0015 % of all research articles in the MEDLINE database have been covered by manual curation; therefore researchers have started to look into automated means of extracting PPIs from literature. Automatic text mining methods can be divided into two major categories: co-occurrence/frequency-based methods and natural language processing (NLP). Co-occurrence text mining purely relies on matching exact elements (tokens) from the specified dictionary without taking into account the sequence in which the tokens appear or—to some extent—if the tokens are located in the same sentence. On the other hand, for NLP these characteristics of the text are essential and the interaction can be mined only when the two proteins are mentioned in the same sentence. Although NLP does not require repeated co-occurrence of same links in order to mine high-confidence interactions, this advantage is counterbalanced by very low sensitivity of NLP methods due to the sentence constrain [47].

8   Databases of Predicted PPIs

GeneMANIA [6] is a database of known and predicted functional associations between proteins that covers seven major organisms (*D. melanogaster, R. norvegicus, S. Cerevisiae, C. elegans, M. Musculus, A. Thaliana,* and *H. Sapiens*). The physical and genetic interactions included in GeneMANIA are assembled from datasets stored in Pathway Commons (which consists of various other resources including IMEx consortium databases and Bio-GRID). The predicted associations include: interologs predictions (mostly from I2D), predictions based expression profiles from experiments stored in GEO [46], and predictions based on shared domain composition. The distinctive feature of GeneMANIA is that every dataset incorporated in GeneMANIA could be regarded as a separate interaction network. The confidence scores (weights) of links from each network are based on the raw score from the particular set. The resulting final network consists of these scores combined and weighted accordingly to the confidence associated with each data source. The weights from all the links for each of the data sources sum to 100 %, and the user has an ability to choose one of seven different weighting methods including query- and annotation-dependent methods. Using a friendly interface, the user also has full control over which dataset or prediction method should contribute to the network and can turn them on/off accordingly. GeneMANIA automatically finds all the Gene Ontology annotations of proteins in the network and assesses their false discovery rate and their network coverage. It also has a function to color-code the network node according to one or more GO terms. The user has the ability to query GeneMANIA with more than one protein or to upload its own interaction sets, along with confidence scores, that will be incorporated with the GeneMANIA network and framework.

Search Tool for the Retrieval of Interacting Genes/Proteins (STRING) [5] is another database that incorporates both known and predicted functional associations between proteins. It covers 1,133 fully sequenced genomes across all three domains of life. It integrates several different kinds of sources: heavily curated interactions stored in pathway databases such as Reactome and KEGG; manually curated low and high-throughput experiments from IMEx consortium databases and BioGRID; interactions retrieved using the co-occurrence method from text mining of 20,000,000 PubMed abstracts; computational prediction of PPI by correlating mRNA expression profiles across various experiments stored in the GEO database; interactions predicted from genomic context, such as, gene-fusion, gene neighborhood, and phylogenetic profiles, which does not depend on prior knowledge and could be generated for any species with a fully sequenced genome; and interolog predictions for all evidence sources and all species in the STRING database. All known and predicted links are annotated with a confidence score (from 0 to 1), and scores originating from different sources are added in a probabilistic manner in order to create a final network. The protein view contains basic information about a protein including sequence, domain composition, 3D structure, and homologs in a selected species. The STRING network can be recomputed based on user-chosen set sources and cut-offs as well as expended based on user-selected proteins or globally with the most confident links connected to the nodes in a visible network. The interactive STRING network view gives the user basic network analysis tools, such as network clustering, protein-protein interaction enrichment, pathway enrichment, protein domains enrichment, and Gene Ontology terms enrichment. The STRING database is closely coupled with its sister databases: STITCH [48], which, in addition to proteins, covers small molecules and eggNOG [49], the database of orthologous relationships between STRING's proteins.

The Human protein-protein Interaction Prediction [50] (PIPs) database is a resource containing human protein–protein functional associations derived by purely computational methods. Each link in the database is annotated with a score representing a likelihood ratio of the particular interaction occurring. This ratio is evaluated based on HPRD interactions as a positive set and a 100 times larger negative set constructed from random associations between HPRD proteins. The interactions were calculated using six different methods: co-expression; interacting domain co-occurrence; interologs mapping; posttranslational modifications co-occurrence; cell compartment protein co-localization; and a network topology method based on the hypothesis that the more interaction partner proteins share, the more likely that they interact with each other. The confidence scores from different prediction algorithms are combined using the Bayesian method, with a score greater than 1 denoting that the interaction is more likely to occur than not. The results of a user query are presented as an HTML table with annotation on fractional scores and if the particular interaction was seen in other databases (BIND, DIP, HPRD, and I2D).

9   PPI Database Federation

Database federation, which is the transparent interoperation of several databases, can be viewed as an alternative to resources like GeneMANIA and STRING that aim to integrate everything in one place. The early collaboration between major PPI database providers, which resulted in the development of the MIMIx curation standard and common data structure (PSI-MI XML and MITAB), grew to become the International Molecular Exchange (IMEx) consortium. As of 2012 the IMEx consortium comprised of the following members and partners: DIP [18], IntAct [3], MINT [24], MatrixDB [51], MPIDB [52], Molecular Connections (http://www.molecularconnections.com), I2D [37], InnateDB [53], SwissProt group [54], and as an observing member BioGRID [4]. In addition to common curation guidelines and data structures, the member databases now use common controlled vocabularies (PSI-MI ontologies) to describe, for example, the type of experiment. All of these efforts make the datasets from different IMEx databases relatively easy to compare in an automated way.

There is also an ongoing endeavor between IMEx members to release the full, nonredundant, set of guideline compliant protein interactions. Though, as of 2012, most of the participating databases still have data to be released to the IMEx consortium dataset, and integrating different databases is still necessary to get all the available interactions (*see* Fig. 2). There is no central repository of the IMEx interaction set, and the associated records exist only in their source databases tagged respectively to indicate classification as being a part of the IMEx dataset. To address this, the IMEx consortium developed the PSI Common Query Interface (PSICQUIC) [55], which gives users an ability to query all databases simultaneously and presents the result as a list with the option to cluster the results in order to create a nonredundant set of interactions. PSICQUIC View incorporated within EBI web services allows a user to query databases that are outside the IMEx consortium, such as STRING and Reactome.

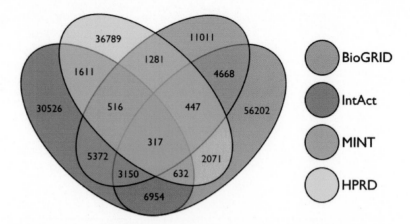

Fig. 2 Overlap of binary human protein interactions between selected IMEx databases. The total count of interactions includes only the interactions between proteins for which identifiers could be mapped onto a common set of human ENSEMBL identifiers

10   Data Integration

The individual interaction sets derived through high-throughput techniques are not only incomplete but also have relatively high false-positive rates [56]. In addition, it has been estimated that the nonfunctional direct biophysical interactions that are found through Yeast 2-Hybrid assays could constitute to as much as 19 % of all interactions from these experiments [57]. These effects alone can explain the often poor overlap between different interaction studies. Not one method could create a complete picture of an interactome; therefore, in order to obtain the closest approximation of it from existing data, we need to integrate different data sources from experimental and computational methods.

The simplest way to integrate different high- and low-throughput datasets is to combine all mined interactions in a list. Especially in the case of manually curated high-throughput interaction sets, this is often done with the accompanying reliability annotation of a particular set either derived from the experiment itself or annotated based on the method used. This method, although simple, does have the advantage of not omitting any supposedly relevant data and is most suitable in situations when a researcher is interested in a very limited set of proteins and where all interaction evidence can thus be assessed manually. The drawback of such an approach is the more limited utility for high-throughput network analysis without extensive postprocessing.

Another common way to integrate different data, employed by several databases, is to compound confidence scores for links for which there are more than one line of evidence. As a result, the final compounded link would have a higher confidence score than any of its contributing evidences individually. This could be done by assigning one fixed confidence score for each link from a particular data source—this is predominantly done for sources for which the links cannot be sorted in a meaningful way that would correlate well with their confidence, e.g., the combined list of interactions from many small-scale experiments. Alternatively, one can calibrate confidence scores based on raw scores from experiments or predictions; these raw scores can, for example, be correlation coefficients between the expression levels of different genes or text mining co-occurrence frequencies.

This compounded confidence integration is predominately done for combining different computational predictions. This is due to the fact that the majority of links have low confidence levels and only when sources are compounded can a more applicable network be formed. If the biases of the prediction methods are not correlated, it is beneficial to have several sources as it would dilute the individual biases of any single source.

11 Bias

The story that is emerging from protein-protein interaction networks forms a consistent view of scale-free networks [58], with essential, well-conserved proteins being the hubs that give rise to robustness in the networks [59]. Indeed, this is undoubtedly true for the data, but it does not necessarily reflect the underlying biological system. This discrepancy happens due to biases connected to the data, as any experiment or prediction method is subject to different unavoidable biases either linked to the experimental method [60] or sampling bias caused by research interest [61].

The majority of research articles report an interaction only when it has been observed (either by high- or low-throughput methods) and do not report when an interaction has been examined but the result was negative. Due to this reporting bias, the integration of the subsequent data can only add new interactions rather than refine the network by eliminating false positives. As a consequence of inherent false-positive rates and imperfect reproducibility of data, it is thus to be expected that the compounded false-positive rate should be higher for well-studied pathways or proteins. Indeed, this appears to be the case as it has been shown that the popularity of a field correlates with a higher false-positive rate of interactions [61]. There is little doubt that to some degree the same effect of selection bias applies, among others, to disease-associated proteins [62], evolutionary conserved proteins [63], and highly expressed proteins [64]. Of course a single high-throughput experiment may not be subjected to selection bias; nonetheless the experiment itself can contain different sorts of biases, i.e., the yeast two-hybrid method forces the two proteins to localize to nucleus, which for some types of proteins (in particular membrane proteins) may cause them to misfold and aggregate, resulting in a lower detection rate [65]. Furthermore, differences in the expression level of proteins may influence the detection rate as well, producing, already mentioned, bias towards higher expressed proteins.

As for now, there is no agreement of how accurate protein-protein interaction networks represent true biological interactomes. There are possibilities to increase reliability of known methods, i.e., by releasing raw, unprocessed data in order to refine networks by incorporating true negatives. Till then, researchers have to rely on their own assessment of biases and take them into account when inferring any knowledge based on protein interaction networks.

References

1. Croft D, O'Kelly G, Wu G, Haw R et al (2011) Reactome: a database of reactions, pathways and biological processes. Nucleic Acids Res 39:D691–D697

2. Kanehisa M, Goto S, Furumichi M, Tanabe M et al (2010) KEGG for representation and analysis of molecular networks involving diseases and drugs. Nucleic Acids Res 38:D355–D360

3. Kerrien S, Aranda B, Breuza L, Bridge A et al (2012) The IntAct molecular interaction database in 2012. Nucleic Acids Res 40:D841–D846

4. Stark C, Breitkreutz B-J, Chatr-Aryamontri A, Boucher L et al (2011) The BioGRID Interaction Database: 2011 update. Nucleic Acids Res 39:D698–D704

5. Szklarczyk D, Franceschini A, Kuhn M, Simonovic M et al (2011) The STRING database in 2011: functional interaction networks of proteins, globally integrated and scored. Nucleic Acids Res 39:D561–D568

6. Warde-Farley D, Donaldson SL, Comes O, Zuberi K et al (2010) The GeneMANIA prediction server: biological network integration for gene prioritization and predicting gene function. Nucleic Acids Res 38:W214–W220

7. Goel R, Harsha HC, Pandey A, Prasad TSK (2012) Human Protein Reference Database and Human Proteinpedia as resources for phosphoproteome analysis. Mol Biosyst 8:453–463

8. Cherry JM, Hong EL, Amundsen C, Balakrishnan R et al (2012) Saccharomyces Genome Database: the genomics resource of budding yeast. Nucleic Acids Res 40:D700–D705

9. Murali T, Pacifico S, Yu J, Guest S et al (2011) DroID 2011: a comprehensive, integrated resource for protein, transcription factor, RNA and gene interactions for Drosophila. Nucleic Acids Res 39:D736–D743

10. Goodman N, McCormick K, Goldowitz D, Hockly E et al (2003) Plans for HDBase—a research community website for Huntington's Disease. Clin Neurosci Res 3:197–217

11. Lechner M, Höhn V, Brauner B, Dunger I et al (2012) CIDeR: multifactorial interaction networks in human diseases. Genome Biol 13:R62

12. Dinkel H, Chica C, Via A, Gould CM et al (2011) Phospho.ELM: a database of phosphorylation sites–update 2011. Nucleic Acids Res 39:D261–D267

13. Caspi R, Foerster H, Fulcher CA, Kaipa P et al (2008) The MetaCyc Database of metabolic pathways and enzymes and the BioCyc collection of Pathway/Genome Databases. Nucleic Acids Res 36:D623–D631

14. Smoot ME, Ono K, Ruscheinski J, Wang P-L et al (2011) Cytoscape 2.8: new features for data integration and network visualization. Bioinformatics 27:431–432

15. Brown KR, Otasek D, Ali M, McGuffin MJ et al (2009) NAViGaTOR: Network Analysis, Visualization and Graphing Toronto. Bioinformatics 25:3327–3329

16. Gene T., Consortium O. (2010) The Gene Ontology Consortium in 2010: extensions and refinements. Nucleic Acids Res 38:D331–D335

17. Hakes L, Robertson DL, Oliver SG (2005) Effect of dataset selection on the topological interpretation of protein interaction networks. BMC Genomics 6:131

18. Salwinski L, Miller CS, Smith AJ, Pettit FK et al (2004) The Database of Interacting Proteins: 2004 update. Nucleic Acids Res 32:D449–D451

19. Salwinski L, Eisenberg D (2007) The MiSink Plugin: cytoscape as a graphical interface to the Database of Interacting Proteins. Bioinformatics 23:2193–2195

20. Deane CM, Salwiński Ł, Xenarios I, Eisenberg D (2002) Protein interactions: two methods for assessment of the reliability of high throughput observations. Mol Cell Proteomics 1:349–356

21. Deng M, Mehta S, Sun F, Chen T (2002) Inferring domain-domain interactions from protein-protein interactions. Genome Res 12:1540–1548

22. Graeber TG, Eisenberg D (2001) Bioinformatic identification of potential autocrine signaling loops in cancers from gene expression profiles. Nat Genet 29:295–300

23. Hastings J, de Matos P, Dekker A, Ennis M et al (2013) The ChEBI reference database and ontology for biologically relevant chemistry: enhancements for 2013. Nucleic Acids Res 41:D456–D463

24. Ceol A, Chatr Aryamontri A, Licata L, Peluso D et al (2010) MINT, the molecular interaction database: 2009 update. Nucleic Acids Res 38:D532–D539

25. Persico M, Ceol A, Gavrila C, Hoffmann R et al (2005) HomoMINT: an inferred human network based on orthology mapping of protein interactions discovered in model organisms. BMC Bioinformatics 6(Suppl 4):S21

26. Chatr-aryamontri A, Ceol A, Peluso D, Nardozza A et al (2009) VirusMINT: a viral protein interaction database. Nucleic Acids Res 37:D669–D673

27. Ceol A, Chatr-aryamontri A, Santonico E, Sacco R et al (2007) DOMINO: a database of domain-peptide interactions. Nucleic Acids Res 35:D557–D560

28. Amberger J, Bocchini C, Hamosh A (2011) A new face and new challenges for Online Mendelian Inheritance in Man (OMIM®). Hum Mutat 32:564–567

29. Kandasamy K, Mohan SS, Raju R, Keerthikumar S et al (2010) NetPath: a public resource of curated signal transduction pathways. Genome Biol 11:R3

30. Breuer K, Foroushani AK, Laird MR, Chen C et al (2013) InnateDB: systems biology of innate immunity and beyond–recent updates and continuing curation. Nucleic Acids Res 41:D1228–D1233

31. Bader GD, Donaldson I, Wolting C, Ouellette BF et al (2001) BIND–The Biomolecular Interaction Network Database. Nucleic Acids Res 29:242–245

32. Royer L, Reimann M, Andreopoulos B, Schroeder M (2008) Unraveling protein networks with power graph analysis. PLoS Comput Biol 4:e1000108

33. Barsky A, Gardy JL, Hancock REW, Munzner T (2007) Cerebral: a Cytoscape plugin for layout of and interaction with biological networks using subcellular localization annotation. Bioinformatics 23:1040–1042

34. Fu W, Sanders-Beer BE, Katz KS, Maglott DR et al (2009) Human immunodeficiency virus type 1, human protein interaction database at NCBI. Nucleic Acids Res 37:D417–D422

35. Resource Coordinators NCBI (2013) Database resources of the National Center for Biotechnology Information. Nucleic Acids Res 41: D8–D20

36. Chen R, Jeong SS (2000) Functional prediction: identification of protein orthologs and paralogs. Protein Sci 9:2344–2353

37. Niu Y, Otasek D, Jurisica I (2010) Evaluation of linguistic features useful in extraction of interactions from PubMed; application to annotating known, high-throughput and predicted interactions in I2D. Bioinformatics 26:111–119

38. Hermjakob H, Montecchi-Palazzi L, Bader G, Wojcik J et al (2004) The HUPO PSI's molecular interaction format–a community standard for the representation of protein interaction data. Nat Biotechnol 22:177–183

39. Hart GT, Ramani AK, Marcotte EM (2006) How complete are current yeast and human protein-interaction networks? Genome Biol 7:120

40. Burns DM, Horn V, Paluh J, Yanofsky C (1990) Evolution of the tryptophan synthetase of fungi. Analysis of experimentally fused Escherichia coli tryptophan synthetase alpha and beta chains. J Biol Chem 265:2060–2069

41. Enright AJ, Iliopoulos I, Kyrpides NC, Ouzounis CA (1999) Protein interaction maps for complete genomes based on gene fusion events. Nature 402:86–90

42. Marcotte EM, Pellegrini M, Ng HL, Rice DW et al (1999) Detecting protein function and protein-protein interactions from genome sequences. Science 285:751–753

43. Dandekar T, Snel B, Huynen M, Bork P (1998) Conservation of gene order: a fingerprint of proteins that physically interact. Trends Biochem Sci 23:324–328

44. Overbeek R, Fonstein M, D'Souza M, Pusch GD et al (1999) Use of contiguity on the chromosome to predict functional coupling. In Silico Biol 1:93–108

45. Eisen MB, Spellman PT, Brown PO, Botstein D (1998) Cluster analysis and display of genome-wide expression patterns. Proc Natl Acad Sci U S A 95:14863–14868

46. Barrett T, Troup DB, Wilhite SE, Ledoux P et al (2011) NCBI GEO: archive for functional genomics data sets–10 years on. Nucleic Acids Res 39:D1005–D1010

47. Hirschman L, Park JC, Tsujii J, Wong L et al (2002) Accomplishments and challenges in literature data mining for biology. Bioinformatics 18:1553–1561

48. Kuhn M, Szklarczyk D, Franceschini A, von Mering C et al (2012) STITCH 3: zooming in on protein-chemical interactions. Nucleic Acids Res 40:D876–D880

49. Powell S, Szklarczyk D, Trachana K, Roth A et al (2012) eggNOG v3.0: orthologous groups covering 1133 organisms at 41 different taxonomic ranges. Nucleic Acids Res 40: D284–D289

50. McDowall MD, Scott MS, Barton GJ (2009) PIPs: human protein-protein interaction prediction database. Nucleic Acids Res 37: D651–D656

51. Chautard E, Fatoux-Ardore M, Ballut L, Thierry-Mieg N et al (2011) MatrixDB, the extracellular matrix interaction database. Nucleic Acids Res 39:D235–D240

52. Goll J, Rajagopala SV, Shiau SC, Wu H et al (2008) MPIDB: the microbial protein interaction database. Bioinformatics 24:1743–1744

53. Lynn DJ, Winsor GL, Chan C, Richard N et al (2008) InnateDB: facilitating systems-level analyses of the mammalian innate immune response. Mol Syst Biol 4:218

54. The UniProt Consortium (2011) Ongoing and future developments at the Universal Protein Resource. Nucleic Acids Res 39:D214–D219

55. Aranda B, Blankenburg H, Kerrien S, Brinkman FSL et al (2011) PSICQUIC and PSISCORE: accessing and scoring molecular interactions. Nat Methods 8:528–529

56. Sambourg L, Thierry-Mieg N (2010) New insights into protein-protein interaction data lead to increased estimates of the S cerevisiae interactome size. BMC Bioinformatics 11:605

57. Nakayama M, Kikuno R, Ohara O (2002) Protein-protein interactions between large proteins: two-hybrid screening using a functionally classified library composed of long cDNAs. Genome Res 12:1773–1784

58. Jeong H, Tombor B, Albert R, Oltvai ZN et al (2000) The large-scale organization of metabolic networks. Nature 407:651–654

59. Wuchty S, Oltvai ZN, Barabási A-L (2003) Evolutionary conservation of motif constituents in the yeast protein interaction network. Nat Genet 35:176–179

60. Von Mering C, Krause R, Snel B, Cornell M et al (2002) Comparative assessment of large-scale data sets of protein-protein interactions. Nature 417:399–403

61. Ioannidis JPA (2005) Why most published research findings are false. PLoS Med 2:e124

62. Tang JL (2005) Selection bias in meta-analyses of gene-disease associations. PLoS Med 2:e409

63. Pál C, Papp B, Hurst LD (2003) Genomic function: rate of evolution and gene dispensability. Nature 421:496–497, discussion 497–8

64. Bloom JD, Adami C (2003) Apparent dependence of protein evolutionary rate on number of interactions is linked to biases in protein-protein interactions data sets. BMC Evol Biol 3:21

65. Brito GC, Andrews DW (2011) Removing bias against membrane proteins in interaction networks. BMC Syst Biol 5:169

Chapter 4

Computational Prediction of Protein-Protein Interactions

Tobias Ehrenberger, Lewis C. Cantley, and Michael B. Yaffe

Abstract

The prediction of protein-protein interactions and kinase-specific phosphorylation sites on individual proteins is critical for correctly placing proteins within signaling pathways and networks. The importance of this type of annotation continues to increase with the continued explosion of genomic and proteomic data, particularly with emerging data categorizing posttranslational modifications on a large scale. A variety of computational tools are available for this purpose. In this chapter, we review the general methodologies for these types of computational predictions and present a detailed user-focused tutorial of one such method and computational tool, *Scansite*, which is freely available to the entire scientific community over the Internet.

Key words Scansite, Protein-protein interaction prediction, Sequence motif, PSSM, Binding motif, Phosphorylation sites, Bioinformatics

1 Introduction

Decades of research in molecular biology have resulted in the availability of vast amounts of data, including genomic sequences, protein sequences, structural data, and protein metadata including functional domain information and interaction data. Unfortunately, the availability of these data types does not necessarily result in a clear understanding of what all the data means in a broader context. The bulk of the available data is single molecule-centric, limiting our ability to understand how molecules are integrated into pathways and networks. With the advent of new experimental techniques, it is possible to enrich these pieces of data with additional information related to protein-protein interactions and enzyme–substrate relationships. One of the most important breakthroughs in this context was the rise of experimental techniques that allowed the rapid and large-scale detection of protein-protein interactions [1]. Since the molecular apparatus of a cell is mainly controlled by protein–protein and protein–nucleic acid interactions, detecting and understanding such events, particularly

Cheryl L. Meyerkord and Haian Fu (eds.), *Protein-Protein Interactions: Methods and Applications*, Methods in Molecular Biology, vol. 1278, DOI 10.1007/978-1-4939-2425-7_4, © Springer Science+Business Media New York 2015

direct interactions, is the first step to a broader view of biological systems. Interaction information has been collected in a number of different public databases [2, 3], and the information stored in these databases mostly contains experimentally verified information, i.e., data from in vivo or in vitro experiments. Unfortunately, given the current trend towards large-scale proteome wide analyses and the fact that these databases are far from complete, this information often proves insufficient for many analyses. The missing pieces in the puzzle that elucidates a more complete view of the cell interactome can be provided by interaction prediction tools. These tools create in silico predictions of protein-protein interactions and kinase–substrate relationships, are typically inexpensive and fast compared to conventional time- and resource-intensive experimental methods, and can provide a focused list of predictions that can then be verified or refuted by further focused experimental testing.

Over the past years, a number of different computational approaches for predicting protein-protein interactions have been developed. These techniques can be divided into those that are based on a single biological feature and those that attempt to use a range of different features and data types and can therefore be categorized based on the types of data that they use. A detailed overview of how each of these approaches works, and what the shortcomings of these methods are, can be found elsewhere [4, 5], but a short summary is provided here, focusing on which general features are used to predict protein-protein interactions.

At one extreme are methods based on a protein's three-dimensional structure, generally referred to as "protein docking" techniques. Given a 3D model (usually based on high-resolution data from X-ray crystallography or NMR experiments, deposited in the Protein Data Bank [1]) for two potentially interacting proteins, the best fit for each potential interaction interface on the surface of these models can be searched for and scored [6]. However, finding low-energy fits is very challenging, often due to the static nature of the PDB structures and the dynamic plasticity that can occur at protein-protein interfaces. Thus, conformational changes, the arrangements of side chains, and the energy levels of a potential conformation combination and interaction, potential posttranslational modifications that may or may not be included in a model, and a number of other factors have to be considered. Because of the large variance in the quality of the prediction of different methods in this field, CAPRI (Critical Assessment of Protein Interactions), a community-based program that regularly evaluates the algorithms to predict protein-protein interactions based on structures in a double-blind manner, has been initiated [7]. Alternatively, machine learning methods can be developed, usually based on databases of experimentally verified interactions and a number of additional biological properties. These points of data are then used to train a prediction engine based on known data [8]. The problem with this approach—as with any other machine learning approach used in this manner—is that the resulting predictor does not provide easily decipherable information about why the proteins are likely to interact. This means that, although it may yield useful results, it is hard to reconstruct and understand exactly why a prediction is made by this black-box predictor. A very specific type of machine learning method tries to find a pattern based on features at the interaction interface of the proteins involved. A method closely related to this approach is described later in this chapter. Other prediction methods are based on genomics. Gene fusion methods predict that discrete proteins are likely to interact if their homologues are fused into single genomic entities in other species. Other techniques based on gene neighborhood conservation are built on the hypothesis that gene pairs within such neighborhoods that are evolutionary conserved across different species are likely to interact.

No matter which method is used, it is important to keep the caveats of the method in mind. First, no prediction can guarantee either biological correctness or relevance. This is especially important if prediction tools are used to design and plan further experiments without first confirming the initial prediction. Failure of a method to predict an interaction may not reflect a fundamental problem with the method but may instead reflect limitations of the data that the method is based on. The data, be it experimentally verified interaction sites, 3D data of proteins, or other information, originates from experiments which are all error-prone, though in some cases the extent of the error may be difficult to estimate. This also applies to methods that use machine learning to train a predictor, as these methods are highly dependent on the quality of the underlying training dataset. Obviously, a large set of training data is necessary to create a good predictor and, indeed, large databases of experimentally verified protein-protein interactions are now available. However, training a predictor also requires a negative dataset that provides information of what interactions are very unlikely to happen. Experimental data of this type are typically not published, at least in part due to difficulties in distinguishing whether the lack of an observed interaction is the result of a technically failed experiment or because there is no biologically relevant interaction [9]. The end result is a lack of reliable negative training data for computational method development. The quality and nature of the training data should therefore be one important consideration in the user's choice of whether to trust a predictor trained on these kinds of data types, including the species of the proteins in the training dataset, the type of experiments used to verify the sites, and of course the number of sites and proteins included. Thus, it is very important for prediction tools to explicitly (1) give information about how the method works and what information it uses, (2) provide some type of quantitative measure that allows users to compare different results and distinguish between good and not-as-good predictions, and (3) provide any additional information that helps the user decide whether to trust the predictions. This information could be incorporated into the prediction method itself but is also very helpful if this type of metadata is simply presented for the user to examine independently of whether this information is explicitly used in the prediction algorithm.

One of the most important features in describing protein-protein interactions is elucidating the exact sites on the proteins where the interaction occurs, either at the detailed atomic/structural level or at the level of specific amino acid sequences. That is, on the surface of the protein, which part of the amino acid sequence directly contacts or indirectly influences the interaction partner? By focusing solely on sequence information, the complexities required for interaction prediction by docking-type simulations (conformational states of side chains, energetic contributions, etc.) are radically reduced.

Since protein-protein interactions are mediated by attractive forces based on the physicochemical properties of amino acids, in many cases it is sufficient to describe potential interaction partners by amino acid sequence patterns alone. This can be clearly shown by considering kinase–substrate interactions: kinases generally only phosphorylate serine (S), threonine (T), or tyrosine (Y) residues based on the ability of their phosphate acceptor hydroxyl groups to nucleophilically attack the γ-phosphate of ATP. However, more than 500 different kinases are known alone in humans, each of which targets a different set of substrates [10]. The specific site of phosphorylation is therefore not the only amino acid that plays a role in substrate recognition. Instead, 4–12 amino acid residues on the substrate flanking the phospho-acceptor likely physically contact a kinase's active site [11] and help to position the substrate for an in-line attack on the phosphate while simultaneously optimizing the geometry of the kinase's catalytic machinery to facilitate stabilization of the resulting transition state. This indicates that this part of the substrate's primary structure may be sufficient to determine whether an acceptor residue is likely to be phosphorylated by a given kinase. Although sequence patterns are only one piece of information in a puzzle of many (secondary structure, tertiary structure, surface accessibility, etc.), an abundance of data suggests that this is one of the most distinguishing factors in describing a kinase–substrate interaction and in many cases it is a sufficient predictive feature [12, 13]. Obviously, this idea does not apply only to kinases but can be used to describe other protein-protein interactions mediated by other types of modular protein domains that recognize short linear sequence motifs on their binding partners in a phospho-dependent or -independent manner, such as SH2 and SH3 domains, FHA and BRCT domains, 14-3-3 proteins, etc.

Specific amino acid preferences can be described in two ways. One is to describe them in a strict combinatorial regular expression-like pattern (Boolean matching model). This approach was originally used in PROSITE [14] to search for patterns in a sequence database. However, these patterns are very inflexible and do not allow for including differently weighted preferences for amino acids. A more flexible and powerful approach is the use of position-specific scoring matrices (PSSMs) to describe patterns/motifs in this form. This approach was implemented in *Scansite* [15, 16], an application to predict short linear sequence motif sites. A PSSM matrix like this contains a probability value for each amino acid (columns) at each position of a sequence window of certain size (rows), where each value in a column and row of the matrix describes the binding partner's preference for that amino acid at that position in the motif. *Scansite* is a web application that uses PSSMs to predict interaction sites that are important in cellular signaling and includes more than 120 kinases and proteins that recognize specific short linear binding motifs. It can be used to show all potential sites in a given protein or all proteins in a database that contain sites for one or more motifs. Directions for both uses are provided in the following sections.

2   Materials

Scansite 3 (http://scansite3.mit.edu/) requires nothing more than a computer with an Internet connection and a modern web browser. Although it works with all popular web browsers, the recommended options are Mozilla Firefox, Google Chrome, and Opera. On some pages that display search results, *Scansite* will show content in pop-up windows so that results can be viewed side by side. Therefore it is recommended that you allow pop-ups in your browser for these pages to work properly. Wherever *Scansite* allows you to choose a sequence database (e.g., when selecting proteins or for searching a database), you can choose from these resources: SwissProt [17], SGD (yeast) [18], Ensembl (human and mouse) [19], NCBI Protein (GenPept) [20], and TrEmbl [21]. *Scansite* uses local mirrors of these databases in order to allow fast queries. Over the past years *Scansite* has also become popular for analyses of whole proteomes or subsets thereof. These are generally not done using the web interface, but computationally. If you are interested in using *Scansite* for this purpose, please *see* **Note 1** for information about *Scansite*'s web service.

To perform *Protein Scans* you need either a protein sequence or a protein identifier (accession number or ID) for the protein you are

*2.1 Scanning a
Protein for Motifs*

interested in from one of *Scansite*'s protein sequence database mirrors.

*2.2 Searching a
Sequence Database for
Motifs*

Database Searches only require information about the motif that is searched for in a particular sequence database. All of the standard *Scansite* matrices for kinases and modular binding domains are available. In addition, you may enter more specific information to restrict the search to a smaller number of proteins.

3 Methods

Scansite's two most important interaction prediction searches will be described in detail in this section: *Protein Scans* that search for motif matches in a given protein and *Database Searches* that find proteins that contain one or more motifs in a protein sequence database. A short overview of *Scansite*'s other features is given in **Note 2**. In the following, you will be guided through the steps necessary to use these features properly. Furthermore, some guidance on how to interpret these searches' results will be given.

*3.1 Scanning a
Protein for Motifs*

The key feature of *Scansite* is the prediction of motif-relevant sites in a given protein. This feature is referred to as *Protein Scan* or *Scan Proteins for Motifs* and allows a range of different inputs.

1. *Navigate to Input Page.* To get to the Protein Scan input screen from anywhere in *Scansite*, click the "Scan Proteins for Motifs" button in the navigation section on the left-hand side of the web page.

2. *Choose the Protein to Scan.* There are two different ways of choosing proteins in *Scansite*: by protein identifier (default option) and by sequence.

   To choose a protein by accession number, select "Protein Accession" from the "Choose Protein by..." drop-down list. Below, select a protein sequence database and enter a protein ID. Links on the right-hand side of the text boxes refer to the different sequence databases that *Scansite* currently supports and where you can search for protein identifiers. After entering at least three characters in the text box entitled "Protein Accession", *Scansite* searches for protein IDs that start with these characters and presents a list of options below the text box. The same happens when the "Check!" button next to the text box is clicked or the Enter key is pressed. You can either continue typing or select an ID from the list. The text box turns green for valid and red for invalid protein identifiers.

   In order to enter a peptide sequence, select "Input Sequence" from the drop-down list. The area below this menu will change accordingly. Then, enter or paste a name and an amino acid sequence. Invalid characters (punctuation marks, white space, digits, etc.) are stripped from the sequence automatically. This means that you can just paste a sequence that is formatted with spaces and line breaks or annotated with numbers. If you paste a FASTA-formatted sequence, make sure not to copy the FASTA header (">...") with the sequence. Otherwise all possible amino acid one letter codes in the header will also become part of the sequence.

3. *Choose Motifs to Consider.* It is possible to search for all motifs of a motif class, for only a selected subset of motifs or motif groups or both, or for a user-defined motif (instructions on how to create your own motif can be found in **Note 3**). You can choose from these options in the drop-down menu entitled "Look for". Again, the area below this menu will change accordingly dependent on your choice, offering a number of additional choices. Begin by selecting a motif class (mammalian or yeast). By default, *Scansite*'s mammalian motifs are displayed. To select more than one motif or motif group, hold down the control key on your keyboard and make selections using your mouse. If you are not sure which motifs belong to which groups, you can either click the link below the list of groups ("Show Group Definitions") or follow the instructions in **Note 4**. When using your own motif, select the motif file from your computer. After the file is uploaded (this happens automatically after you selected a file), you get a chance to make changes to affinity values if you wish to do so.

4. *Select a Stringency Level.* This measure defines how high sites have to score in order to be displayed as results. The setting *high* only displays the very best sites, i.e., the top 0.2 % of sites (sites that have a score less than or equal to the top 0.2 % of motif-specific scores in the reference proteome). *Medium stringency* displays the top 1 %, *low* the top 5 %, and *minimum* displays the top 15 %. These settings apply only for motifs from the *Scansite* database. Since no precompiled reference proteome score distribution (*see* **Note 5**) is available for user-defined motifs, these always display all sites with a score ≤ 5.

5. *Additional Options.* The two additional options that users are given are to decide whether to show predicted domains in the result as supporting information (*see* **Note 6**) and whether to use an alternative reference proteome. At the moment users can use either SwissProt's Vertebrate proteins as a reference (default) or all of SGD's proteins (default for scans using yeast motifs). Domains can also be requested later on from the result page.

6. *Click the Submit Button.*

Protein Scan Results: *P53_HUMAN (swissprot)*

Protein Overview

Protein Scanned: P53_HUMAN (see SwissProt, see PhosphoSite)

Descriptions: RecName: Full=Cellular tumor antigen p53; AltName: Full=Antigen NY-CO-13; AltName: Full=Phosphoprotein p53; AltName: Full=Tumor suppressor p53;

Keywords: Apoptosis, Cell cycle, Host-virus interaction, Endoplasmic reticulum, DNA-binding, Isopeptide bond, Cytoplasm, Zinc, Alternative promoter usage, Transcription regulation, Complete proteome, Ubl conjugation, Polymorphism, Transcription, Metal-binding, Alternative splicing, Glycoprotein, Acetylation, 3D-structure, Phosphoprotein, Disease mutation, Li-Fraumeni syndrome, Activator, Methylation, Reference proteome, Tumor suppressor, Nucleus

Accessions: Q9NP68, Q9NZD0, P53_HUMAN, Q2XN98, Q3LRW2, Q3LRW1, Q3LRW5, Q3LRW4, Q9UQ61, Q9HAQ8, Q16848, Q8J016, Q16808, Q16809, Q16807, Q9NPJ2, Q86UG1, Q9UBI2, Q9BTM4, P04637, Q16810, Q16811, Q99659, Q15087, Q15088, Q16535, Q15086

Molecular Weight: 43658.8

Isoelectric Point: 6.33

Scan Overview

Protein Plot

Predicted PFAM-Domains (from InterProScan): P53_TAD (6 - 29), P53 (95 - 288), P53_tetramer (318 - 358)

Note: The domains' positions are retrieved from InterProScan. For this reason the numbers may differ slightly from PFAM-retrieved domains.

Go to PFAM.

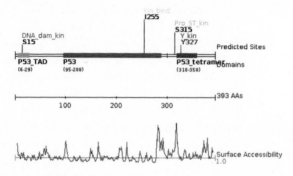

Protein: P53_HUMAN (swissprot)

Predicted Motif Sites (Table)

Please allow popups in your browser settings to make links in the table work properly!

Score	Percentile	Motif	▲ Motifgroup	Site	Sequence	Surface Accessibility	Gene Info	Previously Mapped Site
0.159	0.001%	ATM Kinase (ATM_Kin)	DNA damage kinase group (DNA_dam_kin)	S15	PSVEPPLsQETFSDL	1.6951	ATM	PhosphoELM, Phosphosite
0.400	0.128%	DNA PK (DNA_PK)	DNA damage kinase group (DNA_dam_kin)	S15	PSVEPPLsQETFSDL	1.6951	PRKDC	PhosphoELM, Phosphosite
0.323	0.010%	Erk D-domain (ErkDD)	Kinase binding site group (Kin_bind)	I255	RRPILTIiTLEDSSG	0.3350	MAPK1	
0.269	0.074%	CDK1 motif 2 - [ST]PxxK (CDK1_2)	Proline-dependent serine/threonine kinase group (Pro_ST_kin)	S315	LPNNTSSsPQPKKKP	2.2838	CDK1	PhosphoELM, Phosphosite
0.326	0.068%	Fgr Kinase (Fgr_Kin)	Tyrosine kinase group (Y_kin)	Y327	KKPLDGEyFTLQIRG	0.6342	FGR	Phosphosite
Score	Percentile	Motif	Motifgroup	Site	Sequence	Surface Accessibility	Gene Info	Previously Mapped Site

DISCLAIMER: These results are purely speculative and should be used with EXTREME CAUTION because they are based on the assumption that the peptide library data is correct and sufficient to predict a site!

Also, if an evidence for a site is given ('previously mapped site') it is only site- and protein-specific, meaning that this site is known to be phosphorylated by some kinase, but *not necessarily* by the kinase Scansite associates with this site!

Repeat Scan

Stringency: Medium ∨ | Scan! |

| Repeat Search with Different Parameters |

Download Results

Download results as tab separated file...

Additional Analyses

| Score sites using DisPhos (Disorder-Enhanced Phosphorylation Site Predictor)... |

Fig. 1 The results of a high stringency protein scan for all mammalian motifs using the SwissProt protein *P53_HUMAN* and the default reference proteome are shown. The section entitled "Scan Overview" which summarizes the parameters of the scan of the page is collapsed to better fit this figure on the page

As an example for a Protein Scan result page, the results of a high stringency protein scan are shown in Fig. 1. The result page is split in seven sections (divided by grey bars): Protein Overview, Scan Overview, Protein Plot, Predicted Motif Sites (Table), Repeat Scan, Download Results, and Additional Analyses. Each of these sections is collapsible by clicking on the grey title areas. This allows the user to quickly get to the bottom of the page if a long list of predicted sites is displayed.

In the "Protein Overview" section, some information about the input protein is listed, including alternative identifiers and keywords (only for proteins from *Scansite*'s databases), and the protein's molecular weight and isoelectric point (calculated according to ref. 22). The "Scan Overview" summarizes the input parameters of the search and displays the number of sites that have been detected using these settings. In the next part of the page ("Protein Plot"), a plot of the protein gives a visual overview of the search results displaying the protein sequence as a straight line annotated with some additional information. If domain information about the query protein was requested to be displayed, the predicted domains are listed above the image. The plot displays the predicted sites (annotated with the position and motif group), the protein's domains (if requested) along with their names and positions, and a surface accessibility plot that shows which parts of the protein are likely to be exposed to the surface and which ones are likely to be buried. If domains have not been requested earlier, a button will be displayed below the image that allows the user to request domain prediction at this point. The links in the list of displayed domains refer to these domains' PFAM pages (*see* **Note 6**).

The sites that are outlined in the protein plot are listed in more detail in the table view below ("Predicted Motif Sites"). Most columns can be sorted by clicking on the label in the table's header. Here, each site that was found is displayed along with some motif information (motif, motif group, hyperlink to motif's gene information page), its score and percentile, and the surrounding sequence. In addition, *Scansite-*3 offers hyperlinks to PhosphoSite [23], PhosphoELM [24], and Phosida [25] if a site was reported in one of these databases before (for more information about "Previously Mapped Sites" *see* **Note 7**). The other links displayed in the table, more specifically the columns "Score" and "Sequence," refer to a histogram view of a site in the reference proteome and to a view that shows a site's sequence highlighted in the protein's sequence, respectively. The latter view also offers a link that directly submits the site's sequence (15 amino acids) to NCBI's basic local alignment search tool (BLAST) [26]. More information on BLASTing sites in *Scansite* can be found in **Note 8**.

In the "Repeat Scan" section of the result page, it is possible to either directly rerun the scan with a different stringency setting or to go back to the input page to change other search parameters. This is especially helpful if your search did not return any results. The next part in the page ("Download Results") offers a link to a download-able version of the table shown above (tabulator-separated file). At the bottom of the result page ("Additional Analyses") users can directly submit the current protein's sequence to DisPhos [27], a Disorder-Enhanced Phosphorylation Site Predictor (*see* **Note 9**).

3.2 Searching a Sequence Database for Motifs

The *Scansite* feature *Search Sequence Database for Motifs* or short *Database Search* performs a broader search than single protein scans. Given a motif (or a set of motifs) and a sequence database, it searches the database for sequences that contain motif-relevant sites. One of the most powerful parts of this tool is the option of targeting a search to specific experimental requirements by restricting searches to proteins of a specific organism class, species, molecular weight and isoelectric point range, annotation, and sequence property. For example, this tool can be used to help identify unknown bands in two-dimensional (2D) gel electrophoresis experiments.

1. *Navigate to Input Page.* To get to the Database Search input screen from anywhere in *Scansite*, click the button "Search a Sequence Database for Motifs" in the navigation section on the left-hand side of the web page.

2. *Choose the Search Method.* The area below this drop-down list will change dependent on what you select. Searches for single "Database motifs" from the *Scansite* database are the easiest option to choose. Alternatively, you can search for your own motifs (*see* **Note 3**) or so-called "Quick Motifs" (*see* **Note 10**). It is also possible to search for sequences that match up to five motifs. These searches can include either database motifs, user-defined motifs, or a combination of both. The score of a multi-motif site is the mean (average) of all the scores of the sites involved. Co-occurrences of different motif sites in proteins can be filtered in different ways. First of all, it is possible to penalize gaps between sites of different motifs. Gap penalty settings are either *high, medium, low,* or *none*. Penalties p are then added to the score according to the maximum distance d_{max} between the involved sites (i.e., position of site closest to C-terminus minus position of site closest to N-terminus). The penalty values are calculated as follows: $p_{low} = 0.001 \times d_{max}$; $p_{medium} = 0.01 \times d_{max}$; $p_{high} = 0.1 \times d_{max}$. Secondly, it is possible to define up to three strict minimum and maximum distance bounds between motif-specific sites. This can be used if you know which motifs to expect and how far apart you expect them to be in the protein sequence. If you just want to get an overview of peptides that have multiple motif sites, it is recommended to use a gap penalty. Using distance bounds is the better option for very specific searches.

3. *Select Database to Search* from the drop-down menu.

4. *Restrict Search.* It is recommended that you exclude as many proteins from the search as possible to both target your search as much as possible to what you are looking for and to decrease the runtime of the search. Database searches can take several minutes and the runtime of a search mostly depends on the number of proteins that are searched. You will find useful hints on what kinds of restrictions you can apply in **Note 11**.

5. *Select Number of On-Screen Results.* Since Database Searches may find a very high number of results and visual exploration of a table of thousands of results generally is avoided, the number of sites that are displayed in the web browser is limited. By default, the size of the output list is limited to 50, but users can also choose sizes 100, 200, 500, 1,000, and 2,000. Please note that these are just the numbers of sites that are displayed in the table on the result page. A file containing all the hits that were found in this search can be downloaded from the result page as well.

6. *Click the Submit Button.*

A result page of a Database Search is displayed in Fig. 2. Four sections can be distinguished within the Database Search result page. The "Search Input" section at the top of the page summarizes the preferences defined in the input page. "Search Results" gives an overview of the number of proteins in the entire sequence database, the number of proteins found that matched the given restrictions, and the number of sites found in these proteins. In addition, the median and MAD (median absolute deviation) of these sites' scores is displayed. This part is followed by a table view of the sites found ("Predicted Motif Sites"). The table shows the (combined) site score, some information about the protein that was found (including MW and pI), and displays some site-specific information (site and surrounding sequence). For multi-motif searches a site and sequence column for each motif in the motif's site is given. The first column in the table allows to directly scan the protein for other motifs. This is useful if you want to know what other motifs are found in that protein, if a site has been reported before (previously mapped) in another database, and how the protein is generally composed (domains, surface accessibility). The link in the column labeled "Accession" takes the user to the protein's page in its primary database. The score column links to a histogram that shows the site's score in comparison to all scores found in that search. At the bottom of the page, options for downloading the entire result set and for repeating the search are given.

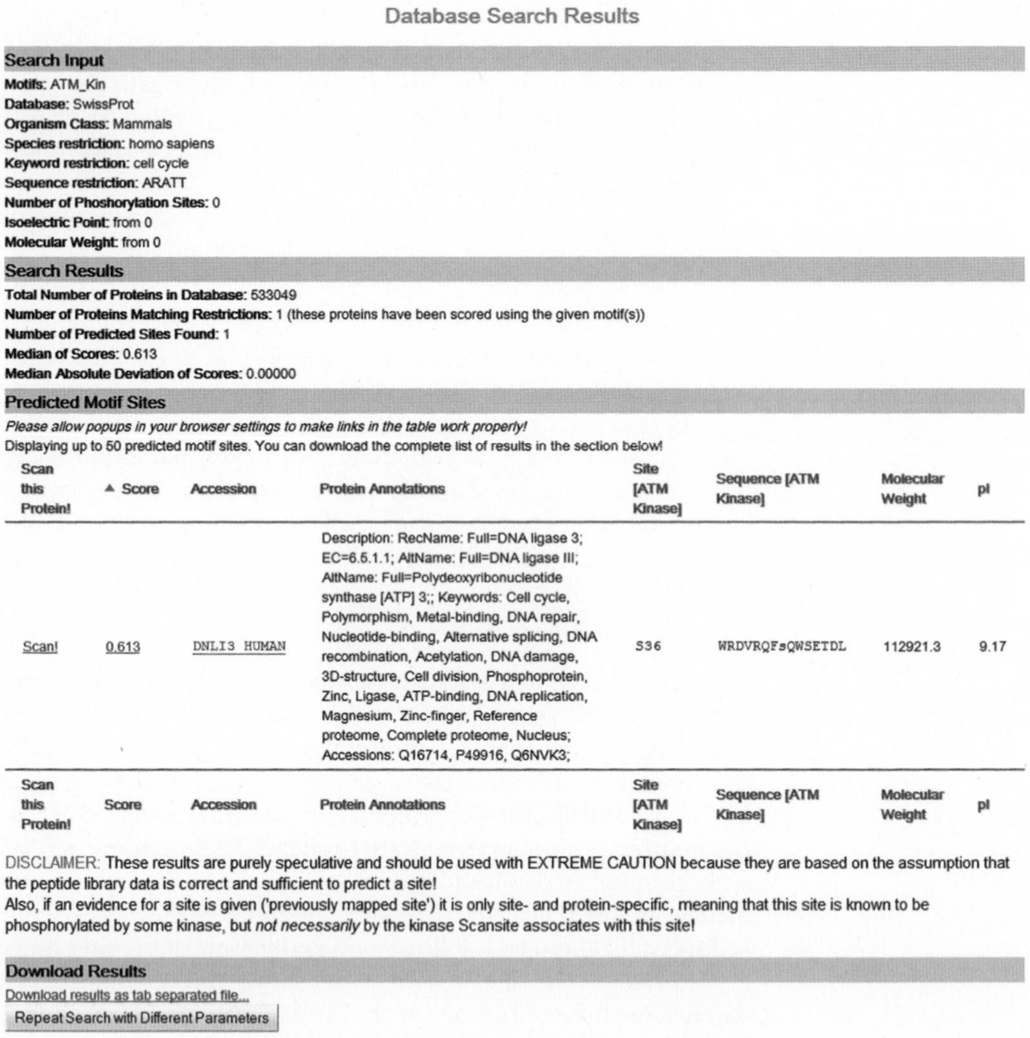

Database Search Results

Search Input

Motifs: ATM_Kin
Database: SwissProt
Organism Class: Mammals
Species restriction: homo sapiens
Keyword restriction: cell cycle
Sequence restriction: ARATT
Number of Phoshorylation Sites: 0
Isoelectric Point: from 0
Molecular Weight: from 0

Search Results

Total Number of Proteins in Database: 533049
Number of Proteins Matching Restrictions: 1 (these proteins have been scored using the given motif(s))
Number of Predicted Sites Found: 1
Median of Scores: 0.613
Median Absolute Deviation of Scores: 0.00000

Predicted Motif Sites

Please allow popups in your browser settings to make links in the table work properly!
Displaying up to 50 predicted motif sites. You can download the complete list of results in the section below!

Scan this Protein!	▲ Score	Accession	Protein Annotations	Site [ATM Kinase]	Sequence [ATM Kinase]	Molecular Weight	pI
Scan!	0.613	DNLI3 HUMAN	Description: RecName: Full=DNA ligase 3; EC=6.5.1.1; AltName: Full=DNA ligase III; AltName: Full=Polydeoxyribonucleotide synthase [ATP] 3;; Keywords: Cell cycle, Polymorphism, Metal-binding, DNA repair, Nucleotide-binding, Alternative splicing, DNA recombination, Acetylation, DNA damage, 3D-structure, Cell division, Phosphoprotein, Zinc, Ligase, ATP-binding, DNA replication, Magnesium, Zinc-finger, Reference proteome, Complete proteome, Nucleus; Accessions: Q16714, P49916, Q6NVK3;	S36	WRDVRQFsQWSETDL	112921.3	9.17
Scan this Protein!	Score	Accession	Protein Annotations	Site [ATM Kinase]	Sequence [ATM Kinase]	Molecular Weight	pI

DISCLAIMER: These results are purely speculative and should be used with EXTREME CAUTION because they are based on the assumption that the peptide library data is correct and sufficient to predict a site!
Also, if an evidence for a site is given ('previously mapped site') it is only site- and protein-specific, meaning that this site is known to be phosphorylated by some kinase, but *not necessarily* by the kinase Scansite associates with this site!

Download Results

Download results as tab separated file...

Repeat Search with Different Parameters

Fig. 2 The results of a Database Search for ATM in human proteins of SwissProt that are annotated with "cell cycle" and contain the sequence "ARATT". Here, only one protein matched the given restrictions and this protein also contains the motif that was searched for

4   Notes

1. *Accessing Scansite Computationally.* The current era of genomics and proteomics often requires analyses of large numbers of proteins. To make tasks like this easier it is now possible to access *Scansite* computationally using a web service. The parameters of protein scans, database searches, and other utility functions are sent to *Scansite* using a URI. The results are then returned in XML format. Detailed instructions and examples are available online at http://scansite3.mit.edu/Scansite3Webservice/. This link can also be found in *Scansite*'s FAQ online.

2. *Getting the Most out of Scansite.* In addition to the features described in detail above, *Scansite* offers some more useful tools. To start with, you can search *Scansite*'s sequence databases for simple wildcard-based sequence patterns or regular expressions. Another tool calculates a sequences molecular weight and isoelectric point for a given number of putative phosphorylations. Last, a tool called "Calculate Amino Acid Composition" visualizes a protein sequence's amino acid composition by highlighting selected sites and displaying the relative abundance of sites (e.g., all tyrosines in a sequence that are followed by leucines two residues downstream). In addition, this tool displays a protein's domain information as calculated by InterProScan [28]. One can also use one of these tools to analyze a protein sequence, copy/paste it to make changes (e.g., introduce mutations), and then use it as an input for protein scans.

3. *Creating Scansite Motifs.* Both main search options in *Scansite* allow the use of user-defined motifs. These motifs have to be in a *Scansite*-specific tabulator-separated file format. All user-defined motifs that are uploaded to *Scansite* are only used for the user's searches and are deleted as soon as the user leaves the site. If you have a clear idea of what motif you want to look for, use the information below to specify your own *Scansite*-specific motif file.

PSSMs in *Scansite* describe amino acid-specific affinity values for a sequence window of 15 residues. Lines correspond to positions in the sequence window, columns (separated by tabulators) to amino acids. It is not necessary to define values for every single amino acid—default values are used for omitted residues. The first line (row 1, header) defines the residue-to-column assignments using amino acid one letter codes. Those amino acids can be in any order. The following lines (rows 2–16) define affinity values for the respective residues; rows 2–8 and 10–16 define the N- and the C-terminal side of the motif, respectively. *Scansite*'s search for sites in a peptide sequence highly depends on the PSSM's central residue (row 9). At least one site in this position needs to be invariant in the motif sequence. For example, the fixed residue should be a Y for motifs recognized by tyrosine-kinases and S and T for serine-/threonine-kinases. To mark a position as invariant, the value 21 has to be used.

In addition to columns of standard amino acids (default values of 1), it is also possible to incorporate special requirements. A motif's preference for a protein sequence's N- or C-terminus can be incorporated by using a column labeled "$" (dollar sign) or "*" (asterisk), respectively. These positions are assigned values of 0 by default. *Scansite 3* also recognizes the rarely occurring amino acids selenocysteine (U) and pyrrolysine (O), which can be added by their one letter code as well. Due to their similar chemical structure, the default numbers for these residues are the values of cysteines and lysines, respectively. Lastly, some wildcard values can be used for very special cases: B (aspartate/asparagine), Z (glutamate/glutamine), J (leucine/isoleucine), and X (any residue). These symbols are included because they occur rarely in public protein databases. Generally speaking, they have no relevance for actual research purposes. The default values for these wildcards are the mean values of the amino acids they encode.

Now that the general structure and default values of motif files were defined, you may wonder what values to use to define affinities. *Scansite*'s scoring system ranges from 0 to roughly 21. Giving an individual amino acid a score of 1 at one position in the motif indicates that no preference exists, positive or negative, for that particular amino acid in that position. Giving all amino acids in one position of the motif a score of 1 (i.e., making all values in a single row of the matrix equal to one) indicates no preference exists for any particular residue type at that position in the motif. The value 21 defines that the amino acid that is given this value in a position is required in this position for the motif to find a match. Values higher than 21 are permitted to indicate very strong affinities. However, negative values are not permitted for defining a strong disfavoring of amino acids. Instead, values between zero and one should be used for that purpose. Beware that the scoring function uses logarithms, so values less than 1, particularly those less than 0.5, strongly penalize for that particular residue in a motif.

Here is a short checklist to avoid the most common pitfalls of creating motifs:

- Is there at least one amino acid with value 21 in the central position?
- Is there a header line defining the columns using amino acid one letter codes?
- Are there 16 lines (1 header and 15 lines with values) in the file?
- Are all column separators in the file tabulators (and not spaces or other characters)?

4. *Learn more about Scansite's Data.* In a section called "Databases and Motifs" in the navigation section of the web page (left-hand side), an overview of *Scansite's* database mirrors (release dates and sizes), motifs, and motif group definitions is presented. In the motifs section you can select a motif and click "Get Info!." Clicking this button will visualize the motif as a sequence logo [29] and display a link that takes you to a web

page that gives information about the gene that recognizes this motif. Mammalian motifs and yeast-specific motifs are supported by information from GeneCards [30] and SGD, respectively.

5. *Interpreting Scansite Scores. Scansite*'s scores range from 0 to (theoretically) ∞. However, you will never see scores higher than 5 because sites with scores that high are discarded in the scoring process. Please be aware that scores in *Scansite* are always motif-dependent. This means that scores for different motifs should not be directly compared to each other. For example, knowing that one motif's optimal score is 0.001 and another motif's best score is 0.4 it is easy to say that these are the best possible scores, so hits with these scores are equally good. However, the only way to extend this knowledge to slightly poorer scores is to know how likely other scores are to occur. To make this possible and allow a comparison among motifs, *Scansite* offers percentile values. The percentiles used in *Scansite* are calculated from the so-called reference proteomes which are proteomes that are commonly used in research. In the process of adding a motif to *Scansite*, it is scored against every single peptide in the reference proteome and the scores are stored to create a score distribution. This distribution is then used to calculate percentile values from scores calculated when users run certain searches. Using these values it is possible to rank sites from different motifs.

6. *Domains in Scansite 3. Scansite* uses InterProScan [28] to predict a protein's PFAM domains [31]. Therefore the domain positions displayed in *Scansite* may vary by a few amino acids from the positional assignments seen on the PFAM homepage. This is mentioned because these variations may cause confusion but do not pose a problem since all these positions are predictions and there is no way to tell which numbers are more correct in the absence of clear structural data from crystallographic or NMR experiments.

7. *Previously Mapped Sites in Scansite.* Displaying previously mapped sites in *Scansite* is only possible for proteins from public protein databases and works best with proteins from SwissProt. Please note that these references are only site-specific but not motif-specific. This means that if a previously mapped site shows up in the list, the site is reported in the linked databases; however, this does not imply that the *Scansite* motif that was found at this site is related to the site reported in the database. It could be that a completely different gene is responsible for this site. Wherever possible, the hyperlinks refer directly to the external databases page about this site. If a database does not support direct linking, the link just takes you to the database's homepage.

8. *BLASTing of Sites. Scansite* allows to directly submit the 15-mers around identified sites to NCBI's BLAST. This is a simple approach to see if a site is conserved in organisms that are expected to be physiologically similar to the one at hand. If the site is also found in similar proteins in other species, the site is more likely to be biologically relevant.

9. *Intrinsically Disordered Proteins.* Disordered regions in proteins are stretches of amino acids that do not have a rigid tertiary structure and are therefore enabled to change conformation. Disordered Proteins are proteins with disordered regions. It has been shown [32] that many posttranslational modifications and binding sites occur in disordered regions because these regions make a protein more flexible, which facilitates binding and interaction processes. DisPhos is a disorder-prediction engine that focuses on potential phosphorylation sites. The results of DisPhos searches can therefore be used as supporting information for phosphorylation sites predicted by *Scansite*.

10. *Using Quick Motifs.* Creating a custom motif only makes sense if enough information about the affinities of the kinase or binding domain is known. This, however, requires a very specific idea about the motif. Often, only very little detail about a motif is known. In cases like these, creating a "Quick Motif" to search a database is the best option. For defining a quick motif, the user can enter a set of primary and secondary preferences for each position of a 15-mer. These preferences are then used to calculate a simple *Scansite* motif. As for actual *Scansite* motifs, the center position needs to be fixed, so it is not possible to enter secondary preferences there. The web page describes a number of wildcards that can be used in this process to easily describe amino acid subsets by their physicochemical properties (e.g., hydrophobic or positive residues). A simplified regular expression-like version of the motif that is entered is displayed below the text boxes (with resolved wildcards) as soon as values in the text boxes are changed.

11. *Restricting Searches.* Searches of protein databases can be restricted in a number of ways to allow better more targeted searches. At the same time applying restrictions reduces the number of proteins that have to be scanned and therefore may significantly reduce the time a query takes. Consequently, users are encouraged to restrict their searches as rigorously as possible. For some on-site information, a short help text about each restriction can be displayed by clicking on the links next to the text boxes.

    • For many searches, you may only be interested in matches from humans or a particular model organism. Searches can be restricted this way by entering the species' name in the text box labeled "Single Species." This feature supports many MySQL-style wildcards (regular expressions) to match species names. For example, if you are tired of writing out "Caenorhabditis elegans", you can use "C.* elegans" instead. In a regular expression, the period (.) matches any single character, and the asterisk extends that match to multiple characters (or even zero characters). This also allows for genus-wide searches, by entering just "Rattus" for example. However, this may yield unexpected results when trying to search for all kinds of mice with "Mus." This expression will accidentally match "Thermus aquaticus" as well, but you can avoid that by entering "^Mus." The caret symbol (^) requires the text to match at the beginning of the entered name. One of the most common pitfalls is when the species entered does not match the organism class specified above (e.g., a search for "yeast" when "Mammals" is selected). Please note that *Scansite*'s organism classes are not taxonomic "classes" in the conventional sense (except for Mammals) but groups of species frequently used for research purposes.

- The molecular weight, isoelectric point, and phosphorylation options are intended for use in conjunction with 2D gel electrophoresis experiments. When you find a few spots appearing reproducibly on a 2D gel under a particular test condition and not under the control, you could use Scansite to find what proteins are expected to be in that region of the 2D gel by putting in ranges for molecular weights and isoelectric points. You could simultaneously constrain the species to match the cell line you used in the experiment. If it is an experiment involving possible phosphorylation events, you can see how much a putative phosphorylation would move the peptides on the gel.

- Matches for "Keywords" are searched in a protein's annotations and are therefore primarily useful for searching well-annotated databases like SwissProt. For example, proteins involved in the cell cycle can be easily identified by entering "cell cycle," novel proteins in GenPept by searching for "hypothetical."

- The "Sequence Contains" text field is a quick way to restrict your search to proteins containing a consensus sequence. It is important to note that the consensus sequence entered here is not required to be part of the motif being searched for. It is merely required to show up somewhere in the sequence. Also, note that regular expressions have to be used here instead of the protein wildcard signs ("." instead of "X", "[ND]" instead of "B", etc.). For example, the sequence "PXXP" is represented as "P..P" in regular expression syntax. More information on regular expressions and how they can be used in *Scansite* is available in *Scansite's* frequently asked questions (FAQ) section online.

References

1. Berman HM, Westbrook J, Feng Z et al (2000) The Protein Data Bank. Nucleic Acids Res 28:235–242
2. Mathivanan S, Periaswamy B, Gandhi T et al (2006) An evaluation of human protein-protein interaction data in the public domain. BMC Bioinformatics 7:S19
3. Turinsky A, Razick S, Turner B, et al. (2010) Literature curation of protein interactions: measuring agreement across major public databases. Database (Oxford) 2010: baq026
4. Shoemaker B, Panchenko A (2007) Deciphering protein-protein interactions. Part II. Computational methods to predict protein and domain interaction partners. PLoS Comput Biol 3:e43
5. Pitre S, Alamgir M, Green J, et al. (2008) Computational methods for predicting protein–protein interactions. In: Advances in biochemical engineering/biotechnology: protein-protein interaction. Springer, Heidelberg, pp 247–267
6. Andrusier N, Mashiach E, Nussinov R et al (2008) Principles of flexible protein–protein docking. Proteins 73:271–289
7. Janin J (2002) Welcome to CAPRI: a Critical Assessment of PRedicted Interactions. Proteins Struct Funct Genet 47:257
8. Rhodes DR, Tomlins SA, Varambally S et al (2005) Probabilistic model of the human protein-protein interaction network. Nat Biotechnol 23:951–959
9. Trost B, Kusalik A (2011) Computational prediction of eukaryotic phosphorylation sites. Bioinformatics 27:2927–2935
10. Hutti J, Jarrell E, Chang J et al (2004) A rapid method for determining protein kinase phosphorylation specificity. Nat Methods 1:27–29
11. Songyang Z, Blechner S, Hoagland N et al (1994) Use of an oriented peptide library to determine the optimal substrates of protein kinases. Curr Biol 4:973–982
12. Kemp BE, Pearson RB (1990) Protein kinase recognition sequence motifs. Trends Biochem Sci 15:342–346
13. Pinna LA, Maria Ruzzene M (1996) How do protein kinases recognize their substrates? Biochim Biophys Acta 13143:191–225
14. Bairoch A (1992) PROSITE: a dictionary of sites and patterns in proteins. Nucleic Acids Res 20(Suppl):2013–2018
15. Yaffe M, Leparc G, Lai J et al (2001) A motif-based profile scanning approach for genome-wide prediction of signaling pathways. Nat Biotechnol 19:348–353
16. Obenauer J, Cantley L, Yaffe M (2003) Scansite 2.0: proteome-wide prediction of cell signaling interactions using short sequence motifs. Nucleic Acids Res 31:3635–3641
17. M. M, UniProt-consortium (2011) UniProt Knowledgebase: a hub of integrated protein data. Database: bar009
18. Cherry J, Hong E, Amundsen C et al (2011) Saccharomyces Genome Database: the genomics resource of budding yeast. Nucleic Acids Res 40(Database issue):D700–D705
19. Flicek P, Amode M, Barrell D et al (2011) Ensembl 2011. Nucleic Acids Res 39(Suppl 1):D800–D806
20. Burks C, Cassidy M, Cinkosky MJ et al (1991) GenBank. Nucleic Acids Res 19:221–225
21. Boeckmann B, Bairoch A, Apweiler R et al (2003) The SWISS-PROT protein knowledgebase and its supplement TrEMBL in 2003. Nucleic Acids Res 31:365–370
22. Bjellqvist B, Hughes G, Pasquali C et al (1993) The focusing positions of polypeptides in immobilized pH gradients can be predicted from their amino acid sequences. Electrophoresis 14:1023–1031
23. Hornbeck P, Kornhauser J, Tkachev S et al (2012) PhosphoSitePlus: a comprehensive resource for investigating the structure and function of experimentally determined post-translational modifications in man and mouse. Nucleic Acids Res 40:D261–D270
24. Dinkel H, Chica C, Via A et al (2011) (2011) Phospho.ELM: a database of phosphorylation sites—update 2011. Nucleic Acids Res 39: D261–D267
25. Gnad F, Ren S, Cox J et al (2007) PHOSIDA (phosphorylation site database): management, structural and evolutionary investigation, and

prediction of phosphosites. Genome Biol 8: R250

26. Altschul SF, Madden TL, Schäffer AA et al (1997) Gapped BLAST and PSI-BLAST: a new generation of protein database search programs. Nucleic Acids Res 25:3389–3402

27. Iakoucheva L, Radivojac P, Brown C et al (2004) The importance of intrinsic disorder for protein phosphorylation. Nucleic Acids Res 32:1037–1049

28. Hunter S, Apweiler R, Attwood T et al (2009) InterPro: the integrative protein signature database. Nucleic Acids Res 37: D211–D215

29. Schneider T, Stephens R (1990) Sequence logos: a new way to display consensus sequences. Nucleic Acids Res 18:6097–6100

30. Stelzer G, Dalah I, Stein T et al (2011) In-silico human genomics with GeneCards. Hum Genomics 5:709–717

31. Punta M, Coggill P, Eberhardt R et al (2012) The Pfam protein families database. Nucleic Acids Res 40:D290–D301

32. Uversky V, Dunker A (2010) Understanding protein non-folding. Biochim Biophys Acta 1804:1231–1264

Chapter 5

Structure-Based Computational Approaches for Small-Molecule Modulation of Protein-Protein Interactions

David Xu, Bo Wang, and Samy O. Meroueh

Abstract

Three-dimensional structures of proteins offer an opportunity for the rational design of small molecules to modulate protein-protein interactions. The presence of a well-defined binding pocket on the surface of protein complexes, particularly at their interface, can be used for docking-based virtual screening of chemical libraries. Several approaches have been developed to identify binding pockets that are implemented in programs such as SiteMap, fpocket, and FTSite. These programs enable the scoring of these pockets to determine whether they are suitable to accommodate high-affinity small molecules. Virtual screening of commercial or combinatorial libraries can be carried out to enrich these libraries and select compounds for further experimental validation. In virtual screening, a compound library is docked to the target protein. The resulting structures are scored and ranked for the selection and experimental validation of top candidates. Molecular docking has been implemented in a number of computer programs such as Auto-Dock Vina. We select a set of protein-protein interactions that have been successfully inhibited with small molecules in the past. Several computer programs are applied to identify pockets on the surface, and molecular docking is conducted in an attempt to reproduce the binding pose of the inhibitors. The results highlight the strengths and limitations of computational methods for the design of PPI inhibitors.

Key words Protein-protein interactions, Molecular docking, Structure-based drug design, Virtual screening, Small molecules, Inhibitors

1 Introduction

Protein-protein interactions (PPIs) control nearly every aspect of normal cellular function. These interactions can also promote a diverse set of cellular processes that lead to pathological processes such as cancer [1]. Protein interactions are typically identified through affinity purification and other pull-down techniques, but modern approaches like yeast-two-hybrid have enabled large-scale mapping of protein interaction networks [2]. Genomic data for cancer cells can be mapped onto these networks to uncover new PPIs. Three-dimensional structures can facilitate the rational design

Cheryl L. Meyerkord and Haian Fu (eds.), *Protein-Protein Interactions: Methods and Applications*, Methods in Molecular Biology, vol. 1278, DOI 10.1007/978-1-4939-2425-7_5, © Springer Science+Business Media New York 2015

of small molecules that bind and modulate these interactions [3]. Protein structures often harbor well-defined binding pockets that may be located at the protein-protein interface, at distal sites outside the interface, or at enzyme active sites. It is expected that small molecules that bind to these pockets will modulate the formation of the PPIs either by directly interfering with binding or through allosteric effects.

Structure-based computational methods have significantly matured over the past two decades. These methods provide tools to identify pockets and sites that are suitable for small-molecule binding. Pockets enable the screening of large chemical libraries to generate hit compounds that can be further optimized to modulate the protein interactions. Here we describe how to search for protein structures and scan their surface for binding pockets. We discuss the use of molecular docking of chemical libraries to these pockets to generate protein–ligand structures that can be scored and ranked. We discuss the process of scoring in virtual screening. We end with a case study of six PPIs that have been successfully targeted with small molecules.

2   Methods

2.1   Identifying Three-Dimensional Structures of Protein Interactions

Starting with the sequence of a gene that encodes for PPI proteins, the sequence of one of the binding partners can be used in a BLAST search using the advanced search tool at the Protein Data Bank (PDB) [4] (*see* **Note 1**). The FASTA format [5] for the protein sequence can be used in the search. A threshold value known as the E-value limits the search to proteins that possess significant sequence identity and coverage of the query sequence. Based on previous studies [6, 7], a value of 10^{-6} is expected to identify protein domains or very close homologs of the query protein (*see* **Note 2**). In addition, the search parameter "Number of Entities" can be used to filter for protein complexes by setting the Entity Type to "Protein" with a minimum bound of 2. Different structures that overlap the same protein sequence can also be removed by checking the "Remove similar sequences at 90 % identity" box.

Similarly, structures of protein complexes can be identified by using the Web server Interactome3D [8]. This tool provides structural annotations of individual proteins as well as protein complexes in a variety of model organisms, which can be queried by using UniProt accession IDs. If the protein complex is known, then the interaction pair can be used in the search. Otherwise, searching for individual proteins results in a network with all the interactions of that protein. Clicking on any of the edges in this network yields a ranked list showing structures from PDB which contain this particular interaction, as well as other information such as: (1) resolution of the structure; (2) chain of each of the binding partners in the protein complex; (3) sequence identity of each of the structures; and (4) how much of the protein sequence is covered by the structure.

2.2 Identifying Pockets on Protein Structures

The presence of a well-defined binding cavity at the surface of a protein can significantly facilitate the design of high-affinity small molecules. Pockets reduce compound exposure to solvent and through van der Waals and electrostatic forces provide additional stability to a ligand [9]. The location of a pocket on the structure of a protein–protein complex determines the impact of a ligand on the interaction. Compounds targeting an interface pocket will likely disrupt a protein interaction. But compounds that bind outside the interface may either stabilize or destabilize the complex through allosteric effects.

Several methods have been developed to identify pockets on protein surfaces. The most common approach uses a three-dimensional grid around the entire protein to determine the van der Waals energies at each grid point as implemented in SiteMap, a module available in the Schrödinger software suite [10]. The van der Waals radius of each grid point is compared to the distance from the grid point to nearby protein atoms to determine whether the point is outside the protein. Points outside the protein are kept if the van der Waals interaction energy is less than a given cutoff. These points are then clustered to define the binding pocket. SiteMap typically returns the top 5 detected binding sites, but a larger number of sites may be necessary on larger proteins. Another method implemented in the freely available program fpocket [11] uses a grid-free approach. Voronoi tessellation, a method for spatial division, is used to determine the location of α spheres on the protein. In this approach, α spheres are placed at the Voronoi intersects formed by the tessellations. Each of these spheres forms contacts with four atoms at its boundary, such that the sphere radii determine whether or not it is on the protein's surface. Pockets are detected by clustering α spheres within a specific radius; a range of 3–6 Å is the default. A third approach implemented in the program Cavbase [12] follows a similar strategy to what is implemented in SiteMap. A Cartesian grid around the protein is defined by the calculation of the van der Waals energies at each grid point. However, CavBase is mainly used for comparing different binding sites based on the definition of pseudocenters [12]. Finally, FTSite [13] is a Web server for pocket detection that uses fast Fourier transforms to calculate the energy of chemical probes (functional groups) on a grid. Thus, identifying the clusters formed by of each of the probes separately and also the overlap of these clusters between probes is used to identify likely binding sites.

In each case, the protein structure is preprocessed to be used for the pocket identification program. The coordinates of the protein must be separated from the protein–protein or protein–ligand complex as a separate file to properly detect binding sites. In addition, noncovalently bound atoms and molecules are also removed; these typically include water molecules, or ions used during the crystallization process. It is often the case that structures lack electron density within specific regions of the protein (*see* **Note 3**). Most commercial packages such as the Prime module in the Schrödinger package can incorporate missing side chains or even missing loops. Once the structure processing is completed, the pocket identification programs are used to identify various pockets on the structure.

2.3 Scoring Pockets

Various metrics can aid in determining whether a pocket can accommodate a small molecule. Two metrics have been developed and implemented in SiteMap for this purpose: SiteScore and DrugScore [14]. SiteScore is a measure of whether a pocket can accommodate a small molecule, while DrugScore provides information about the suitability of the pocket for the development of therapeutics. Both SiteScore and DrugScore use the weighted sums of the same parameters, namely (1) the number of site points in the binding pocket; (2) enclosure score that is a measure of how accessible the pocket is to solvents; and (3) hydrophilic character of the binding pocket (hydrophilic score). Unlike DrugScore, SiteScore limits the impact of hydrophilicity in charged and highly polar sites. On the other hand, fpocket provides only one metric known as the Druggability Score [15]. The Druggability Score is a general logistical model based on the local hydrophobic density of the binding site, the hydrophobicity score, and the normalized polarity score. Generally, a SiteMap SiteScore above 0.8, corresponds to a pocket that can accommodate a high-affinity small molecule. A SiteMap DrugScore above 0.9 [14], or an fpocket Druggability Score above 0.7 [15] correspond to druggable pockets. FTSite identifies and ranks the detected pockets by the number of amino acids that are in contact with the probes, but does not give a quantitative assessment on the druggability of a pocket.

It is often the case that pockets will be identified outside a protein-protein interface or on the surface of a protein whose complex with its binding partner has yet to be solved by crystallography. In many cases, the binding partners may not even be known. Potential binding partners can be identified from various datasets that report PPIs such as BioGRID [16] or MINT [17]. These datasets typically provide experimental data from yeast two-hybrid studies or other techniques in tab-delimited or PSI-MI XML formats, or through their web interface. The data can be readily converted to protein sequence by first retrieving a common identifier, such as the UniProt ID, and then retrieving the FASTA format sequence from UniProt [18]. Pockets may also potentially be an enzyme active site. Binding of a small molecule to an enzyme active site located on a protein–protein complex may potentially act in an allosteric manner and modulate the protein interaction. The location of these enzymatic binding residues can be retrieved from UniProt [18] or Catalytic Site Atlas [19] for the associated protein and compared to the location of the identified binding pockets.

2.4 Virtual Screening of Chemical Libraries to Target Pockets

Structure-based virtual screening is widely used to enrich large chemical databases to generate focused libraries that can be experimentally validated [20–22]. To date, only a handful of cases have been reported whereby compounds that inhibit PPIs emerged directly from virtual screening [23–26]. In most cases, virtual screening provides moderate to low affinity compounds that can serve as a starting point for the development of PPI inhibitors. Structure-based virtual screening consists of two steps, namely docking and scoring. Docking corresponds to the series of computational steps to predict the binding mode of a ligand to a target protein. The resulting protein–ligand complex provides coordinates that can be used to score and rank-order the interactions. In virtual screening, it is not uncommon that hundreds of thousands of compounds are docked to a target, followed by scoring for the selection of the top ~100 compounds.

Several computer programs have been developed for docking, such as DOCK [27, 28], AutoDock [29], GOLD [30, 31], FlexX [32], Glide [33, 34] among others [35–39]. AutoDock is one of the most cited open-source molecular modeling simulation docking programs. It was designed, implemented, and maintained by The Scripps Research Institute. Its latest version, Vina [40], was released in 2010. Vina inherits major ideas and approaches from AutoDock. By using a new source code, scoring function, and algorithms, Vina significantly improves the average accuracy of the binding mode prediction compared to AutoDock.

As described above, docking requires a three-dimensional structure of a protein as well as a pocket on the surface. Typically, a simple visualization of the structure can identify the location of the pocket of interest. For PPIs, pockets located at the interface are the most desirable since compounds that bind to these pockets are expected to disrupt the interaction. Pockets that include hot-spot residues are particularly attractive. The hot-spots can be located within the binding partner that contains the pocket or on the partner that occupies the pocket [25, 41–44]. If the location of the pocket is unknown, it can be identified using programs such as SiteMap or fpocket (*see* above) if necessary. Once a pocket is identified, a grid box based at the center of the pocket is created. The pre-calculated grid maps obviate the need to calculate the interaction energies at each step of the docking process. This results in a significantly faster docking run (*see* **Note 4**). Vina generates the map and carries out the energy calculations.

Typically, virtual screening is carried out on a single crystal structure that is kept rigid during the docking. But in solution, proteins sample multiple conformations. Different compounds will often bind to distinct conformations of the protein. Hence, using a single structure during virtual screening is likely to miss compounds that would otherwise bind to alternative conformations. There are several ways to introduce protein flexibility, such as induced-fit docking [45–47], docking with multiple crystal structures [48–50], NMR structures [51, 52], and structures from molecular dynamics (MD) simulations [23, 53]. AutoDock Vina offers the option of predefining flexible residues in the pocket. The selected side chains in the receptor are treated explicitly in a separate file. The grid maps and interaction energies are also generated and calculated for flexible side chain atoms. In other words, the atoms of flexible side chains are treated like atoms in the ligand. The grid box has to be enlarged when using flexible residues to offer enough space for the movement of the atoms in the flexible side chains.

AutoDock Tools (ADT) in MGLTools package is the interactive UI used to prepare a structure for docking with Vina. There are Python scripts in MGLTools that can convert the files in batch mode. All protein and ligand structure files have to be converted to PDBQT before docking. PDBQT is an extended PDB format coordinate file, which includes atomic partial charges, atom types, and information on the torsional degrees of freedom. The PDBQT files for ligand and receptor, center coordinates, and X, Y, Z dimensions of the search space grid box are the required input parameters for Vina docking. The exhaustiveness parameter controls the length of the docking run. The default value of *exhaustiveness* is 8, which is usually suitable for docking in a cube with an 18 Å edge. While a larger *exhaustiveness* parameter for Vina will increase the length of the docking run, it will not necessarily lead to improved docking results [54]. It is recommended that the *exhaustiveness* is increased when using a search space grid box larger than $30 \times 30 \times 30$ Å.

ZINC [55] is a free database of commercial and annotated compounds that are processed for virtual screening. Over 34 million unique compounds are loaded from 134 commercial supplier catalogs and 36 annotated catalogs. About 40 % of the compounds are "drug-like"; 13 % are "lead-like"; and 1.5 % are "fragment-like." Structure files in SDF, SMILES, and Mol2 format can be accessed from the ZINC Website (http://zinc.docking.org/). The protonation state of compounds is generated using Epik (version 2.1209) [56] in the Schrödinger Suite at four different pH ranges: reference range (pH = 7.1); middle range (pH of 6–8); high range (pH of 7–9.5); and low range (pH of 4.5–6). Atomic charges and desolvation of the compounds are calculated using AMSOL [57, 58]. OEChem software [59] from OpenEye is applied to convert original 2D SDF files into isomeric SMILES. Molecular Networks Corina program

[60] is used to generate the initial 3D conformation of the small molecule. ZINC provides easy access for users to search the biological activity of molecules, or search compounds that are active against a particular target.

Other databases that provide chemical structures of small molecules are also available. These include ChEMBL [61], PubChem [62], DrugBank [63], and BindingDB [64]. ChEMBL is an database that contains 5.4 million bioactivity measurements for more than 1 million compounds and 5,300 protein targets. PubChem is one of the most comprehensive databases that includes over 25 million unique chemical structures and 90 million bioactivity outcomes associated with several thousand macromolecular targets, but many of the compounds are not commercially available. Drug-Bank is a richly annotated database of drug and drug target information. BindingDB provides approximately 20,000 experimentally determined binding affinities of protein–ligand complexes for 110 protein targets and 11,000 small molecule ligands.

It is important to note that docking programs use a fitness function to guide the docking process towards the optimum binding mode. But these functions may not perform well for library enrichment during virtual screening. Re-scoring of protein-compound complexes from the docking step is often carried out for better enrichment performance. There are several types of scoring functions that have been developed over the years that include; (1) force field, such as GBSA [65] and PBSA [66]; (2) empirical, such as ChemScore [67], GlideScore [33], and SVRKB [68]; (3) knowledge-based scoring functions, such as PMF [69], DrugScore [70]; and (4) machine learning algorithms such as our own SVMSP approach [71]. Force field-based scoring functions use potential energy functions that describe various bonded and nonbonded interactions in a molecule; examples include stretch, bend, dihedral, and van der Waals interactions. Knowledge-based scoring functions are developed from statistical pairwise potentials derived from a large number of three-dimensional structures. SVMSP and SVRKB are two newly developed scoring functions that use a support vector machine trained from features obtained from three-dimensional structures [71].

3 Case Studies

A set of protein–protein complexes that have been successfully inhibited with small molecules and described in a previous review article [72] is used to illustrate the above methods. To identify the structure of these proteins, a keyword search was conducted at the PDB. For example, a search for interleukin-2 led to 268 structures. Structures with IL-2 in complex with other proteins were identified by limiting the protein stoichiometry to display only heteromers.

Another approach used the sequence of IL-2 that was obtained from UniProt (UniProt ID: P60568). A search using the FASTA sequence and a more relaxed E-value of 10^{-4} led to 17 structures of IL-2 without any additional filters. Of these 17 structures, one structure contained the IL-2/IL-2RA complex (PDB: 1Z92), another contained the ternary complex of IL-2 with its beta and gamma subunits (PDB: 3QAZ), and two contained the quaternary complex of IL-2 with all three subunits (PDB: 2ERJ and 2B5I). Similarly, the structures of the other five protein–protein complexes can be identified in this manner.

The structure of the binding partners can be used to identify binding pockets and assess whether these pockets are suitable to accommodate small molecules. To identify binding cavities on the six PPIs in Table 1, the monomer structures were first separated from the complex and preprocessed using the Protein Preparation Wizard in the Schrödinger package. The structure of the cytokine IL-2, an essential regulator in T cell growth and proliferation [73], forms an interaction interface with its alpha receptor (PDB ID: 1Z92) that is inhibited by a small molecule (PDB: 1PY2). To prepare the IL-2 monomer structure, the Protein Preparation Wizard tool in Maestro was used to remove chain B and add explicit hydrogen atoms to the structure. While SiteMap was unable to identify the binding pocket of the ligand on the monomer structure, both fpocket and FTSite identified the binding pocket on the IL-2/IL-2RA complex with a Druggability Score of 0.655.

Similarly, the BCL-X_L-BAD complex (PDB: 2BZW) is inhibited by the small molecule ABT-737 (PDB: 2YXJ). In addition to the preparation needed for the IL-2 structure, the structure of BCL-X_L, an apoptosis regulator [74], contains additional water molecules that need to be removed. The Arg-102 residue is also missing heavy atoms on its side chain, which need to be incorporated. All three pocket detection algorithms were able to identify the binding pocket occupied by the ligand in the

Table 1
Pockets identified by the various detection programs

Protein	Target	PPI structure	Protein-compound	SiteScore (SiteMap)	DrugScore (SiteMap)	Druggability (fpocket)	FTSite
IL-2	IL-2RA	1Z92	1PY2	Undetected		0.655	Site 2
BCL-X_L	BAD	2BZW	2YXJ	0.793	0.824	0.365, 0.897	Site 1, 2, 3
HDM2	p53	1YCR	1RV1	0.871	0.923	0.773	Site 1, 2
HPV E2	HPV E1	1TUE	1R6N	Undetected		0.084	Site 1
ZipA	FtsZ	1F47	1Y2F	Undetected		Undetected	Site 3
TNF	TNF	1TNF	2AZ5	Undetected		Undetected	Site 2

protein–protein complex, with SiteScore and DrugScore scores of 0.793 and 0.824, respectively. The fpocket Druggability scores are 0.365 and 0.897, respectively. While SiteMap was able to enclose the real binding site in one pocket, both of the two other algorithms found that the ligand spanned more than one pocket as evidenced by the two fpocket Druggability scores. In addition to the pockets at the interaction interface, SiteMap identified a druggable allosteric pocket near the interaction interface, with a SiteScore of 0.831 and a DrugScore of 0.776.

The mouse homolog of HDM2 was shown to bind and block the tumor-suppressor protein p53 [75]. Crystal structures show that this complex (PDB: 1YCR) have been inhibited by both an imidazoline (PDB: 1RV1) and a benzodiazepine (PDB: 1T4E) inhibitor. Similar to the IL-2 monomer, HDM2 did not have any water molecules or missing side chains. Once again, all three algorithms identified the binding site at the PPI interface. While both SiteMap and fpocket represented the binding site as only one pocket, with a SiteScore of 0.871, DrugScore of 0.923, and Druggability Score of 0.773, FTSite required two pockets to properly enclose the bound ligand.

The HPV E2 protein regulates the transcription and replication of the viral genome with the E1 protein [76, 77]. The HPV E2-E1 complex (PDB: 1TUE) is inhibited by a small molecule at the N-terminal transactivation domain of HPV E2 (PDB: 1R6N). Water molecules were removed from the structure and pocket detection algorithms were applied to the monomer structure. While SiteMap failed to identify the specific binding pocket, fpocket identified the binding site as an unlikely druggable target, with a Druggability Score of only 0.084. FTSite, on the other hand, identified the proper binding site as its highest ranking pocket.

The ZipA-FtsZ complex (PDB: 1F47), which is essential in cell division of bacteria [78], is inhibited by a small molecule (PDB: 1Y2F). To prepare the ZipA structure, water molecules were removed from the ZipA monomer of the protein–protein complex. Both SiteMap and fpocket failed to identify the binding pocket of the ligand, but fpocket did identify two additional druggable pockets on ZipA, with Druggability Scores of 0.631 and 0.721. FTSite also identified the proper binding site, but only as its third ranking pocket.

Finally, the cytokine TNFα forms a homotrimer complex (PDB: 1TNF) that is involved in the inflammatory response [79]. This complex was shown to be disrupted by small molecules (PDB: 2AZ5), thereby inhibiting TNFα activity. Similar to the ZipA-FtsZ complex, both SiteMap and fpocket failed to detect the binding pocket of the protein-protein interface occupied by the small molecule. FTSite successfully identified the pocket as its second ranking site.

In addition to the pockets at the interaction interface, all three programs identified additional pockets outside of the interface, albeit they are unlikely to be druggable pockets based on the pocket scoring metrics. SiteMap identified only one noninterface pocket with a SiteScore or DrugScore greater than 0.7 among all six protein monomers, a pocket near the interaction interface on BCL-X_L with a SiteScore of 0.831 and a DrugScore of 0.776. FPocket on the other hand only identified two additional druggable pockets on ZipA with Druggability Scores of 0.631 and 0.721. FTSite consistently identified three pockets for each of these six proteins, but offered no measurement of its druggability.

Once a pocket is identified, molecular docking can be used to predict the binding mode of a large number of compounds for subsequent scoring and ranking. The reliability of molecular docking depends on the properties of the binding pocket. To illustrate this, all six PPI inhibitors in Table 1 were extracted and re-docked using AutoDock Vina. Default parameters were used. Typically, a docking program will generate multiple binding poses along with a score associated with each pose. The predicted binding pose with the lowest Vina binding energy is selected (Table 2). The root-mean-square deviation (RMSD) comparing the position of heteroatoms of the docked inhibitors to the crystal structure are listed in Table 2. For the first three targets with detectable pockets, AutoDock Vina reliably reproduced the binding pose of the compounds. This is evidenced by an RMSD that was less than 1.5 Å in each case. In contrast, molecular docking of the compounds that bind to HPV E2, ZipA, and TNF resulted in binding modes that did not agree with the crystal structure (RSMD >5 Å). It is of interest to note that the binding mode of the compounds that bind to well-defined pockets (IL-2, BCL-X_L, and HDM2) was reliably predicted by molecular docking. The absence of a well-defined pocket for HPV E2, ZipA, and TNF posed a significant challenge to the docking program. These results highlight the strengths and limitations of using molecular docking for the search of small molecules that bind at PPIs. New scoring functions are needed to guide molecular docking and scoring of small molecules that bind to protein interfaces with shallow pockets.

4 Notes

1. If a large number of protein sequences are being queried, it is often easier to install a local copy of BLAST+ and the PDB amino acid (pdbaa) database and go through the sequences in an iterative manner. The pdbaa database returns amino acid sequence queries as PDB chains, whose metadata can be retrieved using the REST API in PDB through the use of XML queries. This metadata can then be filtered to identify structures that contain dimerization.

Table 2
AutoDock Vina results and RMSD compared to crystal structure

PDB	Compound structure	Compound name	Binding energy (kcal/mol)	RMSD (Å)
1PY2		SP4206	−7.7	1.06
2YXJ		ABT-737	−9.5	1.28
1RV1		IMIDAZOLINE	−6.6	0.89

(continued)

Table 2
(continued)

PDB	Compound structure	Compound name	Binding energy (kcal/mol)	RMSD (Å)
1R6N		Compound 23	−7.5	6.65
1Y2F		Compound 1	−6.4	8.21
2AZ5		SP304	−8.3	5.68

2. The cutoff used in the BLAST search will only identify PDB entries that are nearly identical to the initial sequence query. If structures that you expect to see are not returned, lowering the cutoff will result in a larger number of potential structures. However, a cutoff that is too low will often yield structures that have very low sequence identity to the original.

3. Sometimes there are no atomic coordinates for residues within a PDB entry. Often, these missing residues occur at the beginning or end of the PDB sequence, or in intrinsically disordered regions, which are difficult to crystallize. If these missing residues are within the proximity of a known binding cavity, using an alternative crystal structure from the BLAST search can yield a PDB structure that has these coordinates available. Otherwise, homology modeling can be used to thread the missing residues into a homologous structure. This can be accomplished with software packages such as MODELLER or Prime.

4. During docking, the ligands can dock outside of the specified binding pocket if the grid is too large. In this case, either visual inspection or a computational approach can determine which ligands, if any, are outside the binding pocket. In the computational approach, the distance between the center of mass of the binding pocket and the center of mass of the docked ligand can be used to determine which ligands are outside the binding cavity.

Disclosure of Potential Conflicts of Interest

No potential conflicts of interest were disclosed.

References

1. Vidal M, Cusick ME, Barabasi AL (2011) Interactome networks and human disease. Cell 144:986–998

2. Ngounou Wetie AG, Sokolowska I, Woods AG et al (2013) Protein–protein interactions: switch from classical methods to proteomics and bioinformatics-based approaches. Cell Mol Life Sci 71:205–228

3. White AW, Westwell AD, Brahemi G (2008) Protein–protein interactions as targets for small-molecule therapeutics in cancer. Expert Rev Mol Med 10:e8

4. Berman HM, Westbrook J, Feng Z et al (2000) The protein data bank. Nucleic Acids Res 28:235–242

5. Lipman DJ, Pearson WR (1985) Rapid and sensitive protein similarity searches. Science 227:1435–1441

6. Li L, Bum-Erdene K, Baenziger PH et al (2010) BioDrugScreen: a computational drug design resource for ranking molecules docked to the human proteome. Nucleic Acids Res 38: D765–D773

7. Huang YJ, Hang D, Lu LJ et al (2008) Targeting the human cancer pathway protein interaction network by structural genomics. Mol Cell Proteomics 7(10):2048–2060

8. Mosca R, Ceol A, Aloy P (2013) Interactome3D: adding structural details to protein networks. Nat Methods 10:47–53

9. Li L, Meroueh SO (2008) Receptor-ligand interactions in biological systems. In: Encyclopedia for the life sciences. Wiley, London, p. 19. http://onlinelibrary.wiley.com/book/10.1002/9780470048672/homepage/EditorsContributors.html

10. Halgren T (2007) New method for fast and accurate binding-site identification and analysis. Chem Biol Drug Des 69:146–148

11. Le Guilloux V, Schmidtke P, Tuffery P (2009) Fpocket: an open source platform for ligand pocket detection. BMC Bioinformatics 10:168

12. Kuhn D, Weskamp N, Hullermeier E et al (2007) Functional classification of protein kinase binding sites using cavbase. ChemMed-Chem 2:1432–1447

13. Ngan CH, Hall DR, Zerbe B et al (2012) FTSite: high accuracy detection of ligand binding sites on unbound protein structures. Bioinformatics 28:286–287

14. Halgren TA (2009) Identifying and characterizing binding sites and assessing druggability. J Chem Inf Model 49:377–389

15. Schmidtke P, Barril X (2010) Understanding and predicting druggability. A high-throughput method for detection of drug binding sites. J Med Chem 53:5858–5867

16. Stark C, Breitkreutz BJ, Reguly T et al (2006) BioGRID: a general repository for interaction datasets. Nucleic Acids Res 34:D535–D539

17. Licata L, Briganti L, Peluso D et al (2012) MINT, the molecular interaction database: 2012 update. Nucleic Acids Res 40:D857–D861

18. UniProt C (2012) Reorganizing the protein space at the Universal Protein Resource (Uni-Prot). Nucleic Acids Res 40:D71–D75

19. Porter CT, Bartlett GJ, Thornton JM (2004) The catalytic site atlas: a resource of catalytic sites and residues identified in enzymes using structural data. Nucleic Acids Res 32: D129–D133

20. Leach AR, Gillet VJ, Lewis RA et al (2009) Three-dimensional pharmacophore methods in drug discovery. J Med Chem 53:539–558

21. Hubbard RE (2011) Structure-based drug discovery and protein targets in the CNS. Neuropharmacology 60:7–23

22. Cheng T, Li Q, Zhou Z et al (2012) Structure-based virtual screening for drug discovery: a problem-centric review. AAPS J 14:133–141

23. Khanna M, Wang F, Jo I et al (2011) Targeting multiple conformations leads to small molecule inhibitors of the uPAR·uPA protein–protein interaction that block cancer cell invasion. ACS Chem Biol 6:1232–1243

24. Scheper J, Guerra-Rebollo M, Sanclimens G et al (2010) Protein–protein interaction antagonists as novel inhibitors of non-canonical polyubiquitylation. PLoS One 5:e11403

25. Koes D, Khoury K, Huang Y et al (2012) Enabling large-scale design, synthesis and validation of small molecule protein–protein antagonists. PLoS One 7:e32839

26. Geppert T, Bauer S, Hiss JA et al (2012) Immunosuppressive small molecule discovered by structure-based virtual screening for inhibitors of protein–protein interactions. Angew Chem Int Edit 51:258–261

27. Kuntz ID, Blaney JM, Oatley SJ et al (1982) A geometric approach to macromolecule-ligand interactions. J Mol Biol 161:269–288

28. Makino S, Kuntz ID (1997) Automated flexible ligand docking method and its application for database search. J Comput Chem 18:1812–1825

29. Goodsell DS, Olson AJ (1990) Automated docking of substrates to proteins by simulated annealing. Proteins 8:195–202

30. Jones G, Willett P, Glen RC et al (1997) Development and validation of a genetic algorithm for flexible docking. J Mol Biol 267:727–748

31. Jones G, Willett P, Glen RC (1995) Molecular recognition of receptor sites using a genetic algorithm with a description of desolvation. J Mol Biol 245:43–53

32. Rarey M, Kramer B, Lengauer T et al (1996) A fast flexible docking method using an incremental construction algorithm. J Mol Biol 261:470–489

33. Friesner RA, Banks JL, Murphy RB et al (2004) Glide: a new approach for rapid, accurate docking and scoring. 1. Method and assessment of docking accuracy. J Med Chem 47:1739–1749

34. Halgren TA, Murphy RB, Friesner RA et al (2004) Glide: a new approach for rapid, accurate docking and scoring. 2. Enrichment factors in database screening. J Med Chem 47:1750–1759

35. Pierce BG, Hourai Y, Weng Z (2011) Accelerating protein docking in ZDOCK using an advanced 3D convolution library. PLoS One 6:e24657

36. McGann M (2011) FRED pose prediction and virtual screening accuracy. J Chem Inf Model 51:578–596

37. Pedretti A, Villa L, Vistoli G (2004) VEGA – an open platform to develop chemo-bio-informatics applications, using plug-in architecture and script programming. J Comput-Aided Mol Des 18:167–173

38. Thomsen R, Christensen MH (2006) MolDock: a new technique for high-accuracy molecular docking. J Med Chem 49:3315–3321

39. Abagyan R, Totrov M, Kuznetsov D (1994) ICM – a new method for protein modeling and design: applications to docking and structure prediction from the distorted native conformation. J Comp Chem 15:488–506

40. Trott O, Olson AJ (2010) AutoDock Vina: improving the speed and accuracy of docking

with a new scoring function, efficient optimization, and multithreading. J Comput Chem 31:455–461

41. Obiol-Pardo C, Alcarraz-Vizán G, Cascante M et al (2012) Diphenyl urea derivatives as inhibitors of transketolase: a structure-based virtual screening. PLoS One 7:e32276

42. Dessal AL, Prades R, Giralt E et al (2011) Rational design of a selective covalent modifier of G protein βγ subunits. Mol Pharm 79:24–33

43. Trosset J-Y, Dalvit C, Knapp S et al (2006) Inhibition of protein–protein interactions: the discovery of druglike β-catenin inhibitors by combining virtual and biophysical screening. Proteins 64:60–67

44. Grüneberg S, Stubbs MT, Klebe G (2002) Successful virtual screening for novel inhibitors of human carbonic anhydrase: strategy and experimental confirmation. J Med Chem 45:3588–3602

45. Elokely KM, Doerksen RJ (2013) Docking Challenge: Protein Sampling and Molecular Docking Performance. J Chem Inf Model 53:1934–1945

46. Lill MA, Winiger F, Vedani A et al (2005) Impact of Induced Fit on Ligand Binding to the Androgen Receptor: A Multidimensional QSAR Study To Predict Endocrine-Disrupting Effects of Environmental Chemicals. J Med Chem 48:5666–5674

47. Sherman W, Day T, Jacobson MP et al (2005) Novel procedure for modeling ligand/receptor induced fit effects. J Med Chem 49:534–553

48. Arooj M, Sakkiah S, Kim S et al (2013) A combination of receptor-based pharmacophore modeling & QM techniques for identification of human chymase inhibitors. PLoS One 8: e63030

49. Zhou S, Li Y, Hou T (2013) Feasibility of using molecular docking-based virtual screening for searching dual target kinase inhibitors. J Chem Inf Model 53:982–996

50. Li Y, Kim DJ, Ma W et al (2011) Discovery of novel checkpoint kinase 1 inhibitors by virtual screening based on multiple crystal structures. J Chem Inf Model 51:2904–2914

51. Isvoran A, Badel A, Craescu C et al (2011) Exploring NMR ensembles of calcium binding proteins: perspectives to design inhibitors of protein–protein interactions. BMC Struct Biol 11:24

52. Knegtel RMA, Kuntz ID, Oshiro CM (1997) Molecular docking to ensembles of protein structures. J Mol Biol 266:424–440

53. Carlson HA, Masukawa KM, Rubins K et al (2000) Developing a dynamic pharmacophore model for HIV-1 integrase. J Med Chem 43:2100–2114

54. Kukol A (2011) Consensus virtual screening approaches to predict protein ligands. Eur J Med Chem 46:4661–4664

55. Irwin JJ, Sterling T, Mysinger MM et al (2012) ZINC: a free tool to discover chemistry for biology. J Chem Inf Model 52:1757–1768

56. Greenwood JR, Calkins D, Sullivan AP et al (2010) Towards the comprehensive, rapid, and accurate prediction of the favorable tautomeric states of drug-like molecules in aqueous solution. J Comput Aid Mol Des 24:591–604

57. Cramer CJ, Truhlar DG (1992) An SCF solvation model for the hydrophobic effect and absolute free energies of aqueous solvation. Science 256:213–217

58. Cramer CJ, Truhlar DG (1992) AM1-SM2 and PM3-SM3 parameterized SCF solvation models for free energies in aqueous solution. J Comput Aided Mol Des 6:629–666

59. Hawkins PCD, Skillman AG, Nicholls A (2006) Comparison of shape-matching and docking as virtual screening tools. J Med Chem 50:74–82

60. Tetko IV, Gasteiger J, Todeschini R et al (2005) Virtual computational chemistry laboratory-design and description. J Comput-Aided Mol Des 19:453–463

61. Gaulton A, Bellis LJ, Bento AP et al (2012) ChEMBL: a large-scale bioactivity database for drug discovery. Nucleic Acids Res 40: D1100–D1107

62. Li Q, Cheng T, Wang Y et al (2010) PubChem as a public resource for drug discovery. Drug Discov Today 15:1052–1057

63. Knox C, Law V, Jewison T et al (2011) DrugBank 3.0: a comprehensive resource for "omics" research on drugs. Nucleic Acids Res 39:D1035–D1041

64. Liu T, Lin Y, Wen X et al (2007) BindingDB: a web-accessible database of experimentally determined protein–ligand binding affinities. Nucleic Acids Res 35:D198–D201

65. Still WC, Tempczyk A, Hawley RC et al (1990) Semianalytical treatment of solvation for molecular mechanics and dynamics. J Am Chem Soc 112:6127–6129

66. Luo R, David L, Gilson MK (2002) Accelerated Poisson–Boltzmann calculations for static and dynamic systems. J Comput Chem 23:1244–1253

67. Eldridge MD, Murray CW, Auton TR et al (1997) Empirical scoring functions: I. The development of a fast empirical scoring function to estimate the binding affinity of ligands

in receptor complexes. J Comput Aided Mol Des 11:425–445

68. Li L, Wang B, Meroueh SO (2011) Support vector regression scoring of receptor–ligand complexes for rank-ordering and virtual screening of chemical libraries. J Chem Inf Model 51:2132–2138

69. Muegge I, Martin YC (1999) A general and fast scoring function for protein – ligand interactions: a simplified potential approach. J Med Chem 42:791–804

70. Gohlke H, Hendlich M, Klebe G (2000) Knowledge-based scoring function to predict protein–ligand interactions. J Mol Biol 295:337–356

71. Li L, Khanna M, Jo I et al (2011) Target-specific support vector machine scoring in structure-based virtual screening: computational validation, in vitro testing in kinases, and effects on lung cancer cell proliferation. J Chem Inf Model 51:755–759

72. Wells JA, McClendon CL (2007) Reaching for high-hanging fruit in drug discovery at protein–protein interfaces. Nature 450:1001–1009

73. Malek TR (2003) The main function of IL-2 is to promote the development of T regulatory cells. J Leukoc Biol 74:961–965

74. Willis S, Day CL, Hinds MG et al (2003) The Bcl-2-regulated apoptotic pathway. J Cell Sci 116:4053–4056

75. Moll UM, Petrenko O (2003) The MDM2-p53 interaction. Mol Cancer Res 1:1001–1008

76. Muller M, Demeret C (2012) The HPV E2-host protein–protein interactions: a complex hijacking of the cellular network. Open Virol J 6:173–189

77. Hughes FJ, Romanos MA (1993) E1 protein of human papillomavirus is a DNA helicase/ATPase. Nucleic Acids Res 21:5817–5823

78. Pazos M, Natale P, Vicente M (2013) A specific role for the ZipA protein in cell division: stabilization of the FtsZ protein. J Biol Chem 288:3219–3226

79. Locksley RM, Killeen N, Lenardo MJ (2001) The TNF and TNF receptor superfamilies: integrating mammalian biology. Cell 104:487–501

Chapter 6

Targeting Protein-Protein Interactions for Drug Discovery

David C. Fry

Abstract

Protein-protein interactions are associated with key activities and pathways in the cell, and in that regard are promising targets for drug discovery. However, in terms of small molecule drugs, this promise has not been realized. The physical nature of many protein-protein interaction surfaces renders them unable to support binding of small drug-like molecules. In addition, there are other unique hurdles presented by this class that make the drug development process difficult and risky. Nevertheless, success stories have begun to steadily appear in this field. These experiences are starting to provide general strategies and tools to help overcome the problems inherent in pursuing protein-protein interaction targets. These lessons should improve the rate of success as these systems are pursued in the future.

Key words Protein-protein interactions, Drug discovery, MDM2, Nutlins, Druggability, Protein NMR, NMR screening, Screening library, Fragment-based drug discovery

1 Framing the Issue

Proteins commonly interact with other proteins. Some act as "hubs" and have many partners. In fact, it has been estimated that there are on the order of 600,000 protein-protein interactions that occur in the cell [1]. These interactions serve various purposes, including regulation of function, step-wise transmission of a signal, or attraction of a partner to a particular location. Protein-protein interactions are associated with a wide variety of critical physiological activities and pathways. Accordingly, one would expect that malfunctions with these systems could be linked to a number of important disease states. Beneficial modulation of protein-protein interactions could, therefore, be the mode of action of effective drugs. However, historically, this situation has not been exploited [2]. If one surveys the mechanisms of conventional (i.e., small organic molecule) drugs, it is found that very few of them act by affecting a protein-protein interaction. Why is this so?

The physical nature of most protein-protein interaction surfaces lowers the probability that a small organic compound can

Cheryl L. Meyerkord and Haian Fu (eds.), *Protein-Protein Interactions: Methods and Applications*, Methods in Molecular Biology, vol. 1278, DOI 10.1007/978-1-4939-2425-7_6, © Springer Science+Business Media New York 2015

bind there [3–5]. These surfaces are typically large and flat, and devoid of significant subpockets. Binding affinity is achieved by summing up a large number of weak interactions. The interactions are so widely spaced that a small molecule cannot adequately duplicate them. However, this situation is a generalization, and not all protein-protein systems adopt such a strategy. Pioneering work by Jim Wells and colleagues demonstrated that, in some cases, a limited number of amino acids mediate nearly all of the key interactions that produce binding affinity. These sub-regions have been referred to as "hot spots," and their dimensions can be comparable to the size of a small organic molecule [6].

Therefore, protein-protein interaction systems that are candidates for drug discovery must be judged for druggability on a case-by-case basis. The approaches and techniques for making such a judgment are still being refined and are not fully reliable yet. As a consequence, embarking on a drug discovery program targeting a protein-protein interaction system is still a high-risk undertaking. An unwillingness to take on such risk has meant that protein-protein targets have not received the same degree of active pursuit as other target classes. This is one of the major reasons for the paucity of marketed drugs with a mode of action that involves modulation of a protein-protein interaction.

Another reason for lack of success against this target class is the inability to overcome the unique hurdles that are presented by this class during the drug discovery and development process. The field is gaining an awareness of, and an appreciation for, these hurdles, and is acquiring an understanding of how to overcome them, as success stories steadily appear in this area. It has been helpful to examine, at a molecular level, cases in which a drug-like small molecule has been successfully developed against a protein-protein interaction target, and a collective consideration of these examples has led to important lessons that should improve the means by which these targets are addressed in the future.

2 Recent Success Stories

One of the more recent examples, and a case in which a small molecule inhibitor of a protein-protein interaction has progressed fully to clinical trials, is the development of the "Nutlin" series of compounds, which bind to MDM2 and inhibit its interaction with p53 [7]. Overexpression of MDM2 is one way in which certain cancer cells evade the apoptosis-triggering activity of p53, and the Nutlins have been shown capable of activating cell cycle arrest and apoptosis, and of inhibiting growth of human tumor xenografts. The Nutlin case exemplifies several of the key issues that need to be considered when approaching a protein-protein interaction target, and reveals important attributes of the system that, if understood and exploited correctly, can enhance the chance of success.

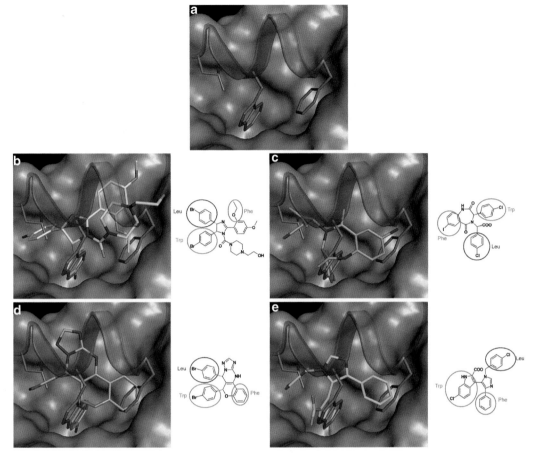

Fig. 1 Series of one single and four superimposed X-ray structures showing how the p53 peptide binds to MDM2 [8] (**a**), and how various small molecule inhibitors [7, 34–36] are able to mimic its binding strategy (**b–e**). MDM2 is depicted as a tan surface. The p53 peptide backbone is shown as a green ribbon, and key side chains (Phe19, Trp23, and Leu26) as green sticks. The inhibitors are shown in stick form. All structural figures in this chapter were prepared using the PyMol Molecular Graphics System (Schroedinger, LLC)

An initial structure of a complex between MDM2 and a peptide representing p53 [8] indicated that the interaction occurs at a single binding cleft on the MDM2 protein. The p53 peptide adopts an alpha helical conformation, and inserts three hydrophobic side chains (Phe19, Trp23, and Leu26) into subpockets of MDM2 (Fig. 1a). An assessment of the druggability of this site, using early computational tools, suggested that the target would be challenging but worth pursuing [9].

A high-throughput screen of MDM2 was conducted, using a library of diverse drug-like small molecule compounds, and numerous active hits were obtained. Protein-observe NMR, using an isotopically labeled version of a stabilized construct of MDM2, was employed to assess the authenticity of the hits [10]. This was a critical step, because all but two of the hits were found to be false positives. Extensive synthetic chemistry was applied, and one of the lead classes (the Nutlin series) was able to be optimized into more potent derivatives, and ultimately into a clinical candidate [11]. During this process, an NMR structure [12], and X-ray structures [7], of Nutlins bound to MDM2 were obtained (Fig. 1b), and they revealed the means by which the small molecules were able to replicate the binding strategy of the crucial segment of p53.

The imidazoline core scaffold, which is the primary feature of the Nutlins, is able to direct substituents into the three subpockets of MDM2 that are normally occupied by hydrophobic side chains of p53. Trp23 and Leu26 are each mimicked by a halogenated phenyl group, and Phe19 is mimicked by an ethoxy group. It is important to note that there is no exact matching of chemotype— for example, an aliphatic group can substitute for an aromatic side chain, and an aromatic group can fulfill the role of an aliphatic side chain. Also, the trajectories by which the small molecule substituents enter the subpockets are completely different from those implemented by the side chains.

The backbone of p53 is not directly replicated at all by the small molecule. While the protein must make use of a series of hydrogen bonds to attain rigidity, in this case along an alpha helix, the imidazoline scaffold possesses inherent rigidity. Further, in terms of geometry, this scaffold is able to economically span a segment of alpha helix that is eight residues in length.

The Nutlins feature an appendage emanating from the N1 atom of the imidazoline core that appears to project out beyond contact with the binding cleft, into the solvent. The chemical composition of this appendage was found to be crucial with respect to influencing potency, yet its precise role cannot be gleaned from the structure. One can speculate that it helps as a general shield keeping solvent away from the binding cleft, or that it may sterically direct a key substituent into the Phe subpocket. A third alternative, that it contributes to a local dipole and in that way influences orientation and affinity, has not been studied.

In the period since the discovery of the Nutlins, MDM2 inhibitors representing other chemical classes have been reported [13]. It is instructive to compare how these various scaffolds bind to the same site, to appreciate the variety of binding interactions that can be made there, and to observe what kinds of substituent groups can participate in these interactions (Fig. 1c–e).

The core scaffolds of the various inhibitors are quite different chemically and with regard to shape. Nevertheless, the compounds share certain binding principles. In every case, the p53 backbone is not directly replicated, and in every case the small molecule inhibitor fills all three subpockets of MDM2 with hydrophobic moieties.

Into the Trp subpocket, the inhibitors insert a chloro-indole or halogenated phenyl group, and these rings all attain the same orientation as the parent Trp side chain. The Leu subpocket is filled by all inhibitors in a similar manner, through use of a chlorophenyl, bromophenyl, or chlorobenzyl group. None of these groups correspond to the branched aliphatic nature of the parent Leu side chain. Also, the trajectories of these inhibitor rings into the subpocket, while similar to each other, are basically perpendicular to that of the Leu side chain. Occupancy of the Phe subpocket is the most varied of all, and is achieved by the following variety of chemotypes: ethoxy, iodophenyl, benzyl edge from a four-ring system, and phenyl. Similar to the situation with the Leu subpocket, the inhibitor substituents enter with different trajectories from that of the parent Phe side chain, and they end up oriented perpendicular to it.

Recently, a potent MDM2 inhibitor from yet another chemical class has been discovered, and this series binds in a completely unique manner [14]. This class emerged from a screen against a protein called MDMX, which is highly similar to MDM2, binds to p53 in a comparable manner, and participates in regulation of p53 function in the cell. The inhibitor binds in an equipotent manner to both proteins, and with a similar strategy. As exemplified by MDM2 (Fig. 2), the inhibitor induces the protein to dimerize, and the result is a symmetric complex consisting of two MDM2 molecules and two inhibitor molecules. Each inhibitor molecule spans the two proximal MDM2 binding sites, by inserting its difluorophenyl group into the Trp subpocket on one side, and positioning its indolyl-hydantoin group into the Phe subpocket of the other MDM2 molecule. This orientation allows a stacking interaction between

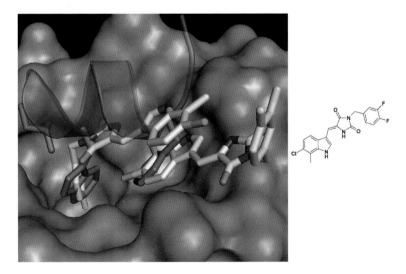

Fig. 2 MDM2/MDMX dual inhibitor that causes dimerization. Depicted is the X-ray structure of the complex of MDM2 with the p53 peptide [7] superimposed with the structure of the inhibitor [14]. The peptide is visualized as described for Fig. 1, and the inhibitor is in stick form with its carbons colored white. Both inhibitor molecules from the dimer are shown, but only one of the MDM2 molecules is shown, as a tan surface

the two indolyl groups, and this energetically favorable situation is reinforced by further stacking with a Tyr63 ring from each protein molecule. The Leu subpocket is basically ignored by this inhibitor. While it is not unusual to encounter a ligand that bridges two binding sites across protein molecules that normally dimerize, the present case is a heretofore unobserved instance of an inhibitor pulling together two molecules of a protein domain that is normally monomeric, and forming a ligand/ligand dimer in the process.

In summary, although at the beginning of the program it was wondered whether any small molecule could be found that would bind tightly to MDM2 and effectively inhibit its interaction with p53, there are now several different classes of molecule which have achieved this goal.

With regard to other protein-protein interaction systems, there have been a number of reports describing successful discovery of small drug-like modulators. These cases have been collected and described in a series of review articles [15–20]. There are also newer reports that have appeared since the latest survey [21, 22]. Some of these modulators bind to a remote site and act allosterically, and some are directly competitive with respect to the partner protein. If one collectively considers all of the successful cases, some general principles emerge:

1. For instances involving direct competition, successful cases usually feature early identification of a peptide that can adequately replicate the binding of the entire partner protein.

2. The protein-protein modulator is usually higher in molecular weight than a typical drug. This is likely due to the need to participate in multiple interactions that are spaced relatively far apart, due to the large surface area typically observed at a protein-protein interface.

3. The modulator is also usually more three-dimensional than a typical drug [23], and often more rigid. The three-dimensionality can be attained by stacking or by the inherent geometric constraints of the central scaffold. As a consequence of having a three-dimensional molecule seated in a shallow cleft, parts of the modulator may stick out beyond the perimeter of the binding cleft. For flexible protein targets, rigidity may help allosteric modulators induce subpocket formation at their binding site.

4. For the modulators that act competitively, there is usually little need to replicate the backbone of the partner protein, just the critical side-chains. An organic scaffold can deliver these substituents in a much more economical way than can a peptide backbone.

5. There is substantial leeway with respect to the chemotypes that can serve as effective core scaffolds, and with respect to

substituents that can mimic the key interacting side chains of the natural partner protein. Also, entry vectors of these substituents into subpockets at the binding site can be completely different from, and even perpendicular to, those displayed by the side chains of the natural partner.

3   Current Best Approaches

Given the experience gained from many attempts over the years at drug discovery involving protein-protein systems, and the recent availability of numerous successful examples in this area, a generalized optimal approach can be laid out for an attack on a protein-protein target:

3.1   Choosing Targets/Assessing Druggability

The target will be chosen based on biology considerations, but then must be quickly put under scrutiny at a molecular level [24]. If a drug-like small molecule modulator is already known for the system in question, then by definition it is druggable. Otherwise, if structural information is available for the target protein, a druggability assessment can be made by computational methods [25–28]. These methods have not yet proven to be completely accurate, but should be a reasonable guide at the extremes—for example, a target with a complete lack of subpockets at the interface is almost certain to be highly challenging. Given the unreliability of the computational methods, an inexpensive experimental method for assessing druggability is advisable. The most established approach is to perform a fragment screen. It has been shown in two retrospective studies that there is a correlation between the inability to find fragment hits for a target and its eventual failure in a small molecule drug discovery program.

3.2   The Role of Peptides

A helpful first step is identification of a peptide that can serve as a fully competent surrogate for the entire partner protein. This can be accomplished via design, based on structural or mutagenesis information; by phage display; or by a screen of a peptide library. Once obtained, a surrogate peptide can serve as a useful tool for assay development and calibration, and for structural studies. One should even consider developing the peptide itself into a drug. The methodologies for doing this represent an area of intense current interest [29], but an exposition is beyond the scope of this article.

3.3   Computational Screening and De Novo Design

Experimental screens are expensive, and any method that can efficiently focus such a screen, or bypass it altogether, is worthwhile. If sufficient structural information is available on the target, a virtual screen can be carried out. This requires the preparation of a virtual compound library, where choices must be made about the conformations of the small molecules that will be sampled. Also, it is desirable, although costly CPU-wise, to allow for flexibility in the target protein during the virtual screen, and to somehow incorporate the influence of solvent. Finally, robust and reliable docking and scoring functions must be available to predict the affinity of compounds during the screening process. The resulting hits can be followed up in experimental assays, or can be the springboard for a focused screen limited to related chemotypes. The current state of affairs is that virtual screening has had sporadic successes, but few would depend solely on it for a high priority target. However, the reliability of virtual screening should improve steadily, and because it is so much more cost efficient, the hope is that it will ultimately be able to fully replace experimental random screening.

Structural information on the target complexed with the partner protein, a peptide, or some other lead molecule, should allow the de novo design of alternate scaffolds that could be more drug-like and possibly more potent. The ability of experts in the field to design an active small organic molecule directly from a bound peptide structure is still not very advanced. More success has been realized by designing alternate scaffolds from bound small molecules. This is another case where improvements in methodologies are expected, and ultimately will lead to major cost savings, but at the moment the de novo design approaches are not robust enough to allow the abandonment of experimental screening.

3.4 Experimental Screening

Screening involves utilization of a compound library and a method for detecting when interaction of a compound with the target protein has occurred. There are a variety of methods available for detection. Some are indirect, observing an activity that is triggered by the protein-protein interaction and monitoring whether added compounds reduce this activity. Others measure binding directly. In this category, labels on the proteins can be used to report that they are bound together, and to monitor whether addition of compound has disrupted the complex. Alternatively, there are label-free methods that employ just the member of the protein pair that is the target of interest, measure an inherent property of the target protein, and detect perturbations caused by binding of a compound. Such methods include SPR, Tm-shift, and NMR. In these cases, hit compounds must be checked in a follow-up assay to see if their binding is consequential—that is, whether they are actually modulating the protein-protein interaction. Another approach is to utilize a labeled peptide ligand and monitor competition by added compounds.

Most of the screening techniques involve some sort of manipulation of the native protein, such as immobilization on a matrix, or attachment of a label or a fusion partner, and any of these can cause falsification of signal. Screening campaigns involving a protein-protein interaction target are notorious for producing a high rate of false positives. Screening results are expressed as an activity curve, and for well-behaved targets, such as kinases and other enzymes, the compounds in the high-activity tail of the curve are usually real—while in the case of protein-protein interactions, these are usually false, and the real hits are buried in the main part of the curve. This high false positive rate probably occurs because protein-protein interactions involve exposed hydrophobic surfaces, and these can allow opportunistic binding—that is, weak transient binding that is insufficient for further chemical optimization, or for formation of co-crystals, but that nonetheless produces a detectable signal. It is essential, therefore, that following a screen of a protein-protein target, some means of sorting the reals from the false positives is applied. This could consist of performing multiple screens using different techniques and looking for hits that are agreed upon by all the methods. An even better approach, if possible, would be to apply one or both of the two "gold standard" hit verification techniques that are essentially free of false positive readouts—protein-observe NMR and X-ray crystallography. One extra advantage of these two techniques is that they reveal the location of binding, allowing a distinction to be made between allosteric effects and a directly competitive mechanism.

Deciding on a screening strategy requires an analysis of numerous parameters, and ultimately some level of compromise on the issues they present. In general, methods that have the advantages of high speed, high sensitivity, and small volumes leading to low protein consumption usually have the disadvantages of high cost of robots and consumables, and the need to format the compound library in a specific way. Methods that are more accurate are usually slower, less sensitive, require more protein, and are more demanding with regard to the purity and physical state of the protein. It should be added that, when considering overall cost and speed for the less accurate "high-throughput" methods, one must add the resources and time needed to sort out the true positives. It has not been unusual for optimization chemistry to begin on a false lead, and for the inability to obtain meaningful SAR to be the basis of making the final judgment, which is a costly mode of operating.

3.5 Protein-Protein Focused Screening Library

It has already been noted that modulators of protein-protein interactions have unique properties, in particular that they tend to be larger and more three-dimensional than typical drugs. We have observed in screens of protein-protein targets that the hits are statistically more three-dimensional than the hits from screens involving other target classes [23]. These extremes of size and shape tend to be underrepresented in typical corporate compound libraries. To remedy this, we have built a "PPI Library," which is a collection of compounds for screening that have substantial three-dimensionality. This library has already shown promise by producing authenticated hits against protein-protein interaction targets that had yielded very low hits rates in earlier screens using our conventional library [30].

3.6 Fragment Approach

Screening libraries have traditionally consisted of compounds in the molecular weight range 200–500 Da. However, a strategy employing libraries comprised only of small compounds, the "fragment-based approach" [31], has been gaining in popularity. Practitioners seem to have settled on a similar set of characteristics for the fragments composing their libraries, featuring a molecular weight range of 100–300 Da. Protein-protein interaction systems, however, represent a unique class of drug target, and it is an open question whether fragments meant to serve as potential leads for protein-protein interaction targets should also have properties distinct from those of conventional fragments. One step toward answering that question would be to deconstruct known successful protein-protein inhibitors into successively smaller fragments, until the smallest active fragment for each is identified, and then derive the key properties that are shared by the successful fragment leads. Such a deconstruction has already been applied to ABT-737, a potent inhibitor of the Bcl-2 family, which has a molecular weight of 813 Da. In a retrospective study, compounds comprising portions of ABT-737 were gathered and were checked for activity, and the smallest piece that still exhibited binding was identified and found to have a molecular weight of 293 Da [32]. We have systematically deconstructed a Nutlin (MW = 728 Da) into successively smaller fragments, and measured the ability of these fragments to bind to MDM2 [33] (Fig. 3). Binding activity has been verified and examined using protein-observe NMR and X-ray crystallography. In this system, the smallest active fragment has a molecular weight of 305 Da. In both cases, where the parent had a high molecular weight typical of protein-protein modulators, the smallest active fragment was found to be at the highest limit of the molecular weight range that normally defines fragments. This suggests that a "PPI Fragment Library" should be a unique entity that contains relatively larger fragments, and possibly other specialized properties. With regard to three-dimensionality, we have observed in fragment screens against protein-protein interaction targets that the hit sets were no more three-dimensional than those obtained against other target classes. So, unlike the final optimized compounds in a protein-protein program, at the fragment level the leads are not exceptionally three-dimensional.

3.7 Optimization

Leads must undergo chemical optimization to acquire greater potency, and to take on the properties that will allow performance as a drug—bioavailability, stability, and lack of toxicity. These requirements are not different when the target is a protein-protein system. However, the unique properties of the compounds that tend to be selected for this target class may present heightened challenges.

Structure-guided design during lead optimization is now a fairly common approach, but may be particularly important in the

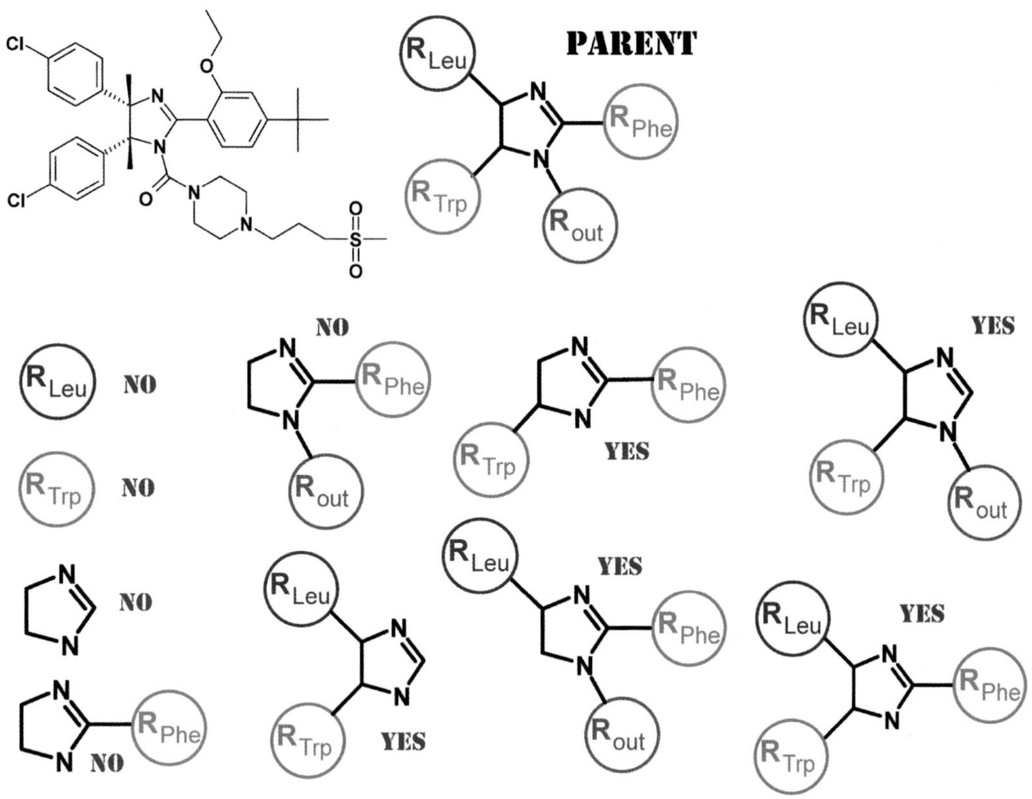

Fig. 3 Deconstruction of a Nutlin into fragments. The parent molecule is shown at the top as a chemical structure and, to the immediate right, as a simplified schematic depiction. The various fragments are shown below in schematic form, and their binding behavior to MDM2 as assessed by protein-observe NMR is given, expressed as "Yes" or "No"

case of a protein-protein target. These binding sites present fewer clear opportunities for making additional interactions—such as hydrogen bond partners to engage with—and fewer obvious nearby subpockets to try to fill. Further, the protein site is likely to be more mobile, presenting a "moving target." So optimization steps are more speculative and exploratory, and it is more critical for the project chemists to have a steady source of feedback to understand what is being accomplished and what next modifications appear promising. Also, the structural work can verify that increasing activity is still grounded in authentic binding, and not due to the introduction of an unwanted mechanism such as higher stoichiometry or unfolding of the protein.

It has been stated that successful compounds in this target class are larger than what is typical. Also, we have observed that in protein-protein projects, increases in potency tend to be accompanied by increases in three-dimensionality. At first glance, higher molecular weight and greater chemical complexity would suggest that problems may be encountered with candidates in these projects, with regard to solubility and intracellular availability. While this expectation would be consistent with historic trends, some protein-protein modulators have been found to be clear exceptions to these trends, so a negative judgment should not be made a priori. Also, formulation methodology has advanced significantly and may be able to deal quite successfully with unusual molecules.

Specificity is always an issue, since a lack thereof may lead to toxicity. For protein-protein targets it is hard to measure, or even predict, the level of specificity of a compound. Although "protein-protein" is a clearly defined class, it does not carry with it a common active site architecture, as does a class like "kinases." There is no pre-established panel for doing assays that adequately represents coverage of the protein-protein universe. One source of worry is that proteins can employ "anti-binding" elements to achieve specificity. That is, the interaction site can feature protrusions that will keep unwanted protein partners away, while the desired partner will have a complimentary surface that can safely avoid or accommodate these protrusions. A small molecule that is already mimicking the interactions that mediate affinity is too small to also mimic anti-binding elements. In the end, there are many potential off-target protein-protein interaction surfaces that a compound could inadvertently bind to, and these cannot be pre-identified, so it is impossible to design against them. All one can do is take a clinical candidate into the stage of in vivo studies and hope that toxicity due to nonspecificity does not appear.

4 Future Outlook

The field of targeting protein-protein interactions for drug discovery holds considerable promise. The numbers alone indicate that many viable targets must exist in this class. The lack of progress in this area has been due largely to an unwillingness to gamble resources on these high-risk targets. However, emerging success stories are starting to indicate ways to reduce the risk.

One overriding lesson is that candidate protein-protein interaction targets must be examined on a case-by-case basis. While many protein-protein surfaces may be too large and featureless to be druggable by a small molecule, this is a blanket characterization that does not apply to all. As the reliability for predicting and measuring druggability improves, the trepidation toward pursuing these targets will be alleviated. Also, as important diseases continue to resist drug discovery, the failures on more traditional target classes will necessitate a move toward more difficult targets.

Following a commitment to invest in protein-protein programs, improved experimental approaches and methodologies will increase the rate of success. Strategies will be acquired for overcoming the special hurdles presented by this class. From reducing the false positive rate in initial screening, to finding ways of formulating optimized candidates for in vivo work, the discovery process will become much more efficient. More reliable computational methods will allow virtual steps to replace experimental steps, resulting in significant cost savings.

As more drug discovery chemistry is performed on protein-protein targets, compound libraries will begin to better match this target class, and cover more extensively the compound space that fits these binding sites, just as occurred historically with other popular target classes. Better-matched libraries will lead to higher success rates in screening, and faster expansion around hits.

In conclusion, the field of protein-protein interactions presents an attractive opportunity for drug discovery, and high optimism is justified for future engagements with this challenging, but captivating, target class.

References

1. Stumpf MPH, Thorne T, de Silva E, Stewart R, An HJ, Lappe M, Wiuf C (2008) Estimating the size of the human interactome. Proc Nat Acad Sci 105:6959–6964

2. Overington JP, Al-Lazikani B, Hoopkins AL (2006) How many drug targets are there? Nat Rev Drug Disc 5:993–996

3. Fry DC (2006) Protein-protein interactions as targets for small molecule drug discovery. Biopolymers 84:535–552

4. Wells JA, McClendon CL (2007) Reaching for high-hanging fruit in drug discovery at protein-protein interfaces. Nature 450:1001–1009

5. Doemling A (2008) Small molecular weight protein-protein interaction antagonists – an insurmountable challenge? Curr Opin Chem Biol 12:281–291

6. Clackson T, Wells JA (1995) A hot spot of binding energy in a hormone-receptor interface. Science 267:383–386

7. Vassilev LT, Vu BT, Graves B, Carvajal D, Podlaski F, Filipovic Z, Kong N, Kammlott U, Lukacs C, Klein C, Fotouhi N, Liu E (2004) In vivo activation of the p53 pathway by small-molecule antagonists of MDM2. Science 303:844–848

8. Kussie PH, Gorina S, Marechal V, Elenbaas B, Moreau J, Levine AJ, Pavletich NP (1996) Structure of the MDM2 oncoprotein bound to the p53 tumor suppressor transactivation domain. Science 274:948–953

9. Fry DC, Graves BJ, Vassilev LT (2005) Exploiting protein-protein interactions to design an activator of p53. In: Golemis EA, Adams PD (eds) Protein-protein interactions: a molecular cloning manual. Cold Spring Harbor Laboratory, Cold Spring Harbor, NY, pp 893–906

10. Fry DC, Graves BJ, Vassilev LT (2005) Development of E3-substrate (MDM2-p53) binding inhibitors: structural aspects. Methods Enzymol 399C:622–633

11. Vu BT, Vassilev L (2011) Small-molecule inhibitors of the p53-MDM2 interaction. In: Vassilev L, Fry D (eds) Small-molecule inhibitors of protein-protein interactions. Springer, Berlin, pp 151–172

12. Fry DC, Emerson SD, Palme S, Vu BT, Liu CM, Podlaski F (2004) NMR structure of a complex between MDM2 and a small molecule inhibitor. J Biomol NMR 30:163–173

13. Wang S, Zhao Y, Bernard D, Aguilar A, Kumar S (2012) Targeting the MDM2-p53 protein-protein interaction for new cancer therapies. Top Med Chem 8:57–80

14. Graves B, Thompson T, Xia M, Janson C, Lukacs C, Deo D, DiLello P, Fry D, Garvie C, Huang K, Gao L, Tovar C, Lovey A, Wanner J, Vassilev L (2012) Activation of the p53 pathway by small-molecule induced MDM2 and MDMX dimerization. Proc Nat Acad Sci 109:11788–11793

15. Arkin MR, Wells JA (2004) Small-molecule inhibitors of protein-protein interactions: progressing towards the dream. Nat Rev Drug Disc 3:301–317

16. Fry DC (2008) Drug-like inhibitors of protein-protein interactions: a structural examination

of effective protein mimicry. Curr Prot Pep Sci 9:240–247

17. Sperandio O, Reynes CH, Camproux AC, Villoutreix BO (2010) Rationalizing the chemical space of protein-protein interaction inhibitors. Drug Disc Today 15:220–229

18. Higueruelo AP, Schreyer A, Bickerton GRJ, Pitt WR, Groom CR, Blundell TL (2009) Atomic interactions and profile of small molecules disrupting protein-protein interfaces: the TIMBAL database. Chem Biol Drug Des 74: 457–467

19. Morelli X, Bourgeas R, Roche P (2011) Chemical and structural lessons from recent successes in protein-protein interaction inhibition (2P2I). Curr Opin Chem Biol 15:1–7

20. Fry DC (2012) Small-molecule inhibitors of protein-protein interactions: how to mimic a protein partner. Curr Pharm Des 18: 4679–4684

21. Cerchietti LC, Ghetu AF, Zhu X, DaSilva GF, Zhong S, Matthews M, Bunting KL, Plol JM, Fares C, Arrowsmith CH, Yang SN, Garcia M, Coop A, MacKerell AD, Prive GG, Melnick A (2010) A small-molecule inhibitor of BCL6 kills DLBCL cells in vitro and in vivo. Cancer Cell 17:400–411

22. Buckley DL, Van Molle I, Gareiss PC, Tae HS, Michel J, Noblin DJ, Jorgensen WL, Ciulli A, Crews CM (2012) Targeting the von Hippel-Lindau E3 ubiquitin ligase using small molecules to disrupt the VHL/HIF-1alpha interaction. J Am Chem Soc 134:4465–4468

23. Fry DC, So S-S (2012) Modulators of protein-protein interactions: the importance of three-dimensionality. In: Doemling A (ed) Protein-protein interactions in drug discovery. Wiley-VCH, Weinhelm, pp 55–62

24. Wanner J, Fry DC, Peng Z, Roberts J (2011) Druggability assessment of protein-protein interfaces. Future Med Chem 3:2021–2038

25. Fuller JC, Burgoyne NJ, Jackson RM (2008) Predicting druggable binding sites at the protein-protein interface. Drug Disc Today 14: 155–161

26. Bourgeas R, Basse MJ, Morelli X, Roche P (2010) Atomic analysis of protein-protein interfaces with known inhibitors: the 2P2I database. PLoS One 5:e9598

27. Meireles LMC, Doemling AS, Camacho CJ (2010) ANCHOR: a web server and database for analysis of protein-protein interaction binding pockets for drug discovery. Nucleic Acids Res 38:W407–W411

28. Sugaya N, Kanai S, Furuya T (2012) Dr. PIAS 2.0: an update of a database of predicted druggable protein-protein interactions. Database 2012:bas034

29. Vlieghe P, Lisowski V, Martinez J, Khrestchatisky M (2009) Synthetic therapeutic peptides: science and market. Drug Disc Today 15:40–56

30. Fry D, Huang K-S, Di Lello P, Mohr P, Mueller K, So S-S, Harada T, Stahl M, Vu B, Mauser H (2013) Design of libraries targeting protein-protein interfaces. Chem Med Chem 8:726–732

31. Murray CW, Rees DC (2009) The rise of fragment-based drug discovery. Nat Chem 1: 187–192

32. Hajduk PJ (2006) Fragment-based drug design: how big is too big? J Med Chem 49:6972–6976

33. Fry DC, Wartchow C, Graves B, Janson C, Lukacs C, Kammlott U, Beluins C, Palme S, Klein C, Vu B (2013) Deconstruction of a nutlin: dissecting the binding determinants of a potent protein-protein interaction inhibitor. ACS Med Chem Lett 4(7):660–665

34. Grasberger BL, Lu T, Schubert C, Parks DJ, Carver TE, Koblish HK, Cummings MD, LaFrance LV, Milkiewicz KL, Calvo RR, Maguire D, Lattanze J, Franks CF, Zhao S, Ramachandren K, Bylebyi GR, Zhang M, Manthey CL, Petrella EC, Pantoliano MW, Deckman IC, Spurlino JC, Maroney AC, Tomczuk BE, Molloy CJ, Bone RF (2005) Discovery and cocrystal structure of benzodiazepinedione HDM2 antagonists that activate p53 in cells. J Med Chem 48:909–912

35. Allen JG, Bourbeau MP, Wohlhieter GE, Bartberger MD, Michelsen K, Hungate R, Gadwood RC, Gaston RD, Evans B, Mann LW, Matison ME, Schneider S, Huang X, Yu D, Andrews PS, Reichelt A, Long AM, Yakowec P, Yang E, Lee TA, Oliner JD (2009) Discovery and optimization of chromenotriazolopyrimidines as potent inhibitors of the mouse double minute 2 – tumor protein 53 protein interaction. J Med Chem 52:7044–7053

36. Popowicz GM, Czarna A, Wolf S, Wang K, Wang W, Doemling A, Holak T (2010) Structures of low molecular weight inhibitors bound to MDMX and MDM2 reveal new approaches for p53-MDMX/MDM2 antagonist drug discovery. Cell Cycle 9:1–8

Part II

Label-Free Biosensors and Techniques

Chapter 7

Studying Protein-Protein Interactions Using Surface Plasmon Resonance

Zaneta Nikolovska-Coleska

Abstract

Protein-protein interactions regulate many important cellular processes, including carbohydrate and lipid metabolism, cell cycle and cell death regulation, protein and nucleic acid metabolism, signal transduction, and cellular architecture. A complete understanding of cellular function depends on full characterization of the complex network of cellular protein-protein interactions, including measurements of their kinetic and binding properties. Surface plasmon resonance (SPR) is one of the commonly used technologies for detailed and quantitative studies of protein-protein interactions and determination of their equilibrium and kinetic parameters. SPR provides excellent instrumentation for a label-free, real-time investigation of protein-protein interactions. This chapter details the experimental design and proper use of the instrumentation for a kinetic experiment. It will provide readers with basic theory, assay setup, and the proper way of reporting this type of results with practical tips useful for SPR-based studies. A generic protocol for immobilizing ligands using amino coupling chemistry, also useful if an antibody affinity capture approach is used, performing kinetic studies, and collecting and analyzing data is described.

Key words Protein-protein interactions, Surface plasmon resonance, Real-time and label-free biosensor applications, Kinetics, Affinity, Association, Dissociation

1 Introduction

Protein-protein interactions (PPIs) are fundamental in regulation of biological processes such as cellular signaling pathways. The "omic" technologies (genomics, transcriptomics, proteomics) have provided important insights into PPIs in biological systems demonstrating that these interactions control the assembly of multi-protein complexes that mediate biological functions. Systematic mapping of interactome networks and identification of the "central nodes" essential for homeostasis has become a major goal of current biomedical research [1, 2]. Changes in PPIs are involved in many diseases, including cancer, neurodegenerative diseases, viral and bacterial infections [3, 4], and such alterations are increasingly attracting attention as a new frontier in drug development.

Cheryl L. Meyerkord and Haian Fu (eds.), *Protein-Protein Interactions: Methods and Applications*, Methods in Molecular Biology, vol. 1278, DOI 10.1007/978-1-4939-2425-7_7, © Springer Science+Business Media New York 2015

Characterization of the protein-protein binding interactions and understanding their function and role in biological processes and diseases is important and necessary for validation of potential therapeutic targets [3, 5–7].

The ultimate goal of studying protein-protein interactions is recognition of the consequences of the interaction for cell function, which depend on the strength of the interaction. Thus, characterization of the PPIs through determination of the interaction affinity, selectivity and binding kinetic parameters, has become an important part of defining and understanding in vivo bimolecular interactions.

One of the most used technologies for studying, analyzing and measuring the binding parameters of PPIs is surface plasmon resonance (SPR). SPR is a powerful and sensitive experimental tool, based on an optical detection technique, which allows quantitative kinetic, equilibrium, and thermodynamic analysis of biomolecular interactions [8, 9]. Since its initial development, SPR has been widely used in molecular biology due to its many advantages which include: (1) high sensitivity, (2) label-free detection, (3) lack of dependence on spectroscopic properties, (4) real-time monitoring, (5) quantitative evaluation, (6) determination of kinetic rate constants, (7) low volume sample consumption, and (8) high degree of automation and throughput. SPR can be used to assess and quantify the kinetic and binding parameters of protein-protein, protein–DNA, and protein–ligand interactions, and in combination with other methods, to identify specific and selective interactions. There are a variety of commercially available SPR systems that can be categorized as laboratory, portable, and imaging SPR. The first SPR system was developed in the early 1990s, by Biacore, subsequently acquired by GE Healthcare, currently the main supplier of SPR instruments [10]. These instruments typically are able to measure association or on-rates (k_a) in the range of 10^3–10^7 M^{-1} s^{-1}, dissociation or off-rates (k_d) of 10^{-5}–1 s^{-1}, affinity at equilibrium (K_D) in the range of 100 μM to 200 pM, and analyte concentration in the range of 1 mM to 10 pM.

This chapter provides an overview of the detection principles, assay design, immobilization techniques and data analysis of SPR, followed by a detailed description of the materials and methods, and guidance for analysis of protein-protein interactions.

1.1 Surface Plasmon Resonance: Principle

SPR is an optical-based real-time detection method for monitoring specific binding events between two or more biomolecules, without the use of labels, and provides quantitative information on the specificity, kinetics and affinity of the biomolecular interactions [11–13]. SPR can be also used for thermodynamic analysis, epitope mapping, and to determine analyte concentration [14–16]. A unique approach to protein investigations is enabled by the combination of SPR and mass spectrometry, which increases limits

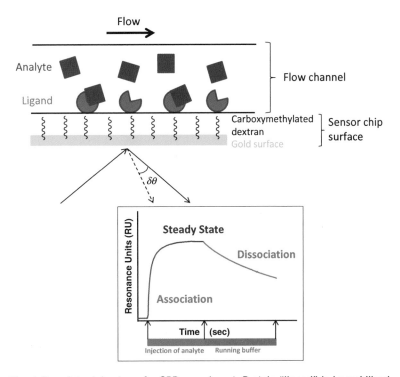

Fig. 1 Experimental setup of a SPR experiment. Protein "ligand" is immobilized on the dextran-coated sensor chip surface. Protein "analyte" is passed over the chip surface. Protein-protein interactions between the "ligand" and the "analyte" increase the density at the sensor chip surface and change the angle of refracted light ($\delta\theta$) generating a real-time binding curve known as a sensorgam. Schematic representation of a sensorgram: Association phase (two or more molecules bind to each other); Steady state-equilibrium (the amount of the molecules that are binding is equal to the amount of molecules that dissociate); Dissociation phase (the bonds between the molecules are breaking and the molecules are dissociating)

of detection, ligand fishing, multi-protein analysis, and protein-complex delineation [17, 18].

In SPR systems, interactions and detection occur in multichannel flow cells on a sensor surface, which is part of the flow system. The SPR-based binding method involves immobilization of a ligand on the surface of a sensor chip which has a monolayer of carboxymethylated dextran covalently attached to a gold surface. The ligand of interest is immobilized on the surface of the sensor chip using well-defined chemistry allowing solutions with different concentrations of an analyte to flow over it and to characterize its interactions to the immobilized ligand (Fig. 1). The SPR signal originates from changes in the refractive index at the surface of the gold sensor chip. The increase in mass associated with a binding event causes a proportional increase in the refractive index, which is observed as a change in response. These changes are measured as changes in the resonance angle ($\delta\theta$) of refracted light when the analyte, flowing in a microfluidic channel, binds to the immobilized ligand and increases in density at the sensor chip. Importantly, for protein-protein interactions the change in refractive index on the surface is linearly related to the number of molecules bound [19]. The response signal is quantified in resonance units (RU) and represents a shift in the resonance angle. Monitoring the change in the SPR signal over time produces a sensorgram, a plot of the binding response (RU) versus time which allows different stages of a binding event to be visualized and evaluated (Fig. 1). During the injection of an analyte, the binding response increase is due to the formation of analyte–ligand complexes at the surface and the sensorgram is dominated by the association phase. After a certain time of injection, a steady state is reached, in which binding and dissociating molecules are in equilibrium. The decrease in response after analyte injection is terminated is due to dissociation of the complexes, defining the dissociation phase. Depending on the dissociation rate of the tested ligand, some assays may require a regeneration step in order to reach the baseline again. Fitting the sensorgram data to an appropriate kinetic binding model allows calculation of kinetic parameters such as the association (k_a) and dissociation (k_d) rate constants, and the binding affinity of the tested interactions.

1.2 Sensor Surface Preparation/ Immobilization

The first step in the interaction analysis is the surface preparation and immobilization of one of the binding partners (the "ligand") on the sensor chip surface. Several factors should be considered for this procedure and include selecting (1) the binding partner for the immobilization, (2) the appropriate immobilization level, and (3) the method to be used for immobilization, and consequently, the sensor chip that will be selected for use.

1.2.1 Selection of Ligand

Several properties can influence the decision as to which ligand will be immobilized on the chip. These include the molecular weight of binding partners, their purity, the number of binding sites, the functionality of the immobilized ligand, the isoelectric point (pI) of the protein(s), the amount of ligand available, and the assay conditions.

When direct coupling is used as an immobilization method, it is important that the purity of the ligand be >95 %. The immobilized ligand should be stable and the binding activity must survive the coupling procedure as well as the regeneration protocol. During SPR experiments, binding is measured as a change in the refractive index on the surface of the sensor chip caused by accumulation of mass within the surface dextran layer and the measured response is related to the mass of the bound analyte. Therefore, when possible, the smallest and most stable molecule in the system should be used as the immobilized component. The larger molecule of the pair should be the analyte thus providing high sensitivity to the assay. Proteins with more than one binding site should be immobilized on the chip. If they are used as the analyte, avidity issues can arise, complicating the kinetic analysis. Less amount of sample is required for coupling and consequently the protein that is less available should be immobilized.

1.2.2 Immobilization Levels

The binding capacity of the sensor chip surface depends on the immobilization level and the optimal amount of the immobilized ligand depends on the experimental goal. For example, higher levels of immobilized ligand are needed in experiments designed to determine the concentration of analytes in solution, while for kinetic analysis the immobilization level of the ligand should be low. The maximum binding capacity (R_{max}) of the immobilized ligand recommended in the literature is typically in the range of 50–150 resonance units (RU), depending on the relative molecular weights and sizes of ligand and analyte. An overloaded surface causes problems of steric hindrance, aggregation, rebinding and diffusional limitations such as mass transport. Mass transport limitations can result in calculated binding kinetics that are slower than the true binding parameters. Problems due to mass transport effects are complex and they are best avoided by use of a low density surface to provide a maximum binding response level (R_{max}) of no more than 50–100 RU, and high flow rates, preferably 100 μL/min.

The maximum response (R_{max}) that can be obtained in experiments depends directly on the molecular weight of the protein-protein complex that will be formed on the surface after flowing the analyte molecule. The theoretical R_{max} describes the maximum binding capacity of the surface, which is usually higher than the experimental R_{max}. The theoretical R_{max} can be calculated using the following equation:

$$R_{max} = \frac{MW_{analyte}}{MW_{ligand}} \times R_L \times S_m$$

R_{max}—maximum capacity of surface
($MW_{analyte}/MW_{ligand}$)—ratio of the mass of the analyte and ligand
R_L—amount of ligand on surface
S_m—stoichiometry

The following rearranged equation can be used to determine an appropriate immobilization level that will generate an R_{max} in the desired range of 50–150 RU for measurement of the binding kinetics of studied PPIs:

$$R_L = R_{max} \left(\frac{1}{S_m}\right) \left(\frac{MW_{ligand}}{MW_{analyte}}\right)$$

This calculation assumes 100 % active analyte and ligand. The capture technique typically retains 100 % activity of the bound ligand whereas the direct coupling technique has 50–80 % of the activity of the bound ligand, although in some cases this can be lower. The quality of the binding data depends on the ligand activity, which is based on the determined experimental R_{max} and should be calculated using the following equation:

$$\% \text{ ligand activity} = \frac{\text{Experimental } R_{max}}{\text{Theoretical } R_{max}} \times 100 \%$$

Poor surface activity can result from denaturation of the ligand as a result of the pre-concentration solution, blocking and inactivating the binding site by immobilization chemistry, or impurities present in the ligand which can also be immobilized, lowering the surface activity. In these cases, changes in the coupling chemistry or ligand capture approach should be considered in order to improve the surface activity.

1.2.3 Immobilization Method

Direct covalent coupling and affinity capturing techniques are two general methods for immobilization of the ligand. Covalent coupling is the most commonly used means of immobilizing the ligand. In this technique, the ligand is directly immobilized on the sensor surface using established coupling chemistry depending on the available reactive groups, for example amine, thiol, or aldehyde groups in the protein. In such cases, direct immobilization does not require any modification of the ligand. The sensor chip, CM5, is the most versatile available chip for covalent coupling of ligands with high binding capacity (Table 1). It gives high response and excellent chemical stability providing accuracy and allowing repeated analyses on the same surface. The immobilization level is easily controlled and ligand consumption is low. The disadvantage of this immobilization method is that it can be associated with multiple attachment sites leading to randomized coupling and heterogeneous orientation of the ligand.

In contrast to direct immobilization, the affinity capturing method offers steric orientation of the immobilized interaction partner for optimum site exposure and assurance that all immobilized molecules will be in the same orientation, resulting in a highly active and homogeneous surface (Fig. 2a). The method relies on non-covalent protein-protein interactions and provides oriented coupling, because binding occurs at a well-defined site on the target. Another advantage of this approach is that it is fully regenerable and allows immobilization from crude protein mixtures, such as cell lysates. A disadvantage of the capturing method is that it may not produce a stable surface, thus complicating the analysis of analyte binding.

Three major coupling classes can be used for the affinity capturing method: artificially introduced affinity tags (e.g., biotin or hexahistidine), antibody–antigen systems (e.g., antibodies against

Table 1
Available sensor chips, their characteristics and applications

Sensor chip	Surface	Characteristics and applications
CM5	Standard surface—Carboxymethylated dextran matrix	Excellent chemical stability; Versatile chip suitable for most applications
CM4	Low carboxylation—Carboxymethylated dextran matrix with lower degree of carboxylation than CM5, i.e., less negatively charged	Reduces nonspecific binding of highly positively charged molecules that may be found, for example, in crude samples Convenient for measurement of low R_{max} needed in kinetic applications
CM3	Short dextran—Carboxymethylated dextran matrix, with the same level of carboxylation as CM5, but shorter matrix	For low immobilization levels and work with cells, viruses, and multicomponent complexes with high molecular weight analytes
CM7	High Carboxylation—Comparable with the CM5 chip but has a higher density of carboxymethylated dextran	Gives higher immobilization capacity; Ideal for applications that use small molecules and fragments
C1	Flat carboxymethylated surface	For work with particles such as cells and viruses, and in applications where dextran matrix is undesirable
SA	Carboxymethylated dextran matrix pre-immobilized with streptavidin	Captures biotinylated ligands such as carbohydrates, peptides, proteins, and DNA fragments
NTA	Carboxymethylated dextran matrix pre-immobilized with NTA (nitrilotriacetic acid)	Designed to bind histidine-tagged ligands via metal chelation; Allows control of steric orientation of ligand component for optimal site exposure
HPA	Flat hydrophobic surface	For studying lipid monolayers interacting with membrane binding biomolecules and membrane-associated interactions
L1	Carboxymethylated dextran matrix modified with lipophilic substances	For rapid and reproducible capture of liposomes with retention of lipid bilayer structure

different tags as capture molecules), and interactions between proteins and naturally occurring sites (e.g., Protein A/IgG). An example of the antibody–antigen system is using anti-Flag antibody, which can be covalently linked to the sensor chip surface. Proteins fused with Flag are readily immobilized on a sensor chip surface through the Flag/anti-Flag antibody interaction. The advantage of this approach is that the antibody serves as a universal adaptor, permitting the capture of different Flag-fusion proteins on the sensor chip surface. As the antibody is covalently bound to the sensor chip, the anti-Flag surface may be regenerated many times

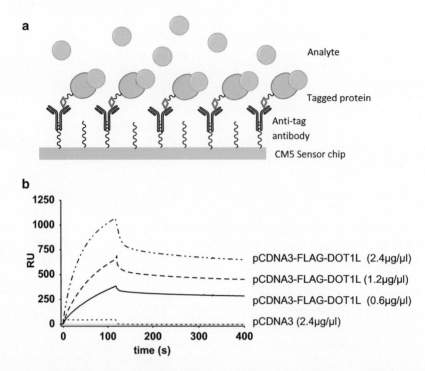

Fig. 2 Affinity based immobilization method. (**a**) Illustration of the capture affinity immobilization method using an antibody–antigen interaction. (**b**) Dose-dependent and specific immobilization of Flag-tagged full length histone methyltransferase, DOT1L, obtained from crude cell lysate of 293 cells transfected with pCDNA3-FLAG-DOT1L, using anti-flag antibody as a capture molecule[20] (pCDNA3—empty vector)

and coated with different Flag-fusion constructs. This approach also allows selective ligand capture when crude samples are used, for example immobilization of proteins that are transiently expressed in cells and can be hard to purify. Using this strategy, the Flag-tagged full length of histone methyltransferase DOT1L, obtained from crude cell lysate, was specifically and dose dependently captured on a CM5 sensor chip through immobilized anti-Flag antibody (Fig. 2b), and used for determination of the kinetic binding parameters of PPIs between DOT1L and the MLL-fusion proteins, AF9 and ENL [20].

Additional affinity capturing strategies involve using biotin and hexahistidine tags for which different sensor chips should be used. The sensor chip SA has carboxymethylated dextran pre-immobilized with streptavidin and is suitable for capture of biotinylated ligands such as proteins, peptides, nucleic acids, or carbohydrates, where controlled biotinylation enables oriented immobilization. The sensor chip NTA is another surface that controls steric orientation for immobilization of histidine-tagged molecules. The carboxymethylated dextran is pre-immobilized with nitrilotriacetic acid (NTA) and His-tagged molecules are immobilized via Ni^{2+}/NTA chelation. A list of the available sensor chips for the Biacore Systems with their surface characteristics as well as applications are presented in Table 1. Recently, it was reported that the negative charge of the carboxymethyldextran matrix on the biosensor surface can influence the binding kinetics and K_D determination of PPIs, suggesting that it is important that these types of matrix-mediated artifacts to be identified early in the measurement process and the most appropriate sensor chip be selected [21].

The choice of immobilization strategy depends on ligand properties. For example, covalent coupling might on occasion result in loss of ligand activity due to direct modification of residues in the binding site or steric hindrance. In such cases, alternative chemistry, such as thiol coupling, or an affinity capturing technique for immobilization should be used. If the ligand is unstable, impure, or the regeneration of the surface is difficult then the capture method is preferable. In order to confirm that the immobilized ligand is in its proper confirmation and its activity is retained, the surface should be tested with a positive control, if available. The time period for which the chips are reusable depends upon the stability and maintenance of the functionality of the attached ligand.

1.2.4   Coupling Chemistry   Different covalent coupling chemistries have been developed for immobilization of ligands: coupling through primary amines, thiol groups, and aldehyde coupling (Fig. 3). Choosing an immobilization method depends mostly on the nature of the ligand.

Amine coupling is the most generally applicable coupling chemistry because of its universality, stability, speed and most macromolecules contain amine groups. This coupling uses primary amine groups (the N-terminus and lysine residues), which are often solvent-exposed due to their hydrophilicity. They react directly with active esters generated by *N*-hydroxysuccinimide–1-ethyl-3-(3-dimethylaminopropyl)carbodiimide (NHS/EDC) activation (Fig. 3). The second type of covalent coupling chemistry, thiol coupling, provides an alternative approach when amine coupling cannot be performed or the immobilization level will not be sufficient, such as immobilization of acidic proteins. Thiol coupling utilizes exchange reactions between thiols and active disulphide groups which can be introduced either on the dextran matrix to exchange with a thiol group on the ligand (ligand thiol approach) or on the ligand molecule to exchange with a thiol group introduced on the dextran matrix (surface thiol approach). A recommended reagent for introducing active disulphide groups is 2-(2-pyridinyldithio)ethaneamine (PDEA). Thiol coupling chemistry is not suitable for experiments where the surface is exposed to reducing agents or high pH, since the coupling bond is unstable under such conditions. Ligands containing aldehyde groups (either native or introduced by oxidation of *cis*-diols) can be immobilized after activating the surface

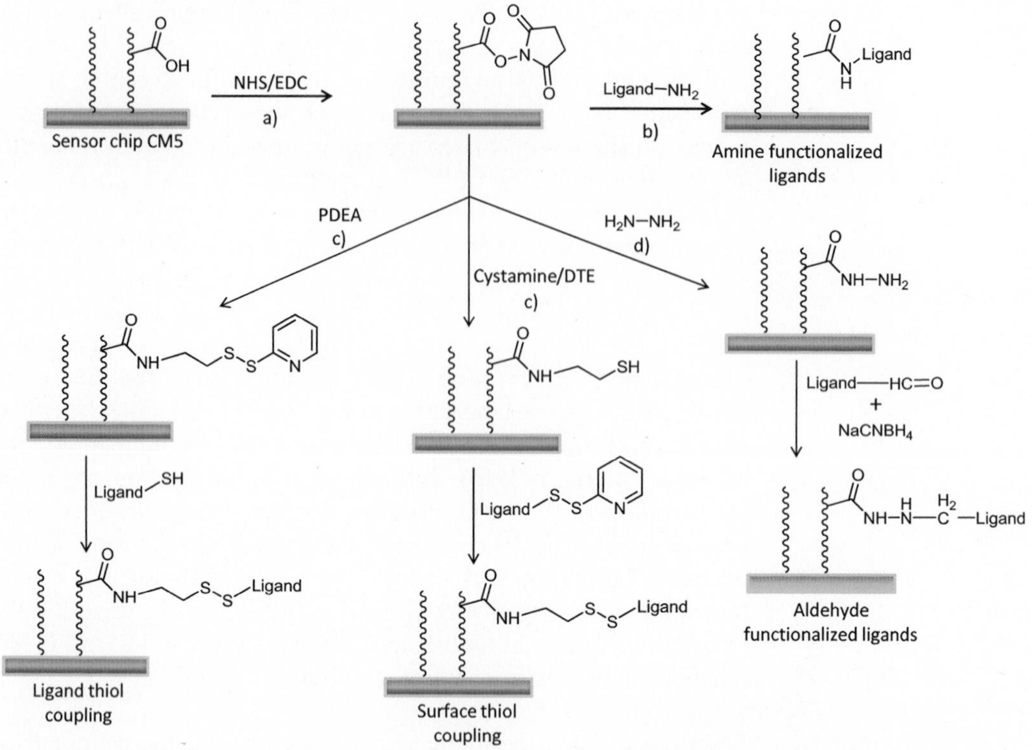

Fig. 3 Steps in coupling chemistry for covalently attaching biomolecules to the sensor surface: (**a**) Activation of the matrix-based carboxyl groups by NHS (*N*-hydroxysuccinimide) and EDC (1-ethyl-3-(3-dimethylamino-propyl)-carbodiimide); (**b**) Direct immobilization of amine functionalized ligands; (**c**) Thiol coupling utilizing disulphide exchange (ligand and surface thiol coupling); (**d**) Immobilization of aldehyde functionalized ligands using reductive ammination. (PDEA Thiol coupling reagent: 2-(2-pyridinyldithio)ethaneamine hydrochloride; H_2N-NH_2 Hydrazine; $NaCNBH_4$ Cyanoborohydride)

with hydrazine or carbohydrazide. Aldehyde coupling provides an alternative approach for immobilizing glycoproteins and other glycoconjugates.

1.2.5 Determining of Optimal Immobilization pH (pH Scouting)

The immobilization process is based upon the principle of electrostatic interaction of the ligand and the activated surface molecules, which results in a covalent binding. In order to reach the highest efficiency of the amine coupling reaction, proteins have to be pre-concentrated on the sensor chip surface. A ligand can be concentrated at the sensor surface by electrostatic attraction. The pK_a of the dextran matrix is 3.5 and at pH > 3.5 the dextran matrix will have a net negative charge. Efficient pre-concentration requires that the pH be between the pK_a of the sensor surface and the isoelectric point (pI) of the ligand (the pH at which there is no net charge on the protein). Such surface attraction will be reached by using immobilization buffer with pH > 3.5, but lower than the pI of the ligand. As a general rule, the pH of the buffer should be at least

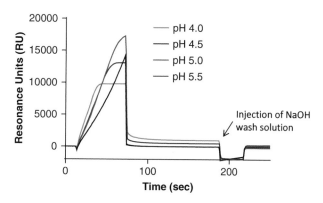

Fig. 4 Pre-concentration profile of DOT1L protein onto the dextran surface of a CM5 sensor chip. Solutions of 50 μg/mL of Mocr-DOT1L (826–1095) protein in 10 mM sodium acetate buffer with various pHs were injected over an untreated chip surface at 10 μL/min. After each injection, 50 mM NaOH was used to completely remove the protein from the surface. Electrostatic interactions of DOT1L protein occur for each tested pH and binding increases as the pH is increased from 4.0 to 5.0. The sodium acetate buffer with pH 5.0 was chosen as optimum to minimize possible denaturation

0.5 units below the pI of the ligand. In addition, the lower the ionic strength of the coupling buffer, the more ligand will be pre-concentrated and immobilized. This will allow strong electrostatic attractions between the negatively charged carboxyl group of the matrix and the positively charged amine groups of the ligand. Therefore, acidic proteins with pI values <3.5 cannot be immobilized by amine coupling. Buffers such as Tris buffer and reagents bearing primary amines should not be used due to possible competition with the amino groups of the protein. To determine the appropriate pH for immobilization, an experimental procedure known as pre-concentration pH-scouting, should be performed. The pre-concentration is driven by electrostatic interactions, which are most pronounced in low ionic strength buffers. For this purpose 10 mM sodium acetate buffer with different pH values, in a range from 4.0 to 5.5, is tested in order to determine the most optimal pH (Fig. 4). The magnitude and slope of the response generated due to the increased mass on the surface provide information about the ligand density. By comparing the pre-concentration responses of each immobilization buffer, the mildest pH solution that can achieve the targeted level for immobilization should be selected in order to minimize possible denaturation. In general, for many proteins, 10 mM sodium acetate buffer (pH 4.5) is optimal for their preconcentration.

1.2.6   Reference Surface

During an SPR experiment the choice of an appropriate reference surface is crucial. Usually the surface on the chip is divided into several flow cells, depending on the instrument, which can be used individually or in a number of combinations, and at least one flow cell of the chip should be used as a reference surface. The use of a reference surface is important because it is a means to: (1) demonstrate that the observed response is not due to nonspecific binding, (2) correct for bulk refractive index change, (3) correct for baseline drift, and (4) in combination with "double referencing", correct for instrument noise due to injections.

Three types of reference surfaces can be used. An unmodified surface is used to check for nonspecific binding to the dextran matrix. A surface that is treated with the same coupling chemistry used to immobilize the ligand is the most commonly used control as a reference surface. This activated–deactivated surface is prepared by treating the surface with the immobilization procedure but omitting the ligand. In this way the negative charge is decreased and therefore nonspecific binding is reduced. The third type of control surface is similar to that used for immobilization but using a different ligand, which can be a nonspecific molecule or an inactive form of the ligand that does not bind the analyte. This inactive protein should be immobilized to approximately the same level as the ligand on the active surface and it should mimic the physical properties of the active surface ligand in terms of, for example, size and charge. If the capture-based assay is used it might be necessary to have an additional control surface onto which the capture ligand will be immobilized at a level similar to that on the active test surface.

Before starting the kinetic measurements, it is very important to demonstrate that the analyte does not show significant binding responses on the control surface as a result of electrostatic or hydrophobic interactions. One of the possibilities for a significant degree of nonspecific binding is a highly positive charge of the analyte in the sample buffer and a high electrostatic attraction to the chip surface. Electrostatic nonspecific binding can be minimized by the addition of NaCl to the sample and running buffers, or using a CM4 chip that has a carboxymethylated dextran matrix with a lower degree of carboxylation, therefore a less negatively charged ligand than CM5 (Table 1), or the analyte could be immobilized on the chip surface. In this way, nonspecific binding of highly positively charged molecules will be reduced. Hydrophobic nonspecific binding usually can be minimized by the addition of a detergent, such as 0.05 % polysorbate 20 or 10 mM CHAPS in the buffer solution.

Each analyte and blank sample should be injected over the ligand and the reference surfaces simultaneously. Differences in the refractive index between the running and sample buffers will give rise to changes in responses that are known as bulk refractive index changes. This is a common phenomenon and in order to eliminate its effect on the measurement of the binding interactions, in particular when the association phase is used for quantification of

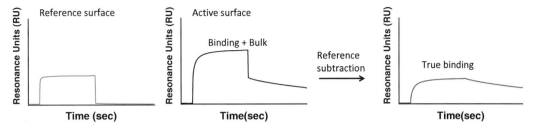

Fig. 5 Bulk effects and reference surface subtraction. Bulk effects are due to differences in the refractive index of the running buffer and sample solution. The bulk contribution should be subtracted using a reference surface

the binding interaction, it is important that the bulk contribution be subtracted using the reference surface responses (Fig. 5).

Furthermore, SPR experiments should be designed in a way that double referencing of the sensorgram data can be performed. Double referencing data helps correct for artifacts such as bulk refractive index changes, nonspecific binding, systematic instrument noise, and baseline drift, all of which are common in almost every SPR experiment. To double reference sensorgram data, both the reference surface responses and the blank responses are subtracted from the analyte sensorgrams during sensorgram processing. The former corrects for refractive index shifts and nonspecific binding, while the latter corrects for systematic instrument noise and baseline drift. Often, the quality of sensorgram data sets cannot be assessed without double referencing, and normally, data cannot be reliably fit for kinetic rate constants without double referencing.

2   Materials

The materials and step-by-step procedures provided in this chapter are universally applicable and can be used for any biomolecular interaction analysis and most Biacore systems, GE Healthcare.

2.1   Instrument Cleaning

1. Maintenance sensor chip (GE Healthcare).
2. Running buffer:
   (a) HBS-P: 10 mM HEPES, pH 7.4, 150 mM NaCl, 0.005 % (v/v) Surfactant P20 or Tween 20.
   (b) HBS-EP is an additional commonly used buffer where the HEPES buffer is supplemented with 3 mM EDTA.

   The buffer must be filtered (0.22 μm), degassed, and stored at 4 °C. It should be allowed to equilibrate to room temperature prior to use.
3. Desorb solution 1: Sodium dodecyl sulfate (0.5 % w/v SDS).

4. Desorb solution 2: 50 mM Glycine, adjusted to pH 9.5 with 5 N NaOH.

5. 1 % AcOH.

6. 0.2 M $NaHCO_3$.

7. 6 M guanidine-HCl.

8. 10 mM HCl.

9. Ultrapure water.

2.2 Surface Preparation, Ligand Pre-concentration, and Amine Coupling

1. Sensor chip CM5 (GE Healthcare).

2. Running buffer HBS-P.

3. 10 mM HCl.

4. 50 mM NaOH.

5. 0.1 % SDS.

6. Sodium acetate (NaOAc) buffers for immobilization: 10 mM NaOAc, pH 4.0, 4.5, 5.0 and 5.5, adjusted with 10 % (v/v) AcOH and filtered (0.22 μm). The immobilization buffer should contain no primary amines such as Tris buffer.

7. Ligand solution (for direct immobilization) or antibody (for capture coupling) diluted with immobilization buffer.

8. NHS amine coupling reagent: 100 mM N-Hydroxysuccinimide in water.

9. EDC amine coupling reagent: 400 mM 1-Ethyl-3-(3-dimethylaminopropyl) carbodiimide hydrochloride in water.

10. Ethanolamine solution: 1 M Ethanolamine hydrochloride, pH 8.5.

11. 20 mM NaOH.

12. Polypropylene tubes with rubber caps (GE Healthcare).

2.3 Regeneration

1. Full range of regeneration solutions:
   (a) 10 mM glycine buffers, pH 1.5–3.0.
   (b) 10–100 mM phosphoric acid, HCl and NaOH.
   (c) 1–5 M NaCl and 2–4 M $MgCl_2$.
   (d) 8 M urea and 6 M guanidine hydrochloride.

2. Analyte solutions prepared in a range of different concentrations including the highest concentration or a concentration above the range that will be used in the application.

2.4 Kinetic Studies

1. Stock solution of the analyte (≥100 μg/mL).

2. Running buffer (HBS-P is the most commonly used).

3. Regeneration solution, as previously determined and optimized.

3   Methods

3.1   Amine Coupling

The coupling procedure consists, in general, of three steps: (1) *activation* resulting in an active group that can be further modified or used to couple the ligand; (2) *coupling*, when the ligand is injected over the activated surface until sufficient ligand is bound; and (3) *deactivation* to quench remaining activated sites and obtain the final level of immobilized ligand (Fig. 6). The covalently immobilized ligand either participates directly in the interaction under study or is used for affinity capture of one of the interacting molecules. The methods presented below are the same if an interacting ligand is immobilized or a corresponding antibody will be used for affinity capturing of the ligand.

3.1.1   Instrument Cleaning

Cleaning and routine instrument maintenance is a key contributor to good quality SPR data. Besides the "desorb" and "super clean" protocols which are described below, it is also important to sanitize the instrument fluidics frequently, especially if the instrument is exposed to crude biological samples.

1. Use a maintenance chip and running buffer to run desorb from working tools (20 min) with Desorb solutions 1 (0.5 % w/v SDS) and 2 (50 mM Glycine-NaOH, pH 9.5) (*see* **Note 1**).

2. Run "Super Clean"—a comprehensive cleaning procedure to clean the system including the needle and tubing. The whole procedure requires about 2 h. Run desorb method with the listed reagents below, followed by priming with water heated to 50 °C:

   (a) 2 × 5 mL vials of 1 % AcOH (use desorb method).

   (b) Prime.

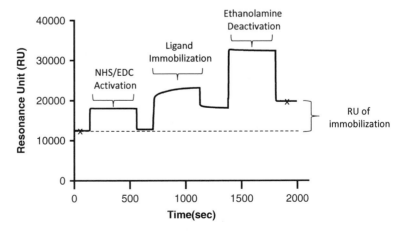

Fig. 6 Sensorgram from a typical amine coupling illustrating the immobilized amount of the ligand

(c) 2×5 mL vials of 0.2 M NaHCO$_3$ (use desorb method).

(d) Prime.

(e) 2×5 mL vials of 6 M guanidine-HCl (use desorb method).

(f) Prime.

(g) 2×5 mL vials of 10 mM HCl (use desorb method).

(h) Prime three times and then put the instrument on standby or continue with preparation of the sensor chip.

3.1.2 Preconditioning of the Sensor Chip CM5

1. Allow the sealed sensor chip pouch to equilibrate at room temperature for 30 min in order to prevent condensation on the chip surface. Insert Sensor Chip CM5 in the instrument and dock the chip.

2. Run three times *Prime* with running buffer (HBS-P) to fill the syringes and compartments.

3. Set the flow rate to 100 μL/min and sequentially using "Quickinject" command inject two times 10 mM HCl, 50 mM NaOH and 0.1 % SDS to precondition the chip surface. This procedure allows washing and hydrating the surface simultaneously, before starting the immobilization procedure.

4. "Rinse," "Flush," and "Prime" to wash the microfluidic system to be ready for the next step of experiments.

3.1.3 Ligand Pre-concentration

1. Start a sensorgram, choose the flow cell where the immobilization will take place and decide on a control, reference surface for blank subtraction. Usually, flow cell 1 is used as the reference surface.

2. Set flow rate to 10 μL/mL.

3. Dilute and prepare the ligand solution (protein sample for immobilization) (*see* **Note 2**) to a constant concentration in a range from 5 to 20 μg/mL using a variety of 10 mM NaOAc immobilization buffers with different pH values (e.g., 4.0, 4.5, 5.0, and 5.5 where the pH is adjusted using 10 % (v/v) acetic acid) (*see* **Note 3**). Place the solutions into propylene tubes and cap the tubes with rubber caps.

4. Inject 10 μL of the prepared protein in NaOAc for 60 s contact time.

5. Inject 50 mM NaOH for 15 s after each pre-concentration as a wash solution to remove the ligand from the surface, if the response does not return to baseline.

6. Analyze the level of RU and the slope of the pre-concentration response to determine the ligand density that can be achieved (Fig. 4). Compare all responses to identify the mildest pH

solution that can achieve the targeted level of immobilization, which depends on the application (*see* **Note 4**).

3.1.4 Ligand Immobilization

1. Start new sensorgram with a 10 µL/min flow rate and select the surface upon which the ligand protein will be immobilized (Fc2, Fc3, and/or Fc4).

2. Prepare fresh solution for activation of the CM5 surface by mixing equal volumes of 100 mM NHS and 400 mM EDC and inject 70 µL (7 min contact time) (*see* **Note 5**).

3. Prepare the ligand in the NaOAc immobilization buffer (with pH determined in the pre-concentration procedure) at a concentration between 10 and 50 µg/mL and inject it over the activated surface until the target levels for ligand immobilization are obtained (*see* **Note 6**).

4. Inject 70 µL of 1 M ethanolamine hydrochloride, pH 8.5, to deactivate excess reactive groups (*see* **Note 7**).

5. Repeat **steps 1–4** until each flow cell has been treated and the ligand has been immobilized.

6. Prepare the reference surface on flow cell 1 (Fc1) to be used for reference-subtracted data necessary for kinetic and affinity measurements. Repeat **steps 1–4**, replacing the ligand for immobilization with an inactive protein or omit immobilization of the control protein so that the control surface is treated only with the activation and deactivation reagents.

7. Wash the surface(s) with a washing solution (i.e., a regeneration solution that is tolerated by the ligand, such as two times 15 s injections of 20 mM NaOH). This allows determination of stably immobilized ligand and the net ligand immobilization level.

8. The sensor chip is now ready to be used or stored. To ensure a stable baseline for the interaction analysis, running a sensor chip overnight in running buffer is recommended (*see* **Note 8**).

3.2 Sample Preparation

Preparation of the samples for testing and injection over a prepared immobilized surface is a critical factor for obtaining high-quality results and accurate kinetic measurements. Similar to the ligand, the analyte should also be pure, homogenous and unaggregated since differently sized binding species will give different responses. The buffer used in experiments can make a significant difference in the binding of the analyte to the ligand and additions such as detergents, chelating agents or denaturing chemicals can influence the binding characteristics and the stability of the complex. The standard running buffer, HBS-P, contains: 10 mM HEPES, pH 7.4, 150 mM NaCl, and 0.005 % P20 (polyoxyethylene sorbitan), with or without EDTA (3 mM). The NaCl is necessary to reduce nonspecific binding, P20 is a non-ionic detergent used to avoid adsorption of the analyte by the flow channels and should be included in the running buffer if possible. Phosphate buffer (10.1 mM Na_2PO_4, 1.8 mM KH_2PO_4, pH 7.4, 137 mM NaCl, 2.7 mM KCl) and Tris buffer (50 mM Tris-HCl, pH 7.4, 150 mM NaCl) are also commonly used buffers. In certain cases if the analyte requires specific additives, for example, when analytes bind in a metal ion-dependent manner, supplementation of the buffers with compatible metal salts will be necessary. During preparation the samples, in order to prevent bulk effects due to the differences in the refractive index of sample solution and running buffer, it is important that the sample buffer match the running buffer as closely as possible. Usually, samples are prepared by serial dilution with running buffer, especially if the dilution factor is sufficiently high so that residual solution components will be negligible, or by dialysis. Samples that contain high refractive index components, such as high-salt solutions, glycerol, or DMSO, should be buffer-exchanged into the running buffer or the components should be added to the running buffer to match the composition of the samples. For example, when small-molecules are tested for their binding to immobilized target proteins, the same concentration of DMSO is added to the running buffer, since DMSO is a commonly used solvent for small molecules. Use of similar buffers will ensure that association (during sample injection) and dissociation (after the end of the injection) occur in the same environment.

Accurate analyte concentration is another important factor since determination of the association rate and affinity constants are dependent on concentration. Thus, it is important that appropriate methods be used to measure the analyte concentration [22]. The ideal value is the concentration of analyte that is active and can bind to the immobilized ligand, which is not necessarily the same as the total analyte concentration.

The concentrations tested in the kinetic experiment should be determined based on the dissociation constant (K_D) of the interaction. Ideally, the analyte samples should be injected at several concentrations ranging from tenfold higher and tenfold less than the expected K_D. If the K_D is unknown, preliminary experiments over a wide range of analyte concentrations should be performed in order to generate complete binding curves and obtain information about appropriate concentrations.

3.3 Regeneration

Regeneration is a process that completely removes the bound analyte from the sensor chip surface after analysis of a sample. Determination of the optimal conditions for efficient surface regeneration is fundamental to efficient and robust protein-protein interaction kinetic studies. It is necessary that the optimal conditions for regeneration be determined empirically, and these conditions will be specific for each studied PPIs. The most important part in this process is that the activity of the surface remains unaffected. Determining suitable regeneration conditions is a two-step process including scouting for the best conditions followed by verification of the suitability of the chosen conditions.

There are several conditions that are generally effective with a wide range of interactants, and these should be used as starting points. These conditions include low pH (10 mM glycine-HCl, pH 3–1.5) [23]; ethylene glycol (50, 75, and 100 %); high pH (1–100 mM NaOH), high ionic strength (up to 5 M NaCl or 4 M MgCl$_2$), detergent (low concentrations of SDS, up to 0.5 %), and denaturant (8 M urea, 6 M guanidine hydrochloride). Additional conditions that have been identified as useful for regeneration of some ligand–analyte interactions include: 10–100 mM HCl, 0.1 % trifluoroacetic acid, 1 M formic acid, 1 M ethanolamine–HCl, pH 9 or higher. In some cases, it is necessary to use regeneration cocktails composed of mixtures of different types of solutions [24]. Regeneration conditions have to be tested experimentally, and whether the regeneration is efficient and removes all bound analyte should be determined. Regeneration reagents of either low pH (e.g., phosphoric acid, glycine–HCl) or high pH (e.g., NaOH) should be injected in relatively short pulses of 10–30 s to minimize exposure time at the ligand surface. Often multiple pulses of a regeneration reagent may be needed. A wash step and an injection of sample buffer should always be performed after the regeneration injections to wash out the microfluidics system, as the regeneration solution can be carried over between binding cycles and affect the next binding reaction and kinetic analysis. The amount of analyte bound to the surface can affect the conditions required for optimal regeneration. Therefore, high analyte concentrations (the highest or a concentration above) that will be tested in the application should be used for the scouting procedure. It is also important that the regeneration solution is fully compatible with the running buffer, so that precipitation will not occur at interfaces between the two solutions.

Once regeneration scouting has identified suitable conditions, it is important to verify the performance of regeneration over a larger number of repeated cycles of analyte injection and regeneration to fully assess the selected conditions. As a general recommendation, good regeneration conditions should give a consistent analyte response, within 10 %, over multiple cycles demonstrating that the ligand is still fully active. If regeneration conditions are too mild, the analyte response will decrease and the baseline response will increase, indicating that the analyte has not been completely removed. On the other hand, conditions that are too harsh might lead to decreases in the response and a constant or decreasing baseline, suggesting that the surface is not fully active and/or has been overstripped. A successful regeneration procedure has been achieved when multiple, properly referenced sensorgrams of identical analyte concentrations can be reproduced. If the off rate is fast, then the regeneration step will not be necessary. In this case, one should just wash the surface until all of the analyte dissociates from it. Recently, a detailed method for a systematic, seven-step experimental approach to efficiently determine the optimal regeneration conditions for SPR surfaces with covalently coupled proteins was reported [25]. Successful selection of a regeneration solution is a very important step ensuring that the surface chip will have intact immobilized ligand and no build-up of bound analyte protein. This will allow multiple uses of the chip and accurate K_D determinations.

3.3.1 Identification of Suitable Regeneration Solution

1. Run sensorgrams using the prepared analyte solutions and various regeneration solutions, under constant contact time and flow rate (*see* **Note 9**).

2. Insert report points and analyze the results through obtained responses to determine that regeneration is within 10 % of the first injection (*see* **Note 10**).

3.4 Measurement of Binding Kinetics and Data Analysis

When designing a kinetic experiment, several experimental parameters should be known or determined as discussed above: immobilization level, injection time and flow rate, dissociation time, and optimal analyte concentration range [26].

Kinetic characterization of interactions can be performed using several different assay formats, which depend on the ligand–analyte interaction kinetics. The most commonly used assay format is the classical format involving multi-cycle kinetics in which each injection of analyte or blank is done in a separate cycle. In this format, the dissociation phase is monitored for the same length of time with each analyte concentration. If the dissociation is very slow, a more efficient approach is the "short and long" experiment in which short and long dissociation times are combined. Usually, the highest tested concentration is used to obtain a long dissociation time. This allows an adequate decrease in response and collection of more dissociation information for calculation of the dissociation rate constant (k_d). Both of these assay formats require regeneration of the ligand surface after each analyte injection in order to remove the bound analyte and prepare the surface for the next concentration to be tested. A kinetic titration or single-cycle kinetics format is useful for studying interactions that are difficult to regenerate or when regeneration is detrimental to the ligand [27]. The analyte is injected from a low to high concentration separated by short dissociation times concluding with a long dissociation time. All the injections are analyzed in one sensorgram with a special equation for kinetic titration. A steady state or equilibrium experiment is possible when the dissociation rate is fast, usually when the k_d is greater than 10^{-3} s^{-1}. This assay format is often used with small compounds, which have a fast dissociation and consequently reach steady state quickly. Generally, determination of the association and dissociation rate using this assay format is difficult because the steady state is reached rapidly. When performing equilibrium analysis to determine the equilibrium constant K_D, it is important to use data in which the responses of all analyte concentrations have reached equilibrium (R_{max}) and plot at a given range of concentrations of analyte.

The order of analyte injection can be important for generation of accurate data. It is better to inject the analyte concentrations in a randomized manner, because injecting the analyte from low to high concentration can hide problems with regeneration and baseline drift. Normally, a single experiment contains several injections of the same analyte at various concentrations to show the stability and reproducibility of the system. Typically, analyte injections are performed at a high flow rate between 50 and 100 µL/min which helps to reduce mass transport effects. In order to calculate accurate binding constants, the association and dissociation phase should be long enough. In particular, higher analyte concentrations should show enough curvature in the association phase of the sensorgrams to allow the fitting model to estimate k_a reliably, and show enough signal decay in the dissociation phase to support a reliable estimate of k_d.

The obtained sensorgrams should be overlaid and used for curve fitting and determination of the kinetic rate constants and binding affinity of the interaction (Fig. 7). Robust curve fitting

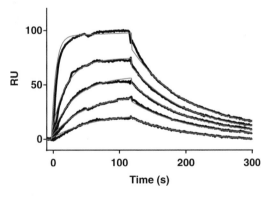

k_a (1/Ms)	k_d (1/s)	K_D (M)	R_{max} (RU)	Chi² (RU)
6.96 x 10⁴	0.0116	1.67 x 10⁻⁷	87.4	1.09

Fig. 7 Plots illustrating the experimental curve-fitting methodology for a simple binding model (1:1 Langmuir). Association and dissociation phases can be seen in each plot. *Black curves* indicate the experimental binding data for each tested concentration (2,000, 500, 250, 125, and 65 nM) of the analyte (ENL protein 489–559) flowing over immobilized Flag-tagged full length DOT1L protein using the capture affinity immobilization method. *Red traces* show the corresponding binding model curves. Calculated binding parameters of the tested protein-protein interactions using global fitting are presented

starts with high quality data from several independent runs or experiments. Many fitting programs (BIAevaluation, Scrubber 2 software) can calculate values for single curves (local fitting) or calculate a single value for all the fitted curves (global fitting). A global value is a more robust outcome because it averages all curves. The most important step is choosing a mathematical model for the fitting, which should be based on an understanding the chemistry and physiology of the studied PPIs.

There are a number of kinetic models that can be applied to the measured data. In general, the simplest model should be used to determine the kinetic constants such as the Langmuir model, which describes a 1:1 interaction between two interactants. This model assumes that the binding reaction is the same and independent at all binding sites and the reaction rate is not limited by mass transport. In this case, analysis of the sensorgram curve in the association phase, in which the binding is measured while the analyte solution flows over the ligand surface, defines the rate of the complex formation and allows calculation of the association constant, k_a (M^{-1} s^{-1}). In the dissociation phase where the injection of the analyte is stopped, the rate of the complex decrease is defined by the dissociation constant, k_d (s^{-1}). In a kinetic analysis the equilibrium constant, K_D (M), is calculated as a ratio from these two kinetic constants: $K_D = k_d/k_a$ (see Note 11).

After the fit is made, the fitted curves should be analyzed with the following questions: (1) how well do the fitted curves follow the measured data; (2) is the dissociation fitted correctly; and (3) are the calculated association constant and R_{max} within the expected range? From the initial 1:1 fitting, a decision should be made regarding whether there is a need for using two other Langmuir kinetic models, which consider the baseline drift and mass transport in order to improve the fit. Langmuir with drift is commonly used in experiments that use a capture surface, where the capture ligand may dissociate from the capture reagent on the chip surface, leading to baseline drift. Adding drift to the fitting can make the fit better but also can skew the results [28]. Therefore, it is very important that the experimental conditions first be optimized before use of different fitting models to give better fit of the data (see Note 12).

When the best fit is chosen, validation of the fitting results is essential and evaluation of the parameters should be performed. For example different zones of the experimental curves should be used for fitting purposes and different fitting should be performed and compared (local versus globally fitted data). If kinetic parameters are consistent throughout all these fits, the chosen kinetic model is probably correct and the determined kinetic parameters are confirmed and reliable. The fitting and experimental curves should be visually inspected and deviations of the fitting from the actual data can be easily identified as either random deviations, which should have a normal distribution, or systematic deviations, which arise because the model is not an adequate description of the experimental data [29]. Chi^2-values and the residual plots for the fitting should be also examined. Residual plots should form a random scattering of the same order of magnitude as the noise level and can reveal systematic differences. Chi^2 is the average of the squared residuals (differences between the measured data points and the corresponding fitted values) and it is an indicator of the fitting confidence. Empirically, this value should be less than 10 % of R_{max}. Calculated values (R_{max}, Chi^2, k_a, k_d) should be inspected and should be within an expected and reasonable range (Fig. 7). For example, in order to achieve a valid calculation of k_d the dissociation should be at least 5 % of the starting value. The value of R_{max}, which is also calculated during the fit and reflects the maximal response when all ligand sites are occupied, should be used to judge if the curves follow a 1:1 kinetic relationship. If R_{max} is very high compared to the response of the curves, this can be an indication that the fit is in error.

In summary, the following factors should be considered in order to obtain robust results, to avoid systematic errors and to validate them: (1) prepare a wide (100-fold) concentration range of the analyte, (2) run replicate experiments, (3) inject the samples in a random order, (4) include buffer injections, (5) use double referencing, (6) apply global analysis, (7) use sensor surfaces with different ligand concentrations or different immobilization techniques, and (8) reverse the analyte and ligand to confirm the results [10, 30].

Most potential sources for generating artifactual data using SPR technology can be avoided with the proper experimental design and data processing, using high-quality protein reagents and proper reporting of the results including all necessary experimental conditions in order to be reproducible. Rich and Myszka have done an excellent work reviewing and analyzing the biosensor literature, and recognizing the lack of proper communication of the use of SPR technology [10, 31–38]. In order to improve the quality, reliability and reproducibility of reported biosensor data, they proposed a list of The Bare Minimum Requirements For An Article Describing Optical Biosensor Experiments (TBMRFAADOBE) [39] (*see* **Note 13**).

3.4.1 Kinetic Analysis of PPIs

1. Run Desorb (**step 1**, Subheading 3.1), then dock the CM5 chip with ligand and prime with running buffer.

2. Prepare a twofold serial dilution of the analyte into running buffer based on the optimum concentration range. The recommended range is 10–0.1 times the expected K_D value. However, it is important to avoid analyte concentrations that can introduce artifacts such as nonspecific binding, steric hindrance or aggregation. Prepare at least one set of duplicate analyte

sample concentrations as well as a buffer sample (zero analyte sample), which will be used for double-reference subtraction during data evaluation prior to curve fitting. Load the dilutions into polypropylene tubes with rubber caps and place them in the instrument.

3. Acceptable flow rates for binding studies are ≥30 μL/min with contact times (i.e., the association phase) ranging from 2 to 4 min. The dissociation period depends on the dissociation rate of the analyte and is usually from 10 to 30 min (*see* **Note 14**).

4. Inject the sample through the control flow cell followed by the flow cell(s) immobilized with the ligand. Using the Kinject command will allow definition of the dissociation period (*see* **Note 15**).

5. Regenerate the surface using two 30 s pulses of optimum regeneration solution determined previously, followed by 1–2 min stabilization period after regeneration.

6. Repeat **steps 4** and **5** with all analyte dilutions, including "zero" analyte concentration, and determine the binding curves. The various dilutions of analyte should be tested in "random order" (*see* **Note 16**).

7. Prepare the obtained sensorgrams for evaluation and curve fitting by aligning all sensorgrams based on the inject start point and set zero on the *x*-axis. Set the baseline at zero on the *y*-axis.

8. Subtract the buffer sample from all sensorgrams. If the subtraction was carried out correctly, the control run will be a flat line. Delete any spikes that may occur at the start or the end of the injection, as well as air spikes.

9. Use the command "Kinetics Simultaneous k_{on}/k_{off}" in the BIAevaluation software to calculate k_{on}, k_{off}, and K_D. This command requires definition of the injection start and end points, checking the alignment of the sensorgrams and selecting the association and dissociation phases on the sensorgram, which will be used for calculation of the kinetic parameters (*see* **Note 17**). The "Kinetics Simultaneous k_{on}/k_{off}" command fits the curves globally and gives the most robust data. It is also possible to perform local fitting, which is usually adopted when the maximum response, R_{max}, during the experiment decreases, allowing the R_{max} parameter to be treated as a local variable across the data set.

10. Insert the tested analyte concentrations and their units (*see* **Note 18**).

11. Selection of the proper binding model for fitting is important. The 1:1 binding model and global fit module should be used unless there is a strong reason to do otherwise (*see* **Note 19**).

12. Identify inappropriately fitted curves or aberrant binding profiles by visual inspection of the fitted curves. The residual

spread and Chi^2 values should fall within the defined acceptance criteria.

4   Summary

SPR technology is a versatile and reliable platform to assess and quantify the kinetics and equilibrium constants of different biological molecules including protein-protein, protein–DNA, protein–RNA, and protein–small molecule interactions and thus separate selective and non-selective interactions. Improved methods for data collection and processing significantly improve the quality of biosensor data. The ability to describe binding data with simple interaction models, together with the option of providing thermodynamic information about the binding reaction, contribute to elucidating and understanding of biomolecular recognition and interaction events. The SPR platform continues to evolve and provide valuable information on binding kinetics, and is expanding beyond these studies towards diagnostic, therapeutic, and drug discovery settings.

5   Notes

1. A Desorb routine should be performed after each use or at least once per week to keep the needle, tubing and microfluidics clean. The method for Desorb is under the Tools tab, Working tools, and the reagents and protocol are described by the software. This procedure is performed using a docked maintenance chip.

2. For immobilization, either direct or via capture, a very small amount of material (2–10 µg) is required. By calculating the R_{max}, one can determine whether the ratio of the molecular masses of the binding partners could limit the response of the interaction and verify which protein should be immobilized. If an antibody–antigen interaction is studied, the antibody should be immobilized as the ligand to avoid binding avidity effects.

3. The pHs of tests are chosen based on the pI of the ligand protein and have to be 1–3 units below the pI, when the net charge of the protein will be positive. There are available bioinformatics programs that can assess the pI of proteins, for example ProtParam available at the SIB Bioinformatics Resource Portal ExPASy. Many proteins have limited stability in the low ionic strength, low pH solutions used for pre-concentration, so dilution of the ligand should be performed just prior to injection. A pH scouting experiment should be performed on the

flow cell that will be used for the immobilization, not on the reference surface.

4. As a general rule, pH values down to about 3.5 can be used and for pH 3.5–4.0 apply citrate buffer. If pre-concentration is inadequate even at pH 3.5, the ligand may be too acidic and a different immobilization approach should be considered. Some ligands can be immobilized at pH values above 5.5; maleate buffers are suitable for immobilization at pH values in the range 5–6. Furthermore, low ionic strength buffer should be used, and the ligand should be sufficiently diluted or desalted from salt-containing stock solutions. The total ion concentration should be 10 mM or less.

5. After 7 min of activation of the sensor chip surface with EDC/ NHS, approximately 40 % of the carboxyl groups will be activated leaving a net negative charge during the immobilization procedure, which enables pre-concentration. Ligand contact should be completed within 15 min after surface activation to ensure coupling before the reactive esters on the surface can be hydrolyzed. If high-density surfaces are required, the injection time of EDC/NHS can be increased.

6. If the immobilization level is low, increase the contact time if the immobilization sensorgram indicates that more ligand can bind. Also, increase the ligand concentration. Make sure that the EDC and NHS are fresh solutions. Immobilization and running buffers should not contain primary amines, for example Tris or sodium azide, since they can compete with the ligand for reactive groups on the surface.

7. Immobilization can be performed on each flow cell of the sensor chip simultaneously or only on the flow cell intended for use. Experience shows that the activation level of each flow cell might not be the same if the activation solution is run over all flow cells at the same time, however, deactivation of all flow cells together is satisfactory.

8. Once the immobilization is finished, the sensor chip with bound ligand can be undocked and kept in a 50 mL conical tube at 4 °C. Before using the sensor chip again and docking it in the instrument, allow the sensor chip to equilibrate to room temperature.

9. Initial regeneration steps should be performed with mild regeneration solutions for a minimal contact time to ensure that the immobilized ligand will not be affected or otherwise damaged. If necessary, the concentration of the regeneration solution can be increased during the procedure. NaOH regeneration solution should be kept in glass vials and should be freshly prepared because its efficiency as a regeneration solution decreases over the time.

10. A larger number of repeated cycles of analyte injection and regeneration (minimum of 20 cycles) should be repeated to verify the regeneration. Usually there is a slight reduction in activity in the first few cycles, thus the assessment should be calculated using the later cycles, for example 6–25. For some particular interactions, it is not always possible to find satisfactory regeneration conditions. Two alternative approaches should be applied: (1) reverse the roles of the ligand and analyte, since regeneration should preserve the activity of the ligand; (2) use a capturing approach instead of directly attaching the ligand to the sensor surface. In this case, regeneration is directed at removing the ligand from the capturing molecule, and any damage to the ligand does not matter.

11. The association rate constant k_a describes the rate of complex formation, i.e., the number of ligand–analyte (LA) complexes formed per second in a one molar solution of L and A. The units of k_a are $M^{-1} s^{-1}$ and its values are typically between 10^3 and 10^7 in biological systems. The dissociation rate constant k_d describes the stability of the complex, i.e., the fraction of complexes that decays per second. The unit of k_d is s^{-1} and is typically between 10^{-1} and 10^{-6} in biological systems. A k_d of $1.10^{-2} s^{-1}$ $(0.01 s^{-1})$ means that 1 % of the complexes decay per second. K_D describes the system at equilibrium but not the dynamics. For the dynamics, refer to the association and dissociation rate constants. The best parameter to compare the strength of the binding is the dissociation rate constant (k_d) because this parameter gives the time an interaction exists.

12. Baseline drift is usually a sign of a suboptimally equilibrated sensor surface. Adding sufficient wash steps after regeneration, longer equilibration times and blank injections in the measurements (double referencing) can improve or eliminate drift. Several buffer injections before the actual experiment can minimize drift during analyte injection. To determine whether a particular interaction is limited by mass transport and whether the Langmuir with mass transport model should be used, the analyte sample should be injected at different flow rates. If the association curves are different, than this interaction is mass-transport limited. In contrast, if the association curves are independent of the flow rate (all binding curves overlap), then diffusion is not the rate-limiting factor, and the simple Langmuir model can be applied.

13. A list of the experimental conditions that should be included in publications in order to facilitate reproducibility by other investigators include: (1) instrument used in analysis; (2) identity, source, and molecular weight of the ligand and analyte; (3) surface type; (4) immobilization condition; (5) ligand density; (6) experimental buffers; (7) experimental temperatures; (8)

analyte concentrations; (9) regeneration conditions; (10) figure of binding responses with fit; (11) overlay of replicate analyses; (12) model used to fit the data; and (13) binding constants with standard errors.

14. Strong binding ($k_d > 10^{-4}\,\text{s}^{-1}$) is difficult to analyze when the dissociation curve is too short. As a rough rule, the dissociation curve should decrease by at least 5 % before analysis is attempted. For a dissociation constant of $10^{-4}\,\text{s}^{-1}$ this will result in a dissociation time of at least 12 min. In addition, sufficient blank injections with the same long "dissociation" times are needed to compensate for possible baseline drift [40]. Long dissociation times are not practical when obtained with every injection. Instead, long dissociation time should be applied for the highest tested analyte concentration, while other concentrations should be analyzed using short dissociation time (200–300 s).

15. Inspect the analyte response on the reference flow cell to identify nonspecific binding. The bulk refractive index has a square-shaped response, while nonspecific binding will typically have an increasing response on the reference, control surface. Problems with nonspecific binding can, in general, be addressed in several ways: (1) experimental conditions: purifying the sample to remove components that interfere with the assay, optimizing the composition of the running buffer (150 mM or higher salt concentration will help to suppress nonspecific electrostatic interactions); (2) sensor surface: different sensor chip types have different characteristics with respect to nonspecific binding (Table 1); (3) sample additives such as soluble carboxymethyl-dextran which can compete for molecules that bind to the dextran on the sensor surface without interfering with the analyte–ligand interaction.

16. Before data evaluation, it is important to analyze and compare the first and last run checking the baseline level, as changes in this level will affect data evaluation. For example, a rise in the baseline level indicates accumulation of the analyte, while a decrease in the baseline level is an indicator of excessive removal of ligand.

17. When a set of binding curves are being evaluated, the region that will be selected for implementing the fit is important. The BIAevaluation software offers a split view function where the original curves and their derivative functions can be examined, which can help in the evaluation of whether the model and the parts of the sensorgram selected are appropriate for data evaluation. The $\ln(Y0/Y)$ tool, for the dissociation phase, and $\ln[\text{abs}(dY/dX0)]$ tool, for the association phase, enable observation

of binding curves exhibiting single-binding kinetics which are linear, while they are curved for more complex systems.

18. It is very important that the analyte is well characterized, with accurate known concentration, high purity (>95 %), functional, no aggregates and monovalent for 1:1 interaction studies, because regardless of the experimental design, the quality of the SPR data directly depends on the quality of the reagents used. In addition, it is important that glycerol is not present in the analyte solution, as glycerol has high refractive index and might interfere with the SPR response.

19. Avoid model shopping until a decent fit has been obtained. Instead, before using fitting models other than the 1:1 Langmuir model be sure that: (1) the system is clean and equilibrated; (2) the ligand is pure and homogenous; (3) the amount of ligand is low in order to minimize mass transport; (4) the analyte is pure and homogenous; (5) the analyte concentration range is wide enough (zero to saturation); (6) the analyte buffer matches the running buffer, minimizing bulk shift; (7) the injection time is long enough to give the association curvature; (8) the dissociation time is long enough to give a reasonable signal decay; (9) blank injections are used for double referencing; and (10) replicate analyte injections are performed to demonstrate system stability.

References

1. Cusick ME, Klitgord N, Vidal M et al (2005) Interactome: gateway into systems biology. Hum Mol Genet 14(Spec No. 2):R171–R181

2. Ge H, Walhout AJ, Vidal M (2003) Integrating 'omic' information: a bridge between genomics and systems biology. Trends Genet 19:551–560

3. Wells JA, McClendon CL (2007) Reaching for high-hanging fruit in drug discovery at protein-protein interfaces. Nature 450:1001–1009

4. Blazer LL, Neubig RR (2009) Small molecule protein-protein interaction inhibitors as CNS therapeutic agents: current progress and future hurdles. Neuropsychopharmacology 34:126–141

5. Ryan DP, Matthews JM (2005) Protein-protein interactions in human disease. Curr Opin Struct Biol 15:441–446

6. Gerrard JA, Hutton CA, Perugini MA (2007) Inhibiting protein-protein interactions as an emerging paradigm for drug discovery. Mini Rev Med Chem 7:151–157

7. Ivanov AA, Khuri FR, Fu H (2013) Targeting protein-protein interactions as an anticancer strategy. Trends Pharmacol Sci 34:393–400

8. Fagerstam LG, Frostell-Karlsson A, Karlsson R et al (1992) Biospecific interaction analysis using surface plasmon resonance detection applied to kinetic, binding site and concentration analysis. J Chromatogr 597:397–410

9. Myszka DG (2000) Kinetic, equilibrium, and thermodynamic analysis of macromolecular interactions with BIACORE. Methods Enzymol 323:325–340

10. Rich RL, Myszka DG (2008) Survey of the year 2007 commercial optical biosensor literature. J Mol Recognit 21:355–400

11. Hunter MC, O'Hagan KL, Kenyon A et al (2014) Hsp90 binds directly to fibronectin (FN) and inhibition reduces the extracellular fibronectin matrix in breast cancer cells. PLoS One 9:e86842

12. Chow CR, Suzuki N, Kawamura T et al (2013) Modification of p115RhoGEF Ser(330)

regulates its RhoGEF activity. Cell Signal 25:2085–2092

13. Pal A, Huang W, Li X et al (2012) CCN6 modulates BMP signaling via the Smad-independent TAK1/p38 pathway, acting to suppress metastasis of breast cancer. Cancer Res 72:4818–4828

14. Day YS, Baird CL, Rich RL et al (2002) Direct comparison of binding equilibrium, thermodynamic, and rate constants determined by surface- and solution-based biophysical methods. Protein Sci 11:1017–1025

15. Mehand MS, Srinivasan B, De Crescenzo G (2011) Estimation of analyte concentration by surface plasmon resonance-based biosensing using parameter identification techniques. Anal Biochem 419:140–144

16. Towne V, Zhao Q, Brown M et al (2013) Pairwise antibody footprinting using surface plasmon resonance technology to characterize human papillomavirus type 16 virus-like particles with direct anti-HPV antibody immobilization. J Immunol Methods 388:1–7

17. Nedelkov D, Nelson RW (2003) Surface plasmon resonance mass spectrometry: recent progress and outlooks. Trends Biotechnol 21:301–305

18. Williams C, Addona TA (2000) The integration of SPR biosensors with mass spectrometry: possible applications for proteome analysis. Trends Biotechnol 18:45–48

19. Davis TM, Wilson WD (2000) Determination of the refractive index increments of small molecules for correction of surface plasmon resonance data. Anal Biochem 284:348–353

20. Shen C, Jo SY, Liao C et al (2013) Targeting recruitment of disruptor of telomeric silencing 1-like (DOT1L): characterizing the interactions between DOT1L and mixed lineage leukemia (MLL) fusion proteins. J Biol Chem 288:30585–30596

21. Drake AW, Tang ML, Papalia GA et al (2012) Biacore surface matrix effects on the binding kinetics and affinity of an antigen/antibody complex. Anal Biochem 429:58–69

22. Pace CN, Vajdos F, Fee L et al (1995) How to measure and predict the molar absorption coefficient of a protein. Protein Sci 4:2411–2423

23. Andersson K, Areskoug D, Hardenborg E (1999) Exploring buffer space for molecular interactions. J Mol Recognit 12:310–315

24. Andersson K, Hamalainen M, Malmqvist M (1999) Identification and optimization of regeneration conditions for affinity-based biosensor assays. A multivariate cocktail approach. Anal Chem 71:2475–2481

25. Drake AW, Klakamp SL (2011) A strategic and systematic approach for the determination of biosensor regeneration conditions. J Immunol Methods 371:165–169

26. Karlsson R, Larsson A (2004) Affinity measurement using surface plasmon resonance. Methods Mol Biol 248:389–415

27. Karlsson R, Katsamba PS, Nordin H et al (2006) Analyzing a kinetic titration series using affinity biosensors. Anal Biochem 349:136–147

28. Rich RL, Papalia GA, Flynn PJ et al (2009) A global benchmark study using affinity-based biosensors. Anal Biochem 386:194–216

29. Cornish-Bowden A (2001) Detection of errors of interpretation in experiments in enzyme kinetics. Methods 24:181–190

30. Cannon MJ, Papalia GA, Navratilova I et al (2004) Comparative analyses of a small molecule/enzyme interaction by multiple users of Biacore technology. Anal Biochem 330:98–113

31. Rich RL, Myszka DG (2000) Survey of the 1999 surface plasmon resonance biosensor literature. J Mol Recognit 13:388–407

32. Rich RL, Myszka DG (2001) Survey of the year 2000 commercial optical biosensor literature. J Mol Recognit 14:273–294

33. Rich RL, Myszka DG (2002) Survey of the year 2001 commercial optical biosensor literature. J Mol Recognit 15:352–376

34. Rich RL, Myszka DG (2003) A survey of the year 2002 commercial optical biosensor literature. J Mol Recognit 16:351–382

35. Rich RL, Myszka DG (2005) Survey of the year 2004 commercial optical biosensor literature. J Mol Recognit 18:431–478

36. Rich RL, Myszka DG (2005) Survey of the year 2003 commercial optical biosensor literature. J Mol Recognit 18:1–39

37. Rich RL, Myszka DG (2006) Survey of the year 2005 commercial optical biosensor literature. J Mol Recognit 19:478–534

38. Rich RL, Myszka DG (2007) Survey of the year 2006 commercial optical biosensor literature. J Mol Recognit 20:300–366

39. Rich RL, Myszka DG (2010) Grading the commercial optical biosensor literature-Class of 2008: 'The Mighty Binders'. J Mol Recognit 23:1–64

40. Katsamba PS, Navratilova I, Calderon-Cacia M et al (2006) Kinetic analysis of a high-affinity antibody/antigen interaction performed by multiple Biacore users. Anal Biochem 352:208–221

Chapter 8

Resonant Waveguide Grating for Monitoring Biomolecular Interactions

Meng Wu and Min Li

Abstract

Label-free detection technologies have been widely used to characterize biomolecular interactions without having to label the target molecules. These technologies exhibit considerable potential in facilitating assay development and enabling new integrated readouts. When combined with high-throughput capability, label-free detection may be applied to small molecule screens for drug candidates. Based on the resonant waveguide grating biosensors, a label-free high-throughput detection system, the Epic® System, has been applied to monitor molecular interactions. Here we describe a generic label-free assay to quantitatively measure phospho-specific interactions between a trafficking signal—phosphorylated SWTY peptide and 14-3-3 proteins or anti-phosphopeptide antibodies. Compared with the solution-based fluorescence anisotropy assay, our results support that the high-throughput resonant waveguide grating biosensor system has shown the capability not only for high-throughput characterization of binding rank and affinity but also for the exploration of potential interacting kinases for the substrates. Hence, it provides a new generic HTS platform for phospho-detection.

Key words Resonant waveguide grating, Label-free, Epic, High throughput, Phospho-specific interactions, SWTY peptide, 14-3-3 proteins, Kinase

1 Introduction

Label-free detection technologies have been widely used to characterize biomolecular interactions without having to label the target molecules [1, 2]. Different transduction techniques such as surface plasmon resonance (SPR), optical ellipsometry, quartz crystal microbalance, Raman scattering, and calorimetry have been developed. In spite of their diverse applications, most of their compatibility with high-throughput screening remains a challenge, although recent developments, i.e., SPR imaging [3], or incidence-angle dependence of optical reflectivity difference imaging [4], have shown potential for increasing the throughput. Resonant waveguide grating (RWG) sensors have been developed into a high-throughput microplate-based biosensor system, the Epic® System, combining the features of

Cheryl L. Meyerkord and Haian Fu (eds.), *Protein-Protein Interactions: Methods and Applications*, Methods in Molecular Biology, vol. 1278, DOI 10.1007/978-1-4939-2425-7_8, © Springer Science+Business Media New York 2015

label-free detection with high-throughput capability [5]. Among the many applications label-free optical biosensors can be used for, the Epic system offers the possibility of developing a generic assay for evaluating protein-protein interactions specific to phosphorylated target sequences.

1.1 Assay Principle of Biosensors

The Epic® system is based on a RWG biosensor, which exploits the evanescent wave generated by the resonant coupling of light into a waveguide via a diffraction grating. The guided light can be viewed as one or more mode(s) of light that all have directions of propagation parallel with the waveguide, due to the confinement by total internal reflection at the substrate–film and medium–film interfaces. The waveguide has a higher refractive index value than its surrounding medium. Because the guided light mode has a transversal amplitude profile that covers all layers, the effective refractive index N of each mode is a weighted sum of the refractive indices of all layers:

$$N = f_N(n_F, n_S, n_C, n_{ad}, d_F, d_{ad}, \lambda, m, \sigma) \tag{1}$$

Here, n_F, n_S, n_C, and n_{ad} is a refractive index of the waveguide, the substrate, the cover medium, and the protein adlayer, respectively. d_F and d_{ad} are the effective thicknesses of the film and the protein adlayer, respectively. λ is the vacuum wavelength of the light used. $m = 0, 1, 2,$ is the mode number; and σ is the mode type number, which equals 1 for TE (transverse electric or s-polarized) and 0 for TM modes (transverse magnetic or p-polarized). Because of its higher sensitivity, generally the TM ($\sigma = 0$) is used for measuring the binding of biomolecules to the probe proteins immobilized on the surface of the waveguide substrate.

When a laser illuminates the waveguide at varying angles or wavelengths, light is coupled into the waveguide only at corresponding specific angles or wavelengths. This coupling is determined by the effective refractive index of the guided mode, denoted as N. The value of N can be calculated numerically from the mode equation for a given mode of a four-layer waveguide configuration:

$$0 \cong \pi m - k\left(n_F^2 - N^2\right)^{0.5}\left(d_F + d_A \frac{n_A^2 - n_C^2}{n_F^2 - n_C^2}\left[\frac{(N/n_C)^2 + (N/n_A)^2 - 1}{(N/n_C)^2 + (N/n_F)^2 - 1}\right]^\sigma\right)$$

$$+\arctan\left[\left(\frac{n_F}{n_S}\right)^2\left(\frac{N^2 - n_S^2}{n_F^2 - N^2}\right)^{0.5}\right]$$

$$+\arctan\left[\left(\frac{n_F}{n_C}\right)^2\left(\frac{N^2 - n_C^2}{n_F^2 - N^2}\right)^{0.5}\right] \tag{2}$$

Here, $k = 2\pi/\lambda$.

Since the laser light is coupled to, and propagates parallel to the surface in the plane of a waveguide film, this creates an electromagnetic field (i.e., an evanescent wave) in the liquid adjacent to the interface. The amplitude (E_m) of the evanescent wave decays exponentially with increasing distance d from the interface:

$$E_m(d) = E_m(0)\exp\left(\frac{-d}{\Delta Z_C}\right) \quad (3)$$

with:

$$\Delta Z_C = \frac{1-\sigma}{k\left(N^2 - n_C^2\right)^{0.5}} + \frac{\sigma\left[\left(N/n_F\right)^2 + \left(N/n_C\right)^2 - 1\right]^{-1}}{k\left(N^2 - n_C^2\right)^{0.5}} \quad (4)$$

ΔZ_C is the penetration depth (also termed as sensing volume; typically ~150 nm) of the evanescent tail of the waveguide mode that extends into the cover medium. This means that a target or complex of a certain mass contributes more to the overall response when the target or complex is closer to the sensor surface, as compared to when it is further from the sensor surface.

1.2 Resonant Waveguide Grating Biosensor Detection

A beta version of the Corning® Epic® System is comprised of three major components for bioassay applications: an Epic® sensor microplate, an RWG detector, and a liquid handling system. The sensor microplate consists of a glass bottom plate attached to a holey plastic 384-well plate, which enables high-throughput screening. Each well in the 384-well Epic® microplate contains an RWG sensor, which consists of an optical grating and a high index of refraction waveguide coating [6, 7]. When illuminated with broadband light at a fixed angle of incidence, these sensors reflect only a narrow band of wavelengths that is a sensitive function of the effective index of refraction of the waveguide. The sensors are coated with a surface chemistry layer that enables covalent attachment (via peptide bond formation) of peptides/proteins or other biomolecules. Binding of molecular recognition partners to the immobilized target induces a change in the effective index of refraction of the waveguide, and this is manifest as a shift in the wavelength of light that is reflected from the sensor. The magnitude of this wavelength shift is proportional to the amount of analyte that binds to the immobilized target. Unlike most commercial SPR biosensors, which generally use a continuous flow system for the determination of the kinetics of binding, the Epic® System does not utilize flow channels. RWG sensors are evanescent in nature which means that the magnitude of the electric field in the medium adjacent to the sensor surface decays exponentially from the sensor surface. The distance from the sensor surface at which the electric field strength has decreased to $1/e$ of its initial value is the penetration depth. For the Epic® System, the penetration depth is ~150 nm. Thus, the system is selective and sensitive to binding events that take place within this penetration. For most of the

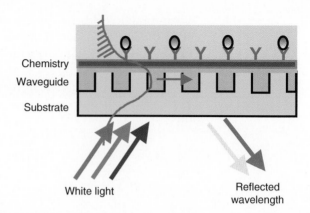

Chemistry

Waveguide

Substrate

White light   Reflected
wavelength

Fig. 1 Detection scheme of RWG biosensor for detecting the binding of biomolecules. The detection scheme of RWG biosensor for detecting the binding of target molecules (●) in a sample to the probe molecules (Y) immobilized onto the surface of a waveguide substrate. The specific binding event is manifested the shift in the wavelength of the reflected light. The waveguide substrate consists of a region within which a grating structure is embedded. The probe molecules are coupled to the derivatized waveguide substrate. From: Wu, M Long S, Frutos AG, Eichelberger M, Li M, Fang Y., *Journal of Receptors and Signal Transduction*, 2009; 29 (3, 4): 202–210, copyright © 2009, Informa Healthcare. Reproduced with permission from Informa

experiments described in this chapter, a well adjacent to the sample well is used as a reference. The most recent version of the Epic® microplate utilizes a self-referencing scheme in which each well of the plate has its own reference region. This is enabled by a plate design that provides protein binding chemistry on only half of the sensor surface so that when a solution of peptide/protein target is added to the well, it only binds to half of the sensor surface, leaving the other half as an in-well or self-reference.

Figure 1 is a schematic drawing for detecting the binding of biomolecules in a sample to the probe molecules immobilized onto the surface of the waveguide substrate. In a particular assay, the probe molecules are pre-coupled to the surface, primarily through covalent-coupling or bio-specific interaction (e.g., biotin–avidin interaction). Alternatively, the immobilization of probe molecules can be monitored in real-time to ensure the efficiency and quality of the coupling. Nonetheless, after the immobilization of probe molecules, the binding of target molecules, in the absence and presence of a modulator, can be directly monitored by the label-free RWG biosensor, as manifested by the shift in wavelengths or angles of the reflected light.

1.3 RWG Biosensor Detection of Phosphorylated Protein-Protein Interactions

Phosphorylation is a key posttranslational process that confers diverse regulation in biological systems involving specific protein-protein interactions recognizing the phosphorylated motifs. Of great interest for ion channels, which represent an important, but underdeveloped class of drug targets, is the surface expression of these membrane proteins modulated by the posttranslational phosphorylation. A common effect of phosphorylation is a change in protein-protein interactions. 14-3-3 proteins were the first protein modules to be identified as binding specifically to phosphorylated substrates. Evidence from structural studies and sequence analyses indicates that the primary function of 14-3-3 proteins lies in their preferential binding to phosphorylated substrates, through their antiparallel bivalent binding sites. In addition to known canonical binding motifs, several earlier reports have identified interactions between 14-3-3 proteins and the C termini of target proteins. This characteristic binding (i.e., SWpTY motif) has high binding affinity that is comparable to that of the canonical binding motifs. Recently, studies have suggested that 14-3-3 proteins, through binding to inducible phosphorylated motifs, regulate protein expression on the cell surface.

Currently, multiple methods are available for detecting phosphorylation, including antibody-based fluorescence detection, radioactive-ATP, and fluorescent-labeled peptide substrates for FRET or fluorescence polarization detection. Although several choices are available for characterization of phosphorylation and potentially for the development of high-throughput screening assays, such label-based methods are prone to artifacts and other detrimental effects (i.e., label effect on antibody interactions). Here we present a protocol for using a high-throughput label-free optical biosensor system–Epic®—for the interrogation of an example phospho-specific interaction, 14-3-3 with SWpTY motif.

2 Materials

2.1 Reagents

1. Inorganic salts of analytical purity.

2. PEG-amine   (O,O′-Bis(2-aminopropyl)polyethylene glycol 1900).

3. Ethanolamine.

4. Boric acid.

5. Dimethylsulfoxide (DMSO).

6. 3-[(3-cholamidopropyl)dimethylammonio]-1-propanesulfonate (CHAPS).

7. Octylphenolpoly (ethyleneglycolether) (NP40).

8. Trition X-100.

9. Dithiothreitol (DTT).

10. PBS buffer: a phosphate buffered saline solution with a phosphate buffer concentration of 0.01 M and a sodium chloride concentration of 0.154 M. The solution pH will be 7.4.

11. Anti-SWpTY antibody produced from rabbit.

12. Peroxidase labeled anti-Rabbit IgG (H + G, made in goat).

13. Anti-14-3-3 antibody (made in rabbit).

14. Supersignal ELISA Femto Maximum Sensitivity substrate kit.

All solutions were prepared in a 10 mM 4-(2-Hydroxyethyl) piperazine-1-ethanesulfonic acid (HEPES) buffer (pH 7.3) unless indicated.

2.2 Preparation and Purification of Recombinant 14-3-3 Proteins

1. 14-3-3ζ protein expressed as a GST-tagged fusion and purified from *E. coli* strain BL21-SI as previously described [8].

2. Recombinant GST-tagged 14-3-3 proteins purified using Glutathione Sepharose 4B beads.

Fluorescence anisotropy measurements were used to determine the binding affinity of 14-3-3ζ for SWpTY [8].

2.3 Synthesis and Preparation of Peptides

1. NH2-SWpTY (NH2AhxAhxFRGRSWpTY-COOH, Ahx: 6-aminohexyl-) peptide and non-phosphorylated NH2-SWTY peptide synthesized by Biomer Technology.

2. SWpTY, RGRSWpTY-COOH; SWTY, RGRSWTY-COOH; SWpTD, RGRSWpTD-COOH; SWpTP, RGRSWpTP-COOH; and SWpTP, RGRSWpTP-COOH peptides from New England Peptides (or another commercial source).

The peptides were prepared as stock solutions by dissolving in water, and if necessary with the addition of a minimal amount of acetonitrile.

3 Methods

3.1 14-3-3 Detection Protocol on the Epic® System

As a common protocol for 14-3-3 detection [9] (Fig. 2), the first step is to immobilize NH2-SWpTY peptide (50 μg/ml, pH 7.5), as a 14-3-3 binding motif, on the Epic® plate. This was followed by the addition of ethanolamine to quench the remaining active sites on the surface. Reference wells were made by the addition of PEG-amine (50 μg/ml, pH 9.2) or the non-phosphorylated peptide NH2-SWTY (50 μg/ml, pH 7.5). The second step is the addition of 14-3-3 protein solutions into both the sample wells and the reference wells. As seen in Fig. 2c, the wavelength shift after the subtraction of the signals from the reference wells is termed the referenced signals for the 14-3-3 binding event. The signal without subtraction of the reference is termed the unreferenced signal. Usually six sample wells and two reference wells were used for each binding reaction unless otherwise described.

To evaluate the specificity of the interaction, 14-3-3 was added into wells modified with either NH2-SWpTY or non-phosphorylated NH2-SWTY (Fig. 3a). Only wells immobilized

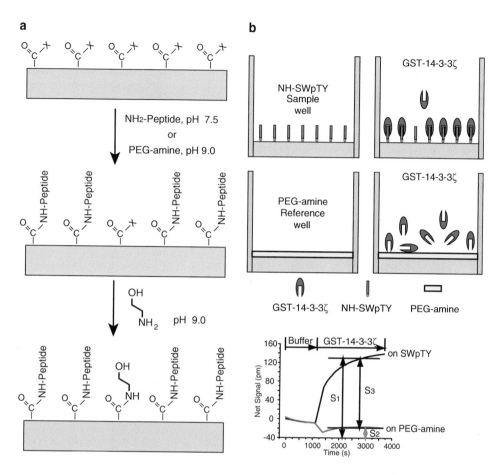

Fig. 2 Scheme of Epic® detection. (**a**) Surface reaction scheme for immobilization of N-terminal amino group modified peptides. X, reactive group; NH2-Peptide, N-terminal amino group modified peptides. (**b**) Binding scheme for 14-3-3. *Left two panels* show the immobilized substrate (NH2-SWpTY) and reference (PEG-amine). *Right two panels* show GST-14-3-3ζ binding. In the sample well, where GST-14-3-3ζ is added, GST-14-3-3ζ binds to the immobilized NH2-SWpTY peptide. In the reference well containing immobilized PEG-amine, there is no binding of GST-14-3-3ζ. (**c**) The time-course charts of the respective responses from sample wells and reference wells, with S1 as an unreferenced signal from GST-14-3-3ζ on NH2-SWpTY (50 μg/ml at pH 5.5. GST14-3-3 at 1 μM) S2 as an unreferenced signal from GST-14-3-3ζ on PEG-amine (50 μg/ml, pH 9.0 GST14-3-3 at 1 μM), and S3 as a referenced signal obtained by subtraction of S1 with S2. From: Wu, M Long S, Frutos AG, Eichelberger M, Li M, Fang Y., *Journal of Receptors and Signal Transduction*, 2009; 29 (3–4): 202–210, copyright © 2009, Informa Healthcare. Reproduced with permission from Informa

with the phosphorylated peptide NH2-SWpTY gave detectable signals. Furthermore, upon addition of competitive 14-3-3 binding peptides, SWpTY or a nonhomologous 14-3-3 binding peptide R18, the binding signals were significantly reduced (Fig. 3b). In contrast, the non-phosphorylated SWTY peptide did not affect the

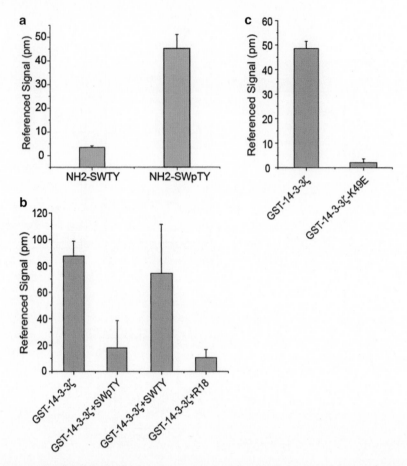

Fig. 3 Specificity of detection. (**a**) The referenced signals for the binding of 14-3-3 to immobilized SWpTY, and SWTY. SWpTY, and SWTY were immobilized at 50 μg/ml, pH 5.5, and GST-14-3-3ζ was added at 0.4 μM. (**b**) Competition assay with competitors and control. 1 μM of competitors (SWpTY and R18) and control peptide (SWTY) were pre-incubated with 0.4 μM GST-14-3-3ζ and applied on the immobilized NH2-SWpTY (50 μg/ml). (**c**) Comparison of the referenced signals from the GST-14-3-3ζ and its validated binding mutant GST-14-3-3ζ (K49E). GST-14-3-3 was used at 0.4 μM

binding of 14-3-3. To be more definitive, the SWpTY peptide binding was tested with wild type 14-3-3 and 14-3-3 K49E, which has a mutation at the binding site [14]. Consistently, the binding signal was obtained only in wild type 14-3-3 proteins (Fig. 3c). Therefore, these experimental results provide the evidence for the specific detection of 14-3-3 interactions with substrate peptides.

3.2 Validation of 14-3-3 Detection by In Situ ELISA

To further validate the 14-3-3 binding event on the Epic® plates, the same Epic® plates above were then subjected to an ELISA assay. After washing five times with PBS buffer, 50 μl of the anti-14-3-3 antibody (1:200 dilution) was added and incubated at 4 °C for 0.5 h. After washing with PBS buffer with 0.1 % Tween, 50 μl of

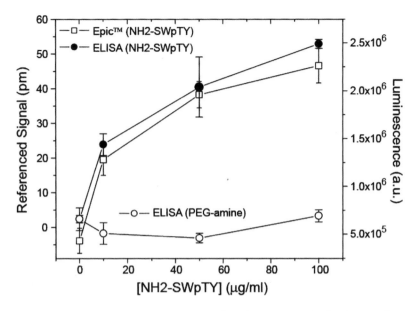

Fig. 4 Binding detection of 14-3-3 to immobilized NH2-SWpTY peptide. NH2-SWpTY in different concentrations (pH 5.5) was immobilized on the Epic™ plate. After addition of GST-14-3-3ζ (0.4 μM), the referenced signals were plotted on the left axis against the concentrations of NH2-SWpTY. The same plate was used for ELISA detection as described in Subheading 3 and the luminescence signals were plotted on the right axis

Peroxidase labeled Anti-Rabbit IgG (1:1,000 dilution) was added and incubated at 4 °C for 0.5 h. After washing with PBS buffer with 0.1 % Tween again, 50 μl of luminescence substrate from Supersignal ELISA Femto Maximum Sensitivity substrate kit was added to each well and subjected to luminescence detection on a multifunctional plate reader.

The ELISA signals proportionally matched those of the Epic™ System (Fig. 4, filled circles). In contrast, reference wells with PEG-amine did not produce any ELISA signal (Fig. 4, open circles). Hence, immobilization of NH2-SWpTY peptide at 50 μg/ml allows for approximately 80 % of the maximal signal and therefore was used for the subsequent experiments.

3.3 Affinity Ranking by Competitive Replacement

The rank order of affinity of the peptide motifs was determined by competition assays. After immobilization of NH2-SWpTY on the Epic® plate, a pre-mixed solution of 14-3-3 and varying concentrations of the competitive peptides were added into the Epic® wells. The referenced signals were applied in Eq. 5 to determine the relative affinities of the corresponding competitive peptides.

$$y = B \times \left\{ \left(\frac{K_{D_{NH\text{-}SWpTY}}}{K_{D_{competitor}}} x + K_{D_{NH\text{-}SWpTY}} + D_0 + P_0 \right) \right.$$

$$\left. - \sqrt{\left(\frac{K_{D_{NH\text{-}SWpTY}}}{K_{D_{competitor}}} x + K_{D_{NH\text{-}SWpTY}} + D_0 + P_0 \right)^2 - 4 \times D_0 \times P_0} \right\}$$

(5)

The fitting was done with Eq. 5 using Origin 7.0 (OriginLab), with y, the referenced binding signal; B as a constant; D_0, the concentration of NH2-SWpTY that conferred 90 % of saturated GST-14-3-3ζ response; P_0, the concentration of GST-14-3-3; $K_{D_{NH\text{-}SWpTY}}$, the K_D (=2.1 μM) for NH2-SWpTY binding to GST-14-3-3ζ as determined by the concentration response of GST-14-3-3ζ with 50 μg/ml of NH2-SWpTY immobilized at pH 7.5; $K_{D_{competitor}}$ as K_D for competitive peptides, and x as the concentration of competitor peptide.

In contrast to the direct detection using the immobilized binding partner, the competition assay uses the binding competitors to measure binding affinity in solution. Affinity detection through competitive replacement has been a routine method [10, 11]. The distinct signals by competitor peptides can be observed with considerably different affinities using the Epic® System. The fitting results for K_D values of the competitors SWpTY, SWTY, SWpTD, and SWpTP are 0.12 ± 0.05 μM, >50 μM, 1.75 ± 0.14 μM, and 6 ± 1.67 μM, respectively. Direct comparison of this method with the previous detection of SWTY-14-3-3 interaction with fluorescence anisotropy suggests a similar rank order of the affinity (Table 1).

Table 1
**Comparison of the 14-3-3–SWpTY interaction detection from fluorescence anisotropy and Epic®
label-free detection**

	Fluorescence Polarization				Epic® System			
Sensitivity (nM)	16				38			
Linear range (nM)	16–700				38–2,000			
Probe K_D (μM) (GST-14-3-3ζ)	1.7 ± 0.3				2.1 ± 0.4			
K_D (μM, competitive)	SWpTY 0.17	SWTY >100	SWpTD 2.2	SWpTP 45	SWpTY 0.12	SWTY >50	SWpTD 1.75	SWpTP 6
Z factor	>0.5				>0.5			
S/N ratio	~8.4				~15			

3.4 Characterization for High-Throughput Screening

To characterize the capability of the Epic® System for high-throughput screening of 14-3-3 phospho-specific interactions, 50 µg/ml of NH2-SWpTY at pH 7.5 was immobilized in one set of rows and PEG-amine was immobilized in a second set of rows on the Epic® plate with PEG-amine immobilized wells used as a reference. GST-14-3-3ζ at 1 µM with varying DMSO amounts was added to test the compatibility of adding small molecule library in DMSO solution. GST-14-3-3ζ at 1 µM with or without pre-incubated 20 µM SWpTY (as a positive inhibition control) was added into the wells of one Epic® plate with intrawell self-referencing for the statistic data. The Z factor, Z' factor, and S/N ratio [12] were calculated from Eqs. 6 and 7, respectively, where Av is the average and SD is the standard deviation of the referenced signals.

$$Z = 1 - 3(SD_{\text{With 14-3-3}} + SD_{\text{Without 14-3-3}})/(Av_{\text{With 14-3-3}} - Av_{\text{Without 14-3-3}})$$

$$(6)$$

$$S/N = \left(Av_{\text{With 14-3-3}} - Av_{\text{With 14-3-3 and SWpTY}}\right)/$$
$$\sqrt{\left(SD_{\text{With 14-3-3}}\right)^2 + \left(SD_{\text{With 14-3-3 and SWpTY}}\right)^2} \qquad (7)$$

The statistic data of responses of GST-14-3-3ζ, with or without pre-incubated SWpTY, have shown a good signal to noise ratio of ~15. The addition of 20 µM competitor SWpTY peptide in the solution resulted in a 94 % decrease of the referenced signals, with a Z factor of 0.8, indicating the robustness and compatibility of the assay for high-throughput screening.

4 Notes

Common to most of label-free systems is the issue of system prone to nonspecific interactions. This justifies the thorough investigation of the specificity of the current assay by including the following components.

1. Non-phosphorylated SWTY peptide: Only the immobilized phosphorylated peptide SWpTY gave detectable signals, while immobilized non-phosphorylated peptide SWTY gave only minimal background signals.

2. Competitive 14-3-3 binding peptides: Only the competitive peptides added in the solution gave the reduction of RWG signals due to the competitive binding. SWpTY peptide or a nonhomologous 14-3-3 binding peptide R18, the binding signals were significantly reduced, where non-phosphorylated peptide SWTY has shown no such effect.

3. Wild type and mutant 14-3-3 proteins: The assay was tested with wild type 14-3-3 and 14-3-3 K49E, which has a mutation at the binding site. Consistently, the binding signal was obtained only in wild type 14-3-3 proteins.

4. Two different SWpTY binding proteins: Different phosphor-recognizing modules, 14-3-3 protein and an anti-SWpTY antibody, have been applied using the same detection format. Both gave detectable signals, where BSA gave only minimal background signals.

5. Validation with alternative assays: Direct and indirect schemes of the Epic assay have been validated by the alternative assays, by in situ ELISA and fluorescent polarization assay respectively. The ranking and affinity data have further quantitatively verified the specificity of the current Epic detection method.

In contrast to peptides, proteins, and nucleic acids, small molecules are quite challenging for label-free detection, because of the small molecular weight and theoretically much smaller signals. In addition, the nonspecific interactions by small molecules, especially those promiscuous hydrophobic ones, require additional control tests to discriminate artifacts from real signals. To overcome this issue, a self-reference Epic® plate has been designed, as shown in Fig. 5. One half of the sensor was coated with active chemical, good for immobilization; while the other half was not coated, and resistant to the immobilization, and consequently serves as the control. The Epic® system can detect two areas and normalize the

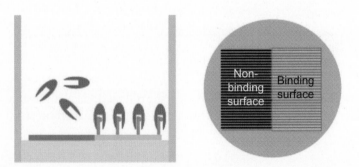

Fig. 5 Schematic representation of the self-referenced biosensor in the well of Epic® plates. The *pink area* denotes the chemical active area good for immobilization; while the *grey area* denotes the area resistant to the immobilization, and consequently serves as the control. The Epic® system can detect two areas and normalize the nonspecific interaction with the surface from the promiscuous small molecules/aggregates. From: Wu, M Long S, Frutos AG, Eichelberger M, Li M, Fang Y., *Journal of Receptors and Signal Transduction*, 2009; 29 (3–4): 202–210, copyright © 2009, Informa Healthcare. Reproduced with permission from Informa

nonspecific interaction with the surface from the promiscuous small molecules/aggregates. For example, as expected, the presence of DMSO did result in a change of bulk refractive index, and, hence, a change in the unreferenced signals of both test wells and reference wells. However, proper use of referencing eliminates this effect, and thus there is no significant difference for the referenced signals between 0 % DMSO up to 10 % DMSO. This is very helpful in the case of competitive format for the detection for screening protein-protein interaction modulators.

5 Conclusion

A label-free assay for the phospho-specific interactions of 14-3-3 proteins has been developed using the resonant waveguide grating Epic® System. When the SWpTY (NH2AhxAhxFRGRSWpTY-COOH) peptide is covalently immobilized to the surface, binding of 14-3-3 proteins can be detected at as low as 38 nM, with adjustable linear ranges depending on the density of immobilized SWpTY. The specificity of detection was validated using competition experiments with a non-phosphorylated SWTY peptide and a binding mutant of 14-3-3 protein. Furthermore, competition assays were performed to determine the rank order of binding affinities of the different peptide motifs. In addition, the assay is compatible with high-throughput screening with a Z factor larger than 0.5 and up to 10 % DMSO tolerance. Therefore the reported assay offers a label-free screening system for phospho-specific interactions applicable at the quantitative level. With minor modifications, the reported assay can be applied for the screening of modulators for 14-3-3 protein-protein interactions. Since the kinase/phosphatase reactions can introduce the phosphorylation and dephosphorylation of the SWTY/SWpTY sequences immobilized on the surface of the sensor plate. The current assay can be further developed into a high-throughput label-free protocol for modulators of kinases and phosphatases. In addition, the assay can be applied into other peptide–protein interactions, i.e., PDZ domains and other extracellular loop sequences of ion channels [13].

The reported assay conditions are tuned for 14-3-3 interactions with the SWpTY peptide. Many aspects of the experimental conditions may be transferable to other interaction systems, especially protein–peptide interactions that are exemplified by phospho-specific antibody to the SWTY peptide. Comparison with solution-based assays such as fluorescence polarization suggests that at least for the interaction between SWpTY and 14-3-3, the Epic® System gave comparable characteristics in terms of sensitivity and dynamic range. Because Epic® allows for quick estimation of binding affinity, it provides an attractive means for quantitative assessment of interactions between different ligands and one protein or between one ligand and different interacting proteins. This could be applicable to a number of assays such as antibody evaluation and detection of interacting components in cell lysates.

References

1. Halai R, Cooper MA (2012) Using label-free screening technology to improve efficiency in drug discovery. Expert Opin Drug Discov 7:123–131

2. Filiou MD, Martins-de-Souza D, Guest PC, Bahn S, Turck CW (2012) To label or not to label: applications of quantitative proteomics in neuroscience research. Proteomics 12:736–747

3. Saito A, Kawai K, Takayama H, Sudo T, Osada H (2008) Improvement of photoaffinity SPR imaging platform and determination of the binding site of p62/SQSTM1 to p38 MAP kinase. Chem Asian J 3:1607–1612

4. Landry JP, Gray J, O'Toole MK, Zhu XD (2006) Incidence-angle dependence of optical reflectivity difference from an ultrathin film on solid surface. Opt Lett 31:531–533

5. Li G, Lai F, Fang Y (2012) Modulating cell-cell communication with a high-throughput label-free cell assay. J Lab Autom 17:6–15

6. Fang Y, Ferrie AM, Fontaine NH, Yuen PK (2005) Characteristics of dynamic mass redistribution of epidermal growth factor receptor signaling in living cells measured with label-free optical biosensors. Anal Chem 77:5720–5725

7. Fang Y, Ferrie AM, Fontaine NH, Mauro J, Balakrishnan J (2006) Resonant waveguide grating biosensor for living cell sensing. Biophys J 91:1925–1940

8. Wu M, Coblitz B, Shikano S, Long S, Spieker M, Frutos AG, Mukhopadhyay S, Li M (2006) Phospho-specific recognition by 14-3-3 proteins and antibodies monitored by a high throughput label-free optical biosensor. FEBS Lett 580:5681–5689

9. Wu M, Long S, Frutos AG, Eichelberger M, Li M, Fang Y (2009) Interrogation of phosphor-specific interaction on a high-throughput label-free optical biosensor system-Epic system. J Recept Signal Transduct Res 29:202–210

10. Huang X (2003) Fluorescence polarization competition assay: the range of resolvable inhibitor potency is limited by the affinity of the fluorescent ligand. J Biomol Screen 8:34–38

11. Dai JG, Murakami K (2003) Constitutively and autonomously active protein kinase C associated with 14-3-3 zeta in the rodent brain. J Neurochem 84:23–34

12. Zhang L, Wang H, Masters SC, Wang B, Barbieri JT, Fu H (1999) Residues of 14-3-3 zeta required for activation of exoenzyme S of *Pseudomonas aeruginosa*. Biochemistry 38:12159–12164

13. Sun H, Li M (2013) Antibody therapeutics targeting ion channels: are we there yet? Acta Pharmacol Sin 34:199–204

14. Zhang L, Wang H, Liu D, Liddington R, Fu H (1997) Raf-1 kinase and exoenzyme S interact with 14-3-3zeta through a common site involving Lysine 49. J Biol Chem 272:13717–13724

Chapter 9

Quartz Microbalance Technology for Probing Biomolecular Interactions

Gabriella T. Heller, Alison R. Mercer-Smith, and Malkiat S. Johal

Abstract

Quartz crystal microbalance with dissipation monitoring (QCM-D) is a useful technique for observing the adsorption of molecules onto a protein-functionalized surface in real time. This technique is based on relating changes in the frequency of a piezoelectric sensor chip, onto which molecules are adsorbing, to changes in mass using the Sauerbrey equation. Here, we outline the cleaning, preparation, and analysis involved in a typical QCM-D experiment, from which one can obtain mass adsorption and kinetic binding information.

Key words QCM-D, Surface, Piezoelectric, Sauerbrey, Frequency, Mass, Adsorption, Deposition, Kinetics

1 Introduction

Quartz crystal microbalance with dissipation monitoring (QCM-D), unlike traditional gravimetric techniques, can measure the binding of molecules to a surface with nanogram sensitivity (Fig. 1). This is achieved by exploiting the piezoelectric properties of crystalline quartz, namely, that applying a precise alternating current to a quartz crystal will cause the crystal to expand and contract, resulting in oscillatory motion. This oscillatory motion is a function of the size and mass of the quartz crystal. Upon altering the mass of the crystal (for example, upon protein binding to the surface) the resonant frequency of the quartz will shift. QCM-D technology works by applying an AC potential to a quartz crystal, which induces the quartz to oscillate at its resonant frequency. By monitoring this resonant frequency over time as molecules are adsorbing to the surface, QCM-D is able to detect these subtle changes in frequency. Because these changes in frequency are related to changes in mass, QCM-D can be thought of as an extremely sensitive balance that measures changes in mass at the surface over time [1].

Cheryl L. Meyerkord and Haian Fu (eds.), *Protein-Protein Interactions: Methods and Applications*, Methods in Molecular Biology, vol. 1278, DOI 10.1007/978-1-4939-2425-7_9, © Springer Science+Business Media New York 2015

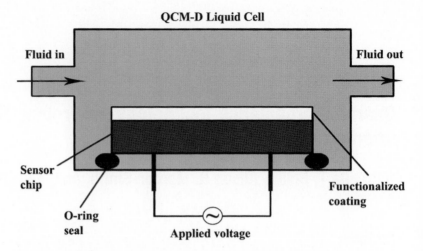

Fig. 1 QCM-D flow cell schematic

Under the assumptions of a rigid surface, the decrease in frequency can then be used to estimate changes in mass using the Sauerbrey equation. Furthermore, information about the viscoelastic properties of the adsorbed film may be obtained using dissipation values, which may be measured by breaking the electrical circuit supplying voltage to the sensor chip. The resulting damping oscillations of the sensor chip are then used to determine dissipation values. If the deposited film is rigid, the energy from the sensor chip oscillations dissipates slowly, and a low dissipation value is reported [1]. Under these assumptions, the Sauerbrey relation may be used. Otherwise, a viscoelastic model must be used to accurately relate frequency changes to changes in mass on the surface. Without examination of dissipation, changes in frequency cannot be accurately correlated to changes in mass.

QCM-D is a useful technique that is comparable to surface plasmon resonance (SPR) and dual polarization interferometry (DPI). While SPR and DPI are optically based methods, QCM-D measurements are based on mechanical frequency changes. As a result, QCM-D measurements account for mass due to trapped solvent within the adsorbent, whereas DPI and SPR do not. Consequently, comparing results from the two methods allows one to decouple the mass effects due to solvent hydration at the surface. Like SPR and DPI, QCM-D can be used to make binding affinity measurements. However, as is the case with any surface technique, a true affinity will not be obtained because the protein's degrees of freedom are constrained due to its immobilization state on a surface [2].

QCM-D experiments are relatively inexpensive and simple to perform. There are three major steps to a typical QCM-D procedure: (1) cleaning, (2) functionalization and data collection, and (3) analysis. As is the case in any surface measurement, a clean system is crucial for reproducible results, and so the sensor chips and flow cells must be thoroughly decontaminated before use. In some cases after cleaning, the sensor chips must be functionalized outside the flow cells; in other cases, this can be accomplished with the sensor chips mounted in the flow cells before the interacting protein is exposed to the sensor chips. In either case, buffer is flowed through the flow cell before taking measurements in order to achieve a stable baseline. Once the measured frequency and dissipation values are stable, the interacting protein or ligand is introduced over the sensor chips. The data collected can then be analyzed to obtain mass deposition values and kinetic information. Overall, the steps involved in the QCM-D procedure are straightforward, making this an approachable and efficient technique for measuring protein–ligand interactions.

2   Materials

When preparing solutions, use ultrapure water (resistivity >18 MΩ cm) and analytical grade reagents. Prepare solutions at room temperature.

2.1   QCM-D Setup (See Note 1)

1. O-rings.
2. SiO_2 sensor chips.
3. Gold sensor chips.
4. Tubing.
5. Gaskets.
6. Tweezers.
7. Teflon sensor holder.
8. Lint-free tissues.
9. Beakers.

2.2   UV/Ozone Treatment Equipment (See Note 2)

1. UV lamp.
2. Safety chamber.
3. Stand.

2.3   Cleaning Chemicals

1. Alkaline liquid detergent solution, specifically designed for sensitive lab equipment cleaning: 2 % (v/v) alkaline liquid concentrate diluted with water.
2. Ethanol, 200 Proof.
3. Compressed nitrogen gas.
4. Oxidizing solution: 1:1:5 volume solution of ammonium hydroxide–hydrogen peroxide–water.
5. Disposable, phosphate-free liquid detergent.
6. Water–ethanol mixture: 1:1 volume solution of water and ethanol.

2.4 Sensor Chip Preparation Materials

1. Phosphate-buffered saline (PBS): Weigh 8 g NaCl, 0.2 g KCl, 1.44 g Na_2HPO_4, and 0.24 g KH_2PO_4. Transfer to a 1,000 mL volumetric flask. Fill the flask with ultrapure water to a few milliliters below the 1,000 mL mark and transfer to a bottled 1 L container. Adjust the pH to 7.4 with HCl or NaOH. Add ultrapure water to obtain a final volume of 1 L as needed.

2. 10 mM salicylic acid solution in water.

3. 10 mM mercaptoundecanoic acid (MUA) solution in methanol.

4. 5 mM 1-ethyl-3-(3-dimethylaminopropyl) carbodiimide/ *N*-Hydroxysuccinimide (EDC/NHS) solution in water.

5. 2 mg/mL metmyoglobin in PBS.

6. 50 mM ethanolamine in ultrapure water.

7. Borate buffer: Weigh 6.18 g boric acid and 1.3 g NaOH. Transfer to a 500 mL graduated cylinder and fill with water to a few milliliters below the 500 mL mark. Transfer to a bottled 500 mL container. Adjust the pH to 9.5 with NaOH and HCl, as needed. Shake to mix thoroughly. Add ultrapure water to obtain a final volume of 500 mL.

8. 0.2, 0.5, 0.7, 1 M sodium azide.

9. 1 mM Nα-(*tert*-Butoxycarbonyl)-L-asparagine, Boc-L-asparagine (Boc-Asn-OH) solution in 200 proof ethanol.

10. 200 mM Nickel ($NiSO_4$) solution in water.

11. 2 mg/mL bovine serum albumin (BSA) in PBS.

3 Methods

Due to QCM-D's high sensitivity, a common source of poor measurements is contamination of the instrument. Thus, it is critical to carefully clean the sensor chips, flow cells, tubing, and handling equipment (tweezers, etc.) before any data is collected. The decontamination procedures outlined below (Subheadings 3.1 and 3.2) should be performed before the experiment (Subheading 3.3) is started to ensure contamination does not confound results obtained using QCM-D.

3.1 Sensor Decontamination

3.1.1 Cleaning SiO₂ Sensor Chips (See Note 3) [3]

1. Assemble UV/Ozone treatment setup (*see* **Note 2**).

2. Expose SiO_2 sensor chips surface side up to UV/Ozone treatment for 10 min.

3. Soak sensor chips in alkaline liquid detergent solution for 10 min.

4. Rinse sensor chips with water and then rinse with ethanol.

5. Dry sensor chips with nitrogen gas (*see* **Note 4**).

6. Expose sensor chips to UV/Ozone treatment for another 10 min.

7. Place sensor chips in decontaminated QCM-D flow cells immediately.

3.1.2 Cleaning Gold Sensor Chips [3]

1. Prepare a 75 °C water bath.

2. Expose gold sensor chips, active surface side up, to UV/Ozone treatment for 10 min.

3. Prepare an oxidizing solution in a beaker (*see* **Note 5**).

4. Place sensor chips in a Teflon sensor holder and submerge holder in the oxidizing solution.

5. Place the beaker in the hot water bath for 5 min.

6. Take the beaker out and let cool.

7. Submerge the Teflon holder with the sensor chips in a beaker containing water.

8. Rinse sensor chips individually with ethanol.

9. Dry sensor chips individually with nitrogen gas.

3.2 Flow Cell Decontamination

3.2.1 Regular Cleaning (To Be Performed Before Each Use) [3]

1. Mount a "cleaning sensor chip" (*see* **Note 6**) in each flow cell, ensuring that a good seal is formed between the sensor chip and the O-ring.

2. Flow approximately 10 mL of alkaline liquid detergent solution (*see* **Note 7**) through the measurement chamber.

3. Flow 20 mL of water through the flow cells.

4. Allow air to flow through the flow cells for 5 min.

5. Remove cleaning sensor chips.

6. Use nitrogen gas to blow extra liquid out of the tubing and dry the flow cells.

3.2.2 Deep Cleaning (To Be Performed as Needed) (See Note 8) [3]

1. Disassemble tubing and measurement chambers. Remove O-rings.

2. Place any component that comes directly into contact with liquid into a beaker filled with disposable, phosphate-free liquid detergent. Sonicate for 1 h.

3. Rinse components with water and place in a beaker containing water–ethanol mixture. Sonicate for at least 15 min.

4. Rinse components with water and place them in a beaker containing water. Sonicate for at least 10 min.

5. Remove all components and let air-dry. Lint-free tissues and nitrogen gas may be used to dry components, but this is not essential.

3.3 Attachment Procedures

When selecting a method for attaching proteins to a surface, it is important to consider orientation of the protein, minimization of secondary interactions, and packing density. Some of these methods of attachment are performed in the QCM-D flow cell, while others are performed outside of the flow cell. In this section we outline three popular methods of attaching proteins to surfaces and discuss the advantages and limitations of each method (*see* **Note 9**).

3.3.1 Electrostatic Protein Layer Formation

It is often possible to form a densely packed layer of protein on the surface due to electrostatic interactions between the protein and a charged surface. This method is very simple as it can be carried out entirely within the QCM-D flow cell. It should be noted, however, that as a layer of protein is formed, the tertiary structure of the protein becomes compromised, making this method of attachment less desirable than the others discussed in this chapter. Below, we outline a procedure for creating a layer of charged protein on a SiO_2 sensor chip (*see* **Note 10**) and measuring the deposition of salicylic acid onto the surface.

1. Mount clean SiO_2 sensors chips in decontaminated flow cells.
2. Flow PBS until the frequency stabilizes (*see* **Note 11**).
3. Flow BSA solution for 10 min.
4. Rinse with PBS until stable.
5. Flow over solution of interacting salicylic acid until stable.
6. Rinse with PBS until stable.

3.3.2 Cross-Linking Protein Attachment

Cross-linking is the process of covalently bonding two or more molecules, and it is commonly used for QCM-D studies. Cross-linking reagents will contain two reactive ends, which chemically attach to specific functional groups on other molecules or proteins. There are several cross-linking reagents that can be selected based on chemical specificity, length between conjugated molecules, and solubility (*see* **Note 12**).

In most of our studies, we employ EDC/NHS cross-linking chemistry (Fig. 2). Unlike procedure 3.3.1, this procedure is relatively time-intensive, and sensor preparation is performed outside of the QCM-D flow cell. Below we summarize the procedure for attaching metmyoglobin to the surface using EDC/NHS cross-linking and measuring its binding interaction with azide.

1. Place clean gold sensor chips in a solution of MUA overnight.
2. Rinse sensor chips with ethanol and dry with nitrogen gas.

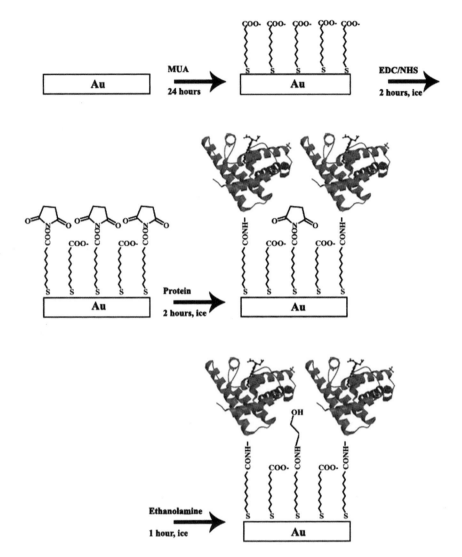

Fig. 2 EDC/NHS functionalization schematic for attaching proteins to a gold surface

3. Submerge in a solution of EDC/NHS for 2 h at 4 °C

4. Rinse with PBS and place in a solution of metmyoglobin for 2 h at 4 °C.

5. Rinse with ultrapure water and place in a solution of ethanolamine for 2 h at 4 °C.

6. Rinse crystals with ultrapure water and dry with nitrogen gas. Mount in a decontaminated liquid flow cell.

7. Obtain a stable baseline by flushing the QCM-D flow cell with borate buffer.

8. Flow 0.2 M sodium azide in borate buffer (pH 9.5) over the surface.

9. Rinse with PBS until stable (*see* **Note 11**).

10. Repeat **steps 1–9** for other concentrations of sodium azide (0.5, 0.7, 1 M).

3.3.3 Histidine Tag Capture

When used appropriately, the His-tag capture method of protein-immobilization is perhaps the most desirable method because of its ability to orient proteins on the surface. When a histidine tag is on the opposite side of a protein than the binding site, this procedure allows the binding site to be exposed during immobilization, thus preventing results from becoming confounded by its inaccessibility. Unfortunately, this method does not apply to all proteins, although some analogs do exist.

1. Rinse Au crystal with ethanol (*see* **Note 13**) until frequency stabilizes (*see* **Note 11**).

2. Flow Boc-Asn-OH solution through the systems for 15 min at 0.1 mL/min (*see* **Note 14**).

3. Flow ethanol until the frequency stabilizes.

4. Flow water until the frequency stabilizes.

5. Charge with nickel solution by flowing nickel solution through the system (*see* **Note 15**).

6. Rinse with water until stable.

7. Rinse with PBS until stable.

8. Flow protein with an exposed histidine residue through the system.

9. Rinse with PBS until stable.

10. Flow ligand through the system.

11. Rinse with PBS until stable.

3.4 Data Analysis

Changes in frequency obtained from QCM-D can be converted to changes in mass using either the Sauerbrey relation or viscoelastic modeling, depending on the viscoelastic properties of the surface. The Sauerbrey equation, Eq. 1, describes a linear relationship between frequency shifts and mass changes in thin films under the assumption that the film is rigidly attached to the sensor.

$$\Delta m = \frac{-C\Delta f}{n} \tag{1}$$

In this equation Δf is the frequency shift (Hz), Δm is the change of mass per area (ng cm^{-2}), C is a constant (17.7 ng Hz^{-1} cm^{-2} for a 4.95 MHz quartz crystal), and n is the overtone number (1, 3, 5, or 7) [3]. For simplicity, here we assume that that dissipation values are minimal, and it is therefore acceptable to proceed using the Sauerbrey relation to convert changes in mass to changes in frequency (*see* **Note 16**). Experimental frequency data and the

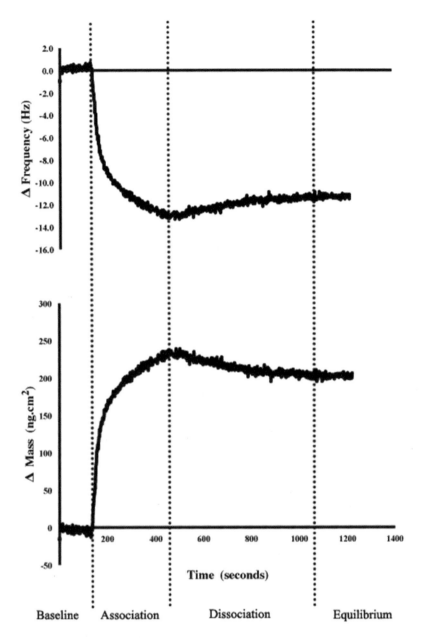

Fig. 3 Sample QCM-D protein-protein binding data. After a stable baseline was achieved for the functionalized sensor chip, interacting protein was introduced to the system at 150 s. Association was terminated at 425 s, and the sensor chip was rinsed with buffer to measure dissociation rates until equilibrium was achieved

corresponding mass data, calculated by the Sauerbrey equation, can be found in Fig. 3.

As is the case with any surface measurement, data can become confounded by unintended interactions between the ligand and the functionalized sensor surface. The method of data analysis presented here assumes that unintended interactions on the surface are minimal.

Procedure for Analyzing Data [2]:

1. Divide data into the following sections (Fig. 3):

   Baseline: a stable frequency before exposure (drift less that 2 Hz per 10 min).

   Association: time during which the sample flows over the sensor and association occurs.

   Dissociation: time during which buffer flows over the sensor and dissociation occurs.

   Equilibrium: steady state is reached.

2. Fit the Dissociation segment of the data according to Eq. 2.

$$f_t = F_0 e^{k_{off}(t-t_0)} \tag{2}$$

   In this case f_t denotes change in frequency at time t (seconds) in Hz, t_0 denotes the start of the dissociation in seconds, F_0 denotes the change in frequency at t_0 in Hz, and k_{off} denotes the dissociation rate constant in inverse seconds (*see* **Note 17**). Because f_t, F_0, and t_0 are all known constants, they can be inserted into Eq. 2 to obtain k_{off}.

3. Fit the Association segment of the data according to Eq. 3.

$$f_t = \frac{k_{on} \, C \, \Delta F_{max}}{k_{on} \, C + k_{on}} \left(e^{-(k_{on} \, C + k_{off})(t-t_0)} - 1 \right) + F_0 \tag{3}$$

   Here f_t denotes change in frequency at time t (seconds) in Hz, k_{on} is the association rate constant in $M^{-1} s^{-1}$, k_{off} is the dissociation constant determined in **step 2**, C is the ligand concentration in mol/l, ΔF_{max} is the frequency shift for a fully saturated surface, t_0 is the time at the start of the association phase, and F_0 is the change in frequency at t_0 in Hz. By inputting all known constants, k_{on} can be obtained.

4. The dissociation constant, K_D, can be obtained using Eq. 4

$$K_D = \frac{k_{off}}{k_{on}} \tag{4}$$

   where K_D is in M^{-1}.

5. Repeat **steps 1–4** for each ligand concentration and average K_D values.

4   Notes

1. Follow manufacturer's assembly instructions. Be sure to check chemical compatibility between solvents and all liquid handling equipment including tubing, O-rings, and gaskets. For simplicity, it is helpful to ensure that all tubing is cut to the same length.

2. UV/Ozone treatment should be performed in a chamber in which sensor surfaces are approximately 5 mm from a lamp, which generates light at 185 and 254 nm. The UV light with the ozone, produced in the breakage of the O-O bond by the 185 nm light, volatilize organic contaminants on the surface and slightly oxidize the surface [4].

3. The same procedure may be followed to treat any tweezers used for handling sensor chips.

4. The sensor chips are dried with nitrogen gas from a pressurized source.

5. Handle the oxidizing solution carefully. Wear appropriate gloves when using this solution. Dispose of the solution down the sink with copious amounts of water.

6. It is common for older, worn out sensor chips that are no longer sensitive enough for measurements to be recycled as "cleaning sensor chips." The presence of the sensor creates a seal within the measurement chamber, so that the pressure within the system can draw up cleaning fluid.

7. Other detergents may also be used such as sodium dodecyl sulfate. Before use, it is crucial to ensure that all liquids are compatible with tubing, O-rings, and gaskets.

8. We typically perform this after 5–10 experiments or if it takes particularly long for the system to achieve a stable baseline (*see* **Note 11**).

9. In all procedures, we generally use flow rates between 100 and 300 µL/min. When switching between solutions, it is crucial to ensure that the pump is not drawing up air, which can cause air bubbles and disrupt measurements. This can be avoided by temporarily stopping the pump before removing it from one solution and only restarting it once the open-end of the tubing is submerged in another solution. The QCM-D sensor used consisted of an AT-cut piezoelectric quartz crystal disk coated with a gold electrode (100 nm thick) on the underside and an active surface layer of gold or SiO_2 (this varies by experiment). All QCM-D sensors described have been optically polished by the vendor with a root-mean-square roughness less than 3 nm.

10. Local positive regions interact with the negatively charged SiO_2 surface, thus allowing for layer formation.

11. We define stable frequency as a change no larger than 2 Hz over 10 min. Sometimes this can take 15 min, but sometimes it can take as long as 3 h.

12. This website provides a great review of cross-linking:

http://www.piercenet.com/browse.cfm?fldID=CE4D6C5C-5946-4814-9904-C46E01232683.

13. Be sure to use tubing that is compatible with ethanol.

14. A decrease in frequency of 12–13 Hz should be observed.

15. A small change in frequency should be observed.

16. Theoretically, as soon as dissipation values are larger than zero, the Voigt model or another viscoelastic model should be used. When a surface displays viscoelastic behavior, the linear nature of the frequency-mass relation fails, and true mass on the surface will be underestimated using the Sauerbrey equation. Data generally require viscoelastic modeling if the dissipation values (in 1E−6) are greater than 5 % of the frequency shifts (in Hz).

17. For best results, this segment should consist of at least 50 points.

Acknowledgements

We would like to acknowledge the Pomona College Chemistry Department and Pomona College Summer Undergraduate Research Program for their continual support.

References

1. Johal MS (2011) Quartz crystal microbalance. In: Press C (ed) Understanding nanomaterials, 1st edn. CRC Press, Boca Raton, FL, pp 101–108

2. Hauck S, Drost S, Prohaska E, Wolf H, Dübel S (2005) Analysis of protein interactions using a quartz crystal microbalance biosensor. In: Golemis EA, Adams PD (eds) Protein-protein interactions: a molecular cloning manual. Cold Spring Harbor Laboratory Press, Cold Spring Harbor, NY, pp 273–284

3. www.qsense.com

4. Vig JR (1985) UV/Ozone cleaning of surfaces. J Vac Sci Technol A 3:1027

Chapter 10

Label-Free Kinetic Analysis of an Antibody–Antigen Interaction Using Biolayer Interferometry

Sriram Kumaraswamy and Renee Tobias

Abstract

Biolayer Interferometry (BLI) is a powerful technique that enables direct measurement of biomolecular interactions in real time without the need for labeled reagents. Here we describe the analysis of a high-affinity binding interaction between a monoclonal antibody and purified antigen using BLI. A simple Dip-and-Read™ format in which biosensors are dipped into microplate wells containing purified or complex samples provides a highly parallel, user-friendly technique to study molecular interactions. A rapid rise in publications citing the use of BLI technology in a wide range of applications, from biopharmaceutical discovery to infectious diseases monitoring, suggests broad utility of this technology in the life sciences.

Key words Label-free, Antibody, Antigen, PSA, Biolayer interferometry, Affinity constant, Kinetic analysis, Ligand, Analyte, Biosensor, Association, Dissociation

1 Introduction

1.1 Label-Free Technology for Analysis of Molecular Interactions

Label-free biosensors are in routine use for the analysis of the kinetics of binding interactions between two biomolecules [1]. In biosensor-based analysis, one of the binding partners is immobilized on the solid surface of the biosensor (ligand) while the other molecule is present in solution (analyte). While traditional techniques for measuring binding activity and affinity rely on a similar format, they typically require enzymatic or fluorescent molecular labeling. Generating labeled biomolecules not only consumes time and material, but can lead to altered protein activity or steric blocking of binding sites. Unlike standard endpoint assays such as ELISA, label-free biosensor technology enables monitoring of binding interactions in real time. Real-time kinetic measurements provide more information on mechanisms of interaction, including association rates (k_a), dissociation rates (k_d), and affinity constants (K_D) [2]. Label-free technology has greatly advanced in recent years, enabling rapid, sensitive and accurate measurement of binding kinetics, affinity and activity of biomolecular complex formation

Cheryl L. Meyerkord and Haian Fu (eds.), *Protein-Protein Interactions: Methods and Applications*, Methods in Molecular Biology, vol. 1278, DOI 10.1007/978-1-4939-2425-7_10, © Springer Science+Business Media New York 2015

that minimizes artifacts or issues associated with traditional end-point techniques [3].

1.2 BioLayer Interferometry

BLI is an optical technique that utilizes disposable fiber-optic biosensors for measurement of biomolecular interactions. In BLI, white light is directed down the length of the biosensor fiber toward two interfaces separated by a thin layer at the tip: the biocompatible surface of the tip, and an internal reference layer. Light is reflected back to the detector from each of the two layers, and the two reflected beams interfere constructively or destructively at different wavelengths in the spectrum. When the tip of a biosensor is dipped into a sample, analyte binds to immobilized ligand on the biosensor surface. This binding forms a molecular layer which increases in thickness as more analyte molecules bind to the surface. As the thickness at the biosensor tip increases, the surface layer effectively moves away from the internal reference layer and the detector, creating a shift in the interference pattern of the reflected light (Fig. 1a). The spectral pattern changes as a function of the optical thickness of the molecular layer, i.e., the number of molecules bound to the biosensor surface. This shift is monitored at the detector, and reported on a sensorgram as a change in wavelength (nm shift) (Fig. 1b). Multiple layers can be bound sequentially to the surface, providing accurate, comprehensive kinetic data on the interactions between molecules [4].

1.3 Biosensors

BLI biosensors are made of glass, and coated with a proprietary biocompatible matrix which minimizes nonspecific binding to the surface. This matrix is pre-coated with one of a wide selection of capture chemistries for specific binding of analyte molecules in a sample. Using a dip-and-read format, biosensors are moved

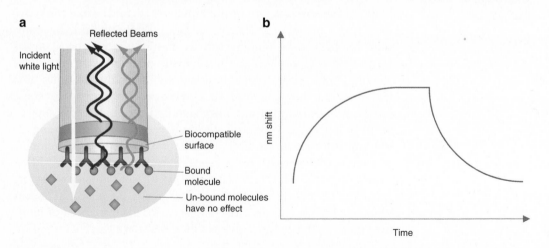

Fig. 1 BLI diagram. (**a**) As more molecules bind to the surface of the biosensor, the interference pattern of the reflected light changes, creating a wavelength shift that is reported in real time on a sensorgram (**b**)

between samples held in standard 96- or 384-well microplates or in a micro-volume drop holder, eliminating the need for microfluidics. Ease of use combined with the ability to measure multiple samples simultaneously allows for higher throughput measurement of binding kinetics and affinity, rapid determination of analyte concentration, and screening of biomolecular interactions. Only molecules binding to or dissociating from the surface of the biosensor shift the spectral interference pattern and generate a response. Unbound molecules in the surrounding solution do not affect the interference pattern, enabling measurements in crude samples such as cell lysates or culture supernatants. Changes in refractive index only minimally affect BLI signals, enabling analysis in solutions containing high refractive index components such as glycerol or DMSO. BLI biosensors are cost-effective and can be disposed of after a single use, or regenerated and reused using optimized conditions suitable for the binding pair under study. Samples are not consumed or destroyed in BLI analysis, and can be recovered when the assay is complete.

1.4 Binding Kinetics with BLI

An example of a binding kinetics experiment on a BLI biosensor is shown in Fig. 2a. The experiment begins with immobilization (loading) of a ligand molecule, such as an antibody, on the surface of the biosensor. The ligand-loaded biosensor is dipped into a solution containing the analyte (association) followed by dipping into buffer (dissociation). Measurements are plotted in real time on a sensorgram (Fig. 2b). Binding association and dissociation rate constants (k_a and k_d, respectively) and affinity constant (K_D) are calculated by fitting the binding and dissociation curves using mathematical equations. The simplest model used to describe an interaction between two biomolecules is

$$A + B \underset{k_d}{\overset{k_a}{\rightleftharpoons}} AB \qquad (1)$$

where A represents the ligand, and B is the analyte [5]. This binding model assumes a 1:1 interaction, where one ligand molecule interacts with one analyte molecule, and binding is independent and of equal strength for all binding sites. The rate of complex (AB) formation during the association step of an assay is a function of the association constant, k_a, which is expressed in $M^{-1} s^{-1}$. When the biosensor is dipped into buffer in the dissociation step, the complex dissociates back to A and B. The dissociation constant, k_d, identifies the fraction of complexes decaying per second, and is expressed in units of s^{-1}. The dissociation constant is a measure of the stability of the interaction; the smaller the k_d, the more stable the complex. The affinity constant, K_D, is a measure of how tightly a ligand binds to its analyte. It is the ratio of the on-rate and off-rate:

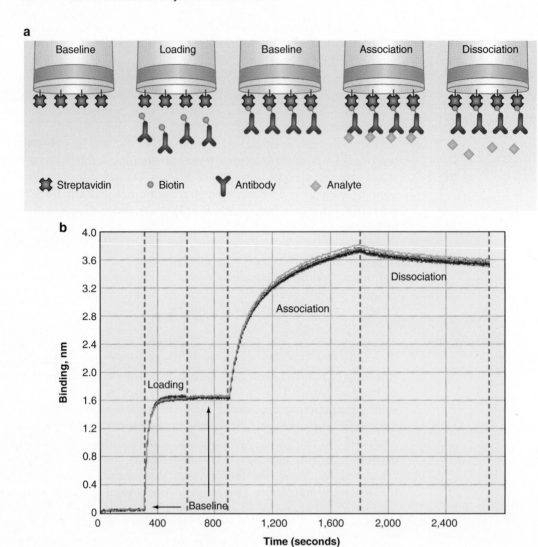

Fig. 2 Binding kinetics experiment on Streptavidin biosensors. (**a**) Biotinylated ligand is immobilized on the streptavidin-coated surface. After a baseline step, an association step is performed where analyte binding occurs, followed by a dissociation step in buffer. (**b**) A typical sensorgram trace showing loading, baseline, association, and dissociation assay steps for replicate samples

$$K_{\mathrm{D}} = \frac{[\mathrm{A}] \cdot [\mathrm{B}]}{[\mathrm{AB}]} = \frac{k_{\mathrm{d}}}{k_{\mathrm{a}}} \qquad (2)$$

K_{D} is represented in molar units (M).

Knowing the concentration of analyte in a sample is required for calculating certain kinetic parameters. To calculate K_{D} and k_{a}, the concentration of analyte must be known. By contrast, k_{d} is concentration-independent, and can therefore be calculated without knowing how much analyte is present. The dissociation rate constant, k_{d}, is useful for ranking sets of analytes of unknown concentration, such as unpurified proteins, based on off-rate.

1.5 Considerations for Designing a Successful Assay

1.5.1 Choosing an Immobilization Strategy

Maintaining the structure and activity of the immobilized ligand is the most important consideration for biosensor selection. Direct immobilization of a target protein to a biosensor can be accomplished by (1) covalent linkage to free lysine residues in the target protein via a carboxyl group on the biosensor or by (2) biotinylated ligand binding to streptavidin-coated biosensors. These methods are compatible with most proteins and enable creation of "custom" biosensor surfaces with virtually any protein. However, these approaches require purified protein and creation of a covalent bond, either directly to the surface or to biotin. Site-directed or capture-based biosensors provide an alternative immobilization strategy that can be used to maximize activity of the ligand on the surface without covalent bonding. Capture biosensors are pre-immobilized with a high affinity capture antibody or protein which binds to the protein ligand via a known motif or tag, enabling favorable orientation of the ligand and improved homogeneity on the surface. Because of the high specificity of these interactions, ligand protein can be captured directly from crude samples such as culture media without need for purification.

1.5.2 Assay Orientation

The choice of which molecule in a binding pair to use as ligand or analyte depends on several factors [6]. Primary factors include (1) binding valency (use the lower valency molecule as analyte to reduce avidity effects and simplify curve fitting; Fig. 3), (2) availability of purified protein to use as ligand, (3) stability of the protein when immobilized (if unstable, use the other molecule as ligand) and (4) sensitivity of detection (BLI signal is a function of molecular size and packing density, so use the larger molecule among the binding pair as analyte to maximize detection sensitivity). Regardless of orientation or assay format, proper assay development is a necessity for obtaining reliable kinetic data with any system.

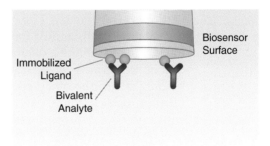

Fig. 3 Avidity effects with multivalent molecules. When a multivalent molecule, such as an antibody, is used as analyte in solution, there is potential for a single analyte molecule to bind multiple immobilized ligand molecules causing avidity effects

1.5.3   Assay Optimization

The amount of ligand to immobilize needs to be optimized for each assay and reagent lot. Typically, increasing the loading density will lead to increased signal in the analyte association step. However, it is a common observation that loading biosensors with very high density of ligand may lead to artifacts and secondary binding effects, such as analyte rebinding, that alter observed binding rates [7]. Hence, a "scouting" experiment must be performed to find the balance between maximizing sensitivity for analyte detection and minimizing secondary binding effects. As a rule, the lowest concentration of immobilized ligand that yields an acceptable signal in the analyte association step should be used.

Analyte concentration is another important consideration for obtaining accurate kinetic and affinity constants. A dilution series of at least four analyte concentrations should be measured in a kinetic assay. The analyte dilution series ideally should range from a concentration of about tenfold higher to tenfold lower than the expected affinity constant. For screening purposes or qualitative analyses, a single concentration is often sufficient. If the approximate K_D of the interaction is not known, performing an analyte concentration scouting step is recommended. In this case, choose a few concentrations that span a wide range to obtain an approximation of the affinity constant [8].

1.5.4   Nonspecific Binding

Biological molecules typically have many points of hydrophobicity and charge that can lead to nonspecific interactions with solid surfaces. Nonspecific binding is commonly observed on microplates and other sample containers, as well as the biosensor tips [9]. Label-free assay technologies generate signal upon binding of biomolecules, so nonspecific binding will contribute to signal. It is thus important to optimize assay conditions to avoid such binding.

The biocompatible layer on BLI biosensors greatly mitigates nonspecific binding; however, a reference sample must be included with every experiment. The reference sample typically consists of the solution medium used for the assay minus the analyte. Double referencing with both a reference sample and a reference biosensor can be performed when background signal due to nonspecific binding is an issue, or in assays where the signal is very small in relation to background. A reference biosensor should be loaded with a non-active protein similar to the active ligand.

1.5.5   Biosensor Regeneration

In many cases, biosensors can be regenerated by removing bound analyte under conditions that do not irreversibly disrupt the binding capacity of the ligand. Regeneration allows for reuse of the same ligand-loaded biosensor to measure binding to multiple analytes or various analyte concentrations [10, 11]. However, reuse of biosensors does require that significant time, cost, and effort be spent on optimizing the conditions of regeneration. While biosensor regeneration is an óption with BLI technology, using a fresh biosensor for every sample is convenient and cost-effective. BLI biosensors are 20–40 times less expensive than those used in other techniques and are unique among label-free biosensors for enabling single use.

1.6 BLI to Study an Antibody–Antigen Interaction

Here we describe how to perform kinetic characterization of the interaction between an antibody and antigen using the BLI-based Octet analytical system. The reagents used in this example are purified human prostate-specific antigen (PSA) and purified anti-PSA monoclonal IgG. This is a previously characterized binding pair of low nanomolar affinity [12]. Anti-PSA antibody is first biotinylated for loading onto streptavidin biosensors. The PSA antigen in solution then is bound to the antibody in an association step, followed by a dissociation step in buffer. The kinetics of the binding interaction are monitored in real time. Data processing and curve fitting parameters are described.

2 Materials

2.1 Reagents

1. Biotin-PEG$_4$-NHS reagent. NHS-PEG$_{12}$-Biotin or Sulfo-NHS-LC-LC-Biotin can alternatively be used (*see* **Note 1**).

2. Distilled water.

3. Phosphate buffered saline (1× PBS): 138 mM NaCl, 2.7 mM KCl, 10 mM Na$_2$HPO$_4$, 2 mM KH$_2$PO$_4$, pH 7.4.

4. Purified mouse IgG monoclonal anti-PSA antibody (e.g., Fitzgerald Industries, clone M612166), 100 μg minimum at a concentration of at least 100 μg/mL (*see* **Note 2**).

5. PSA purified protein antigen (can also be purchased from Fitzgerald Industries).

6. Kinetics Buffer: (Pall ForteBio Part No. 18-5032) 1× PBS, 0.1 % BSA, 0.2 % Tween 20 (*see* **Note 3**).

2.2 Materials and Instrumentation

1. Desalting spin columns or dialysis membrane/cartridge.

2. Streptavidin (SA) biosensors (Pall ForteBio, Part No. 18-5020).

3. 96-well flat-bottom polypropylene microplates, black (*see* **Note 4**).

4. Pall ForteBio's BLI-based Octet RED96, Octet RED384, Octet QKe, or Octet QK384 instrument and software [13].

3   Methods

3.1   Biotinylation of Anti-PSA Monoclonal Antibody Ligand [14]

1. Pipet a minimum of 100 µg purified antibody into a microcentrifuge tube.

2. Follow instructions from the manufacturer to prepare a fresh concentrated stock of biotin-PEG_4-NHS reagent. Dilute this stock in distilled water to make a 1 mM solution of biotin reagent (*see* **Note 5**).

3. Calculate the volume of 1 mM biotin reagent to add to your antibody to achieve a 1:1 molar coupling ratio (MCR) of biotin to protein (*see* **Note 6**).

4. Add the appropriate volume of biotin reagent as calculated in **step 3**. Mix immediately.

5. Incubate the biotinylation mixture for 30 min at room temperature.

6. Remove excess biotin reagent by desalting with a size-exclusion spin column or by dialysis (*see* **Notes 7 and 8**).

7. For dialysis: Dialyze sample 1:1,000 in 1× PBS. Allow to stir gently for a minimum of 3 h before changing PBS buffer. Perform at least four exchanges of buffer before extracting biotinylated protein.

3.2   Optimization of Antibody Loading Concentration on Biosensors

3.2.1   Hydrate Biosensors

1. Pipet 200 µl per well of 1× Kinetics buffer into wells of a 96-well black flat-bottom microplate corresponding to the number and position of biosensors to be used. The buffer used for hydration should be the same as that used throughout the assay.

2. Insert the hydration plate into the biosensor tray. Align the biosensor rack over the hydration plate and lower the biosensors into the wells, taking care not to scrape or touch the tips of the biosensors. Allow biosensors to hydrate for at least 10 min.

3.2.2   Prepare Ligand and Analyte Reagents and Assay Plate

1. Dilute biotinylated anti-PSA antibody to 50 µg/mL in 1× Kinetics Buffer. Equilibrate reagents and samples to room temperature prior to sample preparation. For frozen samples, thaw and mix thoroughly prior to use.

2. Perform dilutions of biotin–anti-PSA in 1× Kinetics buffer to yield antibody concentrations of 50, 25, 10, 5, and 1 µg/mL (*see* **Notes 9 and 10**).

3. Dilute PSA antigen to a concentration of 200 nM in 1× Kinetics buffer (*see* **Note 11**).

4. Pipet 200 µL of each biotin–anti-PSA dilution into Column 2, Rows A–E of a black 96-well polypropylene microplate (*see* **Note 12**).

5. Pipet 200 µL 1× Kinetics buffer into Column 2, Row F.

6. Pipet 200 µL diluted PSA antigen into Column 4, Rows A–F.

7. Pipet 200 μL 1× Kinetics buffer into Column 1, Rows A–F, and Column 3, Rows A–F. The plate layout will be as shown in Fig. 4.

3.2.3 Load Sample Plate and Biosensors onto Octet Instrument and Run Assay

1. Ensure that the Octet instrument lamp is warmed up for at least 40 min prior to starting the assay.

2. Set the sample plate temperature to 30 °C in the Octet software (*see* **Note 13**).

3. Place the sample plate on the sample plate stage inside the Octet system with well A1 toward the back right corner. Place the biosensor hydration assembly on the biosensor stage. Ensure that both the tray and the sample plate are securely in place.

4. Equilibrate the plates in the instrument for 10 min prior to starting the experiment. The delay timer can be used to automatically start the assay after 10 min (600 s).

5. Set up the assay in the instrument software. For details, refer to the *Octet Data Acquisition User Guide*. Table 1 shows an example of the settings for a kinetic assay, which can be used for the ligand loading experiment. Steps consist of equilibration, ligand loading, baseline, association, and dissociation steps (*see* **Note 14**).

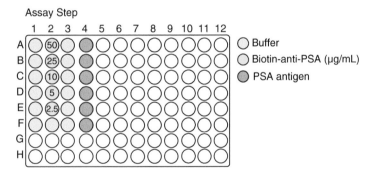

Fig. 4 Plate map diagram for loading optimization experiment

Table 1
Recommended settings for kinetic assay

Step#	Step name	Time (s)	Flow (RPM)	Sample plate column
1	Equilibration	60	1,000	1
2	Loading	300–600	1,000	2
3	Baseline	180–600	1,000	3
4	Association	300–600	1,000	4
5	Dissociation	300–3,600	1,000	3

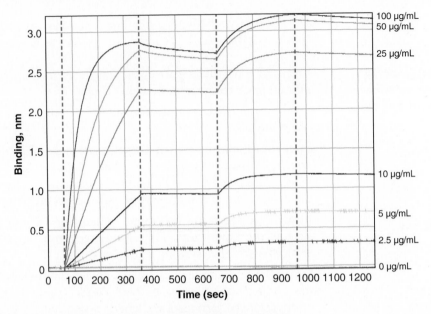

Fig. 5 Sensorgram output for ligand loading optimization experiment

6. Run the assay. An example of the sensorgram output for the ligand loading optimization is shown in Fig. 5. The shape of individual binding curves can be observed. Note that the initial slope of the binding curve corresponds to concentration of ligand.

7. In Octet Data Analysis software, process the data so that the association step is aligned to the baseline. In this view, the relative signal of the analyte binding at each corresponding ligand concentration can be clearly observed (Fig. 6). The optimal concentration of biotin–anti-PSA for a binding kinetics experiment can be selected based on these data. In this example, the optimal loading concentration is estimated to be 25 μg/mL (*see* **Note 15**).

8. Check for nonspecific binding of analyte to the biosensor. The non-ligand loaded sample should show a flat response even in the presence of analyte.

3.3 Run Kinetics Experiment

Hydrate biosensors as described in Subheading 3.2.1.

3.3.1 Hydrate Biosensors

3.3.2 Prepare Assay Samples

1. Equilibrate reagents and samples to room temperature and mix thoroughly.

2. Dilute the biotin–anti-PSA in 1× Kinetics Buffer to a working concentration of 25 μg/mL, based on results from the ligand loading experiment above.

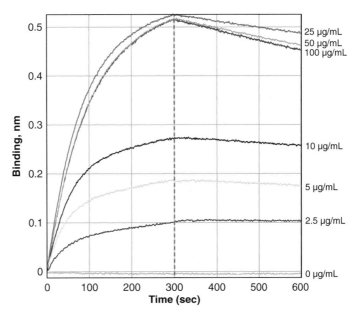

Fig. 6 Aligned sensorgram traces showing association and dissociation steps

3. Dilute the PSA analyte sample to 300 nM in 1× Kinetics Buffer. Perform serial threefold dilutions in 1× Kinetics Buffer to generate 300 nM, 100 nM, 33 nM, 11 nM, and 3.7 nM samples (*see* **Note 16**).

4. Transfer 200 μL of 1× Kinetics Buffer into Columns 1 and 3 of a black polypropylene 96-well microplate (*see* **Notes 17** and **18**).

5. Pipet 200 μL of biotin–anti-PSA ligand into wells in Rows A–F, Column 2.

6. Pipet 200 μL of each dilution of PSA antigen into Column 4, Rows A–E.

7. Pipet 200 μL of 1× Kinetics Buffer into Column 4, Row F (*see* **Note 19**). An example plate setup for a kinetic assay is shown in Fig. 7.

3.3.3 Load Sample Plate and Biosensors onto Octet Instrument and Set Up Kinetic Assay in Octet Software

1. Ensure that the instrument lamp is warmed up for at least 40 min prior to starting the assay.

2. Set the sample plate temperature in the Octet software.

3. Open instrument door and place the sample plate on the sample plate stage with well A1 toward the back right corner. Place the biosensor hydration assembly on the biosensor stage. Ensure that both the tray and the sample plate are securely in place. Close the instrument door.

4. Equilibrate the plates in the instrument for 10 min prior to starting the experiment. The delay timer can be used to automatically start the assay after 10 min (600 s).

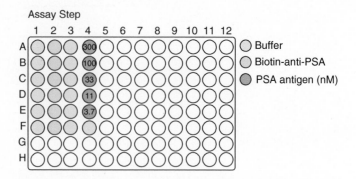

Fig. 7 Plate map diagram for PSA antibody–antigen kinetics

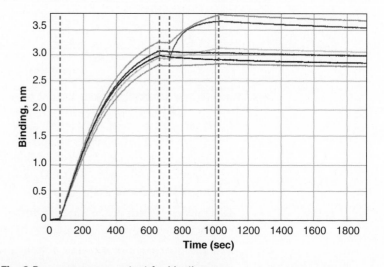

Fig. 8 Raw sensorgram output for kinetic assay

5. Set up the assay in the instrument software. For details, refer to the *Octet Data Acquisition User Guide*. Refer to Table 1 for an example of the settings for the binding kinetics experiment using the plate described in Fig. 7, consisting of equilibration, ligand loading, baseline, association, and dissociation steps (*see* **Notes 20** and **21**).

6. Run the binding assay (*see* **Note 22**).

3.4 Data Analysis

1. Load data into the Octet Data Analysis software. A sample of the raw sensorgram output is shown in Fig. 8 (*see* **Note 23**).

2. Select parameters for data processing. Recommended parameters for a standard protein-protein interaction are:

   (a) Y-axis alignment (select last 5–10 s)

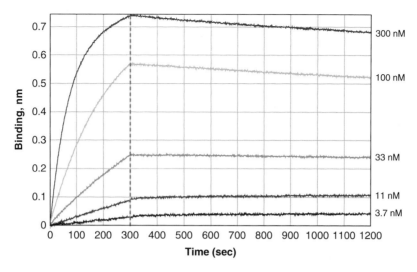

Fig. 9 Processed data for anti-PSA–PSA binding, showing aligned association and dissociation steps

    (b) Use inter-step correction, as long as baseline and association steps were performed in the same well (*see* **Note 24**)

    (c) Reference subtraction using Rows G and H as reference samples

    (d) Savitzky–Golay filtering

3. Process data. Processed data output is shown in Fig. 9, with *X* and *Y* axes aligned.

4. Analyze data. In the analysis tab, select parameters for curve fitting. Several curve fitting models are available (*see* **Note 25**), to represent different types of binding interactions:

    (a) 1:1 binding

    (b) 2:1 heterogeneous ligand

    (c) Mass transport

    (d) 1:2 bivalent analyte

The anti-PSA–PSA interaction shows well-behaved single-analyte to single-ligand molecule binding stoichiometry, so select 1:1 under Model. Fitting should be global, with sensors grouped by color, and R_{max} unlinked by sensor (*see* **Note 26**). Use entire step times. For details on analysis parameters, refer to the *Octet Data Analysis User Guide*.

5. Fit curves (*see* **Note 27**). The ideal fit traces will appear on the graph (Fig. 10), and data table will update with kinetic and affinity constants, rates, errors, and statistics. For guidelines on determining accuracy of fit, *see* **Note 28**.

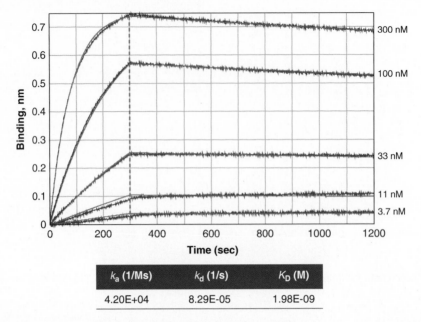

k_a (1/Ms)	k_d (1/s)	K_D (M)
4.20E+04	8.29E-05	1.98E-09

Fig. 10 Analyzed data for anti-PSA–PSA binding using 1:1 binding model. Curve fit overlays are shown as thin lines over traces

4   Notes

1. Biotin-PEG4-NHS is a biotinylation reagent that reacts specifically with primary amine groups, such as the side chains of lysine residues in a polypeptide. The PEG spacer arm confers increased solubility and prevents aggregation of labeled molecules. The spacer also provides a long and flexible linker to minimize steric hindrance when bound via streptavidin to the solid surface of the biosensor.

2. Ensure purified antibody is carrier free and is not in a buffer containing primary amines, such as Tris or glycine. If the protein is suspended in a buffer containing primary amines, perform buffer exchange into 1× PBS either by dialysis or desalting spin columns. It is recommended that the starting concentration be at least 1 mg/mL if buffer exchange is required.

3. Streptavidin biosensors are compatible with a wide range of buffers. Kinetics Buffer is provided from ForteBio as a 10× stock to be diluted 1:10 with PBS, pH 7.4. Other buffers can be used for the assay. Best results are obtained when all buffers/matrices used in the assay are closely matched.

4. For Octet RED384 and QK384 instruments, a 384-well microplate can be used for increased throughput and smaller sample volumes (down to 40 μL).

5. Prepare the biotinylation reagent fresh and use immediately to ensure efficient biotin incorporation. The NHS reagent is moisture-sensitive and typically loses activity in 15–30 min in aqueous solution.

6. We recommend using a 1:1 molar coupling ratio when biotinylating proteins for immobilization on streptavidin biosensors. Over-biotinylation does not improve biosensor loading and has the potential to reduce protein activity.

7. Biotinylation reactions must be desalted to remove excess unincorporated biotin reagent, which will compete for binding sites on the streptavidin surface. Desalting spin columns filled with size-exclusion resin provide a convenient and effective method for reaction buffer exchange and desalting. Dialysis into PBS buffer with an appropriate molecular weight cutoff membrane or cartridge can be used for gentle buffer exchange of more sensitive proteins (100 kDa MW cutoff for antibodies).

8. When using desalting spin columns for buffer exchange, be sure to use the appropriate molecular weight cutoff and select a column size that is suited to the sample volume being applied. This will prevent sample loss.

9. A dilution series of biotinylated ligand is recommended, in general starting around 50 μg/mL and titrating down, to optimize loading concentration. The higher concentrations will be expected to quickly saturate the sensor, while low concentrations may require longer loading times to reach equilibrium. A typical immobilization concentration for a biotinylated antibody lies between 5 and 25 μg/mL.

10. If the ligand concentration is low, e.g., below 5 μg/mL, a longer loading time may be required for sufficient immobilization signal. Overnight incubation in ligand solution may also be performed at 4 °C. Overnight incubation is beneficial in cases with capture biosensors where a ligand molecule is being captured from a dilute supernatant or cell culture sample and can greatly improve results.

11. Bind the analyte at a high concentration during the ligand optimization procedure. Use of a high analyte concentration will enable selection of the lowest ligand concentration that gives an acceptable signal during the association step. It is useful to run a zero-ligand biosensor as a control in this step to assess whether there is nonspecific binding of the analyte to the biosensor.

12. In a standard 384-well plate, use 80–100 μL sample volume per well. In a 384-well Tilted-Bottom plate (Pall ForteBio Part No. 18-5080), use 40 μL or more.

13. We recommend that assays be run at 30 °C for optimal results. Binding to the biosensor is sensitive to fluctuations in temperature. By working at a few degrees above ambient, a consistent temperature can be maintained through the course of the assay.

14. Octet software allows the user to extend a step or skip to the next step, if more or less time is needed. This feature may be used during assay development. When measuring binding kinetics in a run involving multiple concentrations of analyte, use of this feature is not recommended.

15. For best results in a kinetic binding experiment, select the ligand concentration that does not saturate the biosensor, but still provides a strong analyte signal. Ideally, for a 150 kDa antibody, the signal in the loading step should reach about 1.0 nm after 5–10 min loading.

16. For accurate kinetic analysis, it is recommended to run a two-fold or threefold dilution series of at least four concentrations of the analyte. The selected dilutions should ideally range from tenfold above to tenfold below the K_D. If the K_D is not known to even an approximation, a scouting experiment to approximate the K_D is recommended before performing final kinetic analysis.

17. The same sample wells should be used for the baseline and dissociation steps (column 4 in the example above). This will enable use of the inter-step correction feature in the data analysis software, which corrects for steps in data associated with minor changes in microplate well artifacts that occur between sample wells.

18. The buffer used for prehydration, baseline, and dissociation must match the matrix of the analyte sample in the association step.

19. The buffer-alone sample in the analyte column (Column 4, Row F) will serve as a reference sample, or negative control, for subtracting background signal.

20. Shaking speed and assay step lengths can be optimized, depending upon the strength and speed of the interaction. Increasing the shake speed will increase the sensitivity of the assay, and is recommended for weaker binders or lower concentrations of reagent. The association step time can be increased for a slow interaction or a weak binder, to enhance binding signal. The dissociation step should be longer for higher affinity binders. In general, step lengths of 1 min for baseline, 5 min for association and 10 min for dissociation are good starting points, however some optimization may be required.

21. For a high affinity binding pair ($K_D < 1$ nM), a dissociation time of 30 min or more may be required. For accurate calculation of binding and affinity constants, at least 5 % of the complex must dissociate. Do not run the assay longer than 3 h, as sample evaporation from the microplate wells may begin to impact results.

22. When running the assay, be sure the baseline is stable before proceeding to association step. If needed, use the Extend Step function in the software to run the baseline for longer periods. With biotin-streptavidin interactions, there should be minimal signal drift.

23. Some variation in loading levels between individual streptavidin biosensors during the loading step is normal. This variability will not affect calculation of kinetic constants in a properly designed experiment. As long as several concentrations of analyte are run and global curve fitting is performed, small differences in ligand loading level are compensated for.

24. Avoid using inter-step correction for binding pairs with very fast association and dissociation rates.

25. "Model surfing", or determining the type of interaction based on the curve-fitting model that best fits your data, is not recommended. Unless the interaction is known to be more complex, use 1:1 binding model. If the interaction being studied is predicted to follow 1:1 binding, and the data do not fit well, this indicates that further assay development is required.

26. R_{max} may be linked if the same biosensor is used for every sample concentration in the series. This selection is typically made in small molecule analyses, where dissociation is rapid and complete and allows for reuse of the biosensor in a new sample.

27. Note that steady state, or equilibrium analysis can also be performed, but should be used only if the response at the association step for each concentration has reached equilibrium.

28. To determine the quality of the fit and accuracy of the calculated constants, consider the following general guidelines:

    (a) Visually inspect the fit: do the fit lines conform well to the data traces?

    (b) Look at residuals, which are plotted below the data traces. Residual values should not be greater than 1 % R_{max}.

    (c) The k_a error and k_d error values should be no greater than one order of magnitude below the reported constant.

    (d) R^2 should be above 0.95 to indicate goodness of fit.

    (e) Chi-squared value should be below 3.

References

1. Rich RL, Myszka DG (2010) Grading the commercial optical biosensor literature - Class of 2008: 'The Mighty Binders'. J Mol Recognit 23:1–64

2. Rich RL, Myszka DG (2007) Higher-throughput, label-free, real-time molecular interaction analysis. Anal Biochem 361:1–6

3. Cooper MA (2006) Optical biosensors: where next and how soon? Drug Discov Today 11:1061–1067

4. Concepcion J et al (2009) Label-free detection of biomolecular interactions using BioLayer Interferometry for kinetic characterization. Comb Chem High Throughput Screen 12:791–800

5. Elwing H (1998) Protein absorption and ellipsometry in biomaterial research. Biomaterials 19:397–406

6. Markey F (2009) Macromolecular interactions. In: Cooper MA (ed) Label-free biosensors. Cambridge University Press, Cambridge, pp 143–158

7. Myszka DG (1999) Improving biosensor analysis. J Mol Recognit 12:279–284

8. Pall ForteBio Application Note 8: optimizing protein-protein and protein-small molecule kinetics assays. (2012). http://www.fortebio.com/literature.html

9. Karlsson R, Falt A (1997) Experimental design for kinetic analysis of protein-protein interactions with surface plasmon resonance biosensors. J Immunol Methods 200:121–133

10. Pall ForteBio Technical Note 8: regeneration strategies for amine reactive biosensors on the octet system. (2007). http://www.fortebio.com/literature.html

11. Pall ForteBio Technical Note 14: regeneration strategies for streptavidin biosensors on the octet platform. (2009). http://www.fortebio.com/literature.html

12. Katsamba P et al (2006) Kinetic analysis of a high-affinity antibody/antigen interaction performed by multiple Biacore users. Anal Biochem 352:208–211

13. More information may be found at www.fortebio.com

14. Pall ForteBio Technical Note 28: biotinylation of protein for immobilization onto streptavidin biosensors. (2011). http://www.fortebio.com/literature.html

Chapter 11

Characterization of Protein-Protein Interactions by Isothermal Titration Calorimetry

Adrian Velazquez-Campoy, Stephanie A. Leavitt, and Ernesto Freire

Abstract

The analysis of protein-protein interactions has attracted the attention of many researchers from both a fundamental point of view and a practical point of view. From a fundamental point of view, the development of an understanding of the signaling events triggered by the interaction of two or more proteins provides key information to elucidate the functioning of many cell processes. From a practical point of view, understanding protein-protein interactions at a quantitative level provides the foundation for the development of antagonists or agonists of those interactions. Isothermal Titration Calorimetry (ITC) is the only technique with the capability of measuring not only binding affinity but the enthalpic and entropic components that define affinity. Over the years, isothermal titration calorimeters have evolved in sensitivity and accuracy. Today, TA Instruments and MicroCal market instruments with the performance required to evaluate protein-protein interactions. In this methods paper, we describe general procedures to analyze heterodimeric (porcine pancreatic trypsin binding to soybean trypsin inhibitor) and homodimeric (bovine pancreatic α-chymotrypsin) protein associations by ITC.

Key words Protein-protein interaction, Thermodynamics, Calorimetry, Titration, Binding, Dimerization, Dissociation

1 Introduction

1.1 Protein-Protein Interactions

Protein-protein interactions play a critical role in biological signaling. Many pathological conditions including cancer, inflammation, autoimmune diseases, diabetes, osteoporosis, infection, etc. are associated with specific protein-protein interactions and consequently have defined targets for drug development. The number of targets of interest is continuously increasing and range from a vast number of cell surface receptors, such as EGFR, TNFR, and IGFR, to other proteins involved in signaling and regulation [1, 2]. Biologics, i.e., monoclonal antibodies or recombinant versions of ligand proteins and/or soluble regions of the receptors, define the therapeutic arsenal aimed at targeting those interactions. In fact, biologics have become the fastest growing segment of the pharmaceutical industry.

Cheryl L. Meyerkord and Haian Fu (eds.), *Protein-Protein Interactions: Methods and Applications*, Methods in Molecular Biology, vol. 1278, DOI 10.1007/978-1-4939-2425-7_11, © Springer Science+Business Media New York 2015

The development of new biologics requires a precise characterization of the interaction of the target proteins as well as the biologic itself with the selected target.

A survey of protein-protein interactions indicates that their binding affinities span a wide range, from the micromolar to the high picomolar level [3–5], and references therein). In particular, protein ligand–receptor interactions may bind with affinities in the nanomolar and high picomolar level [6–16]. Examples of sub-nanomolar interactions are the binding of IL-4 and erythropoietin to their respective receptors with K_d values of 0.2 nM [8, 10]. The binding affinity is not the only value that shows significant variation. The enthalpy and entropy changes associated with binding also exhibit a significant spread, reflecting the magnitude and nature of the conformational changes coupled to binding. For example, the binding of proteins characterized by intrinsically disordered domains is usually associated with large favorable binding enthalpies and equally large unfavorable binding entropies, as the binding process results in the folding or structuring of the disordered domains [17–19].

ITC provides a unique opportunity to measure ΔG, ΔH, and ΔS simultaneously and, therefore, to develop a complete characterization of the binding process [20–23]. In addition, performing experiments at different temperatures provides access to the change in heat capacity, ΔC_P. The experimental guidelines presented below can be extended to most situations.

1.2 ITC of Protein-Protein Interactions

Two situations are possible when characterizing intermacromolecular interactions in binding reactions: the binding partners are (a) different (heterodimeric complex) or (b) identical (homodimeric complex). Even though experimentally they require different methodologies, the underlying principles are the same in both cases, that is, they follow the same chemical scheme based on a reversible association equilibrium:

$$M_1 + M_2 \leftrightarrow M_1 M_2 \tag{1}$$

where M_1 and M_2 are the interacting macromolecules. The strength of the interaction is described by the association constant, K_a, or the dissociation constant, K_d:

$$K_a = \frac{[M_1 M_2]}{[M_1][M_2]} = \frac{1}{K_d} \tag{2}$$

where $[M_1]$ and $[M_2]$ are the concentrations of the free reactants and $[M_1 M_2]$ is the concentration of the complex. These constants are related to the Gibbs energy of association, ΔG_a, and dissociation, ΔG_d, and can be expressed in terms of the enthalpy, ΔH, and entropy, ΔS, changes in the process:

$$\Delta G_a = -RT \ln K_a = \Delta H_a - T\Delta S_a$$
$$\Delta G_d = -RT \ln K_d = \Delta H_d - T\Delta S_d \tag{3}$$

where R is the gas constant (1.9872 cal/K mol) and T is the absolute temperature (kelvins).

Both enthalpic and entropic contributions to the Gibbs energy reflect different types of interactions underlying the overall process. Accordingly, complexes predominantly stabilized enthalpically or entropically will be preferentially stabilized by specific (hydrogen bonds, electrostatics, van der Waals) or unspecific (hydrophobicity) interactions and will respond differently to environmental changes or mutations in the binding species. A number of reports have shown the importance and consequences of the distribution of the binding affinity into the enthalpic and entropic contributions regarding the identification of binding mechanisms, the optimization of ligand affinity and selectivity, the minimization of ligand susceptibility to target mutations in drug design, as well as the assessment of conformational changes coupled to binding and protein activation and signaling in drug design [17–19, 24–40].

1.2.1 Heterodimeric Interactions

Typical ITC experiments allow for the characterization of the interaction between two different binding partners. A solution of one of the reactants is placed in a syringe and a solution with the other interacting macromolecule is located in a calorimetric cell (Fig. 1). The stepwise addition of the macromolecule from the injection syringe solution triggers the binding reaction, leading the system through a sequence of equilibrium states, the composition of each one being dictated by the association constant. Given the total concentrations of reactants in the calorimetric cell, $[M_1]_T$ and $[M_2]_T$, the association constant, K_a, determines the partition between the different chemical species:

$$K_a = \frac{[M_1 M_2]}{[M_1][M_2]} \tag{4}$$

ITC directly measures the heat, q_i, associated with each change of state after each injection, which is proportional to the increment in the concentration of complex in the calorimetric cell after the injection i:

$$q_i = V \Delta H_a \left([M_1 M_2]_i - [M_1 M_2]_{i-1}\right) \tag{5}$$

where ΔH_a is the enthalpy of binding and V is the calorimetric cell volume. The sequence of injections proceeds until no significant heat is detected, that is, the macromolecule in the cell is saturated and the concentration of complex reaches its maximum.

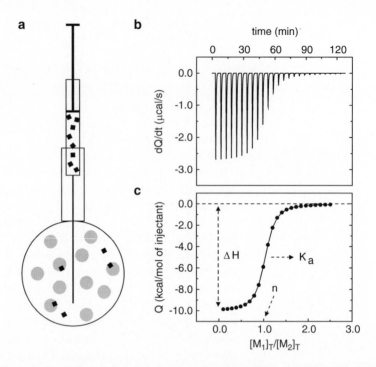

Fig. 1 (**a**) Illustration of the configuration of an ITC reaction cell. The cell volume is 1.4 mL and initially is filled with the macromolecule solution (*gray*). The injection syringe is filled with the ligand solution (*black*). At specified time intervals (400 s), a small volume (10 μL) of the ligand solution is injected into the cell triggering the binding reaction and producing the characteristic peak sequence in the recorded signal (**b**). After saturating the macromolecule, the residual heat effects (the so-called dilution peaks), if any, are due to mechanical and dilution phenomena. After integration of the area under each peak (and subtraction of the dilution heat effects and normalization per mol of injected ligand) the individual heats are plotted against the molar ratio (Panel **c**) from which, through nonlinear regression, it is possible to estimate the thermodynamic parameters n, K_a, and ΔH

Throughout the experiment the total concentrations of reactants, $[M_1]_T$ and $[M_2]_T$, are the known independent variables. Nonlinear regression analysis of q_i, the dependent variable, allows estimation of the thermodynamic parameters (K_a and ΔH_a and, therefore, ΔS_a) (*see* Subheading 3.3).

1.2.2 Homodimeric Interactions

The self-association of a protein leading to the formation of homodimers has been scarcely studied by ITC. From a practical point of view, it is impossible to isolate individual partners and perform a standard mixing assay. However, the strength of the interaction can be measured in dilution experiments [41–43]. A solution of reactant is placed in the syringe and a buffer solution is located in the calorimetric cell. The stepwise addition of the solution in the injection syringe, with the subsequent dilution of the macromolecule, triggers the dissociation reaction, leading the system through a sequence of equilibrium states, the composition of each one being dictated by the dissociation constant, K_d. Given the total macromolecule concentration in the calorimetric cell, $[M]_T$, the dissociation constant determines the partition between the different chemical species, monomer $[M]$ and dimer $[M_2]$:

$$K_d = \frac{[M]^2}{[M_2]} \quad (6)$$

Again, the heat, q_i, associated with each injection is proportional to the increment in the concentration of monomer, $[M]$, in the calorimetric cell after the injection i:

$$q_i = V \Delta H_d \left([M]_i - [M]_{i-1} - F_0 [M]_0 \frac{v}{V} \right) \quad (7)$$

where ΔH_d is the enthalpy of dissociation (per monomer), V is the calorimetric cell volume, v is the injection volume, $[M]_0$ is the total concentration of macromolecule (per monomer) in the syringe, and F_0 is the fraction of monomer in the concentrated solution placed in the syringe. The last term in the parenthesis is a correction that accounts for the increment of monomer concentration in the cell due to the injection of monomers from the syringe and, therefore, not contributing to the heat. As the protein concentration in the calorimetric cell progressively increases, the dissociation process is less favored and the sequence of injections proceeds until no significant heat is detected. Throughout the experiment the total concentration of reactant, $[M]_T$, is the known independent variable. Nonlinear regression analysis of q_i, the dependent variable, allows the estimation of the thermodynamic parameters (K_d and ΔH_d and, therefore, ΔS_d) (*see* Subheading 3.3).

1.3 Information Available by ITC and Experimental Design

1.3.1 Simultaneous Determination of the Association Constant, the Enthalpy of Binding

Every equilibrium binding technique requires the reactant concentrations to be in an appropriate range in order to obtain reliable estimations of the association constant. A practical rule of thumb for ITC is given by the parameter $c = K_a \times [M_2]_T = [M_2]_T / K_d$, which must lie between 0.1 and 1,000, thus, imposing a limit to the lowest and largest association constant measurable at a given macromolecule concentration [44]. This phenomenon is illustrated in Fig. 2. As the c value increases, the transition from low to high total titrant concentration is more abrupt. In the case of very high association constants (macromolecule concentration much higher than the dissociation constant), all the titrant added in any injection will bind to the macromolecule until saturation occurs and, therefore, all the peaks, except the last ones after saturation, exhibit the same heat effect. For low association constants (macromolecule

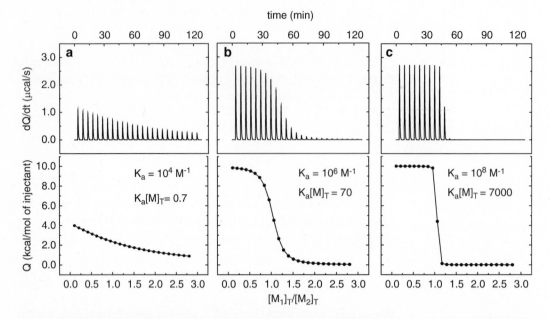

Fig. 2 Heterodimer formation. Illustration of the effect of the association constant value on the shape of a titration curve. The plots represent three titrations simulated using the same parameters (concentrations of reactants and association enthalpy), but different association constants. Low (**a**), moderate (**b**), and high affinity (**c**) binding processes are shown. In order to obtain accurate estimates of the association constant, an intermediate case is desirable $(1 < c = [M_2]_T \times K_a < 1,000)$

concentration far below the dissociation constant), from the very first injection only a fraction of the titrant added will bind, producing a less steep titration in which saturation is hardly reached and often lacking the inflection point. In order to obtain accurate estimates of the association constant, an intermediate case is desirable.

To obtain a satisfactory titration curve, the concentration of titrant should be enough to exceed the stoichiometric binding after completion of the injection sequence (i.e., for a cell volume and an injection volume of ~1.5 mL and ~10 μL, respectively, the reactant concentration in the syringe should be 10–20 times the concentration of macromolecule in the cell). In case of poor solubility, the reactant with lower solubility should be placed in the cell.

The constraints dictated by the parameter c impose an experimental limitation, i.e., for very high association constants optimal concentrations are too small to be practical (low signal-to-noise ratio), and for very low association constants the concentrations may be prohibitively high (possibility of aggregation or economic consideration). While very low c values allow determining the association constant and the binding enthalpy as long as the concentrations of reactants is precisely known, very high c values only allow determining the binding enthalpy.

Sometimes it is possible to change the experimental conditions (temperature or pH, without compromising stability against aggregation or unfolding) in order to modify the association constant toward accessible experimental values [45–47]. To extrapolate to the original conditions appropriate equations will need to be applied.

However, there is an extension of the ITC protocol aimed to overcome such drawbacks. Without changing experimental conditions at all, displacement experiments implemented in ITC extend the range for the association constant determination [27, 48–50]. Basically, a displacement experiment consists of a titration of the high-affinity ligand into a solution of the macromolecule prebound to a weaker ligand, therefore decreasing the apparent affinity of the potent ligand. The thermodynamic parameters for the binding of the high-affinity ligand are calculated from the apparent binding parameters of the displacement titration and the known binding parameters for the weak ligand. This approach can be used to measure extremely high affinity binding processes, as well as very low affinity binding reactions [51, 52].

1.3.2 Simultaneous Determination of the Dissociation Constant and the Enthalpy of Dissociation

For self-associating systems that can be measured by ITC dissociation experiments, another dimensionless parameter can be used to define a practical limit for the accurate measurement of the dissociation constant for a dimeric system. Now, $c = [M]_T/K_d$, which must lie between 10 and 10,000, imposes a limit to the lowest and largest dissociation constant measurable at a given macromolecule concentration. This phenomenon is illustrated in Fig. 3.

As the c value increases, the transition from low to high total protein concentration in the cell is more abrupt. In the case of very low dissociation constants (macromolecule concentration much higher than the dissociation constant) most of the macromolecule in the syringe is forming dimers and, when diluted in the cell, the strength of the interactions determines that only a small fraction of the dimers will dissociate. For that reason the observed heats are very small. For very high dissociation constants (macromolecule concentration far below the dissociation constant), the weak interaction within the dimer makes it possible for a large fraction of the dimers to dissociate upon dilution; however, because the monomer–monomer interaction is weak, the population of dimers in the syringe will be small. In order to obtain accurate estimates of the dissociation constant, an intermediate case is desirable: the curvature of the plot should be enough for a reliable estimation of the dissociation constant and the size of the peaks should give an acceptable signal-to-noise ratio.

1.3.3 Determination of the Enthalpy of Binding

Binding enthalpy is usually determined from titrations in the optimal range of reactant concentrations. However, a greater accuracy can be achieved if enthalpy is measured by performing injections of

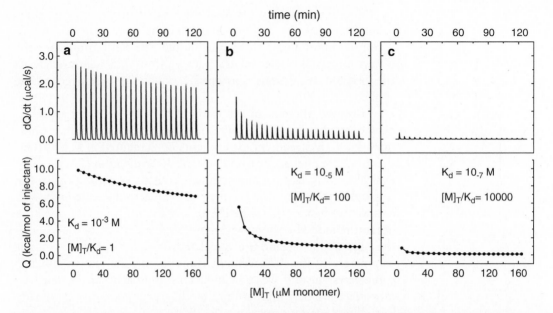

Fig. 3 Homodimer dissociation. Illustration of the effect of the dissociation constant value on the shape of a dissociation curve. The plots represent three titrations simulated using the same parameters (concentration of reactant and dissociation enthalpy), but different dissociation constants. Low (**a**), moderate (**b**), and high (**c**) dissociation constant processes are shown. In order to obtain accurate estimates of the association constant, an intermediate case is desirable ($10 < c = [M]_T/K_d < 10{,}000$)

titrant into a large excess of macromolecule (far above the dissociation constant). Therefore, although there will be no saturation, all the titrant injected will be fully associated giving a good measurement of the binding reaction heat. Blank experiments (see below) must be considered in order to eliminate the contribution of different phenomena (dilution of reactant, solution mixing, and other nonspecific effects) occurring simultaneously during the injections and to obtain the heat effect associated with the binding process.

1.3.4 Blank Experiment   In order to estimate the heat effect associated with dilution, mixing, and other phenomena different from binding, blank experiments can be performed. In this case, the experiment is similar to the one described in Subheading 1.3.3: titrant is injected into a buffer solution without macromolecule. However, this may be not the ideal way to estimate such heat effects, since the degree of solvation of the titrant and the chemical composition of the cell solution would be different compared to the real titration with the other interacting macromolecule present. For that reason, some researchers usually consider the average effect of the last injection peaks as the reference heat effect. If during the experiment it is not possible to reach complete saturation, precluding such averaging, it is possible to include in the fitting function a term accounting for such dilution heat effects. Blank experiments, also called "heat of dilution experiments," are required to measure binding enthalpies under excess macromolecule concentration, as mentioned previously.

1.3.5 Determination of Heat Capacity Change of Binding

The temperature derivative (at constant pressure) of the enthalpy, i.e., the heat capacity change in the process (association or dissociation), ΔC_P, is defined as

$$\Delta C_P = \left(\frac{\partial \Delta H}{\partial T}\right)_P \qquad (8)$$

and can be determined by performing the same experiment at several temperatures. The slope of the enthalpy vs. temperature plot gives the heat capacity of the reaction.

ΔC_P has been shown to originate from surface desolvation upon binding or solvation upon dissociation and, to a less extent, from the difference in vibrational modes between the complex and the free species [53, 54]. It provides information about the nature of the interactions driving the binding, in addition to the main thermodynamic functions (Gibbs energy, enthalpy, and entropy of binding).

1.3.6 Determination of Coupled Proton Transfer Process

Strong dependency of the thermodynamic parameters of the association (dissociation) process on pH is an indication of proton exchange between the binary complex and the bulk solution upon complex formation (dissociation). This is due to changes in the pK_a of some ionizable groups located in any of the binding partners: the microenvironment of these groups is altered upon binding or dissociation, and so do their pK_a and their proton saturation fraction.

ITC is one of the more suitable techniques for the assessment of protonation/deprotonation processes coupled to binding. When a binding process is coupled to proton transfer between the bulk solution and the bound complex, the enthalpy of binding will depend on the ionization enthalpy of the buffer molecule, ΔH_{ion}:

$$\Delta H = \Delta H_0 + n_H \Delta H_{ion} \qquad (9)$$

where ΔH_0 is the buffer-independent enthalpy of binding and n_H is the number of protons being exchanged. Therefore, by repeating the titration under the same conditions, but using several buffers with different ionization enthalpies it is possible to estimate n_H (slope) and ΔH_0 (intercept with the y-axis) from linear regression analysis [55–57]. If n_H is zero, there is no net proton transfer; if n_H is positive, there is a protonation, i.e., a proton transfer from the solution to the complex; if n_H is negative, there is a deprotonation. After determination of these two parameters, n_H and ΔH_0, linkage equations can be used to couple the binding or dissociation process to the proton transfer process and allow the estimation of the thermodynamic parameters for the proton exchange event (enthalpy of ionization and pK_a for each ionizable group involved) [46, 58, 59]. Analogous linkage equations couple the binding or dissociation reaction to the transfer of other ions (e.g., metals or salts [60, 61]).

---

2   Materials

2.1 Reagents and Supplies

1. Porcine pancreatic trypsin.

2. Soybean trypsin inhibitor.

3. Bovine pancreatic α-chymotrypsin.

4. Dialysis membrane of 10,000 molecular weight cutoff (MWCO).

5. Filters (0.22 μm) with low protein binding properties.

6. Pyrex tubes for ITC syringe: 6 × 50 mm.

7. Potassium acetate–calcium chloride buffer: 25 mM potassium acetate, pH 4.5, 10 mM calcium chloride.

8. Sodium acetate–sodium chloride buffer: 20 mM sodium acetate, pH 3.9, 180 mM sodium chloride.

2.2 Isothermal Titration Calorimeter

Modern ITC instruments are simple to use, compact, and computer-controlled. The calorimeter unit consists of two cells, the reference cell and the sample cell embedded in an adiabatic chamber. The system holds the reference cell at a constant temperature. Initially, constant power is applied to the sample cell in order to activate a feedback control mechanism whose purpose is to maintain the temperature difference between the two cells as close to zero as possible. As the reaction, initiated with the injection of titrant, occurs in the sample cell, the system adjusts the power applied to the sample cell up or down depending on whether an endothermic or exothermic reaction is taking place. This power is recorded by the computer and corresponds to the signal observed in the characteristic form of a peak. The area under each peak corresponds to the heat released or absorbed during the reaction after each injection.

2.3 Biological Systems

Two different systems, but functionally and structurally closely related, have been selected to illustrate the two main types of association reactions that can be studied by ITC: formation of heterocomplexes (trypsin–trypsin inhibitor) and homocomplexes (chymotrypsin dimer). These complexes have been characterized enzymatically, energetically, and structurally.

1. Porcine pancreatic trypsin, PPT, (EC. 3.4.21.4) is a 23.8 kDa protein consisting of a single polypeptide chain with six disulfide bonds. It is a serine protease that hydrolyzes peptide bonds with basic side chains (Arg or Lys) on the carboxyl end of the bond.

2. Soybean trypsin inhibitor, STI, is a 20.0 kDa protein consisting of a single polypeptide chain with two disulfide bonds. It inhibits competitively the peptidase activity of trypsin.

3. Bovine pancreatic α-chymotrypsin, BP-α-CT, (EC. 3.4.21.1) is a 25.2 kDa protein consisting of three polypeptide chains interconnected and linked by two of the existing five disulfide bonds. It is a serine protease that hydrolyzes peptide bonds with aromatic or large hydrophobic side chains (Tyr, Trp, Phe, Met, Leu) on the carboxyl end of the bond. This enzyme exhibits a monomer–dimer equilibrium in solution.

The experiments were performed under slightly acidic conditions to minimize the autocatalysis of both enzymes. In addition, at high pH (approx 8) the interaction between trypsin and its inhibitor is so strong that the affinity is above the practical limits ($K_a \sim 10^{10}$–10^{11} M^{-1}).

3 Methods

3.1 Sample Preparation

1. PPT: Prepare PPT at 400 μM in potassium acetate–calcium chloride buffer (pH 4.5) dissolving the protein in buffer (approx 1 mL) and dialyze overnight at 4 °C against 4 L of the same buffer. Filter the solution after dialysis. Measure concentration after dilution (1:20) in the same buffer using an extinction coefficient of 35,700 M^{-1} cm^{-1} at 280 nm.

2. STI: Prepare STI at 30 μM in potassium acetate–calcium chloride buffer (pH 4.5) dissolving the protein in buffer (approx 5 mL) and dialyze overnight at 4 °C against 4 L of the same buffer. Filter the solution after dialysis. Measure concentration without dilution using an extinction coefficient of 18,200 M^{-1} cm^{-1} at 280 nm.

3. BP-α-CT: Prepare BP-α-CT at 200 μM in sodium acetate–sodium chloride buffer (pH 3.9) dissolving the protein in buffer (approx 5 mL) and dialyze overnight at 4 °C against 4 L of the same buffer. Filter the solution after dialysis. Immediately prior to the experiment concentrate the sample to about 1 mM. Measure concentration after dilution (1:50) in the same buffer using an extinction coefficient of 50,652 M^{-1} cm^{-1} at 280 nm.

3.2 Experimental Procedure

This general protocol was designed for most experiments. Depending on the calorimeter or the particular biological system, the protocol may require some adjustments.

What follows is a step-by-step protocol for running an ITC experiment to study heterodimer formation. When studying homodimer dissociation the only differences are in **step 5** (cell is filled with buffer solution only), **step 6** (syringe is filled with BP-α-CT solution) and **step 12** (data analysis (*see* Subheading 3.3)).

1. For typical experiments, allow the equipment to equilibrate one degree below the experimental temperature.

2. Prepare 2.2 mL of STI solution, 0.5 mL of PPT (in a glass tube, 6 × 50 mm) and 10 mL of buffer solution. All solutions should be degassed for 10 min with a vacuum pump (*see* **Note 1**).

3. Meanwhile, thoroughly clean the calorimetric reaction cell. The cell can be washed with a 5 % Contrad 70™ solution or 0.5 M NaOH if necessary, and rinsed thoroughly with water.

4. Fill the reference cell with water. Usually, water should be replaced every week.

5. Rinse the reaction cell with buffer. Slowly load the STI solution into the reaction cell, and carefully remove bubbles. The concentration of the sample should be determined again after loading because some dilution can take place due to residual buffer in the cell.

6. Fill the 250-μL injection syringe with PPT solution (*see* **Note 2**). Rinse the syringe tip with buffer or water and dry.

7. Carefully insert the injection syringe into the reaction cell. Avoid bending the needle or touching any surface with the needle tip.

8. Equilibrate the calorimeter at the experimental temperature.

9. Set the running parameters for the experiment: number of injections (28), temperature (e.g., 25 °C), reference power (10 μcal/s), initial delay (180 s), concentration in syringe (approx 300 μM), concentration in cell (approx 20 μM), stirring speed (490 rpm), file name (*.itc), feedback mode (high), equilibration options (fast), comments, injection volume (first injection 3 μL, the rest 10 μL), duration (automatically set according to the injection volume), spacing between injections (400 s), filter (2 s) (*see* **Notes 3–7**).

10. Start the experiment (*see* **Note 8**). There will be thermal and mechanical equilibration stages. Initiate the injection sequence after a stable no-drift noise-free baseline (as seen in the 1 μcal/s scale) (*see* **Note 9**).

11. At the end of the experiment the system should be cleaned thoroughly with water and the syringe rinsed and dried.

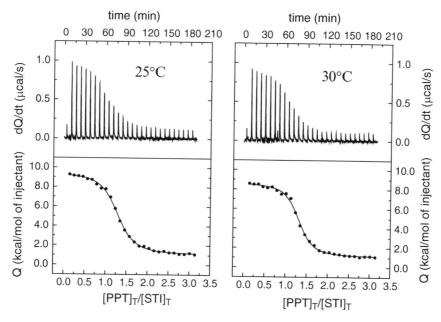

Fig. 4 Titrations of STI with PPT. The experiments were performed in 25 mM potassium acetate, pH 4.25, 10 mM calcium chloride, at 25 °C (*left*) and 30 °C (*right*). The concentrations of reactants are 21 μM STI (in cell) and 312 μM PPT (in syringe). The inhibitor was placed in the calorimetric cell due to its low solubility. The *solid lines* correspond to theoretical curves with $n = 1.26$, $K_a = 1.5 \times 10^6$ M^{-1}, and $\Delta H_a = 8.4$ kcal/mol (*left*) and $n = 1.28$, $K_a = 2.2 \times 10^6$ M^{-1}, and $\Delta H_a = 7.6$ kcal/mol (*right*)

12. Use the software provided by the calorimeter manufacturer for data analysis according to the equations described previously (*see* **Notes 10** and **11**). Results should be consistent with the information shown in Fig. 4.

3.3 Data Analysis

Equilibrium equations are very simple when using free concentrations of reactants (Eqs. **4** and **6**). However, they become more complex when expressed in terms of total concentrations of reactants, the known independent variables in ITC. Data processing through transformation of the experimental data in order to linearize the equilibrium equations must be avoided, since it often provokes a systematic propagation of errors and uneven distribution of statistical weights over the binding curve. Nonlinear fitting procedures should be implemented to directly analyze the raw calorimetric data.

The analysis should be performed with the individual or difference heat plot, not with the cumulative or total heat plot. The analysis in terms of the individual heat (heat associated with each injection) is more convenient since it eliminates error propagation associated with cumulative data. *See* **Notes 12–16** for additional considerations.

The heat, q_i, released or absorbed during injection i is proportional to the change in concentration of binary complex after that injection and is given by Eq. 5. The 1:1 stoichiometric model permits an explicit analytical solution for the concentration of complex (1, 13):

$$[M_1M_2]_i = \frac{1 + [M_2]_{T,i}\,K_a + K_a[M_1]_{T,i} - \sqrt{\left(1 + [M_2]_{T,i}\,K_a + K_a[M_1]_{T,i}\right)^2 - 4[M_2]_{T,i}\,K_a^2[M_1]_{T,i}}}{2K_a}$$

(10)

where $[M_1]_{T,i}$ and $[M_2]_{T,i}$ are the total concentrations of macromolecule M_1 and M_2 in the cell after the injection i.

Even when assuming a model corresponding to 1:1 stoichiometry, it is useful to introduce a parameter n representing the number (or fraction) of binding sites, as in the general 1:n model. When estimating the parameters, n should be equal to unity, within the experimental error, otherwise the following statements would hold:

$n > 1$: There is an error in the determination of the reactants concentration (actual titrating molecule M_1 concentration is lower and/or actual titrated macromolecule M_2 concentration is higher).

There is more than one binding site (specific binding or not) per macromolecule M_2.

$n < 1$: There is an error in the determination of the reactants concentration (actual titrating macromolecule M_1 concentration is higher and/or actual titrated macromolecule M_2 concentration is lower).

There is less than one binding site per macromolecule M_2, that is, the sample is not chemically or conformationally homogeneous, or there is more than one binding site (specific binding or not) per macromolecule M_1.

The 1:n stoichiometric model (n identical and independent binding sites) also permits an explicit analytical solution for the concentration of complex (1):

$$[M_1M_2]_i = \frac{1 + n[M_2]_{T,i}K_a + K_a[M_1]_{T,i} - \sqrt{\left(1 + n[M_2]_{T,i}K_a + K_a[M_1]_{T,i}\right)^2 - 4n[M_2]_{T,i}K_a^2[M_1]_{T,i}}}{2K_a}$$

(11)

Therefore, knowing the total concentrations $[M_1]_T$ and $[M_2]_T$ in the cell after the injections i and i − 1 it is possible to evaluate the heat q_i according to Eq. 5.

The dilution effect on the concentrations, due to the addition of the syringe solution, must be considered. The experiment proceeds at constant cell volume, i.e., when injecting a certain volume

v, the same volume of liquid is expelled from the cell. Therefore, the total concentrations of reactants after the injection i are given by:

$$[M_2]_{T,i} = [M_2]_0 \left(1 - \tfrac{v}{V}\right)^i$$
$$[M_1]_{T,i} = [M_1]_0 \left(1 - \left(1 - \tfrac{v}{V}\right)^i\right) \qquad (12)$$

where $[M_2]_0$ is the initial concentration of macromolecule M_2 in the cell, $[M_1]_0$ is the concentration of macromolecule M_1 in the syringe, v is the injection volume, and V is the cell volume. Because the dilution of reactants lowers the effective concentration of macromolecules in the cell, it will also affect the heat signal measured in each injection. Therefore, a correction to Eq. 5 is used:

$$q_i = V \Delta H_a \left([M_1 M_2]_i - [M_1 M_2]_{i-1}\left(1 - \frac{v}{V}\right)\right) \qquad (13)$$

Nonlinear fitting of q_i as a function of $[M_1]_{T,i}$ or $[M_1]_{T,i}/[M_2]_{T,i}$ (the so-called molar ratio) provides n, K_a, and ΔH_a as adjustable parameters. It is possible to avoid performing blank experiments to estimate the dilution heat effect by including in Eq. 13 a constant (but floating in the fitting procedure) term, q_d, representing such contribution.

In general, models with different sets of binding sites require a numerical approach in which the model parameters (n_j, $K_{a,j}$, and ΔH_j, where j stands for each set of binding sites) are determined iteratively by numerical solution of the binding equations.

For the experiment of porcine pancreatic trypsin and soybean trypsin inhibitor, the data obtained at two temperatures were analyzed using a model that allows the heat of dilution to be fitted simultaneously (Fig. 4). The enthalpy was plotted against temperature and the Gibbs energy dependence on temperature is shown in Fig. 5.

3.3.2 Homodimeric Interaction

The heat, q_i, released or absorbed during injection i is proportional to the change in concentration of monomer after that injection and is given by Eq. 7. As already stated, the third term in the parenthesis accounts for a contribution to the increase in monomer concentration in the cell that does not contribute to the heat measured.

The monomer–dimer equilibrium model permits an explicit analytical solution for the monomer concentration (12):

$$[M]_i = \frac{K_d}{4}\left(\sqrt{1 + \frac{8[M]_{T,i}}{K_d}} - 1\right) \qquad (14)$$

where $[M]_{T,i}$ is the total concentration of macromolecule M (per monomer) in the cell after the injection i. Therefore, knowing the total concentration $[M]_T$ in the cell after the injections i and $i-1$ it is possible to evaluate the heat q_i according to Eq. 7.

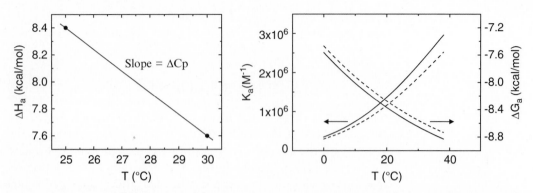

Fig. 5 (*Left panel*) Estimation of the change of heat capacity upon association. Once the enthalpy of association is determined at different temperatures (*see* Fig. 4), these values can be plotted vs. temperature. The slope of this plot corresponds to the heat capacity change. In the case of the PPT/STI interaction, the heat capacity was determined to be −160 cal/K mol. This negative value reflects the desolvation of molecular surfaces upon binding. (*Right panel*) Outline of the dependency of the association constant and the Gibbs energy of association on the temperature according to Eq. 3 and $\Delta G_a(T) = \Delta H_a(T_0) + \Delta C_{P,a}(T - T_0) - T[\Delta S_a(T_0) + \Delta C_{P,a}\ln(T/T_0)]$, where T_0 is a given reference temperature at which the enthalpy and the affinity have been simultaneously measured. The *curves* were calculated using the thermodynamic parameters obtained at 25 °C (*dashed*) and 30 °C (*continuous*). The estimation of the heat capacity change allows the determination of the thermodynamic parameters (affinity, Gibbs energy, enthalpy, and entropy) at any temperature

Considering the dilution effect on the concentration, the total concentration of reactant after the injection i is given by:

$$[M]_{T,i} = [M]_0 \left(1 - \left(1 - \frac{v}{V} \right)^i \right) \tag{15}$$

where $[M]_0$ is the concentration of macromolecule M in the syringe, v is the injection volume, and V is the cell volume. Again, a correction due to the dilution of reactants has to be included in Eq. 7:

$$q_i = V \Delta H_d \left([M]_i - [M]_{i-1} \left(1 - \frac{v}{V} \right) - F_0 [M]_0 \frac{v}{V} \right) \tag{16}$$

where F_0 has been defined above and can be calculated as:

$$F_0 = \frac{K_d}{4[M]_0} \left(\sqrt{1 + \frac{8[M]_0}{K_d}} - 1 \right) \tag{17}$$

Nonlinear fitting of q_i as a function of $[M]_{T,i}$ provides K_d and ΔH_d as adjustable parameters. Blank experiments to estimate the dilution heat effect are obviously not possible; therefore, a constant (but floating in the fitting procedure) term, q_d, representing such contribution must be included in Eq. 16.

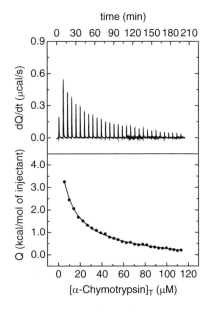

Fig. 6 Dissociation experiment of BP-α-CT. The experiment was performed in 20 mM sodium acetate, pH 3.9, 180 mM sodium chloride, at 25 °C. The concentration of BP-α-CT in the syringe is 608 μM. The *solid line* corresponds to theoretical curve with $K_d = 53 \times 10^{-6}$ M^{-1} and $\Delta H_d = 5.5$ kcal/mol of monomer. The value of the dissociation constant is in agreement with that determined by ultracentrifugation [62]. However, in the ITC experiment the possibility of autocatalysis is minimized since it takes only 2 h and the experiment by ultracentrifugation lasted for 2 days. The short time required to do an experiment is one of the advantages of ITC compared to other binding techniques (e.g., dialysis or ultracentrifugation). Another experiment performed at 30 °C (not shown) yielded a dissociation constant and a dissociation enthalpy of 63×10^{-6} M^{-1} and 6.2 kcal/mol. The estimated heat capacity of dissociation is 140 kcal/K mol. A positive value reflects the solvent exposure of molecular surfaces upon dissociation

For the experiment of bovine pancreatic α-chymotrypsin the data were analyzed using a dimer dissociation model that allows the heat of dilution to be fitted simultaneously (Fig. 6).

4 Notes

1. When performing experiments below room temperature, samples should be kept on ice or one degree below the experimental temperature while degassing.

2. The available total injection volume may be larger than the nominal volume of the syringe. Filling the syringe incompletely would generate nonuniform injections.

3. If the reaction is expected to be so exothermic and/or the concentration of ligand is so high that the heat effect in the first injections exceeds -200 µcal, the reference power should be raised or the injection volume lowered in order to prevent the signal to go below zero. Alternatively, the concentrations could be lowered accordingly.

4. In the equilibration options, fast equilibration is more convenient, compared to the other choices, because it permits simultaneous thermal and mechanical equilibration and it allows the user to manually start the injection sequence whenever the baseline is good.

5. The first injection is usually erroneous because there is some mixing of the solutions inside and outside of the needle when inserting the injection syringe and/or during the time required to reach thermal and mechanical equilibration. Therefore, the first injection is usually set smaller in order to waste the minimal amount of ligand possible. In the data analysis, the first injection will be taken into consideration for the calculation of the total concentrations in the cell. However, its associated heat effect will not be included in the fitting procedure.

6. The feedback should be set to high feedback gain. Otherwise a performance drop will occur: given an injection heat effect, the corresponding peak would be smaller and broader (sensitivity loss and increase of the time for running an experiment).

7. If possible, the stirring speed should be set at a sufficient level in order to ensure proper mixing. Occasionally, lower stirring speeds are required to avoid mechanical denaturation or destabilization of the macromolecules.

8. The experimental conditions and buffer solutions should be carefully determined for each system, depending on the particular characteristics (thermal stability, isoelectric point, solubility, propensity to interact with organic molecules, etc.) of the interacting macromolecules.

9. The time in between injections may need to be adjusted if the system is slow to reach equilibrium.

10. The data analysis software provided by the calorimeter manufacturer is straightforward and sufficient for simple systems (e.g., 1:1 binding reaction). However, for the analysis of more complex systems, such as the dissociation reaction presented here, or for including dilution heat effects, it might be necessary to create additional fitting routines.

11. In data analysis, the heat associated to each injection is usually normalized per mol of macromolecule injected. Therefore, Eqs. 13 and 16 are modified including a normalizing factor: $Q_i = q_i/(v \times [M_1]_0)$ and $Q_i = q_i/(v \times [M]_0)_i$, respectively.

12. Given the values of enthalpy and association (or dissociation) constant, it is possible to determine the optimal range of concentrations for reactants in an experiment. It is encouraged to carry out simulations in the experimental design stage.

13. The user may find different expressions for the necessary corrections in the concentrations and the heat effect due to dilution after injection. However, all those expressions are equivalent as long as the injection volume is small compared to the cell volume.

14. The user should be aware that a correct integration of the heat effect is instrumental for successful data analysis. The baseline may need to be adjusted independently for each injection.

15. Special care must be taken when determining the concentration of the reactants. Uncertainties in the concentration of any protein (e.g., due to contaminants or defective folding, which would decrease the fraction of binding-competent protein) will cause errors in the parameters. Active site titrations and/or reliable spectrophotometric methods should be employed. Colorimetric methods very often provide invalid results.

16. Performing direct and reverse titrations is the best criterion for discriminating between 1:1 and more complex stoichiometries. When normalizing the heat signal per mol of reactant injected, both titrations will provide similar results if the stoichiometry is 1:1 (that is, there is a single binding site). Otherwise, the two titrations will lead to different results if the system exhibits higher stoichiometries [63].

Acknowledgments

This work was supported by grants from the National Science Foundation (MCB0641252) and the National Institutes of Health (GM56550 and GM57144).

References

1. Wells JA, McClendon CL (2007) Reaching for high-hanging fruit in drug discovery at protein-protein interfaces. Nature 450:1001–1009

2. Zinzalla G, Thurston DE (2009) Targeting protein-protein interactions for therapeutic intervention: a challenge for future. Future Med Chem 1:65–93

3. Falconer RJ, Penkova A, Jelesarov I, Collins BM (2010) Survey of the year 2008: Applications of isothermal titration calorimetry. J Mol Recognit 23:395–413

4. Falconer RJ, Collins BM (2011) Survey of the year 2009: Applications of isothermal titration calorimetry. J Mol Recognit 24:1–16

5. Ghai R, Falconer RJ, Collins BM (2012) Applications of isothermal titration calorimetry in pure and applied research - Survey of the literature from 2010. J Mol Recognit 25:32–52

6. Banner DW, D'Arcy A, Janes W, Gentz R, Schoenfeld HJ, Broger C, Loetscher H, Lesslauer W (1993) Crystal structure of the soluble human 55 kd TNF receptor-human TNF

beta complex: implications for TNF receptor activation. Cell 73:431–445

7. Shibata H, Yoshioka Y, Ohkawa A, Minowa K, Mukai Y, Abe Y, Taniai M, Nomura T, Kayamuro H, Nabeshi H, Sugita T, Imai S, Nagano K, Yoshikawa T, Fujita T, Nakagawa S, Yamamoto A, Ohta T, Hayakawa T, Mayumi T, Vandenabeele P, Aggarwal BB, Nakamura T, Yamagata Y, Tsunoda S, Kamada H, Tsutsumi Y (2008) Creation and x-ray structure analysis of the tumor necrosis factor receptor-1-selective mutant of a tumor necrosis factor-alpha antagonist. J Biol Chem 283:998–1007

8. Hage T, Sebald W, Reinemer P (1999) Crystal structure of the interleukin-4/receptor alpha chain complex reveals a mosaic binding interface. Cell 97:271–281

9. Rickert M, Wang X, Boulanger MJ, Goriatcheva N, Garcia KC (2005) The structure of interleukin-2 complexed with its alpha receptor. Science 308:1477–1480

10. Philo JS, Aoki KH, Arakawa T, Narhi LO, Wen J (1996) Dimerization of the extracellular domain of the erythropoietin (EPO) receptor by EPO: one high-affinity and one low-affinity interaction. Biochemistry 35:1681–1691

11. Syed RS, Reid SW, Li C, Cheetham JC, Aoki KH, Liu B, Zhan H, Osslund TD, Chirino AJ, Zhang J, Finer-Moore J, Elliott S, Sitney K, Katz BA, Matthews DJ, Wendoloski JJ, Egrie J, Stroud RM (1998) Efficiency of signalling through cytokine receptors depends critically on receptor orientation. Nature 395:511–516

12. Kelekar A, Chang BS, Harlan JE, Fesik SW, Thompson CB (1997) Bad is a BH3 domain-containing protein that forms an inactivating dimer with Bcl-XL. Mol Cell Biol 17:7040–7046

13. Kussie PH, Gorina S, Marechal V, Elenbaas B, Moreau J, Levine AJ, Pavletich NP (1996) Structure of the MDM2 oncoprotein bound to the p53 tumor suppressor transactivation domain. Science 274:948–953

14. Sundstrom M, Lundqvist T, Rodin J, Giebel LB, Milligan D, Norstedt G (1996) Crystal structure of an antagonist mutant of human growth hormone, G120R, in complex with its receptor at 2.9 Å resolution. J Biol Chem 271:32197–32203

15. Walsh ST, Jevitts LM, Sylvester JE, Kossiakoff AA (2003) Site2 binding energetics of the regulatory step of growth hormone-induced receptor homodimerization. Protein Sci 12:1960–1970

16. Abbate EA, Berger JM, Botchan MR (2004) The X-ray structure of the papillomavirus helicase in complex with its molecular matchmaker E2. Genes Dev 18:1981–1996

17. Myszka DG, Sweet RW, Hensley P, Brigham-Burke M, Kwong PD, Hendrickson WA, Wyatt R, Sodroski J, Doyle ML (1997) Energetics of the HIV gp120-CD4 binding reaction. Proc Natl Acad Sci U S A 97:9026–9031

18. Gift SK, Zentner IJ, Schön A, McFadden K, Umashankara M, Rajagopal S, Contarino M, Duffy C, Courter JR, Zhang MY, Gershoni JM, Cocklin S, Dimitrov DS, Smith AB 3rd, Freire E, Chaiken IM (2001) Conformational and structural features of HIV-1 gp120 underlying the dual receptor antagonism by cross-reactive neutralizing antibody m18. Biochemistry 50:2756–2768

19. Kwong PD, Doyle ML, Casper DJ, Cicala C, Leavitt SA, Majeed S, Steenbeke TD, Venturi M, Chaiken I, Fung M, Katinger H, Parren PW, Robinson J, Van Ryk D, Wang L, Burton DR, Freire E, Wyatt R, Sodroski J, Hendrickson WA, Arthos J (2002) HIV-1 evades antibody-mediated neutralization through conformational masking of receptor-binding sites. Nature 420:678–682

20. Freire E, Mayorga OL, Straume M (1990) Isothermal titration calorimetry. Anal Chem 62:950A–959A

21. Doyle ML (1997) Characterization of binding interactions by isothermal titration. Curr Opin Biotechnol 8:31–35

22. Jelessarov I, Bosshard HR (1999) Isothermal titration calorimetry and differential scanning calorimetry as complementary tools to investigate the energetics of biomolecular recognition. J Mol Recognit 12:3–18

23. Leavitt S, Freire E (2001) Direct measurement of protein binding energetics by isothermal titration calorimetry. Curr Opin Struct Biol 11:560–566

24. Parker MH, Lunney EA, Ortwine DF, Pavlovsky AG, Humblet C, Brouillette CG (1999) Analysis of the binding of hydroxamic acid and carboxylic acid inhibitors to the stromelysin-1 (matrix metalloproteinase-3) catalytic domain by isothermal titration calorimetry. Biochemistry 38:13592–13601

25. Velazquez-Campoy A, Todd MJ, Freire E (2000) HIV-1 protease inhibitors: enthalpic versus entropic optimization of the binding affinity. Biochemistry 39:2201–2207

26. Todd MJ, Luque I, Velazquez-Campoy A, Freire E (2000) Thermodynamic basis of resistance to HIV-1 protease inhibition: calorimetric analysis of the V82F/I84V active site resistant mutant. Biochemistry 39:11876–11883

27. Velazquez-Campoy A, Kiso Y, Freire E (2001) The binding energetics of first- and second-generation HIV-1 protease inhibitors: Implications for drug design. Arch Biochem Biophys 390:169–175

28. Velazquez-Campoy A, Freire E (2001) Incorporating target heterogeneity in drug design. J Cell Biochem S37:82–88

29. Ward WH, Holdgate GA (2001) Isothermal titration calorimetry in drug discovery. Prog Med Chem 38:309–376

30. Velazquez-Campoy A, Muzammil S, Ohtaka H, Schön A, Vega S, Freire E (2003) Structural and thermodynamic basis of resistance to HIV-1 protease inhibition: implications for inhibitor design. Curr Drug Targets Infect Disord 3:311–328

31. Vega S, Kang LW, Velazquez-Campoy A, Kiso Y, Amzel LM, Freire E (2004) A structural and thermodynamic escape mechanism from a drug resistant mutation of the HIV-1 protease. Proteins 55:594–602

32. Ohtaka H, Muzammil S, Schön A, Velazquez-Campoy A, Vega S, Freire E (2004) Thermodynamic rules for the design of high affinity HIV-1 protease inhibitors with adaptability to mutations and high selectivity towards unwanted targets. Int J Biochem Cell Biol 36:1787–1799

33. Ohtaka H, Freire E (2005) Adaptive inhibitors of the HIV-1 protease. Prog Biophys Mol Biol 88:193–208

34. Ruben AJ, Kiso Y, Freire E (2006) Overcoming roadblocks in lead optimization: a thermodynamic perspective. Chem Biol Drug Des 67:2–4

35. Lafont V, Armstrong AA, Ohtaka H, Kiso Y, Amzel LM, Freire E (2007) Compensating enthalpic and entropic changes hinder binding affinity optimization. Chem Biol Drug Des 69:413–422

36. Freire E (2008) Do enthalpy and entropy distinguish first in class from best in class? Drug Discov Today 13:869–874

37. Freire E (2009) A thermodynamic approach to the affinity optimization of drug candidates. Chem Biol Drug Des 74:468–472

38. Ladbury JE, Klebe G, Freire E (2010) Adding calorimetric data to decision making in lead discovery: a hot tip. Nat Rev Drug Discov 9:23–27

39. Kawasaki Y, Freire E (2011) Finding a better path to drug selectivity. Drug Discov Today 16:985–990

40. Schön A, Madani N, Smith AB, Lalonde JM, Freire E (2011) Some binding-related drug properties are dependent on thermodynamic signature. Chem Biol Drug Des 77:161–165

41. Burrows SD, Doyle ML, Murphy KP, Franklin SG, White JR, Brooks I, McNulty DE, Scott MO, Knutson JR, Porter D, Young PR, Hensley P (1994) Determination of the monomer-dimer equilibrium of interleukin-8 reveals it is a monomer at physiological concentrations. Biochemistry 33:12741–12745

42. Czypionka A, de los Paños OR, Mateu MG, Barrera FN, Hurtado-Gomez E, Gomez J, Vidal M, Neira JL (2007) The isolated C-terminal domain of Ring1B is a dimer made of stable, well-structured monomers. Biochemistry 46:12764–12776

43. Bello M, Perez-Hernandez G, Fernandez-Velasco DA, Arreguin-Espinosa R, Garcia-Hernandez E (2008) Energetics of protein homodimerization: effects of water sequestering on the formation of beta-lactoglobulin dimer. Proteins 70:1475–1487

44. Wiseman T, Williston S, Brandts JF, Nin LN (1989) Rapid measurement of binding constants and heats of binding using a new titration calorimeter. Anal Biochem 179:131–137

45. Doyle ML, Louie GL, Dal Monte PR, Sokoloski TD (1995) Tight binding affinities determined from linkage to protons by titration calorimetry. Methods Enzymol 259:183–194

46. Baker BM, Murphy KP (1996) Evaluation of linked protonation effects in protein binding using isothermal titration calorimetry. Biophys J 71:2049–2055

47. Doyle ML, Hensley P (1998) Tight ligand binding affinities determined from thermodynamic linkage to temperature by titration calorimetry. Methods Enzymol 295:88–99

48. Sigurskjold BW (2000) Exact analysis of competition ligand binding by displacement isothermal titration calorimetry. Anal Biochem 277:260–266

49. Velazquez-Campoy A, Freire E (2005) ITC in the post-genomic era…? Priceless. Biophys Chem 115:115–124

50. Velazquez-Campoy A, Freire E (2006) Isothermal titration calorimetry to determine association constants for high-affinity ligands. Nat Protoc 1:186–191

51. Zhang Y-L, Zhang Z-Y (1998) Low-affinity binding determined by titration calorimetry using a high-affinity coupling ligand: a thermodynamic study of ligand binding to protein tyrosine phosphatase 1B. Anal Biochem 261:139–148

52. Velazquez-Campoy A, Ohtaka H, Nezami A, Muzammil S, Freire E (2004) Isothermal

titration calorimetry. Curr Protoc Cell Biol Chapter 17, Unit 17.8

53. Murphy KP, Freire E (1992) Thermodynamics of structural stability and cooperative folding behavior in proteins. Adv Protein Chem 43:313–361

54. Gomez J, Hilser VJ, Freire E (1995) The heat capacity of proteins. Proteins 22:404–412

55. Hinz HJ, Shiao DDF, Sturtevant JM (1971) Calorimetric investigation of inhibitor binding to rabbit muscle aldolase. Biochemistry 10:1347–1352

56. Biltonen RL, Langerman N (1979) Microcalorimetry for biological chemistry: experimental design, data analysis and interpretation. Methods Enzymol 61:287–319

57. Gomez J, Freire E (1995) Thermodynamic mapping of the inhibitor site of the aspartic protease endothiapepsin. J Mol Biol 252:337–350

58. Baker BM, Murphy KP (1997) Dissecting the energetics of a protein-protein interaction: the binding of ovomucoid third domain to elastase. J Mol Biol 268:557–569

59. Velazquez-Campoy A, Luque I, Todd MJ, Milutinovich M, Kiso Y, Freire E (2000) Thermodynamic dissection of the binding energetics of KNI-272, a potent HIV-1 protease inhibitor. Protein Sci 9:1801–1809

60. Wyman J, Gill SJ (1990) Binding and linkage: functional chemistry of biological macromolecules. University Science, Mill Valley, CA

61. Edgcomb SP, Baker BM, Murphy KP (2000) The energetics of phosphate binding to a protein complex. Protein Sci 9:927–933

62. Patel CN, Noble SM, Weatherly GT, Tripathy A, Winzor DJ, Pielak GJ (2002) Effects of molecular crowding by saccharides on α-chymotrypsin dimerization. Protein Sci 11:997–1003

63. Freire E, Kawasaki Y, Velazquez-Campoy A, Schön A (2011) Characterisation of ligand binding by calorimetry. In: Podjarny A, Dejaegere A, Kieffer B (eds) Biophysical approaches determining ligand binding to biomolecular targets. Detection, measurement and modeling. RSC Publishing, Cambridge

Chapter 12

Sedimentation Equilibrium Studies

Ian A. Taylor, Katrin Rittinger, and John F. Eccleston[†]

Abstract

The reversible formation of protein-protein interactions plays a crucial role in many biological processes. In order to carry out a thorough quantitative characterization of these interactions it is essential to establish the oligomerization state of the individual components first. The sedimentation equilibrium method is ideally suited to perform these studies because it allows a reliable, accurate, and absolute value of the solution molecular weight of a macromolecule to be obtained. This technique is independent of the shape of the macromolecule under investigation and allows the determination of equilibrium constants for a monomer–multimer self-associating system.

Key words Analytical ultracentrifugation, Sedimentation equilibrium, Solution interaction, Quaternary protein structure, Protein-protein interaction

1 Introduction

Analysis of macromolecular interactions, in particular protein-protein interactions, is an increasingly common goal in modern biological research. Methods vary widely, from qualitative techniques aimed at the detection of interactions in vivo to quantitative in vitro analyses that attempt to produce a detailed thermodynamic description of an interacting system. A necessary part of any quantitative in vitro method is that an initial characterization of the oligomerization state of the systems components be undertaken. This characterization is essential in order to correctly interpret binding data produced from subsequent titration experiments monitored, for example, by optical and magnetic resonance spectroscopy or isothermal titration calorimetry. Sedimentation equilibrium studies are unique in that they can be used at all stages of these quantitative studies. The technique allows a reliable, accurate and absolute value of the solution molecular weight of a macromolecule to be obtained making it invaluable in an initial

[†](Deceased)

Cheryl L. Meyerkord and Haian Fu (eds.), *Protein-Protein Interactions: Methods and Applications*, Methods in Molecular Biology, vol. 1278, DOI 10.1007/978-1-4939-2425-7_12, © Springer Science+Business Media New York 2015

characterization of a system. Furthermore, the technique can be extended to look at self-associating systems and heterologous equilibria making it complementary to spectroscopic and calorimetric methods (some good reviews about this technique include refs. [1–7].

1.1 Determination of Solution Molecular Weight

Popular methods for the determination of molecular weights are size exclusion chromatography and dynamic laser light scattering. In these methods, estimates of the molecular weight of unknowns are obtained by interpolation, using a curve generated by plotting an experimentally determinable parameter against the molecular weight of a set of standards. In size exclusion chromatography K_{av}, the partition coefficient between a porous matrix and free solution, is often used. In dynamic laser light scattering, a translational diffusion coefficient D_T, derived from the measured autocorrelation function, is often the parameter of choice. Along with the fact that using these methods molecular weights have to be obtained by interpolation, the main drawback is that no account is taken of molecular shape. Consequently, there are often large errors in the values obtained. The effects of molecular shape are not accounted for because the experimentally measured parameter reported by both methods is a function of D_T rather than molecular weight. Although D_T is correlated with molecular weight, it is directly related, through Eq. 1, to an important molecular parameter, the frictional coefficient, f [8].

$$D_T = \left(\frac{RT}{N_0 f} \right) \tag{1}$$

The value of f can be regarded as what determines inherent capacity of a molecule to undergo translational motion, and both molecular size and shape contribute to its value. Taking this into account it is easy to see how measurements of molecular weights are so shape dependent, since it is the frictional coefficient that directly determines both the D_T measured in dynamic light scattering and the partition coefficient observed in size exclusion chromatography. The complications due to molecular shape are serious enough in single component systems. When dealing with protein complexes and interacting systems it should be apparent that these problems are further exacerbated because of the need to account for the effects of equilibrium constants.

1.2 Sedimentation Equilibrium Theory

To overcome problems associated with solution molecular weight determinations it is best to examine a phenomenon where the shape of the molecule does not contribute to the measurement. This occurs when a concentration gradient is established by a macromolecular species sedimenting in a gravitational field, referred to as "sedimentation equilibrium." This method enables accurate, shape-independent, and absolute measurements of molecular weights to be obtained. Furthermore, sedimentation equilibrium studies can be employed to analyze interacting systems, allowing the determination of equilibrium constants and stoichiometries alongside molecular weights.

In order to understand how this works it is necessary to introduce the concept of flux. Equation 2 is a general expression for flow and can be likened to Ohm's law. Where the flux (J_i), or the flow of material, is related to a term for a generalized conductivity (L_i) and ($\delta U_i/\delta r$) a generalized gradient of potential.

$$J_i = -L_i \left(\frac{\delta U_i}{\delta r} \right) \tag{2}$$

In analytical ultracentrifugation, we are concerned with the flow of mass. The total potential in this case is comprised from a component due to the applied field, the centrifugal potential energy, together with a component from the chemical potential gradient generated from the solute. The conductivity term (L_i) in this case is the manifestation of the frictional coefficient and contributes to both flow due to sedimentation and flow due to diffusion.

$$J = \left[\frac{M(1-\nu\rho)}{N_0 f} \right] \omega^2 r C - \left[\frac{RT}{N_0 f} \right] \left(\frac{\delta C}{\delta r} \right) \tag{3}$$

M, molecular weight; ν, partial specific volume; ρ, solute density; N_0, Avogadro's number; ω, angular velocity; r, radial distance from center of rotation; R, gas constant; T, absolute temperature).

Equation 3 provides a full description of the transport process occurring in the ultracentrifuge cell. The terms $[M(1-\nu\rho)/N_0 f]$ and $[RT/N_0 f]$ correspond to the sedimentation coefficient (s) and translational diffusion coefficient (D_T), respectively, and chemical potential, U_i has been replaced by concentration C. It should be noted that the term for the strength of the applied gravitational field ($\omega^2 r$) will dominate at high rotor speeds, diffusion then only manifests itself as the boundary spreading observed in sedimentation velocity experiments. At lower rotor speeds, back diffusion due to the establishment of the chemical potential gradient counteracts transport from the applied gravitational field. Under these conditions equilibrium is established where the net transport in the system vanishes to zero at all points. Equation 3 can then be written as Eq. 4

$$\left[\frac{M(1-\nu p)}{N_0 f} \right] \omega^2 r C = \left[\frac{RT}{N_0 f} \right] \left(\frac{dC}{dr} \right) \tag{4}$$

An important result of this is that the shape term ($N_0 f$) cancels, meaning that whilst the concentration gradient established by the macromolecular species is dependent on the molecular weight, it is completely independent of shape. The useful form of the expression involves rearrangement to give Eq. 5 and then integration with respect to C and r, between the limits C_0 and C_x and r_0 and r_x to give Eqs. 6a and 6b.

$$\int_{C_0}^{C} \left(\frac{1}{C}\right) dC = \frac{M\omega^2(1-\nu\rho)}{RT} \int_{r_0}^{r} r\,dr \tag{5}$$

$$\ln\left(\frac{C_x}{C_0}\right) = \frac{M\omega^2(1-\nu p)}{2RT}\left(r_x^2 - r_0^2\right) \tag{6a}$$

or

$$C_x = C_0 e^{\frac{M\omega^2(1-\nu\rho)}{2RT}\left(r_x^2 - r_0^2\right)} \tag{6b}$$

1.3 Analysis of Molecular Weight Data

There are currently two models of analytical ultracentrifuges produced by Beckman-Coulter: the XL-A and the XL-I. Both contain an absorbance optical system able to measure the absorbance, at a chosen wavelength, at many points in the cell between r_0 and r_x [9]. Providing data is collected within the usable linear range of the optical system, absorbance can replace concentration in Eq. 6b simply using the Beer–Lambert law, Eq. 7.

$$A_x^\lambda = \varepsilon^\lambda c_x l \tag{7}$$

In addition, the XL-I also has Rayleigh interference optics that measures the displacement of fringes (Δj) in an interference pattern produced from combining two monochromatic laser light beams, one passed through the sample channel and the other through the reference. The number of fringes displaced in crossing from a point x_1 to a point x_2 in the sample cell is then proportional to the difference in weight concentration between these points, Eq. 8. Knowledge of the refractive index increment ($dn/dc = 0/186$ g ml^{-1} for proteins) and laser wavelength ($\lambda = 670$ nm) then allows Δj to replace c in Eq. 6b.

$$\Delta j = \frac{c\left(\dfrac{dn}{dc}\right)l}{\lambda} \tag{8}$$

Molecular weights are then simply extracted by direct nonlinear least squares fitting of the data to Eq. 6b between r_0 and r_x with substitution of c by either A or Δj. An offset is also included in the fitting procedure for contributions to the profile from non-exact matching of the cells, but should be treated with care. In the simplest case, a single sedimenting species, the data should fit to a single exponential curve in terms of r^2. Deviations in the data from a single exponential are indicative of sample heterogeneity, nonideality or an associative system. Often this is revealed by inspection of a plot of the fit residuals. A more complex model to account these effects can then be applied.

1.4 *Chapter Outline*	The instrumentation necessary to carry out analytical ultracentrifugation studies is described in the Materials section followed by a description of sample requirements and potential practical limitations of this technique. The Subheading 3 describes general considerations concerning experimental conditions for equilibrium runs and contains a detailed protocol for data collection and analysis (*see* Subheading 3.3 and 3.4).

2 Materials and Equipment

2.1 *Reagents*	1. Baseline buffer: 10 mM Tris–HCl, 100 mM NaCl, 1 mM TCEP, pH 7.5.
	2. Dialysis units (such as a D-Tube Dialyzer, Novagen).
	3. Purified proteins (~1–2 mgs).
	4. Fluorocarbon oil FC-42.
2.2 *Equipment*	1. Optima XL-A or XL-I analytical ultracentrifuge (Beckman-Coulter).
	2. Six channel centerpieces for analytical ultracentrifuge cells.
	3. Scanning (UV/Visible) spectrophotometer.
2.3 *Description of the Instrument*	Although there are two optical systems available for Beckman-Coulter analytical ultracentrifuges, in this chapter we will concentrate mainly on the use of absorbance optics. There are some advantages to the use of Rayleigh interference optics, but these largely surround the speed, the quality, and the density of data that can be collected and are more relevant to sedimentation velocity rather than equilibrium experiments. A good comparison of the advantages and disadvantages of the two optical systems can found in [9].

The basic feature of the absorbance optics is to allow absorbance, as a function of radial distance, to be measured while the solution is being centrifuged. This is achieved by the use of cells in which the solution is contained within quartz windows. A monochromator placed above the cell allows it to be illuminated with monochromatic light between 190 nm and 800 nm, while a slit mechanism below the cell scans radially at 0.001 cm or greater step sizes (Fig. 1).

Two types of centrifuge rotors that hold the cells are available, the An-50 Ti rotor (maximum speed 50,000 rpm) and the An-60 Ti rotor (maximum speed 60,000 rpm). The An-50 rotor has eight rotor holes, seven for sample cells and one for a counterbalance, whereas the An-60 has four rotor holes, three for sample cells and one for a counterbalance. The counterbalance is required in all centrifuge runs in order to coordinate the flash of the lamp with the cell position when it is above the detector. |

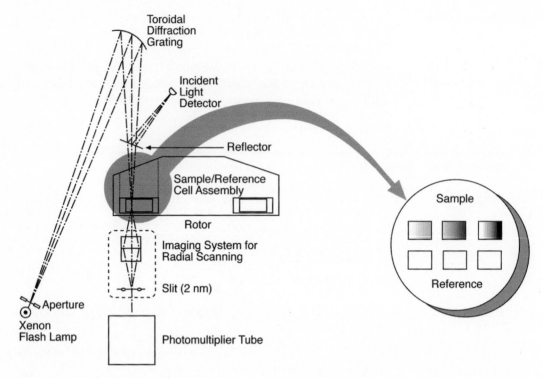

Fig. 1 A Schematic diagram of the optical system of the Beckman Optima XL-A analytical ultracentrifuge. Figure courtesy of Beckman-Coulter

2.4 Software and Data Analysis

The Beckman Optima XL-A/XL-I is supplied with analysis software based on the Origin® software package (MicroCal). This software has been used to fit all the data described in this chapter. The calculations of buffer density (ρ) and the partial specific volumes (ν) have been carried out using SEDNTERP [10] (*see* **Note 1**). This and several other useful programs for the analysis of equilibrium data have been developed by experts in the field and are available by download over the Internet, in most cases as Freeware. The Reversible Associations in Structural and Molecular Biology group (RASMB) maintains a website that provides links to most of these programs, some of which are available for different platforms (http://www.bbri.org/RASMB/rasmb.html). Simulations to determine optimal run conditions can be carried out using the Beckman XL-A software and the freely available Sedfit [11] and Sedphat [1] while the relative amounts of monomer and *n*-mers for self-associating systems can be calculated using the Ultrascan software (http://www.ultrascan.uthscsa.edu/).

2.5 Sample Preparation

When preparing a sample for sedimentation equilibrium analysis various points should be taken into account concerning buffer composition and protein concentration.

| 2.5.1 *Sample Buffer* | The sample buffer should not absorb more than 0.3 OD, referenced against water at the wavelength chosen for the experiment as this may otherwise interfere with the range and linearity of absorbance measurements (*see* **Note 2**). Charge repulsion between macromolecules will lead to problems with non-ideality, which is best addressed by using a buffer with ionic strength >0.1 M and by carrying out the experiments close to the pI of the protein in order to minimize the charge on the surface of the molecule. Furthermore, it should be borne in mind that samples might be in the centrifuge for long periods of time (>60 h), and therefore, buffer components that have significant time-dependent changes in their absorbance spectrum, for instance DTT, should be avoided. |

| 2.5.2 *Protein Concentration* | The sample concentration should not exceed 0.7 OD when measured against the sample buffer, otherwise the linear range of the detector (approximately 1.5 OD) will be exceeded when the absorbance rises towards the bottom of the cell during the run (*see* Fig. 1). If no information about the association state of the protein is available, 0.5 OD is a good starting point. If self-association is present, the loading concentration should be chosen such that there are detectable amounts of monomers and multimers (*see* Subheading 2.4). The protein solution and sample buffer have to be in equilibrium, this is best achieved by gel filtration (using for example NAP-5 columns, GE Healthcare) or exhaustive dialysis (2–3 buffer changes). We routinely use D-Tube Dialyzer, Novagen which is available for a range of different volumes starting at 10 μl. |

Before every ultracentrifugation run, the sample should be checked for aggregation. This might be done by a separate sedimentation velocity experiment or by static and/or dynamic light-scattering analysis if an instrument is available. Otherwise, a simple test is to spin the sample in a microcentrifuge for 15–20 min at maximum speed and to recheck the absorbance (*see* **Note 3**). If the sample OD has significantly decreased, the sample might not be suitable for equilibrium analysis at this time.

| **2.6 Limitations of Measurement for Protein–Protein Complexes** | The range of molecular weights tractable by sedimentation equilibrium is extremely large ($5 \times 10^2 – 10^7$), governed only by the rotor speed attainable. However, the analysis of an interacting system is limited by the absorbance range where reliable concentration measurements can be made within the confines of the equilibrium constant under investigation. Figure 2 shows the fraction of protein monomers expected for a set of monomer–dimer equilibria (K_a $10^3 – 10^9$ M^{-1}) over a concentration range of 1 nM to 1 mM. Superimposed on this figure, in grey, is the concentration range over which reliable absorbance measurements can be made using the XL-A/XL-I optical system. |

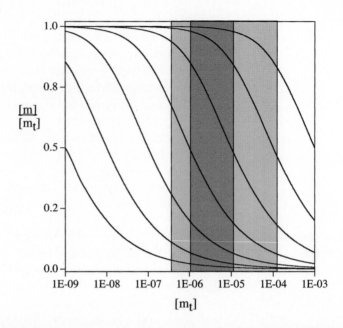

Fig. 2 Equilibrium binding isotherms for a homodimeric interaction. The fraction of monomer, [m]/[mt] is plotted as a function of the total monomer concentration [mt]. Curves from left to right are log order decreases in the association constant ranging from $K_a = 10^9$ M^{-1} to $K_a = 10^3$ M^{-1}. The *grey shaded* area represents the concentration range over which reliable measurements can be made. The *dark grey* area represents measurements made at 280 nm using standard 12 mm cells. The *lighter region* is where data has to be collected off peak, at shorter wavelength or using short pathlength cells

It should be apparent that for interactions with $K_a > 10^8$ M^{-1} the sedimentation experiment will give an accurate value for the molecular weight of the complex and therefore the stoichiometry. For interactions much weaker than 10^4 M^{-1} or where a heterogeneous mixture is present, sedimentation equilibrium will provide the weight averaged molecular weight of the mixture showing no concentration dependency. For interactions between these two limits $K_a > 10^4$ M^{-1} and $<10^8$ M^{-1} sedimentation experiments will also provide a weight averaged molecular weight, but in this case with a concentration dependence allowing the equilibrium constant for the system to be evaluated. Figure 2 is representative of a 50 kD protein with a molar extinction coefficient of 50,000 at 280 nm. Using a cell with a 1.2 cm optical path length, a concentration of 13.3 μM gives a starting absorbance of 0.8 and a concentration of 1.66 μM has an absorbance of 0.1. These represent the limits in absorbance for which reliable data can be collected. However, there are a number of ways to extend this range. One example is by collecting at a shorter wavelength where peptide bond absorbance will contribute to the extinction coefficient. The absorbance optics can measure at 230 nm and this will usually increase the extinction coefficient by about three to four times, extending the absorbance range down to 0.5 µM (*see* **Note 2**). To extend the range upwards shorter pathlength cells (3 mm) are available extending the range up to 50 µM. Finally, data can be collected away from the peak of the absorbance in the 250 nm trough or at 295 nm to give around a two to threefold decrease in sensitivity, expanding the range up to 150 µM.

3   Methods

3.1   Cell Assembly

The cells themselves are assembled for each centrifugation run. There are several different types of cell but all consist of the following main components: a cylindrical cell housing that fits into the rotor, two window assemblies consisting of a window holder, a quartz window, a white Vinylite window gasket, and a Bakelite window liner. These two window assemblies form a sandwich around a centerpiece that contains the samples. This sandwich is assembled in the cell housing and a screw ring and gasket is then screwed into one or both ends of the cell housing and tightened with a torque wrench (*see* **Note 4**). Full details of the cell assembly are provided in the manufacturer's instructions and should be followed very carefully (https://www.beckmancoulter.com/wsrportal/techdocs?docname=LXLA-TB-003).

There are many types of centerpieces constructed of different materials and dimensions. The simplest centerpieces are the double sector centerpieces, which allow for a sample solution and a buffer blank. These have a path length of 1.2 cm and are made in aluminum, aluminum filled Epon or charcoal filled Epon. The aluminum double sector centerpiece has the advantage of having a maximum speed of 60,000 rpm while the others are limited to 42,000 rpm. The charcoal filled Epon centerpiece should be used if there is any possibility of aluminum ions interacting with the sample, otherwise all of the centerpieces are compatible with aqueous solutions but softening of the Epon can occur if strong acids or organic solvents are used (*see* **Note 4**). A 3 mm path length charcoal filled Epon centerpiece is also available and requires the use of two spacers. Charcoal filled Epon centrerpieces that contain six channels (three samples and three buffer blanks) have a maximum speed of 48,000 rpm whereas the eight channel cells (four samples and four buffer blanks) can be operated up to 50,000 rpm. The double sector centerpieces that hold 450 µl per sector are generally used for sedimentation velocity measurements because of their longer column length. However, they can also be used for equilibrium runs if the volume is reduced to 100 µl. This is particularly useful if the 3 mm path length centerpiece is required since this is not available in six channel centerpieces. In this case, the volume is reduces to 33 µl. The volumes required to fill a 6-sector cell are 110 µl of protein and 120 µl of sample buffer.

3.2 Speed and Duration of the Experiment

Before starting a sedimentation equilibrium experiment, two important parameters need to be determined. First, the optimal speed of centrifugation needs to be known. Beckman provides a chart relating optimal speed to expected molecular weight (see https://www.beckmancoulter.com/wsrportal/bibliography?docname=362784.pdf). Alternatively, Eq. 6b can be solved for rpm [rpm $= (30/\pi)\,\omega$] for a given value of C_x/C_0. A good guide is that at equilibrium, around a fivefold increase in concentration over the column length will produce an exponential distribution suitable to get a good fit to the data ($C_x/C_0 = 5$). Regardless of this calculation, it is advisable to centrifuge at speeds lower and higher than this optimal speed (allowing equilibrium to be reached first at the lower speed, then optimal speed and finally at higher speed). The lower and higher speeds can be calculated from Eq. 6b, say for (C_x/C_0) = 3 and 7. Another way of finding optimal experimental conditions in terms of protein concentration or rotor speed is to simulate equilibrium data for a particular set of parameters (see Note 5). This can be done with the Origin XL-A analysis software under 'Utilities/Data Simulator'.

The other important parameter that needs to be determined is the time taken for equilibrium to be reached, T_{eqm}. This can be calculated from Eq. 9 where, T_{eqm} (seconds) is directly proportional to the square of the solution height, h (cm) and inversely proportional to D_T.

$$T_{eqm} = \frac{0.7(h)^2}{D_T} \tag{9}$$

From Eq. 9 it is apparent that smaller molecules reach equilibrium faster than larger ones. A 20 kD protein in about 16 h and a 100 kD protein in around 27 h. A value for D_T can be determined experimentally, but for these purposes it is reasonable to estimate a value by assuming the macromolecule is spherical and applying Eq. 10.

$$D_T = \frac{3 \times 10^{-5}}{\sqrt[3]{M}} \tag{10}$$

Whichever method is used, it is still necessary to show experimentally that the system has reached equilibrium. The best way to do this is to overlay or subtract successive scans, taken at two hourly intervals. When equilibrium is established no further change in the absorbance profile should occur, the only differences being due to noise in the data.

3.3 Sedimentation Equilibrium Studies of a Monomeric Species

1. Prepare around 2 L of a suitable buffer for the sedimentation study to be carried out in. For instance, 10 mM Tris–HCl, 100 mM NaCl, 1 mM TCEP pH 7.5 (r/t) (see Note 2).

2. Dialyze a stock protein solution against at least three changes of 500 ml of this baseline buffer. If necessary concentrate the

sample after dialysis. Ensure that there is enough of the stock protein to be able to prepare around 700 μl of a solution with an optical density of 0.7 at 280 nm, in a 1 cm pathlength cell (*see* **Note 6**).

3. Using this working solution prepare a dilution series of protein solutions, 120 μl in each, ranging from the highest optical density down to about 0.1 in a 1 cm cell. Because of the setup of the cells, it is convenient to use either three or nine samples. If sample is limiting then three is adequate; however, nine allows duplicates and to cover a larger concentration range.

4. Place 10 μl of Fluorocarbon oil FC-42 in each channel of the pre-assembled bottom sections of the six channel centrifuge cells and add 110 μl of each protein solution into the sample channels and 120 μl of baseline buffer into the buffer channels (*see* **Note 7**). Finally, insert the top window assembly and tighten down the top section of each cell.

5. Balance the cells and place into the analytical rotor as described in the user manual. Carefully install the monochromator, set the run temperature, usually 20 °C, and apply the vacuum (*see* **Note 8**).

6. Set the rotor speed to an initial speed of 3,000 rpm. When the centrifuge has reached the required speed, temperature and vacuum, collect a single radial scan at $\lambda = 280$ nm of each cell using a step size of 0.001 cm and two averages per scan. Refer to *see* **Note 9** for expected appearance of this pre-scan.

7. Set the centrifuge speed to the lowest of the three speeds to be collected during the run, for instance about 10,000 rpm for a 90 kDa protein. If association equilibria are expected or suspected set the rotor speed to optimize the gradient in favor of the higher molecular weight species (the slowest).

8. After 16 h collect radial scans every 2 h using a step size of 0.001 cm and five averages per scan. Assess whether equilibrium has been reached by overlaying or subtracting successive scans. Do this until there is no further change in the absorbance profile. The final scan in this dataset should be suitable for molecular weight analysis.

9. Set the centrifuge to the next speed in the set of three, wait for about 12 h before again collecting radial scans every 2 h using a step size of 0.001 cm and five averages per scan. Again, assess when equilibrium has been reached and collect a scan to use in the molecular weight analysis.

10. Finally, set the centrifuge to the highest speed where data will be collected. After 12 h, collect the radial scan data, assess for equilibrium as before, and collect a final scan suitable for use in data analysis.

11. When data has been collected at all the necessary speeds, set the centrifuge to 42,000 rpm. After approximately 16 h, collect a radial scan using a 0.001 cm step size with two replicates. This "overspeed" scan gives a depleted boundary that should be used to provide a reasonable estimate of the cell offset during the data fitting procedure (*see* **Note 10**).

3.4 Analyzing the Data

All data analysis described in the following paragraph has been carried out using the Origin® software package. This program allows the direct fitting of the equilibrium concentration gradients to mathematical models for single species or associating systems using nonlinear least squares algorithms [12, 13]. Furthermore, it allows global fitting of multiple datasets covering a range of concentrations and rotor speeds. This is particularly important for self-associating systems.

1. Once a whole dataset is collected, select the files that will be used in the molecular weight analysis. Load the relevant "RA" files into Origin, then cut out the data corresponding to the sample absorbance profiles using the select subset command (*see* **Note 11**).

2. First, use the Origin software to individually fit each "cut" file assuming an ideal single species model. In this fitting procedure, include the offset value obtained from the "overspeed" scan along with the calculated values for ρ and ν (*see* **Note 10**).

3. Plot the molecular weights obtained against the initial protein concentration. Is there any correlation between initial concentration and apparent molecular weight? If so, this may be indicative of an interacting system.

4. Plot ln(C) against r^2 (under Utilities/Plot) (*see* **Note 12**).

5. If the fit of the individual data files provided any indication of self-association carry out a simultaneous (global) fit, still assuming a single species. Under these conditions there should be clear systematic deviations in the residuals plot indicating that the protein is not an ideal monomeric species.

6. Carry out global fits to monomer–dimer, monomer–trimer, and monomer–tetramer equilibria and carefully inspect for deviations between the fitted curve and the experimental data as well as for systematic deviations in the residuals plot. At this stage it is of great importance to have some previous knowledge of the monomer molecular weight of the protein under investigation as this allows the monomer molecular weight to be fixed (*see* **Note 13**).

7. Convert the association constant expressed in absorbance units into the commonly used association constant K_a (*see* **Note 14**).

3.5 Examples of Fits to Experimental Data

1. The Swi6 transcriptional regulator is a 90 kD protein. Analysis of the molecular weight using size exclusion chromatography and dynamic light scattering provided estimates of the molecular weights of 250 kD and 280 kD, respectively, suggesting that the protein is either dimeric or trimeric. Subsequent analysis of the Swi6 solution molecular weight using sedimentation equilibrium, Figure 3, reveal the data are fit well with an ideal monomer molecular mass of 90 kD [14] and that lnC vs. r^2 plots and the application of associative models provide no indication of oligomerization. The comparative analysis demonstrates how erroneous estimates of molecular weight arise when using techniques that do not account for the effect of molecular shape.

2. Sedimentation equilibrium runs of bovine dynamin I, a 100 kDa protein that oligomerizes in vivo and in vitro [15, 16]. The runs

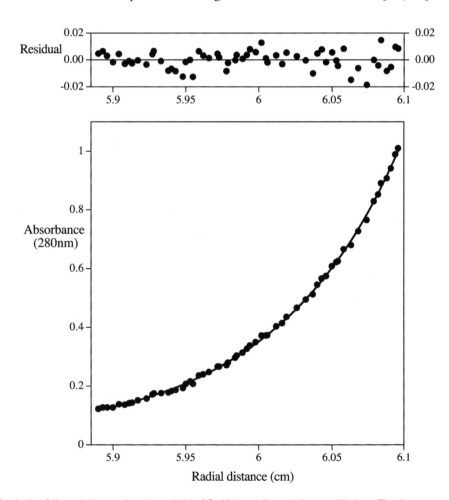

Fig. 3 Analysis of the solution molecular weight of Swi6 by sedimentation equilibrium. The *lower panel* shows the absorbance profile produced at equilibrium by Swi6 at a rotor speed of 12,000 rpm, $T = 293$ K, $\nu = 0.727$, $\rho = 1.003$. The *upper panel* shows the residuals to the plot

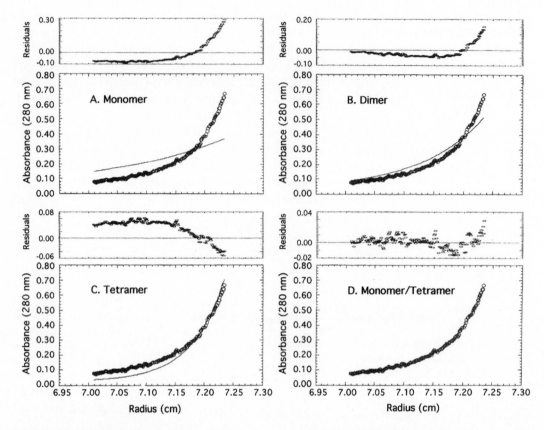

Fig. 4 Sedimentation equilibrium data of dynamin I fitted to different models: (**a**) monomer alone; (**b**) dimer alone; (**c**) tetramer alone; (**d**) monomer/tetramer. Reproduced with permission from Journal of Protein Chemistry (1999) 18, 277–290

shown in Fig. 4 have been carried out at 4 °C at a speed of 7,000 rpm and protein concentration of 0.33 mg/ml. Figure 4a–c show fits of the data, along with residuals, assuming a monomeric, dimeric, and tetrameric species, respectively. The fit to a single species gave a molecular weight of 285 kD, significantly bigger than that of a monomeric protein. Furthermore, the non-random distributions of the residuals of all three fits clearly indicate that none of these models represents the correct association behavior. Only a monomer–tetramer model gave a satisfactory fit with a random distribution of residuals as shown in Fig. 4d, allowing the calculation of an association constant of $K_{1,4} = 1.67 \times 10^{17}$ M^{-3}. Assuming a monomer–dimer–tetramer model resulted in the same quality of fit with no improvement over the monomer–tetramer model. The association constant $K_{1,2}$ for the monomer–dimer equilibrium calculated from this fit is so small relative to $K_{1,4}$, that there would be hardly any dimer present under the experimental conditions, thereby justifying the assumption of a monomer–tetramer model.

3.6 Heterologous Systems

Sedimentation equilibrium methods could be especially useful for studying the heterologous interactions that are important in almost every biological process. However, the analysis of these data is complicated by the fact that the absorbance along the cell has contributions from both the free proteins and the complex(es). If the two proteins are of similar molecular weight and extinction coefficient, they can be treated as a homologous associating system [17]. However, this is usually not the case and so more complex strategies must be used. These are beyond the scope of this chapter, and interested readers are referred to a review article by Minton and colleagues [3] that discusses different methods for the quantitative characterization of heterologous interactions and to Philo [18] who has discussed how problems can be alleviated through the use of numerical constraints during data analysis.

4 Notes

1. In fact, it is the buoyant molecular weight, M^* that is directly obtainable from the experimental data. The absolute molecular weight has to be calculated from Eq. 11 using the solvent density ρ and the partial specific volume, ν of the particle.

$$M = \frac{M^*}{(1 - \nu\rho)} \tag{11}$$

   Good values for the solvent density ρ can be obtained from tabulated values. An approximate value for the partial specific volume of a protein ν is 0.73 ml/g. In some cases, it may be determinable experimentally using densitometry. A reasonably accurate alternative is to calculate partial specific volumes based on the amino acid composition of the protein. The SEDNTERP program does this using Eq. 12, where, n_i, Mr_i, and ν_i are the number, molecular mass, and partial specific volume of the ith residue.

$$\nu = \frac{\sum n_i M_i \nu_i}{\sum n_i M_i} \tag{12}$$

   The need to obtain accurate values for these parameters stems from the $(1 - \nu\rho)$ relationship and because of this a 3 % error in ν results in close to a 10 % error in molecular weight.

2. The optical system of the XL-A/XL-I is capable of measuring data at wavelengths between 800 and 190 nm, although below 230 nm there is very little lamp intensity and so this region is not generally useful. In all measurements it is important to try to keep the buffer absorbance to a minimum because low transmittance of the buffer blank will decrease the range, linearity, and sensitivity of any measurements. If measurements are

carried out at a wavelength in the ultraviolet region, a non-absorbing buffer has to be selected. At short wavelengths NaCl should be substituted by NaF to reduce buffer absorbance. 232 nm is a good wavelength because there is a peak in the lamp spectrum. However, the intensity of the light source of the instrument decreases with time, in particular in the UV region, due to deposits of oil on the lamp and a scan of the light intensity against wavelength should be done regularly to determine if the lamp needs cleaning. This problem has been partially relieved on newer machines by incorporation of a window at the base of monochromator stem that reduces the oil buildup.

3. A UV spectrum taken from 230 to 350 nm can provide a good indication of sample quality. A sloping baseline between 310 and 350 nm is indicative of protein aggregation.

4. Before assembling the cells make sure everything is clean because salt deposits can lead to leakage or cracking of the windows. Furthermore, it is necessary to check if the buffer components are compatible with the cell components. Chemical compatibility tables are available at (https://www.beckmancoulter.com/wsrportal/techdocs?docname=LXLA-TB-003).

5. The 'Simulator Setup' window allows the user to define all necessary run parameters. Note please that the loading concentration is given in mg/ml and will be converted to absorbance based onto the concentration conversion factor. A detailed description of the program can be found at (https://www.beckmancoulter.com/wsrportal/techdocs?docname=LXLA-TB-009).

6. The optical pathlength of the ultracentrifuge cells is 12 mm, rather than 10 mm used in most standard spectrophotometric cuvettes. Bear this in mind when making up the sample dilution series.

7. Load the reference column slightly higher than the sample so the reference meniscus does not interfere with the sample. Fluorocarbon oil FC-42 has the same refractive index as water but is denser and forms a boundary at the bottom of the cell and thereby reduces artifacts caused by light at the bottom of the cell.

8. Do not forget to equilibrate the rotor at the experimental temperature or best at a slightly lower temperature. Also, the experimental parameters, ρ and ν are temperature dependent and usually hydrodynamic data is referenced to 20 °C. If the experiment is not carried out at 20 °C this needs to be accounted for.

9. The 3,000 rpm pre-scan should have a constant absorbance along the whole of the cell. It is worth writing these values down for any subsequent analysis of concentration dependency. The presence of a boundary or the appearance of upward curvature in the absorbance profile towards the bottom of the cell is indicative of high molecular weight or aggregated material in the sample. This probably means the sample is not suitable for further analysis. The lack of any absorbance indicates cell leakage and requires the run to be repeated.

10. The offset value should be used with caution and only if it can be justified in some experimental way. It is highly correlated with A_0 and "adjustment" of its value will dramatically alter the molecular weight values obtained. The use of the over-speed at the end of a run is a reasonable estimation of the value, but may not be correct. The best way to avoid offset problems is to avoid buffers with absorbing components and to take care when preparing the cells for a run, avoid scratches and fingerprints.

11. If the equilibrium run has been carried out using six or eight channel centerpieces, all the data for the sample channels will be in a single file and have to be separated into individual files before any fitting procedures can be carried out. This is done via the "Select subset" option in the in the Utilities menu. This feature allows the selection of a subset of data-points from a whole profile to be saved as an individual file.

12. A simple, model-independent method to check for heterogeneity is a plot of $\ln(C)$ versus r^2. A straight line for this plot suggests the presence of a single species, with the slope equalling M*, the buoyant molecular weight of the protein (*see* Eq. 11 and **Note 1**). An upward curving line is indicative of an associating system. However, this method can be rather insensitive to heterogeneity and should only be taken as a guide.

13. The molecular weight of the protein can easily be calculated based on the amino acid composition if the amino acid sequence of the protein is known, see for example (http://www.expasy. ch/proteomics/protein_characterisation_and_function). Otherwise, the molecular weight has to be experimentally determined, for example by mass spectrometry. Alternatively, sedimentation equilibrium runs can be carried out under denaturing conditions.

14. Equilibrium constants determined in the Beckman origin software have units expressed in absorbance units. Equations 13a, 13b, and 13c are used to convert them into concentration units for monomer–dimer, monomer–trimer, and monomer– tetramer equilibria, respectively.

$$K_a = \frac{K_a^{abs}(\varepsilon_m l)}{2} \qquad (13a)$$

$$K_a = \frac{K_a^{abs}(\varepsilon_m l)^2}{3} \qquad (13b)$$

$$K_a = \frac{K_a^{abs}(\varepsilon_m l)^3}{4} \qquad (13c)$$

Acknowledgement

This work was supported by the Medical Research Council, UK.

References

1. Vistica J, Dam J, Balbo A et al (2004) Sedimentation equilibrium analysis of protein interactions with global implicit mass conservation constraints and systematic noise decomposition. Anal Biochem 326:234–256

2. Howlett GJ, Minton AP, Rivas G (2006) Analytical ultracentrifugation for the study of protein association and assembly. Curr Opin Chem Biol 10:430–436

3. Rivas G, Stafford W, Minton AP (1999) Characterization of heterologous protein-protein interactions using analytical ultracentrifugation. Methods 19:194–212

4. Laue TM, Stafford WF 3rd (1999) Modern applications of analytical ultracentrifugation. Annu Rev Biophys Biomol Struct 28:75–100

5. Cole JL, Lary JW, PMoody T et al (2008) Analytical ultracentrifugation: sedimentation velocity and sedimentation equilibrium. Methods Cell Biol 84:143–179

6. Cole JL, Hansen JC (1999) Analytical ultracentrifugation as a contemporary biomolecular research tool. J Biomol Tech 10:163–176

7. Liu J, Shire SJ (1999) Analytical ultracentrifugation in the pharmaceutical industry. J Pharm Sci 88:1237–1241

8. Cantor CR, Schimmel PR (1980) Ultracentrifugation. In: biophysical chemistry, Part II. Freeman

9. Laue TM (1996) Choosing which optical system of the Optima XL-I analytical centrifuge to use. Beckman-Coulter Technical Application Information Bulletin A-1821-A

10. Laue TM, Shah BD, Ridgeway TM et al (1992) Computer-aided interpretation of analytical sedimentation data for proteins. In: Harding SE, Rowe AJ, Horton JC (eds) Analytical Ultracentrifugation in Biochemistry and Polymer Science. The Royal Society of Chemistry, Cambridge United Kingdom, pp 90–125

11. Brown PH, Schuck P (2006) Macromolecular size-and-shape distributions by sedimentation velocity analytical ultracentrifugation. Biophys J 90:4651–4661

12. Johnson ML, Faunt LM (1992) Parameter estimation by least-squares methods. Methods Enzymol 210:1–37

13. Johnson ML, Correia JJ, Yphantis DA et al (1981) Analysis of data from the analytical ultracentrifuge by nonlinear least-squares techniques. Biophys J 36:575–588

14. Sedgwick SG, Taylor IA, Adam AC et al (1998) Structural and functional architecture of the yeast cell-cycle transcription factor swi6. J Mol Biol 281:763–775

15. Binns DD, Helms MK, Barylko B et al (2000) The mechanism of GTP hydrolysis by dynamin II: a transient kinetic study. Biochemistry 39:7188–7196

16. Binns DD, Barylko B, Grichine N et al (1999) Correlation between self-association modes and GTPase activation of dynamin. J Protein Chem 18:277–290

17. Silkowski H, Davis SJ, Barclay AN et al (1997) Characterisation of the low affinity interaction between rat cell adhesion molecules CD2 and CD48 by analytical ultracentrifugation. Eur Biophys J 25:455–462

18. Philo JS (2000) Sedimentation equilibrium analysis of mixed associations using numerical constraints to impose mass or signal conservation. Methods Enzymol 321:100–120

Chapter 13

Detecting Protein-Protein Interactions by Gel Filtration Chromatography

Yan Bai

Abstract

Upon protein-protein interaction, the formed complex is subject to a change in size. A number of methods can be utilized to detect such a change. Gel filtration technology is well recognized for its ability to monitor and separate protein species of different sizes, and can greatly facilitate functional studies of protein complexes. In addition, gel filtration can be performed in any buffer system that preserves the protein complex formation and function. Therefore, it can be a significantly useful method for studying protein interactions. In this chapter, a protocol for performing gel filtration is described in detail.

Key words Gel filtration, FPLC, Molecular weight estimation, Protein dimerization

1 Introduction

Gel filtration separates substances by their difference in size as they pass through a medium packed in a column, and therefore this method is also known as size-exclusion chromatography (SEC). Gel filtration medium is a porous matrix made up of cross-linked polymer spherical particles. There are many kinds of media that differ in their chemical and physical properties, and are designed to suit a variety of separation tasks. When a solution composed of different sized molecules is applied to a column, the molecules that are too large to diffuse into the pores will be eluted from the column all together. These large molecules are said to be excluded. The smaller molecules diffuse into the pores of the medium particles, take routes with different lengths based on their sizes, and elute in the order of their sizes with the largest molecule first and the smallest last (Fig. 1). The terminology for the size measurement is "Stokes radii" [1], which is the radius of a hard sphere that diffuses at the same rate as the molecule. Since a molecule's Stokes radius is often unavailable, as for globular molecules, their molecular weights are used to represent their size. Commercially available gel filtration media cover a wide molecular weight range from 100

Cheryl L. Meyerkord and Haian Fu (eds.), *Protein-Protein Interactions: Methods and Applications*, Methods in Molecular Biology, vol. 1278, DOI 10.1007/978-1-4939-2425-7_13, © Springer Science+Business Media New York 2015

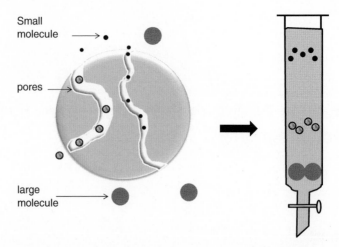

Fig. 1 Illustration of the gel filtration principle. Protein molecules pass through the gel medium via different routes based on their size. Small molecules take longer to elute because they can get into small pores and big pores; big molecules take less time because they are too large to travel through the pores

to 80,000,000 Da, from peptides to large proteins and protein complexes. Any packed column, based on its physical properties, has its own selectivity. Therefore, in practice, an effective separation range is always the primary consideration when choosing an appropriate gel column for a specific task.

Compared with other chromatographic separation methods, gel filtration has several advantages: (1) the solute recovery rate is high, approaching 100 % [2], (2) high reproducibility is expected for repeated runs, (3) various buffer conditions can be used for many types of samples. Separation can be performed in the presence of essential ions or cofactors, therefore proteins can be preserved in their active conformation. The tolerance of high concentrations of detergents and wide range of ionic strength of gel filtration are greatly appreciated when working with less soluble proteins, such as membrane proteins and transcription factors.

The fundamental usage of gel filtration is for protein purification. However, because the technique provides size information for biological substances, it has been used extensively for other biological events that result in a molecule size change or molecule complex formation, such as protein refolding [3], the generation of amyloid fibrils [4], and, especially, functional proteomics by revealing protein-protein interactions under conditions that preserve native protein complexes [5–7]. This chapter reviews important concepts in gel filtration and describes, how to utilize gel filtration technology to study protein-protein interactions, including sample preparation, instrument operation, and data analysis. The procedure for gel filtration is as follows: (1) Set up the gel filtration system and a specific gel filtration method. (2) Equilibrate the

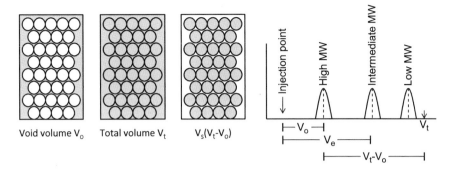

Fig. 2 Common terms in gel filtration. V_o (the void volume): space taken by solvent surrounding the gel beads, usually about 30 % of the total bed volume. V_t: the total volume of a gel bed. V_s: space occupied by solvent inside the medium particles

column and prepare protein standards and samples. (3) Run protein standards to establish a calibration curve. (4) Run protein samples. (5) Analyze collected protein fractions on an SDS-PAGE gel.

1.1 General Terms and Concepts in Gel Filtration

A set of terminology is employed to describe the gel filtration process. As shown in Fig. 2, the total volume of a gel bed, V_t, includes space occupied by solvent inside the medium particles V_s, and space taken by solvent surrounding the gel beads called V_o (the void volume), which is about 30 % of the total bed volume.

$$V_t = V_o + V_s \qquad (1)$$

The elution volume, V_e, is the volume of mobile phase entering the column between the start of the elution and the emergence of the peak maximum [8].

1.2 Gel Filtration Resolution

Detection of protein-protein interaction by size change requires high-resolution gel filtration. Several factors influence resolution: column dimensions, particle size and distribution, pore size of the particles, medium packing density, sample volume, flow rate, and viscosity of the sample and buffer. These factors contribute to two effects that determine the final resolution, the selectivity and the efficiency of the medium. The selectivity of the medium, which solely depends on its pore size distribution, refers to the molecular weight range over which a gel filtration medium can effectively separate molecules. The medium selectivity is described by a selectivity curve, in which a partition coefficient, K_{av}, is plotted against the logarithm of the molecular weight for a set of standard proteins.

$$K_{av} = (V_e - V_o)/(V_t - V_o) \qquad (2)$$

$$K_{av} = a\log(MW) + b \qquad (3)$$

The steeper the curve, the higher the resolution reached.

The efficiency of a packed column defines its ability to produce narrow symmetrical peeks. The uniformity of the column packing and particle size contributes the most to the efficiency. Increasing medium bedding height and decreasing the medium particle size can increase the resolution; however, this slows down the flow rate. Besides the selectivity of the medium, the sample volume and column dimensions are also crucial for the resolution. Usually, the recommended sample volume is within 0.5–4 % of the column bed volume, and for high-resolution results, the sample volume should not exceed 2 % of the total bed volume.

For studying protein-protein interaction by gel filtration, the running buffer is another critical matter. The buffer should be able to maintain the interactions of interest, such as appropriate pH, ion strength, detergent concentration, and optimal buffering range should be carefully chosen.

2   Materials

1. Chromatography system (e.g., AKTApurifier FPLC system (*see* **Note 1**) and AKTA design fraction collector, Frac-920).

2. Gel filtration columns. Superdex 75 (Superdex 75 10/300) (*see* **Note 2**).

3. Gel filtration molecular weight markers. 3.5 mg albumin bovine serum (66 kDa), 1 mg carbonic Anhydrase from bovine erythrocytes (29 kDa), 1 mg aprotinin (6,512 Da).

4. Purified protein sample (e.g., 2 mL His-14-3-3 zeta (~30 kDa) (*see* **Note 3**). The protein was expressed in an *E. coli* system and purified using Ni-NAT beads. The protein sample was filtered through a 0.2 μm, 25 mm syringe filter (Nylon, sterile) to remove any insoluble materials).

5. Distilled water (500 ml), filter through a 0.2 μm filter (non-pyrogenic, sterile) and degas.

6. 20 % ethanol in distilled water (500 ml), filter through a 0.2 μm filter (nonpyrogenic, sterile) and degas.

7. Running buffer (500 ml): 50 mM Tris–HCl, pH 7.5, 100 mM NaCl (*see* **Note 4**), filter through a 0.2 μm filter (nonpyrogenic, sterile) and degas.

3   Methods

3.1   Chromatography System Setup

It is advisable to read the manufacture's user manual for detailed instrument operation and maintenance. A simplified outline for the AKTApurifier FPLC system is provided here.

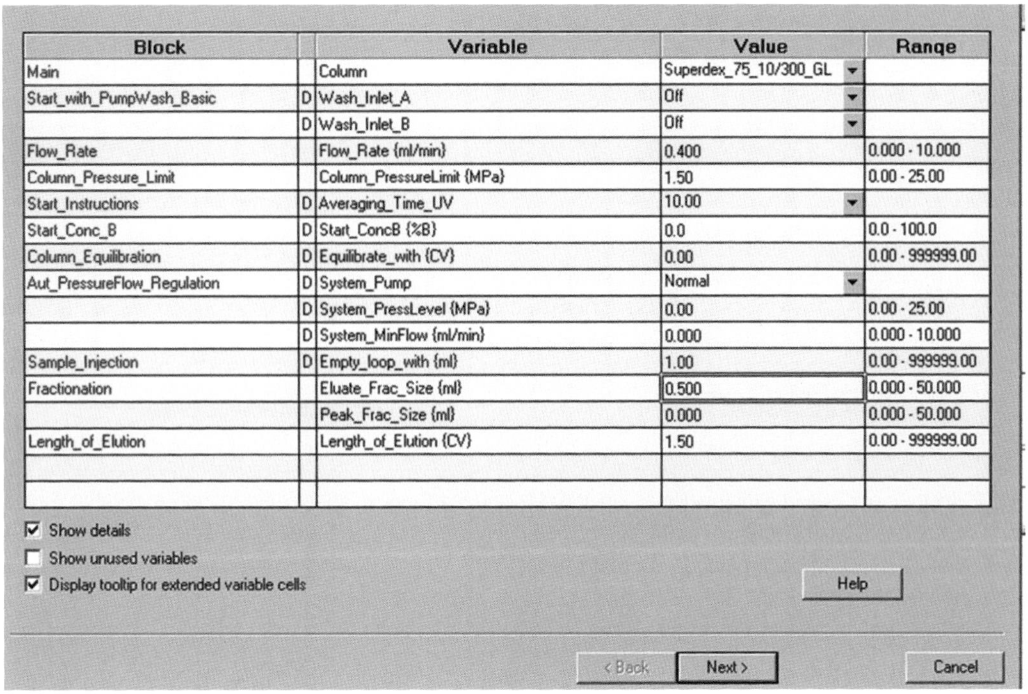

Block		Variable	Value	Range
Main		Column	Superdex_75_10/300_GL ▼	
Start_with_PumpWash_Basic	D	Wash_Inlet_A	Off ▼	
	D	Wash_Inlet_B	Off ▼	
Flow_Rate		Flow_Rate {ml/min}	0.400	0.000 - 10.000
Column_Pressure_Limit		Column_PressureLimit {MPa}	1.50	0.00 - 25.00
Start_Instructions	D	Averaging_Time_UV	10.00 ▼	
Start_Conc_B	D	Start_ConcB {%B}	0.0	0.0 - 100.0
Column_Equilibration	D	Equilibrate_with {CV}	0.00	0.00 - 999999.00
Aut_PressureFlow_Regulation	D	System_Pump	Normal ▼	
	D	System_PressLevel {MPa}	0.00	0.00 - 25.00
	D	System_MinFlow {ml/min}	0.000	0.000 - 10.000
Sample_Injection	D	Empty_loop_with {ml}	1.00	0.00 - 999999.00
Fractionation		Eluate_Frac_Size {ml}	0.500	0.000 - 50.000
		Peak_Frac_Size {ml}	0.000	0.000 - 50.000
Length_of_Elution		Length_of_Elution {CV}	1.50	0.00 - 999999.00

☑ Show details
☐ Show unused variables
☑ Display tooltip for extended variable cells   Help

< Back   Next >   Cancel

Fig. 3 Parameter settings to conduct a typical gel filtration run

The AKTApurifier is an automated liquid chromatography system. It has two major modules: Pump P-900 and monitor UPC-900, a highly precise online monitor to measure UV absorption at 280 nM, conductivity and pH. Each module has its own power switch. The system is controlled and monitored by a software package, UNICORN, which is installed on a PC and runs under the Microsoft Windows operating system.

1. Turn on the chromatography system, turn on the computer and launch UNICORN. Four interfaces will pop up: UNICORN manager, Method Editor, System Control and Evaluation (*see* **Note 5**).

2. Create methods. New methods are created through the Method Wizard. All major parameters for a gel filtration run, such as column type, pressure limit, flow rate, sample volume, fraction volume and length of run can be entered step by step (Fig. 3) (*see* **Note 6**). By clicking *Finish* in the last dialog, the Run Setup Window appears, which allows the user to review all information about the method. The user can save the method by selecting *File:Save* and saving it with an appropriate name.

3. Set up the column and the sample loop. Check the tubing connections and the sample loop. The tubing on the top of the column is connected to port 1 of the injection valve, and the bottom tubing goes to the UV cell port. The sample loop

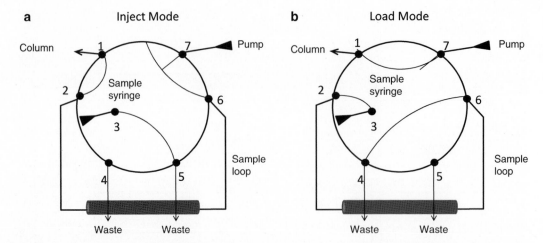

Fig. 4 Illustration of the injection valve at inject mode and load mode. After being manually injected through port 3, the sample will be retained in the sample loop between port 2 and port 6. (**a**) Upon sample loading on the column, the system automatically switches to the "Inject" mode, in which the flow path is 7-6–2-1. (**b**) When elution starts, the liquid flows directly from 7 to 1, bypassing the sample loop

connects between valve port 2 and port 6. The waste tubing is connected to port 4 and 5 (Fig. 4).

4. Set up the fraction collector. Insert collection tubes into the tube holder starting at position 1 and ending at position 70. Gently lift the delivery arm up and slide it to the outer stop, then place the tube rack over the central spindle and pull the spring-loaded drive sleeve out and snap in the rack. Lift and lower the delivery arm, and move it in so the tube sensor touches the outer wall of the tube at position 1 and the eluent tubing is above the center of the collection tube (*see* **Note 7**).

3.2 Sample Loading and Fractionating

1. Equilibrate the column (*see* **Note 8**). Wash the column using at least 1CV (column volume) of filtered distilled water, and equilibrate the column using at least 1CV of running buffer. By the end of the equilibration, the conductivity trace should show an "S" shape and level off in the end. The operation can be set up and executed through **Manual** on the System control module.

2. Wash the sample loop. The sample loop (0.5 ml) should be well cleaned using running buffer. Use a needle syringe to inject 1 ml of running buffer into the fill port three times (*see* **Note 9**).

3. Load the sample. Go to the UNICORN System Control interface, under Manual/ flow path, set **"Inject valve"** to **"Inject"** before taking out the syringe. Take out the syringe and fill the syringe with 180–200 μl of protein sample (*see* **Note 10**). Insert the syringe needle into the fill port. Set **"Inject valve"**

to "**load**", and then slowly inject the sample into the loop. Leave the syringe in position during the entire run. Go to the System Control interface, click "**End**" on the operation tool bar. Ensure that 1 mL of buffer volume is used to empty the sample loop in order to load all of the sample onto the column.

4. Start a run. Right click the "**Run**" button on the top tool bar to select a method for the run. Several dialog windows will be displayed sequentially, on which the user can review/edit the method settings and set up how the results file should be generated. Click "**Start**" on the last page of the dialog window to initiate the run.

5. Monitor the run. The gel filtration progress can be monitored in the **System Control** module. The ongoing experiment can be pulsed or ended at any time by clicking the "**Pulse**" or "**End**" button on the top tool bar.

3.3 Create the Standard Curve and Estimate the Molecular Weight

Studies [9, 10] have shown that, for globular proteins, their partition coefficients, K_{av}, are related to their corresponding molecular weights in a sigmoidal fashion on gel filtration columns. Therefore, a calibration curve, also called a selectivity curve, can be created by plotting a group of known proteins' partition coefficients against the logarithm of their molecular weight. Other elution parameters, such as V_e, V_e/V_o, K_d, have been used in the calibration curve, however the use of K_{av} is recommended because it is insensitive to errors that may be generated by column preparation and dimension variations.

An unknown protein sample's molecular weight can be estimated according to its elution volume using a calibration curve. It is important to make sure that the calibration curve's linear range covers the size of the protein of interest since only the linear region of the plot is reliable [10]. When gel filtration is used to detect or confirm protein complex formation, the protein complex's molecular weight is usually considered as the sum of each individual protein. However, when the complex exhibits an irregular shape, the molecular weight cannot be simply extrapolated from the calibration curve. For example, an elongated protein complex will behave like a bigger molecular on the gel chromatogram than it actually is.

1. Determine the void volume (V_o). Make 2 mg/ml blue dextran in running buffer (*see* **Note 11**).

2. Prepare the protein standards. Dissolve 3.5 mg of Albumin bovine serum, 1 mg of Carbonic Anhydrase and 1 mg of aprotinin in 500 μl of running buffer (*see* **Note 12**).

3. Equilibrate the column using running buffer and run the protein standards as described in Subheading 3.2.

4. Calculate K_{av} values for each protein maker.

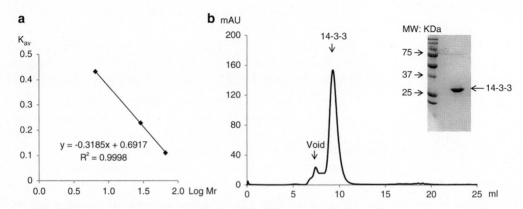

Fig. 5 (a) A standard curve was created based on the experimentally determined K_{av} of the standard proteins. (b) Chromatogram of the protein sample. The experimental V_e is 9.39 mL. Therefore, the estimated molecular weight of the protein sample is 61.74 kDa, indicating a dimer form of the protein. The collected fractions were pooled and run on 12.5 % SDS-PAGE, the protein monomer was observed on the gel at approximately 30 kDa

5. Plot the K_{av} values (Y axis) versus the logarithm of the corresponding molecular weights (X axis). Use linear fitting to generate the equation $K_{av} = a\log(\mathrm{MW}) + b$, and obtain the values for a and b (Fig. 5a).

6. Calculate the molecular weight of the protein sample based on the calibration curve once its K_{av} has been obtained experimentally. Plug the experimental values for $K_{av(\mathrm{sample})}$, a and b in the equation in Fig. 5a, and obtain the MW value for the protein sample.

7. Analyze 1–2 µg of each fraction collected by SDS-PAGE to confirm the size of the protein (Fig. 5b).

4 Notes

1. An HPLC or FPLC system is commonly used for gel filtration; in this chapter we use an FPLC system as an example.

2. There are many commercially available gel filtration columns, choose an appropriate column according to experimental requirements. For high-resolution results, Superdex medium is the first choice. All columns should be well sealed at both ends and stored in 20 % ethanol at 4 °C.

3. Usually 0.1–0.8 ml of protein sample (0.5–4 % column volume) is used for one run. Use the right size sample loop according to the protein sample volume.

4. Running buffers should contain at least 50–100 mM salt to prevent protein aggregation. However, the maximum salt concentration should be empirically determined since high salt concentration can break down protein complexes.

5. The user can control the UNICORN displays and manage files through the UNICORN manager module. Method Editor is for creating new methods. System Control is used to monitor and control runs in real time. The results for each run will be displayed and evaluated in Evaluation module.

6. Correct maximum pressure setup is the most critical parameter to protect columns. Ensure that the back-pressure over the column does not exceed the maximum recommended pressure (1.8 M for a Superdex 75). The recommended maximum pressure for a selected column can be found in manufacturer's instruction sheet. Pressure could gradually build up with increasing system usage. One of the common reasons is that impurities build up on the in-line filter. The filter can be cleaned by soaking it in 0.1 N NaOH in a beaker with gentle agitation overnight and rinsing it thoroughly with filtered deionized water. In addition, the columns have to be washed completely after each use. Once column clogging is observed, a stringent cleaning operation has to be performed immediately according to manufacturer's instructions.

7. The process of setting up a fraction collector varies depending on the type of collector used.

8. The column equilibration step can be incorporated into the method through the Method Wizard so that the equilibration will automatically end and switch to the elution step. However, it may be more advantageous to have the equilibration run separately through manual control, by which the user can alter the equilibration volume as needed to make sure that the column is perfectly equilibrated before starting a run.

9. If partial filling is used (*see* **Note 10**), at the end of the last loop wash, do not remove the syringe from the fill port, otherwise the buffer inside the sample loop will drain out.

10. There are two ways to load the sample, partial filling and complete filling. When a high recovery rate is required, partial filling is used. The maximum sample volume loaded cannot exceed 50 % loop volume, 1/3 of the loop volume is usually loaded. Make sure the sample loop is completely filled with buffer before sample loading. Complete filling requires about 2–3 loop volumes of the sample to achieve 95 % maximum loop volume. For example, in order to fill a 0.5 mL sample loop with 95 % protein sample without diluting it, 1–1.5 mL of protein sample is required to be injected into the loop. When complete filling is used, a buffer volume at least five times the sample loop volume should be used to flush the loop during emptying the loop.

11. 5 % glycerol can be contained in the running buffer to increase the density of the solution, but it is optional.

12. If more than three protein markers are used, dissolving all markers together may cause unassignable peaks because BSA may form a complex with other protein markers. It is advisable to prepare multiple protein marker groups, 2–3 proteins for each group.

References

1. Ackers GK (ed) (1975) The proteins, vol 1. Molecular sieve methods of analysis. Academic, New York, NY

2. Cooper TG (ed) (1977) Gel permeation chromatography. The tools of biochemistry. Wiley, New York

3. Freydell EJ, van der Wielen LA, Eppink MH et al (2010) Size-exclusion chromatographic protein refolding: fundamentals, modeling and operation. J Chromatogr A 1217:7723–7737

4. Hall D, Huang L (2012) On the use of size exclusion chromatography for the resolution of mixed amyloid aggregate distributions: I. Equilibrium partition models. Anal Biochem 426:69–85

5. Monti M, Cozzolino M, Cozzolino F et al (2009) Puzzle of protein complexes in vivo: a present and future challenge for functional proteomics. Expert Rev Proteomics 6:159–169

6. Pedro-Roig L, Camacho M, Bonete MJ (2013) Regulation of ammonium assimilation in *Haloferax mediterranei*: interaction between glutamine synthetase and two GlnK proteins. Biochim Biophys Acta 1834:16–23

7. Kuo WY, Huang CH, Liu AC et al (2013) CHAPERONIN 20 mediates iron superoxide dismutase (FeSOD) activity independent of its co-chaperonin role in Arabidopsis chloroplasts. New Phytol 197:99–110

8. Wechsler DS, Dang CV (1992) Opposite orientations of DNA bending by c-Myc and Max. Proc Natl Acad Sci U S A 89:7635–7639

9. Porath J, Flodin P (1959) Gel filtration: a method for desalting and group separation. Nature 183:1657–1659

10. Abate C, Luk D, Gentz R et al (1990) Expression and purification of the leucine zipper and DNA-binding domains of Fos and Jun: both Fos and Jun contact DNA directly. Proc Natl Acad Sci U S A 87:1032–1036

Chapter 14

Using Light Scattering to Determine the Stoichiometry of Protein Complexes

Jeremy Mogridge

Abstract

The stoichiometry of a protein complex can be calculated from an accurate measurement of the complex's molecular weight. Multiangle laser light scattering in combination with size exclusion chromatography and interferometric refractometry provides a powerful means for determining the molecular weights of proteins and protein complexes. In contrast to conventional size exclusion chromatography and analytical centrifugation, measurements do not rely on the use of molecular weight standards and are not affected by the shape of the proteins. The technique is based on the direct relationship between the amount of light scattered by a protein in solution, and the product of its concentration and molecular weight. A typical experimental configuration includes a size exclusion column to fractionate the sample, a light scattering detector to measure scattered light, and an interferometric refractometer to measure protein concentration. The determination of the molecular weight of an anthrax toxin complex will be used to illustrate how multiangle laser light scattering can be used to determine the stoichiometry of protein complexes.

Key words Light scattering, Stoichiometry, Molecular weight, Size exclusion chromatography, Interferometric refractometer, Protein complex

1 Introduction

Elucidating the stoichiometry of a protein complex can provide insights into its structure and its molecular mechanisms of action. Stoichiometry can be calculated from the measurement of the molecular weight of a complex, but this can be challenging if there are several copies of more than one protein species. Techniques such as conventional size exclusion chromatography and analytical centrifugation have been used to estimate molecular weights, but these techniques are limited by their reliance on protein standards and by the influence of protein shape on the measurement. In contrast, the combination of size exclusion chromatography, laser light scattering, and interferometric refractometry is an absolute method for the determination of molecular weight [1, 2]. It is based on the theory that the amount of light scattered by a protein

Cheryl L. Meyerkord and Haian Fu (eds.), *Protein-Protein Interactions: Methods and Applications*, Methods in Molecular Biology, vol. 1278, DOI 10.1007/978-1-4939-2425-7_14, © Springer Science+Business Media New York 2015

in solution is directly proportional to the product of the protein's concentration and molecular mass. Scattered light is measured by a light scattering (LS) detector, protein concentration is measured by an interferometric refractometer (IR), and the molecular weight is calculated by computer software. The accuracy of this technique allows for precise molecular weight measurements that are required to calculate the stoichiometry of a multi-subunit protein complex. The determination of the stoichiometry of the *Bacillus anthracis* edema toxin complex will be used as an example to describe how multiangle laser light scattering can be used to probe protein structure [3]. Edema toxin consists of two proteins, edema factor (EF) and protective antigen (PA), that are secreted from *Bacillus anthracis* and assemble on the surface of mammalian cells [4]. After PA is cleaved on the cell surface into two fragments, the remaining cell-associated fragment, PA_{63}, oligomerizes into heptamers that can bind edema factor to form a toxic complex. The molecular weight of the saturated complex measured in solution corresponded to a stoichiometry of three molecules of EF per PA_{63} heptamer.

2   Materials

1. HPLC.
2. DAWN EOS 18-angle light scattering detector (Wyatt Technology).
3. OPTILAB DSP interferometric refractometer (Wyatt Technology).
4. ASTRA software (Wyatt Technology).
5. Superdex 200 HR 10/30 column.
6. Column buffer: 20 mM Tris–HCl, pH 8.2, 200 mM NaCl.
7. 0.2 μm filter units.
8. 0.02 μm Anotop 10 filters.

3   Methods

3.1 System Setup

Before measurements are taken, the LS detector must be calibrated with a pure solvent that has a known Rayleigh ratio, such as toluene. Calibration relates the voltage detected by the 90° detector to the measured intensity of scattered light. If the LS detector was calibrated by the manufacturer, it needs to be re-calibrated only if any changes to the detector that affects the 90° signal, such as realignment of the laser, have been made.

The HPLC is connected in series to the LS detector and the IR. The IR should be placed after the LS detector to avoid subjecting the IR to high back-pressure. Connections should be made with tubing of low inner diameter $(0.01'')$ to minimize the dead volume between the instruments.

A solution of 4 mg/mL of bovine serum albumin (an isotropic scatterer) in column buffer can be injected onto the size exclusion column and used to normalize the LS detector, a process that relates the measured voltages at each detector to the 90° detector in order to compensate for slight differences in electronic gain among the detectors. Normalization must be performed each time a different solvent is used.

A *dn/dc* (refractive index increment) value of 0.185 mL/g is used for non-glycosylated proteins in the molecular weight calculations made by ASTRA software.

3.2 Preparation

1. Turn on the IR approximately 12 h before measurements are taken. Set the temperature to approximately 10 °C above room temperature.

2. Turn on the LS detector 2 h before measurements are taken.

3. Connect the size exclusion column to the HPLC (*see* **Note 1**) and run buffer through the column while the light scattering detector is warming (*see* **Note 2**).

4. Once the column has pre-equilibrated with buffer (approximately 2 column volumes), connect the waste line from the HPLC to the LS detector and the output from the LS detector to the IR (*see* **Note 3**).

5. Equilibrate the LS detector and IR with column buffer (*see* **Note 4**).

6. Flow buffer simultaneously through the reference and sample cells of the IR and set the baseline to zero. After the reference has been set, program the IR so that the column buffer bypasses the reference cell.

3.3 Injecting a Sample

Do not stop the HPLC pumps after the equilibration step or between measurements because starting and stopping the pumps may cause particles from the column to dislodge and interfere with the measurements. Start data collection by ASTRA software and inject each component of the protein complex (EF and PA_{63}; *see* **Note 5**) individually and then a mixture of the proteins (*see* **Note 6**). The results of a chromatography run in which a molar excess of EF was mixed with PA_{63} are shown in Fig. 1.

3.4 Analyzing the Results

The results can be analyzed using a computer program, such as ASTRA.

1. Set the baseline voltages of the signals detected by the LS detector and IR.

2. When selecting values for the light scattering and refractive index signals, use data relating to the middle of the protein

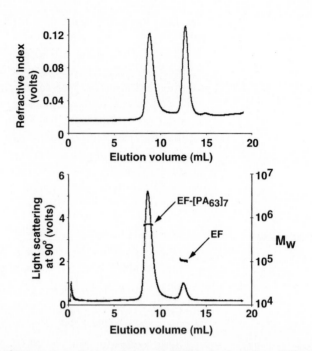

Fig. 1 Multiangle laser light scattering of EF:[PA$_{63}$]$_7$. A mixture (120 µL) of EF (260 µg) and [PA$_{63}$]$_7$ (93 µg; approximately twofold molar excess of EF over PA$_{63}$ monomer) was chromatographed over a Superdex 200 size exclusion column, which was connected to a light scattering detector (*lower panel*) and an interferometric refractometer (*upper panel*). The values of molecular mass determined in volume increments across each peak are shown (*arrows*). Reprinted with permission from *Biochemistry* 41: 1079–1082. Copyright 2002 American Chemical Society

Table 1
Molecular mass determinations using multiangle laser light scattering

	Measured molecular mass (kDa)	Theoretical molecular mass (kDa)
EF	93 ± 0.3	91
[PA$_{63}$]$_7$	460 ± 3	444
EF:[PA$_{63}$]$_7$	720 ± 20	717[a]

Reprinted with permission from *Biochemistry* 41: 1079–1082. Copyright 2002 American Chemical Society
[a]Theoretical value for saturated complex is based on a stoichiometry of three molecules of EF per PA$_{63}$ heptamer

peak where the signal-to-noise ratio is the highest (signal-to-noise ratio should be at least 2).

3. Use ASTRA software to calculate the molecular weights (Table 1; *see* **Note 7**).

4   Notes

1. The HPLC system should have a pulse dampener since small fluctuations in flow rate can affect the measurements. The Superdex 200 size exclusion column was chosen for this study because EF and the EF:$[PA_{63}]_7$ complex are separated by this column.

2. It is important to make fresh buffer with high quality reagents. Filter the buffer through a 0.2 μm filter and degas it thoroughly before use. Pre-equilibrate the column at the flow-rate (e.g., 0.5 mL/min) that will be used to take measurements to avoid changes in flow-rate that might cause particles to dislodge from the column.

3. New columns should be equilibrated for approximately 48 h before measurements are taken. The LS detector and IR are stored in methanol, which should be replaced with filtered water before introducing buffer (solvents that are injected directly into the LS detector or IR should be filtered through a 0.02 μm filter).

4. Equilibrate the LS detector and IR until the signals have stabilized. Baseline noise for the LS detector should be in the 10–20 mV range. The lower angle LS detectors (the Wyatt EOS LS detector has 18 detectors) will have a higher level of noise, but these detectors can be turned off without adversely affecting the measurements. Detectors 8–18 were used in this study.

5. EF and $[PA_{63}]_7$ were purified as described previously [5, 6]. Small amounts of aggregated protein in a sample can interfere with the light scattering measurements because aggregates are very efficient at scattering light. An aggregate will produce a high signal from the LS detector and a very low signal from the IR. Recently prepared protein samples that have not been frozen may contain less aggregate than a sample that has been frozen and thawed.

6. The maximum injected volume is determined by the column specifications (250 μL for the Superdex 200 column), but smaller volumes can be used. The amount of protein required to make an accurate measurement depends on the molecular weight of the protein—more sample is required for low molecular weight proteins (~50 kDa) than for high molecular weight proteins (>100 kDa) because high molecular weight proteins scatter more light. Accurate measurements of the amino terminal domain of *Bacillus anthracis* lethal factor (30 kDa) were made with 200 μg of protein [3]. Accurate measurements of transferrin (75 kDa) and ovalbumin (43 kDa) were made with less than 10 μg of protein [7]. An additional consideration

when measuring protein complexes is that the proteins should be at a high enough concentration to ensure that they fully associate. In general, the protein concentration should exceed the dissociation constant of the interaction by approximately tenfold (taking into account that the protein may be diluted by severalfold on the size exclusion column). The ratios of the proteins in the samples should be varied to ensure that saturation has occurred. The amount of one protein is held constant and increasing amounts of the other protein are added. Saturation has been achieved when the addition of protein does not change the molecular weight of the complex peak and a free protein peak is observed.

7. ASTRA software calculates molecular weights at multiple points across the selected section of an elution peak and uses this data to calculate the average molecular weight of the protein. These points can be displayed in a graph of molecular weight versus elution volume to aid in the analysis of the experiment (Fig. 1, lower panel). Calculated molecular weights across a homogeneous peak will be similar at each point, but these points may form a frowning pattern, which is indicative of peak broadening between the LS detector and IR (see EF: $[PA_{63}]_7$ complex peak in Fig. 1). Slight peak broadening will not adversely affect the molecular weight measurements. If the protein peak is not homogeneous, one might observe a linear pattern with higher molecular weights at lower elution volumes and lower molecular weights at higher elution volumes. If this is the case, a size exclusion column that better separates the species can be used.

References

1. Wyatt PJ (1993) Light scattering and the absolute characterization of macromolecules. Anal Chim Acta 272:1–40
2. Gell DA, Grant RP, Mackay JP (2012) The detection and quantitation of protein oligomerization. Adv Exp Med Biol 747:19–41
3. Mogridge J, Cunningham K, Collier RJ (2002) Stoichiometry of anthrax toxin complexes. Biochemistry 41:1079–1082
4. Young JA, Collier RJ (2007) Anthrax toxin: receptor binding, internalization, pore formation, and translocation. Annu Rev Biochem 76:243–265
5. Zhao J, Milne JC, Collier RJ (1995) Effect of anthrax toxin's lethal factor on ion channels formed by the protective antigen. J Biol Chem 270:18626–18630
6. Miller CJ, Elliott JL, Collier RJ (1999) Anthrax protective antigen: prepore-to-pore conversion. Biochemistry 38:10432–10441
7. Folta-Stogniew E, Williams KR (1999) Determination of molecular masses of proteins in solution: implementation of an HPLC size exclusion chromatography and laser light scattering service in a core laboratory. J Biomol Tech 10:51–63

Chapter 15

Circular Dichroism (CD) Analyses of Protein-Protein Interactions

Norma J. Greenfield

Abstract

Circular dichroism (CD) spectroscopy is a useful technique for studying protein-protein interactions in solution. CD in the far ultraviolet region (178–260 nm) arises from the amides of the protein backbone and is sensitive to the conformation of the protein. Thus, CD can determine whether there are changes in the conformation of proteins when they interact. Changes in the conformation of the protein complexes as a function of temperature or added denaturants, compared to the individual proteins, can be used to determine binding constants. CD bands in the near ultraviolet (350–260 nm) and visible regions arise from aromatic amino acid side chains and prosthetic groups. There are often changes in these regions when proteins bind to each other. Because CD is a quantitative technique, these changes are directly proportional to the amount of the protein–protein complexes formed and thus also can be used to estimate binding constants.

   Key words Conformation, Secondary structure, Binding constants, Thermodynamics of folding

1 Introduction

Circular dichroism (CD) is a valuable spectroscopic technique for studying protein-protein interactions in solution. There are many review articles in the literature on the theory and general applications of CD [1–8], so this chapter will focus on how CD can be used to follow protein-protein interactions. CD is best known as a method of determining the secondary structure of proteins in solution. It thus can be used to determine whether there are changes in protein conformation when proteins interact with each other. However, more importantly, CD can be used to estimate binding affinities for protein interactions. Since CD is a quantitative technique, changes in CD spectra are directly proportional to the concentrations of the protein–protein complexes formed, and these changes can be used to estimate binding constants. One can monitor changes in the "intrinsic" CD arising from the conformational transitions of the amide backbone and amino acid side chains or

Cheryl L. Meyerkord and Haian Fu (eds.), *Protein-Protein Interactions: Methods and Applications*, Methods in Molecular Biology, vol. 1278, DOI 10.1007/978-1-4939-2425-7_15, © Springer Science+Business Media New York 2015

"extrinsic" CD of the proteins arising from bound ligands or prosthetic groups. Changes in the thermodynamics of folding upon protein-protein interaction can also be used to determine binding constants.

1.1 Spectra of the Peptide Backbone

The intrinsic CD spectrum of a protein, arising from its amide backbone, is sensitive to the secondary structure of the protein. The amide backbone bonds absorb in the far UV. Most commercial CD machines can collect far UV spectra between 260 and 178 nm. To a first approximation, increases in negative ellipticity at 222 and 208 nm and positive ellipticity at 193 nm usually indicate an increase in α-helical content, while increases in a single negative band near 218 nm and positive band at 195 nm indicates an increase in β-structure. Figure 1a shows typical CD curves obtained for a polypeptide, Poly-L-lysine in the 100 % α-helical, 100 % β-sheet, and 100 % denatured states [9]. Figure 1b illustrates the increase in α-helix that occurs when a peptide from titin, a Z-repeat, interacts with a fragment of the C-terminal domain of α-actinin [10]. Figure 1c illustrates the increase in β-structure when two cell cycle control proteins, Arf and Hdm2, interact [11]. Both of these proteins are mainly disordered in the absence of the interactions.

1.2 Side Chain and Extrinsic CD Spectra

Non-backbone intrinsic CD bands are due to the chromophores of the aromatic amino acids, tryptophan, tyrosine, and phenylalanine and disulfide bonds. Extrinsic bands are due to noncovalently bound chromophores such as hemes, aromatic cofactors, and colored metal ions. Figure 2a illustrates a change in the aromatic spectrum of the catalytic domain of a cAMP-dependent protein kinase when it binds to an artificial peptide substrate [12], and Fig. 2b illustrates the change in the heme spectrum of cytochrome C when it binds to cytochrome C oxidase [13].

1.3 Binding Constants

There are four methods that can be used to extract binding (or dissociation) constants from CD data. These include:

1. Direct titration of one protein with another.
2. Serial dilution of a protein complex.
3. Changes in susceptibility to denaturants or osmolytes of the protein complex compared to the unbound components.
4. Changes in thermal stability of the complex compared to the unbound components.

If formation of the complex changes the protein secondary structure or causes changes in the environment of optically active aromatic or prosthetic groups of one or both of the interacting proteins, one protein can be titrated by the other. The change in CD upon binding is directly proportional to the amount of complex formed, and thus the binding constant can be directly

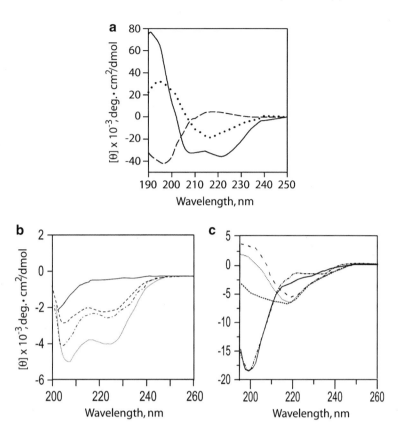

Fig. 1 Examples of conformational changes upon protein-protein interactions. (**a**) CD spectra of a polypeptide, poly-L-Lysine, showing spectra typical for the 100 % α-helical (*solid line*), 100 % antiparallel β-sheet (*dotted line*), and 100 % denatured (*short dashed line*) conformations. Data from Greenfield and Fasman [9] used with permission from the American Chemical Society. (**b**) Gain of signal for the Act-EF34-Zr7 complex (complex of peptides of titin, Zr7, and actinin, EF34) upon complex formation. The spectra of the isolated components, Zr7 (*solid line*), Act-EF34 (*short dashed line*), and the complex (*dotted line*), are shown. The sum of the spectra of the two isolated components is also reported for comparison (*dotted dashed line*). Data from Joseph et al. [10] used with permission from the American Chemical Society. (**c**) The complex of two cell cycle control proteins, Arf and Hdm2, are comprised of β-strands and are thermally stable. CD spectra for Hdm2$_{210-304}$ alone (*solid line*) and mArfN$_{37}$ alone (*dashed dotted line*) and Hdm2$_{210-304}$ mixed with 1 (*short dashed line*), 1.7 (*dotted line*), and 2 (*long dashed line*) molar equivalents of mArfN37. Data of Bothner et al. [11] used with permission from Elsevier

determined. However, even if there are no conformational changes, if the interactions cause a change in stability, one can determine the thermodynamics of folding of the individual and complexed proteins and determine the binding constant by assuming that all of the change in free energy of folding is due to the binding. The binding constant can then be determined using the relationship

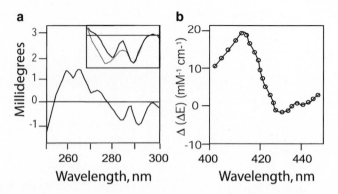

Fig. 2 Examples of changes in the CD spectra arising from nonamide chromophores upon protein-protein interactions. (**a**) Near UV CD of the catalytic subunit of adenosine cyclic 5′ monophosphate-dependent protein kinase. (*Inset*) CD spectrum of the catalytic subunit from 275 to 300 nm (*solid line*) enzyme alone; (*dashed line*) enzyme plus 250 μM Kemptide, a synthetic peptide substrate (buffer plus Kemptide base line subtracted). Data from Reed and Kinzel [12] used with permission from the American Chemical Society. (**b**) CD difference spectra of cytochrome c upon binding to cytochrome c oxidase. Data from Michel et al. [13] used with permission from the American Chemical Society

$k = \exp(-\Delta\Delta G/RT)$, where $\Delta\Delta G$ is the difference in free energy of folding of the mixture versus the unmixed proteins, R is the gas constant, T is the temperature, and k is the association constant.

2 Materials

2.1 Proteins

CD can be performed on proteins ranging in concentration from 0.005 to 2 mg/ml depending on the path lengths of the cuvettes used (see below). CD depends on concentration of chromophores, rather than molar protein concentrations. CD is usually reported either in units of mean residue ellipticity, $[\theta]$, in deg cm^2/dmol, or the difference between the absorption of left- and right-handed circularly polarized light, ΔE. $\Delta E = [\theta]/3298$. For quick calculations, $[\theta] = \theta \times \mathrm{MRW}/c$, where θ is the ellipticity in millidegrees, MRW is the mean residue weight (the molecular weight of the protein divided by the number of amino acids), and c is the protein concentration in mg/ml. When one is studying the aromatic spectrum of a protein, the results are usually reported as molar ellipticity (mean residue ellipticity times the number of residues).

2.2 Buffers

For the most precise estimates of protein secondary structure from CD spectra, it is important to collect data to low wavelengths; 178 nm has been recommended [3, 14, 15]. In this case one should use buffers with very low absorption, such as 10 mM potassium phosphate. If salt is necessary to stabilize the protein, 100 mM KF or 50 mM K_2SO_4 is preferred as they are relatively transparent. NaCl should be avoided. However, for most purposes an adequate estimate of protein conformation can be obtained using data collected between 260 and 200 nm [5]. For these measurements, one can use buffers such as phosphate buffered saline or 20 mM Tris–HCl, or 2 mM Hepes containing low concentrations of EDTA or EGTA, e.g., 1 mM, and 1 mM Dithiothreitol and up to 500 mM NaCl if necessary.

For folding experiments, e.g., those monitored at 222 nm, almost any buffers can be utilized provided that the total absorption of the sample (protein plus buffer) does not exceed 1 at the wavelength of interest.

2.3 Cuvettes

CD measurements are usually performed in cuvettes with pathlengths of 0.1 or 1 cm. Both standard rectangular and cylindrical cuvettes are available for CD machines, and the choice depends on the instrument utilized. For measurements between 260 and 200 nm for concentrations ranging from 0.005 to 0.02 mg/ml, one needs 2–3 ml of solution, as the samples will be studied in 1-cm cuvettes, with a total volume capacity of 3.5 ml. For measurements between 178 and 260 nm at higher concentrations, e.g., 0.05–1 mg/ml, one would use 0.1- to 0.2-cm cuvettes with volume capacities of 0.3 and 0.6 ml, respectively. If one needs to use samples with high concentrations or buffers with high absorbance, 0.1-cm cuvettes are available.

For measurements in the near UV, for following changes in aromatic groups of proteins, or visible regions, e.g., following the CD of prosthetic groups such as flavins or hemes, higher concentrations of 1–2 mg/ml in a 1-cm cell are generally necessary, depending on the number and asymmetry of the chromophores.

2.4 Instrumentation

The following is a list of companies that currently produce commercially available CD machines. All are adequate to perform thermal denaturations and titration experiments. They are listed alphabetically:

1. Applied Photophysics Ltd, 21 Mole Business Park, Leatherhead, Surrey KT22 7BA, UK.

2. Aviv Biomedical, 750 Vassar 750 Vassar Avenue—Suite 2 Lakewood, NJ 08701-6929, USA.

3. JASCO Inc., 28600 Mary's Court, Easton, MD 21601, USA.

4. Olis, Inc., 130 Conway Drive Suites A, B & C, Bogart, GA 30622, USA.

3 Methods

3.1 Analyzing Changes in Conformation Accompanying Protein-Protein Interactions

While changes at 222 or 218 nm can give an estimate of increases in α-helical or β-structure content upon protein complex formation, more precise estimates of changes in structure accompanying protein-protein interactions can be made utilizing computer programs. Computer programs for analyzing CD spectra are not available commercially, but many are available from their authors without charge (*see* **Note 1** for sources of software).

The use of the various methods to determine protein conformation from CD spectra has been reviewed, with a detailed comparison of the advantages and disadvantages of each [5], so only a brief description of each method is given below.

3.1.1 Constrained Multilinear Regression (LINCOMB)

Constrained linear regression programs deconvolute CD spectra into component spectra characteristic of specific secondary structures [9] by fitting the data to a set of reference spectra using the methods of least squares. Standards include spectra of polypeptides with known conformations [9, 16] or reference spectra deconvoluted from the spectra of proteins with known conformations by the method of least squares [17, 18] or the convex constraint algorithm [19, 20]. The sum of the fractions of each component must equal 1. The LINCOMB program of Perczel et al. [19] uses constrained linear regression to determine the percent α-helix, β-sheet, and β-turns contributing to a spectrum. The method uses invariant standards, so it is useful for quantifying changes in each type of secondary structure when one protein is titrated with another. The best method is to fit the mean residue ellipticities of the individual proteins and the protein–protein complexes and calculate the number of residues in each conformation by multiplying the fraction of each conformation by the number of residues. One can then subtract the number of residues in each conformation of the individual proteins from the number of residues in the complex to estimate the total number of residues that undergo conformational changes upon complex formation.

3.1.2 Nonconstrained Multilinear Regression

Nonconstrained multilinear regression [16] analyzes the CD spectrum of a protein by fitting the spectrum to those of standards using the method of least squares. The sum of the components is not constrained to equal 100 %. The method is independent of the intensity of the spectrum, so it is the only method that can be used if the protein concentration isn't known precisely. The method can be used to analyze difference spectra, obtained when the sum of the spectra of the unmixed components is subtracted from the spectrum of the mixture, because one doesn't need to know the number of residues that change conformation. The method is much less

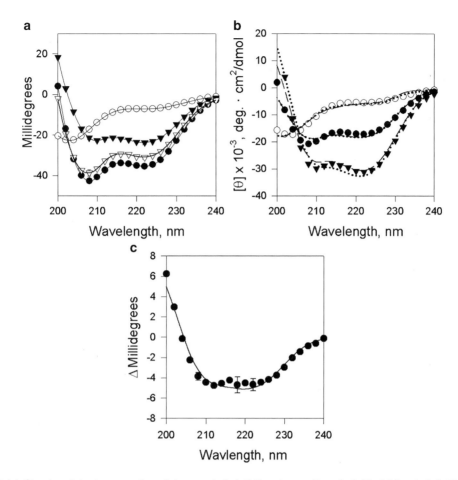

Fig. 3 (**a**) Circular dichroism spectra of (*open circles*) E-Tmod$_{1-130}$, (*inverted filled triangles*) AcTM1bZip, (*inverted open triangles*) the sum of the spectra of E-Tmod1-130 and AcTM1bZip, and (*filled circles*) the spectrum of the mixture E-Tmod1-130 and AcTM1bZip. Data are the average and standard error of four to seven measurements. (**b**) The mean residue ellipticity of (*open circles*) Tmod1$_{1-130}$, (*inverted filled triangles*) AcTM1bZip, and (*filled circles*) the mixture of the two peptides. The data are fit by (*dashed dotted line*) constrained and (*dotted line*) nonconstrained least squares fits. (**c**) (*Filled circles*) Difference between the spectrum of the mixture of the components and the sum of the unmixed components in (**a**), (*Solid line*) nonconstrained least squares fit of the data. The difference data are the average of four measurements. The data are from Greenfield and Fowler [21] used with permission from Elsevier

accurate than the constrained method, however, because one loses the information of the intensities of each CD band.

Figure 3 illustrates the change in CD that occurs when a peptide fragment of E-tropomodulin, now called tropomodulin 1 (E-Tmod$_{1-130}$), binds to a chimeric two-chained coiled-coil peptide containing the N-terminus of a short α-tropomyosin (AcTM1b-Zip). Tmod$_{1-130}$ contains residues 1–130 of chicken E-Tmod plus 12 residues of a glutathione S-transferase linker [21]. The conformations of the proteins were fit by both constrained and nonconstrained multilinear regression using the spectrum of α-tropomyosin [22] as the reference for a coiled-coil α-helix and the data of Brahms and Brahms [16] for the spectra of peptides in the β-pleated sheet, β-turn, and disordered conformations. Figure 3a shows the raw spectra of the two unmixed peptides, the sum of the spectra of the unmixed peptides, and the spectrum of the complex. Figure 3b shows the mean residue ellipticities of the individual proteins and the complex and the best fits using constrained and nonconstrained multilinear regression. Figure 3c shows the difference spectrum of the complex minus the sum of unmixed peptides and the best fit to the curves using non-multilinear regression with the same peptide references. The percentage of the conformations for the nonconstrained fits were normalized to a sum of 100 %. The results are compared in Table 1. The results for the TM1bZip peptide are also compared to the structures determined by X-ray crystallography [24] and NMR [25]. The number of residues of each protein fragment in each conformation was calculated by multiplying the percentage of each conformation by the number of residues of each protein fragment, 74 for AcTM1bZip, 142 for E-Tmod (Tmod1$_{1-130}$), and 216 for the complex. When the tropomyosin peptide binds to the tropomodulin fragment, there is an increase in ellipticity of about 10 % relative to the unmixed peptides. When the data in Fig. 3b is analyzed using LINCOMB, the results suggested that most of the change is due to an increase in helical content of about 22 residues, ~10 %, and a small increase in β-sheet residues, 3, ~1 %. When the nonconstrained least squares fitting program MLR was used, the results suggested that there was an almost equal increase in both helical and β-sheet content of approximately 6 %, 14 and 13 residues, respectively. When the difference spectrum (Fig. 3c) was analyzed directly, the results suggested that the increase in ellipticity upon binding was 44 % due to an increase in helix and 55 % increase in β-sheet. NMR studies of a similar E-Tmod peptide containing residues 1–92 [23] show only that 12 residues [21–35] of tropomodulin are α-helical in the absence of tropomyosin peptides, but residues 1–35 become ordered when tropomodulin binds to the N-terminus of tropomyosin, an increase of 23 residues, in reasonable agreement with the calculations from the CD data. Mutation experiments suggest the TMod and tropomyosin peptides bind via formation of a multichain α-helical coiled coil, so the constrained least squares fits of the CD data are more accurate than the nonconstrained fits.

3.1.3 Ridge Regression (CONTIN)

CONTIN, developed by Provencher and Glöckner [26], fits the CD of unknown proteins by a linear combination of the spectra of a large database of proteins with known conformations. In this method, the contribution of each reference spectrum is kept small, unless it contributes to a good agreement between the theoretical best fit curve and the raw data. The method usually gives very good estimates of β-sheets and turns and gives good fits with polypeptides [5].

Table 1
Characterization of the secondary structural changes induced by AcTM1bZip-Tmod1$_{1-130}$ complex formation

					Number of residues in each conformation			
AcTm1bZip	**LINCOMB**	**MLR**	**X-ray**	**NMR**	**LINCOMB**	**MLR**	**X-ray[a]**	**NMR[b,c]**
α-Helix	90.53 %	73.44 %	81.00 %	79.00 %	67.0	54.3	60	60
β-Sheet	9.47 %	26.56 %	0.00 %	0.00 %	7.0	19.7	0	0
β-Turn	7.36E-09	0.00 %	0.00 %	0.00 %	0.0	0.0	0	0
Remainder	8.08E-08	0.00 %	19.00 %	21.00 %	0.0	0.0	14	16
Total sum	100.00 %	100.00 %	100.00 %	100.00 %	74.0	74.0	74	76
E-Tmod$_{1-130}$	**LINCOMB**	**MLR**			**LINCOMB**	**MLR**		
α-Helix	19.42 %	26.16 %			27.6	37.1		
β-Sheet	18.66 %	8.50 %			26.5	12.1		
β-turn	2.54E-10	0.00 %			0.0	0.0		
Remainder	61.92 %	65.34 %			87.9	92.8		
Total sum	100.00 %	100.00 %			142.0	142.0		
Mixture	**LINCOMB**	**MLR**			**LINCOMB**	**MLR**		
α-Helix	53.99 %	48.81 %			116.6	105.4		
β-Sheet	16.72 %	20.56 %			36.1	44.4		
β-turn	0.00 %	0.00 %			0.0	0.0		
Remainder	29.29 %	30.63 %			63.3	66.2		
Total sum	100.00 %	100.00 %			216.0	216.0		

(continued)

Table 1
(continued)

AcTm1bZip					Number of residues in each conformation			
	Change in number of residues				% Change in number of residues			
	LINCOMB	MLR	X-ray	NMR	LINCOMB	MLR	X-ray[a]	NMR[b,c]
	LINCOMB	MLR			LINCOMB	MLR		
α-Helix	22.0	13.9			10.21 %	6.45 %		
β-Sheet	2.6	12.7			1.21 %	5.88 %		
β-Turn	0.0	0.0			0.00 %	0.00 %		
Remainder	−24.7	−26.6			−11.42 %	−12.33 %		
Total sum	0.0	0.0			0.00 %	0.00 %		

MLR fit of difference spectra

	Raw	Normalized to 100 %
α-Helix	1.11E-04	44.04 %
β-Sheet	1.40E-04	55.96 %
β-Turn	5.64E-13	0.00 %
Remainder	4.77E-13	0.00 %
Total sum	2.51E-04	100.00 %

[a]Meshcheryakov et al. [24]
[b]Greenfield et al. [25]
[c]The NMR studies were performed on a peptide were a glycine residue replaced the N-terminal acetyl group

3.1.4 Singular Value Decomposition (SVD, VARSLC, CDSSTR, SELCON)

The application of singular value decomposition to determine the secondary structure of proteins by comparing their spectra to the spectra of a large number of proteins with known structure was first developed by Hennessey and Johnson [14]. This method extracts basis curves, with unique shapes, from a set of spectra of proteins with known structures. Each basis curve is then related to a mixture of secondary structures, which are then used to analyze the conformation of unknown proteins. The method is excellent for estimating the α-helical content of proteins but is poor for sheets and turns unless the data is collected to very low wavelengths, at least 184 nm [14].

Several newer programs have improved the method of singular value decomposition by selecting references that have spectra that closely match the protein of interest. They include VARSLC [15], SELCON [27–31], and CDSSTR [32]. These give good estimates of β-sheet and turns in proteins and work with data collected only to 200 nm but give poor fits to polypeptides with high β-structure content because reference sets for such compounds are not in their database [5].

3.1.5 Neural Network Programs (CDNN and K2D)

A neural network is an artificial intelligence program which can detect patterns and correlations in data. Two widely used programs are CDNN [33] and K2D [34]. A neural network is first trained using a set of known proteins so that the input of the CD at each wavelength results in the output of the correct secondary structure. The trained network is then used to analyze unknown proteins. The method works very well and the fits seem to be relatively independent of the wavelength range that is analyzed [5].

3.1.6 Convex Constraint Algorithm (CCA)

The CCA algorithm [19, 20, 35] deconvolutes a set of spectra into basis spectra that when recombined generate the entire data set with a minimum deviation between the original data set and reconstructed curves. It is very useful for determining whether there are intermediate states in thermal and denaturant induced unfolding. The method has also been used to estimate protein conformation but is poorer than least squares, SVD, or neural net analyses [5].

3.1.7 Recommendations

1. For determination of globular protein conformation in solution and evaluating the conformation of protein–protein complexes: SELCON, CDSSTR, CDNN, CONTIN, and K2D.

2. For evaluating conformational changes upon protein-protein interactions: LINCOMB

3. For deconvoluting sets of CD spectra, e.g., to follow the effects of binding, denaturants, ligands, or changes in temperature on protein and peptide conformation: the CCA algorithm.

Table 2 compares the results when the CONTIN, SELCON, K2D, and MLR programs were used to analyze the spectra (Fig. 4)

Table 2
Secondary structure analysis of fragments and cleaved and uncleaved thioredoxin

Protein	Method	α-Helix	β-Sheet	Turns	Other
N-terminal fragment	K2D	6.0	29.0	[b]	65.0
	CONTIN	4.0	14.0	10.0	72.0
	MLR[a]	2.5	28.7	10.3	58.5
C-terminal fragment	K2D	2.0	16.0	[b]	82.0
	CONTIN	3.0	11.0	1.0	86.0
	MLR[a]	0.0	15.7	0.0	84.3
Peptide complex	K2D	30.0	20.0	[b]	50.0
	CONTIN	20.0	40.0	7.0	34.0
	MLR[c,d]	22.8	23.7	9.2	21.8, 22.5[d]
	SELCON	24.66	21.5	16.9	22.5
Thioredoxin	K2D	27.0	25.0	[b]	48.0
	CONTIN	18.0	37.0	8.0	37.0
	MLR[c,d]	31.8	22.9	7.2	16.9, 21.2[d]
	SELCON	28.4	24.8	24.4	21.4
	X-ray data[e]	40.0	28.3	14.0	17.8
	X-ray data[f,g]	33.3	27.8	9.3	26.8
	X-ray data[h]	37.0	27.7	14.8	20.5
	NMR data[g,i]	36.1	25.9	9.2	28.8

Data from Georgescu et al. [36] used with permission from the American Chemical Society
[a]Polypeptide ref. [16]
[b]Undetermined
[c]Polypeptide ref. [19]
[d]Estimate of aromatic and disulfide bond contributions
[e]Structure calculations of Georgescu et al. [36] based on molecule A of the PDB file 2TRX [37]
[f]Based on molecule A of the PDB file 2TRX [37]
[g]Percentage of each structure calculated using MolMol [46]
[h]Based on PDB file 1SRX [42]
[i]Based on PDB file 1XOA [45]

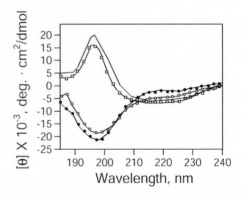

Fig. 4 Circular dichroism of an artificial heterodimer of cleaved peptides of thioredoxin from *E. coli* versus the intact protein. The protein concentration was 20 μM in KPi. Far-UV CD spectra of the isolated N- and C-fragments (*open circles*) N; (*filled circles*), C; (*open squares*), cleaved; and (*solid line*) uncleaved Trx. Data from Georgescu et al. [36] used with permission from the American Chemical Society

of N- and C-terminal fragments of thioredoxin, the binary complex of the two thioredoxin fragments, and the intact protein [36]. All of the methods gave reasonable fits to the data. Note that the amount of helix in the intact protein appeared to be underestimated by all of the programs. This may have occurred because approximately 50 % of the helix in thioredoxin is 3–10 helix rather than α-helix [37].

3.2 Determining Protein–Protein Association Constants Using CD Data

3.2.1 Determination of Binding Constants by Direct Titration of One Protein by Another at Constant Temperature (Isothermal Titrations)

When two proteins interact, there are often changes in either intrinsic or extrinsic circular dichroism. CD is a quantitative technique and the change in CD, relative to the ellipticity of the unmixed components, is directly proportional to the amount of complex formed. The change can then be used to determine the association or dissociation constant of the complex. Ideally, when performing titrations, the concentration of the peptide being titrated, e.g., protein A, should remain constant. Thus, one should titrate protein A with aliquots of a solution containing the starting concentration of A and a large excess of protein B. In this case one does not have to correct for dilution of protein A when determining the change in ellipticity upon complex formation.

When one titrates A with B to form a complex AB, the association constant is K.

$$K = [AB]/[A][B] = [AB]/(([A_o] - [AB])([B_o] - [AB])) \quad (1)$$

where $[A_o]$ and $[B_o]$ are the initial concentrations of A and B and $[AB]$ is the amount of complex formed. The saturation fraction is s.

$$s = [AB]/[A_o] \quad (2)$$

If there is a change in the intrinsic or extrinsic CD accompanying binding, the change, $\Delta[\theta]$, is proportional to the amount of complex formed.

$$\Delta[\theta] = \varepsilon[AB] \quad (3)$$

where ε is the proportionality constant. When all of protein A is bound by protein B,

$$\Delta[\theta]_{max} = \varepsilon[A_o] \text{ and } s = \Delta[\theta]/\Delta[\theta]_{max} = \Delta[\theta]/\varepsilon[A_o] \quad (4)$$

Weak Binding Constants

If the binding of one protein to the other is relatively weak, under conditions where the dissociation constant, $1/k$, of the complex is more than 100-fold the concentration of A, one can assume that the concentration of unbound B is approximately equal to the total amount of B added to A. In this case it is easy to determine the binding constant and one can use linear plots to determine $\Delta[\theta]_{max}$ and K_D.

For example one can fit the data to the Scatchard equation [38] where

$$\Delta[\theta]/[A_o][B] = k(\varepsilon - \Delta[\theta]/[A_o]) \qquad (5)$$

One plots $\Delta[\theta]/[B]$ versus $\Delta[\theta]$. The y intercept $= k\varepsilon[A_o]$ and the slope $= -k$.

The Scatchard equation [38] described above works well for studying interactions where there is one or multiple equivalent binding sites of one protein for another. However, when there are multiple sites, and the binding is cooperative, it is much more difficult to estimate binding constants from spectroscopic data, and usually only apparent constants can be estimated. The Hill equation [39] may be used to fit the change in CD as a function of titrant in a phenomenological fashion to give an estimate of the apparent binding affinity. Here, the first protein A is titrated with a second protein B, where $[B_o]$ is the total concentration of added protein giving an observed change in ellipticity $\Delta[\theta]$.

$$\Delta[\theta] = \left(\Delta[\theta]_{max}k^h[B_o]^h / \left(1 + k^h[B_o]^h\right)\right) + C \qquad (6)$$

where k is the binding constant, $[\theta]_{max}$ is the maximal changes in ellipticity upon protein-protein interaction, h is a constant describing the apparent cooperativity of the interaction, and C is a constant correcting for baseline offsets. When $h = 1$, the binding is noncooperative.

It should be noted that in the case of weak binding, titrations are only feasible using data obtained at wavelengths where one component has very low absorption and ellipticity, such as in the case when an aromatic protein or a protein containing a heme is titrated with a protein with low ellipticity in the aromatic (near UV) or Soret band (visible) regions. These methods are very useful, however, for measuring the interactions of small molecules, such as chromophoric ligands or metals, to proteins.

Tight Binding Constants

If the dissociation constant is close to the concentrations of the proteins being studied, one cannot assume that the free concentration of the added protein, B, is the same as the total protein. One must correct the concentration of the added protein B for the amount that is bound. In this case $[B] = [B_o]-[AB]$. From Eq. 4, the fraction of protein A which is bound

$$[A]_{bound} = [A_O]\left(\Delta[\theta]/\Delta[\theta]_{max}\right) \qquad (7)$$

Therefore, the free concentration of B, $[B]$ at any added concentration of B_o

$$[B] = [B_o] - [A_o](\Delta[\theta]/\Delta[\theta]_{max}) \tag{8}$$

Solving for the change of circular dichroism, $\Delta[\theta]$, at any added concentration of protein B, $[B_o]$,

$$\Delta[\theta] = \Delta[\theta]_{max}\Big\{((1 + k[B_o] + k[A_o])/2\,k[A_o])$$
$$- \Big(((1 + k[B_o] + k[A_o])/2k[A_o])^2\Big) - ([B_o]/[A_o])^{1/2}\Big\} \tag{9}$$

as described by Engel [40], who originally developed the equation for determining binding constants of enzyme–ligand complexes from fluorescence titrations.

Equation 9 can be fit by many available commercial curve fitting programs, e.g., SigmaPlot, Microcal, and Origin among others, that find the best fits to nonlinear equations. To use these programs one inputs starting values of the parameters to be fit, here $\Delta CD_{max,}$ and k, and has the program find the values of A_{max} and k that best fit the data.

Figure 5 illustrates the binding of 130 residue fragments of E- and Sk-Tropomodulins (now called tropomodulin 1 and 4, respectively) to peptides containing the N-terminus of two tropomyosin isoforms [21], where the binding data were fit to Eq. 9.

Estimating Binding Constants When Association of the Proteins Is Coupled with Folding

When two unfolded or partially unfolded proteins fold upon binding to each other, a simple method of determining the dissociation constant is to follow the mean residue or molar ellipticity of a 1:1 mixture of the proteins as a function of concentration. In this case

$$[\theta]_{obs} = [\theta]_F + ([\theta]_U - [\theta]_F)\Big[\Big(-k_D + (k_D^2 + 4P_t k_D)^{1/2}\Big)/(2P_t)\Big] \tag{10}$$

where k_D is the dissociation constant of the two proteins, $[\theta]_{obs}$ is the ellipticity of the mixture of the proteins, at any total concentration, P_t, $[\theta]_F$ is the ellipticity of the fully folded complex, and $[\theta]_U$ is the sum of the ellipticity of the fully unfolded individual proteins. Figure 6 illustrates the use of this equation to determine the dissociation constant of two fragments of the immunoglobulin binding domain B1 of streptococcal protein G [41].

3.2.2 Estimating Protein–Protein Association Constants from the Thermodynamics of Folding

One can determine the thermodynamics of folding of protein–protein complexes by measuring the unfolding of the complex either by thermal denaturation or by chemical denaturation using guanidine-HCl or urea. In some cases, the proteins are only folded when associated. In these cases, determining the ΔG of unfolding directly gives the dissociation constants of the protein–protein complexes. In other cases the proteins are both folded, but

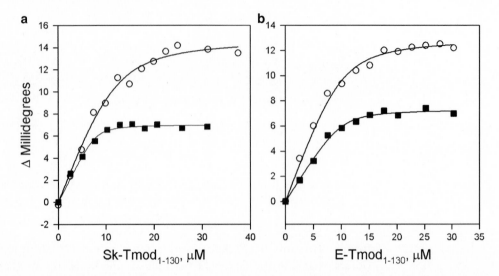

Fig. 5 Increase in ellipticity at 222 nm, 30 °C, when fragments of SK- and E-tropomodulin, (**a**) Sk-Tmod$_{1-130}$ and (**b**) E-Tmod$_{1-130}$ bind to chimeric peptides containing the N-termini of long and short α-tropomyosins, (*open circles*) AcTM1aZip and (*filled squares*) AcTM1bZip, respectively [21]. 10 nmol of the TMZip peptides in 0.5 ml of 100 mM NaCl, 10 mM sodium phosphate, pH 6.5 in 2 mm pathlength cells were titrated with concentrated solutions of the Tmod peptides. The ellipticity changes were corrected for the ellipticity of the Tmod peptides alone and for dilution. The data were fit (solid line) to the equation

$$\Delta[\theta] = \Delta[\theta]_{max}\left\{\left((1 + k[B_o] + k[A_o])/2\,k[A_o]\right) - \left(((1 + k[B_o] + k[A_o])/2k[A_o])^2 - [B_o]/[A_o]\right)^{1/2}\right\}$$

using the Levenberg–Marquardt algorithm [43] implemented in the commercial program SigmaPlot to yield dissociation constants for Sk-TMod$_{1-130}$ for AcTM1aZip and AcTM1bZip of 1.2 ± 0.9 and 0.3 ± 0.3 μM, respectively and for E-TMod$_{1-130}$ for AcTM1aZip and AcTM1bZip of 2.8 ± 1.2 and 0.3 ± 0.2 μM, respectively. Data from Greenfield and Fowler [21] and used with permission from Elsevier

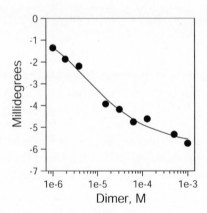

Fig. 6 Association of two complementary fragments of the immunoglobulin binding domain B1 of streptococcal protein G, PGB1(1–40) and PGB1(41–56), evaluated by molecular ellipticity at 222 nm. Equimolar concentrations of the fragments were dissolved in 50 mM phosphate buffer (pH 5.5) at 298 K. The *filled circles* and *solid line* show observed data and best curve fitted to the equation

$$[\theta]_{obs} = [\theta]_F + ([\theta]_U - [\theta]_F)\left[\left(-k_D + (k_D^2 + 4P_tk_D)^{0.5}\right)/(2P_t)\right]$$ to determine the apparent dissociation

constant of the two fragments (Kapp = 9×10^{-6} M). Data from Honda et al. [41]. Used with permission from the American Chemical Society

complex formation increases their stability. In this case, the free energy of binding can be determined by subtracting the free energies of unfolding of the individual proteins from the free energy of unfolding of the complex. The difference, $\Delta\Delta G$, can then be used to calculate the dissociation constant of the complex.

Chemical Induced Unfolding

In the simplest case, two proteins A and B form a complex with 1:1 stoichiometry. The isolated proteins are unfolded but fold to form the complex. When this complex is treated with denaturants, the complex (native form) dissociates to give the disordered proteins. At any concentration of denaturant, [D], the association constant is:

$$k = [AB]/[A][B] = [P_t]\alpha/([P_t](1-\alpha))^2 \qquad (11)$$

where $[P_t]$ is the total concentration of complex and α is the fraction folded. The fraction folded expressed in terms of the protein concentration and association constant is:

$$\alpha = \left\{ 2P_t^2 k + P_t - \left[(-2P_t^2 k - P_t)^2 - 4P_t^4 k^2 \right]^{1/2} \right\} / (2P_t^2 k) \qquad (12)$$

If one assumes that the binding constant is equal to the folding constant, then ΔG can be calculated from the equation:

$$\Delta G = -RT\ln k \qquad (13)$$

It has been shown for many systems that the ΔG of folding of a protein is linearly dependent on the concentration of denaturant, see, e.g., Santoro and Bolen [42].

$$\Delta G_D = \Delta G_o + m[D] \qquad (14)$$

where ΔG_D is the free energy of folding in the presence of denaturant, ΔG_o is the free energy of folding in the absence of the denaturant, [D] is the concentration of denaturant, and m is the slope.

To calculate the binding constant, one then studies the unfolding of the protein complex by denaturants using several different protein concentrations, by following the changes in molar or mean residue ellipticity $\Delta[\theta]_{obs}$ as a function of denaturant concentration. The fraction folded can be calculated at any concentration of denaturant using the equation

$$\alpha = \left([\theta]_{obs} - ([\theta]_U - m_U[D]) \right) / \left(([\theta]_F - m_F[D]) - ([\theta]_U - m_U[D]) \right) \qquad (15)$$

where $[\theta]_U$ and $[\theta]_F$ are the ellipticities of the fully unfolded and folded complexes, respectively, and m_U and m_F are any necessary linear baseline corrections for the change in ellipticity of the fully unfolded and folded peptides with denaturant observed before and after the folding transition.

The k at any denaturant concentration can be calculated using Eqs. 11 and 15, and ΔG can be calculated from Eq. 13 and can be plotted as a function of total denaturant concentration. By linear extrapolation ΔG can be determined at zero denaturant, and the dissociation constant K_D for the protein–protein complex can then be calculated by rearranging Eq. 13, $K_D = \exp(-\Delta G_o/RT)$. One also can fit the data directly using nonlinear regression techniques [41]. Protocols for the direct fitting of data for the unfolding of a heterodimeric peptide are given below (*see* **Note 2**). Georgescu et al. [36] used this method to study the guanidine denaturation of the complex of the two fragments of thioredoxin, shown in Fig. 7. The free energy of folding of the protein complex was -10 ± 0.4 kcal/mol, which agreed with the value determined by direct titration of one protein with another of -9.8 ± 0.2 kcal/mol.

Thermally Induced Unfolding

A change in either the intrinsic or extrinsic CD as a function of temperature can be used to determine the van't Hoff enthalpy of folding since changes in ellipticity are directly proportional to the changes in concentration of the native and denatured forms.

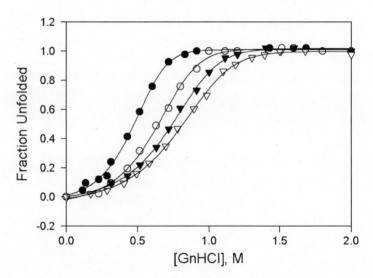

Fig. 7 Chemical denaturation of cleaved Thioredoxin (Trx). Concentration dependence of GnHCl-induced unfolding of cleaved Trx at (*filled circles*) 2.33 μM, (*open circles*) 5 μM, (*filled inverted triangles*) 15 μM, and (*open inverted triangles*) 25 μM in KPi at 20 °C. The fraction of unfolded cleaved Trx was calculated according to a two-state transition process using the intrinsic fluorescence at 350 nm and ellipticity at 222 nm. The data represent the average of three to five experiments. Redrawn from data in Georgescu et al. [36] and used with permission from the American Chemical Society

When a protein is fully folded $[\theta]_{obs} = [\theta]_F$ and when it is fully unfolded $[\theta]_{obs} = [\theta]_U$. In the simplest case of a monomeric protein, the equilibrium constant of folding, $k =$ folded/unfolded. If we define α as the fraction folded at a given temperature, T, then

$$k = \alpha/(1 - \alpha) \tag{16}$$

$$\Delta G = -RT\ln k \tag{17}$$

where ΔG is the free energy of folding and R is the gas constant. From the Gibbs–Helmholtz equation:

$$\Delta G = \Delta H + (T - T_M)\Delta Cp - T[\Delta S + \Delta Cp(\ln(T/T_M)] \tag{18}$$

where ΔH is the van't Hoff enthalpy of folding and ΔS is the entropy of folding, T_M is the observed midpoint of the thermal transition, and ΔCp is the change in heat capacity for the transition. At the T_M, $k = 1$, therefore $\Delta G = 0$ and $\Delta S = \Delta H/T_M$. Rearranging these equations one obtains:

$$\Delta G = \Delta H(1 - T/T_M) - \Delta Cp[(T_M - T) + T(\ln(T/T_M))] \tag{19}$$

$$k = \exp(-\Delta G/RT) \tag{20}$$

$$\alpha = k/(1 + k) \tag{21}$$

$$[\theta]_{obs} = ([\theta]_F - [\theta]_U)\alpha + [\theta]_U \tag{22}$$

To calculate the values of ΔH and T_M that best describe the folding curve, initial values of ΔH, ΔCp, T_M, $[\theta]_F$, and $[\theta]_U$ are estimated, and Eq. 22 is fitted to the experimentally observed values of the change in ellipticity as a function of temperature, by a curve fitting routine such as the Levenberg–Marquardt algorithm [43]. Since at $k = 1$, $\Delta G = 0$, the entropy of folding can therefore be calculated using Eq. 18 where $\Delta S = \Delta H/T_M$. The free energy of folding at any other temperature can then be calculated using Eq. 19. Similar equations can be used to estimate the thermodynamics of folding of proteins and peptides that undergo folded multimer to unfolded monomer transitions (*see* **Note 3**). The association constant, k, of protein–protein complexes can be obtained by subtracting the free energies of folding of the sum of the uncomplexed proteins from the free energy of folding of the complex and then $k = \exp(-\Delta\Delta G/nRT)$.

Figure 8 illustrates the increased stability when a fragment of E-tropomodulin binds to a peptide containing the N-terminus of

258   Norma J. Greenfield

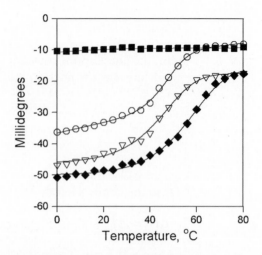

Fig. 8 Thermal denaturation curves for (*filled squares*) E-Tmod$_{1\text{-}130}$, (*open circles*) AcTM1bZip, (*open inverted triangles*) the sum of the spectra of E-Tmod$_{1\text{-}130}$ and AcTM1bZip, and (*filled diamonds*) the spectrum of the mixture E-Tmod$_{1\text{-}130}$ and AcTM1bZip. The lines are the best fits of the data assuming two-state transitions between folded and unfolded conformations. The data were used to estimate dissociation constants for the complex of the peptides of 0.2 ± 0.2 μM. Data from Greenfield and Fowler [21] and used with permission from Elsevier

an isoform of α-tropomyosin. The change in stability was used to estimate a dissociation constant of 0.2 ± 0.2 μM, which is in experimental agreement with the direct titration method of 0.3 ± 0.2 μM (duplicate measurements).

4 Notes

1. Sources of programs for determining secondary structure from circular dichroism data:
   Circular dichroism software can be downloaded from the following sites:

   (a) Variety of Microsoft DOS programs [5] including SEL-CON, VARSLC, K2D, MLR, LINCOMB CONTIN, and CCA algorithm are available from: http://www.nature.com/nprot/journal/v1/n6/suppinfo/nprot.2006.202_S1.html. These programs can be run under Windows 7 in a DOS emulation program: http://www.dosbox.com/download.php?main=1. In addition, equations for evaluating constrained and nonconstrained least squares fits using programs that use the Levenberg–Marquardt algorithm [43] such as Sigmaplot are available at this site.

(b) DICHROPROT [47]: A variety of programs including MLR, VARSLC, SELCON, and CONTIN in a Windows environment: http://dicroprot-pbil.ibcp.fr/.

(c) CDPro [31]: Versions of CONTIN, SELCON, and CDSSTR which calculate regular and distorted helices and sheets: http://lamar.colostate.edu/~sreeram/CDPro/main.html.

(d) Online Data Analysis is available at the Dichroweb web site: http://dichroweb.cryst.bbk.ac.uk/html/home.shtml. Programs include versions of CONTIN, SELCON, CDSSTR, VARSLC, and K2D.

(e) The latest versions of the individual programs: CCA algorithm [19, 20, 35]: A version that runs under 32 bit or better versions of Windows is available from http://www.chem.elte.hu/departments/jimre/.

CDNN [33]: CDNN is available from Applied Photophysics on request. http://www.photophysics.com/tutorials/cdnn-secondary-structure-analysis.

CDSSTR [32]: http://biochem.science.oregonstate.edu/dr-johnsons-software-download-instructions (This program must be run in a command box, CMD.exe, when running Windows.)

CONTIN [26]: http://s-provencher.com/index.shtml and http://lamar.colostate.edu/~sreeram/CD.

K2D [34]: http://www.embl-heidelberg.de/%7Eandrade/k2d.html. Two updated versions named SOMCD [48] and K2D3 [49] are available respectively from http://geneura.ugr.es/cgi-bin/somcd/index.cgi and http://www.ogic.ca/projects/k2d3/.

LINCOMB [19]: A version of Lincomb that runs under most versions of Windows is available from http://www.chem.elte.hu/departments/jimre/. (Note when using the current version of the program, the reference set should be in the form of a comma separated file with the wavelength as the first file, separated by two commas from the columns of the conformational references, e.g., nm, helix, beta sheet, beta turn, random.)

SELCON [31]: http://lamar.colostate.edu/~sreeram/CD.

CAPITO [50] is a new web server-based analysis and plotting tool for circular dichroism data. It is available at http://capito.nmr.fli-leibniz.de.

2. Protocol for fitting the unfolding of a heterodimeric protein complex by denaturants from CD data.

One can fit the change in ellipticity, $[\theta]_{obs}$, as a function of denaturant concentration, $[D]$, to obtain the free energy of folding of a protein complex and the binding affinity in the absence of denaturant. The variables are [D] and $[\theta]_{obs}$. The known constant is the protein concentration, $[P_t]$. The parameters to be fit are ΔG_o, m, $[\theta]_F$, $[\theta]_U$, and m_U and m_F, where ΔG_o is the free energy of folding in the absence of denaturant, m is the constant relating the linear change in free energy to the concentration of denaturant, $[\theta]_F$ and $[\theta]_U$ are the ellipticities of the fully folded and unfolded protein complexes, respectively, and m_F and m_U are the slope describing any linear changes in ellipticity of the folded and unfolded proteins before and after the folding transition.

The equations that are used in the curve fitting procedure for the unfolding of a dimer are:

$$\Delta G = \Delta G_o + m[\mathrm{D}]$$

$$k = \exp(-\Delta G / 1.987\ T)$$

$$\alpha = \left\{2\,P_t^2 k + P_t - \mathrm{sqrt}\left(\left(-2P_t^2 k - P_t\right)^2 - 4P_t^4 k^2\right)\right\} / \left(2P_t^2 k\right)$$

$$[\theta]\mathrm{obs} = \alpha\left([\theta]_F - m_F[\mathrm{D}] - [\theta]_U - m_U[\mathrm{D}]\right) + [\theta]_U$$

In these equations, P_t is the total concentration of the protein, input as a constant, ΔG is the free energy of folding in the presence of denaturant, T is the absolute temperature at which the unfolding experiment is conducted, and 1.987 is the gas constant in kcal/mol, k is the folding/association constant, and α is the fraction folded. The difference between the calculated change in ellipticity and the observed change in ellipticity is minimized to give the best values of the fitting parameters using a nonlinear least squares curve fitting programs. Such curve fitting algorithms are usually included in most commercially available graphics plotting programs, e.g., Kaleidagraph, Origin, or Sigmaplot. If one knows the values of $[\theta]_F$, $[\theta]_U$, and m_U and m_F, they may be input as constants rather than parameters to be determined. Once ΔG_o is determined, the binding constant is calculated using the equation $k = \exp(-\Delta G / RT)$. A curve fitting routine in SigmaPlot format is available: http://www.nature.com/nprot/journal/v1/n6/suppinfo/nprot.2006.229_S1.html.

3. Protocols for fitting circular dichroism data to determine the thermodynamics of folding of monomers, dimers, heterodimers, trimers, and heterotrimers from changes in CD as a function of temperature using nonlinear least squares analysis programs.

Unless otherwise noted, in all of these equations, ΔG is the Gibbs free energy of folding, ΔH is the van't Hoff enthalpy of folding, ΔCp is the change in heat capacity going from the folded to the unfolded form, T is the absolute temperature, Kelvin, T_M is the temperature where $k = 1$, $[\theta]_{obs}$ is the observed ellipticity at any temperature T, α is the fraction folded, $[\theta]_F$ is the ellipticity of the fully folded protein and $[\theta]_U$ is the ellipticity of the unfolded protein, and p is the concentration of the folded complex. The parameters to be fit are ΔH, ΔCp, T_M, $[\theta]_F$, and $[\theta]_U$. One inputs the protein concentration, P_t, as a constant. For complicated fits, to reduce the number of parameters, ΔCp may be set to 0 and the values of $[\theta]_F$ and $[\theta]_U$ may be set as constants rather than parameters to be minimized if they are known.

(a) Monomers

$$\Delta G = \Delta H(1 - T/T_M) - \Delta Cp[(T_M - T) + T(\ln(T/T_M))]$$

$$K = \exp(-\Delta G/RT)$$

$$\alpha = K/(1 + K)$$

$$[\theta]_{obs} = ([\theta]_F - [\theta]_U)\alpha + [\theta]_U$$

(b) Homodimers

$$\Delta G = \Delta H + (\Delta Cp(T - T_m)) \\ - T((\Delta H/T_m) + (\Delta Cp(\ln(T/T_m))))$$

$$k = \exp(-\Delta G/(1.987\ T))$$

$$a = 4kP_t^2$$

$$b = -8kP_t^2 - P_t$$

$$c = 4kP_t^2$$

$$\alpha = \{-b - \text{sqrt}(b^2 - 4ac)\}/2a$$

$$[\theta]_{obs} = \alpha([\theta]_F - [\theta]_U) + [\theta]_U$$

(c) Heterodimers

Here the equations are set up in terms of unfolding and P_t is the total concentration of protein in terms of monomers, and the ΔH and ΔCp values are those of unfolding rather than folding.

$$\Delta G = \Delta H(1 - T/T_m) + Cp(T - T_m - T(\ln(T/T_m)) - 1.987 T\ln(P_t/4)$$

$$k_D = \exp(-\Delta G/1.987\ T)$$

$$a = 2\ k$$

$$b = -2P_t - 2\ k$$

$$c = P_t$$

$$\alpha = \{-b - \text{sqrt}(b^2 - 4\ ac)\}/2a$$

$$[\theta]_{obs} = \alpha([\theta]_F - [\theta]_U) + [\theta]_U$$

This expression is set up to evaluate the apparent T_m at $\alpha = 0.5$. To evaluate it at $k = 1$, the equation to calculate the free energy of folding simplifies to:

$$\Delta G = \Delta H(1 - T/T_m) + Cp(T - T_m - T(\ln(T/T_m))$$

Note that the equations give exactly the same T_M and enthalpy of folding if the unfolding data was analyzed as if the complex was a homodimer, above.

(d) Homotrimers

$$k = \exp\{(\Delta H/(1.987\ T)[(T/T_m) - 1)] - \ln(.75)P_t^2)\}$$

$$z = 3P_t^2$$

$$q = ((3kz) + 1)/(kz)$$

$$e = q - 3$$

$$d = q - 3$$

$$A = \left\{[(-1)(d/2)] + [(d^2/4) + (e^3/27)]^{(1/2)}\right\}^{(1/3)}$$

$$B = (-1)\left\{(d/2) + [(d^2/4) + (e^3/9)]^{(1/2)}\right\}^{(1/3)}$$

$$X = A + B$$

$$\alpha = X + 1$$

$$[\theta]_{obs} = \alpha([\theta]_F - [\theta]_U) + [\theta]_U$$

Note that in this treatment ΔCp is set at 0 and T_m is the observed midpoint of the unfolding transition where $\alpha = 0.5$.

(e) Heterotrimers

$$k = \exp\left((\Delta H/(1.987\ T))(T/T_m - 1)) - \ln\left(P_t^2\right)\right)$$

$$Q = -1/\left(12\ P_t^2 k\right)$$

$$R = -1/\left(8\ P_t^2 k\right)$$

$$A = -\left(R + \mathrm{sqrt}\left((R^2) - (Q^3)\right)\right)^{(1/3)}$$

$$B = Q/A$$

$$y = A + B$$

$$\alpha = 1 - y$$

$$[\theta]_{obs} = \alpha([\theta]_F - [\theta]_U) + [\theta]_U$$

Note that in this treatment ΔCp is set at 0 and T_m is the observed midpoint of the unfolding transition where $\alpha = 0.5$. The equations give the same enthalpy and T_M of folding as the equations for analyzing the unfolding of a homotrimer, above [21]. All of these routines are available in SigmaPlot format from: http://www.nature.com/nprot/journal/v1/n6/suppinfo/nprot.2006.204_S1.html.

References

1. Adler AJ, Greenfield NJ, Fasman GD (1973) Circular dichroism and optical rotatory dispersion of proteins and polypeptides. Methods Enzymol 27:675–735

2. Johnson WC Jr (1988) Secondary structure of proteins through circular dichroism spectroscopy. Annu Rev Biophys Biophys Chem 17:145–166

3. Johnson WC Jr (1990) Protein secondary structure and circular dichroism: a practical guide. Proteins 7:205–214

4. Sreerama S, Woody RW (1995) Computation and analysis of protein circular dichroism spectra. Methods Enzymol 383(Part D):318–351

5. Greenfield NJ (1996) Methods to estimate the conformation of proteins and polypeptides

from circular dichroism data. Anal Biochem 235:1–10

6. Greenfield NJ (2006) Using circular dichroism spectra to estimate protein secondary structure. Nat Protoc 1:2876–2890

7. Greenfield NJ (2006) Using circular dichroism collected as a function of temperature to determine the thermodynamics of protein unfolding and binding interactions. Nat Protoc 1:2527–2535

8. Greenfield NJ (2006) Determination of the folding of proteins as a function of denaturants, osmolytes or ligands using circular dichroism. Nat Protoc 1:2733–2741

9. Greenfield N, Fasman GD (1969) Computed circular dichroism spectra for the evaluation of protein conformation. Biochemistry 8:4108–4116

10. Joseph C, Stier G, O'Brien R, Politou AS, Atkinson RA, Bianco A, Ladbury JE, Martin SR, Pastore A (2001) A structural characterization of the interactions between titin Z- repeats and the alpha-actinin C-terminal domain. Biochemistry 40:4957–4965

11. Bothner B, Lewis WS, DiGiammarino EL, Weber JD, Bothner SJ, Kriwacki RW (2001) Defining the molecular basis of Arf and Hdm2 interactions. J Mol Biol 314:263–277

12. Reed J, Kinzel V (1984) Near- and far-ultraviolet circular dichroism of the catalytic subunit of adenosine cyclic 5′-monophosphate dependent protein kinase. Biochemistry 23:1357–1362

13. Michel B, Proudfoot AE, Wallace CJ, Bosshard HR (1989) The cytochrome c oxidase-cytochrome c complex: spectroscopic analysis of conformational changes in the protein–protein interaction domain. Biochemistry 28:456–462

14. Hennessey JP Jr, Johnson WC Jr (1981) Information content in the circular dichroism of proteins. Biochemistry 20:1085–1094

15. Manavalan P, Johnson WC Jr (1987) Variable selection method improves the prediction of protein secondary structure from circular dichroism spectra. Anal Biochem 167:76–85

16. Brahms S, Brahms J (1980) Determination of protein secondary structure in solution by vacuum ultraviolet circular dichroism. J Mol Biol 138:149–178

17. Saxena VP, Wetlaufer DB (1971) A new basis for interpreting the circular dichroic spectra of proteins. Proc Natl Acad Sci U S A 68:969–972

18. Chang CT, Wu C-SC, Yang JT (1978) Circular dichroic analysis of protein conformation: inclusion of β-turns. Anal Biochem 91:13–31

19. Perczel A, Park K, Fasman GD (1992) Analysis of the circular dichroism spectrum of proteins using the convex constraint algorithm: a practical guide. Anal Biochem 203:83–93

20. Perczel A, Park K, Fasman GD (1992) Deconvolution of the circular dichroism spectra of proteins: the circular dichroism spectra of the antiparallel beta-sheet in proteins. Proteins 13:57–69

21. Greenfield NJ, Fowler VM (2002) Tropomyosin requires an intact N-terminal coiled coil to interact with tropomodulin. Biophys J 82:2580–2591

22. Greenfield NJ, DeGregori H (1993) Conformational intermediates in the folding of a coiled-coil model peptide of the N-terminus of tropomyosin and αα-tropomyosin. Protein Sci 2:1263–1273

23. Greenfield NJ, Kostyukova A, Hitchcock-DeGregori SE (2005) Structure and tropomyosin binding properties of the N-terminal capping domain of tropomodulin 1. Biophys J 88:372–383

24. Meshcheryakov VA, Krieger I, Kostyukova AS, Samatey FA (2011) Structure of a tropomyosin N-terminal fragment at 0.98 Å resolution. Acta Crystallogr D Biol Crystallogr 67:822–825

25. Greenfield NJ, Huang YJ, Palm T, Swapna GV, Monleon D, Montelione GT, Hitchcock-DeGregori SE (2001) Solution NMR structure and folding dynamics of the N terminus of a rat non-muscle alpha-tropomyosin in an engineered chimeric protein. J Mol Biol 312:833–847

26. Provencher SW, Glockner J (1981) Estimation of globular protein secondary structure from circular dichroism. Biochemistry 20:33–37

27. Sreerama N, Woody RW (2000) Estimation of protein secondary structure from circular dichroism spectra: comparison of CONTIN, SELCON, and CDSSTR methods with an expanded reference set. Anal Biochem 287:252–260

28. Sreerama N, Woody RW (1994) Poly(pro)II helices in globular proteins: identification and circular dichroic analysis. Biochemistry 33:10022–10025

29. Sreerama N, Woody RW (1994) Protein secondary structure from circular dichroism spectroscopy. Combining variable selection principle and cluster analysis with neural network, ridge regression and self-consistent methods. J Mol Biol 242:497–507

30. Sreerama N, Woody RW (1993) A self-consistent method for the analysis of protein

secondary structure from circular dichroism. Anal Biochem 209:32–44

31. Sreerama N, Woody RW (2000) Estimation of protein secondary structure from CD spectra: comparison of CONTIN, SELCON and CDSSTR methods with an expanded reference set. Anal Biochem 282:252–260

32. Johnson WC (1999) Analyzing protein circular dichroism spectra for accurate secondary structures. Proteins 35:307–312

33. Bohm G, Muhr R, Jaenicke R (1992) Quantitative analysis of protein far UV circular dichroism spectra by neural networks. Protein Eng 5:191–195

34. Andrade MA, Chacon P, Merelo JJ, Moran F (1993) Evaluation of secondary structure of proteins from UV circular dichroism spectra using an unsupervised learning neural network. Protein Eng 6:383–390

35. Perczel A, Hollosi M, Tusnady G, Fasman GD (1991) Convex constraint analysis: a natural deconvolution of circular dichroism curves of proteins. Protein Eng 4:669–679

36. Georgescu RE, Braswell EH, Zhu D, Tasayco ML (1999) Energetics of assembling an artificial heterodimer with an alpha/beta motif: cleaved versus uncleaved Escherichia coli thioredoxin. Biochemistry 38:13355–13366

37. Katti SK, LeMaster DM, Eklund H (1990) Crystal structure of thioredoxin from Escherichia coli at 1.68 A resolution. J Mol Biol 212:167–184

38. Scatchard G (1949) The attractions of proteins for small molecules and ions. Ann NY Acad Sci 51:660–672

39. Hill AV (1910) The possible effects of the aggregation of the molecules of haemoglobin on its dissociation curves. J Physiol (Lond) 40: iv–vii

40. Engel G (1974) Estimation of binding parameters of enzyme-ligand complex from fluorometric data by a curve fitting procedure: seryl-tRNA synthetase-tRNA Ser complex. Anal Biochem 61:184–191

41. Honda S, Kobayashi N, Munekata E, Uedaira H (1999) Fragment reconstitution of a small protein: folding energetics of the reconstituted immunoglobulin binding domain B1 of streptococcal protein G. Biochemistry 38:1203–1213

42. Santoro MM, Bolen DW (1988) Unfolding free energy changes determined by the linear extrapolation method. 1. Unfolding of phenylmethanesulfonyl alpha-chymotrypsin using different denaturants. Biochemistry 27:8063–8068

43. Marquardt DW (1963) An algorithm for the estimation of non-linear parameters. J Soc Indust Appl Math 11:431–441

44. Holmgren A, Soderberg BO, Eklund H, Branden CI (1975) Three-dimensional structure of Escherichia coli thioredoxin-S2 to 2.8 A resolution. Proc Natl Acad Sci U S A 72:2305–2309

45. Jeng MF, Campbell AP, Begley T, Holmgren A, Case DA, Wright PE, Dyson HJ (1994) High-resolution solution structures of oxidized and reduced Escherichia coli thioredoxin. Structure 2:853–868

46. Koradi R, Billeter M, Wüthrich K (1996) MOLMOL: a program for display and analysis of macromolecular structures. J Mol Graph 14 (51–5):29–32

47. Deléage G, Geourjon C (1993) An interactive graphic program for calculating the secondary structure content of proteins from circular dichroism spectrum. Comput Appl Biosci 9:197–199

48. Unneberg P, Merelo JJ, Chacón P, Morán F, SOMCD (2001) Method for evaluating protein secondary structure from UV circular dichroism spectra. Proteins 42:460–470

49. Louis-Jeune C, Andrade MA, Perez-Iratxetal C (2012) Prediction of protein secondary structure from circular dichroism using theoretically derived spectra. Proteins 80:374–381

50. Wiedemann C, Bellstedt P, Görlach M (2013) CAPITO – a web server-based analysis and plotting tool for circular dichroism data. Bioinformatics 29:1750–1757

Chapter 16

Protein-Protein Interaction Analysis by Nuclear Magnetic Resonance Spectroscopy

Peter M. Thompson, Moriah R. Beck, and Sharon L. Campbell

Abstract

Nuclear magnetic resonance (NMR) has continued to evolve as a powerful method, with an increase in the number of pulse sequences and techniques available to study protein-protein interactions. In this chapter, a straightforward method to map a protein–protein interface and design a structural model is described, using chemical shift perturbation, paramagnetic relaxation enhancement, and data-driven docking.

Key words Nuclear magnetic resonance (NMR), Protein-protein interaction, Chemical-shift perturbation (CSP), Paramagnetic relaxation enhancement (PRE), Docking

1 Introduction

Nuclear magnetic resonance (NMR) continues to be a valuable technique for scientists wishing to study protein-protein interactions. These noncovalent interactions drive much of biology through their presence in signaling pathways, molecular complexes, biomolecular processing, and protein degradation. NMR and X-ray crystallography are the two most common techniques for determining the structure of proteins, RNA, and DNA at a high resolution. However, NMR can be used for more than structure determination. NMR provides, in a site-specific manner, information on conformational dynamics of proteins and nucleic acids in solution, the protonation state of functional groups, and, of relevance to this article, both the identity of atoms involved in protein–ligand interactions and the kinetics of said interaction.

NMR is a unique tool for the study of protein-protein interactions primarily because it allows for the study of these interactions in physiological or near-physiological conditions. Additionally, NMR provides detailed, atomic-level information for even weak interactions ($K_d > 100$ μM). Furthermore, NMR is a very flexible technique, with many different pulse sequences and variations in

Cheryl L. Meyerkord and Haian Fu (eds.), *Protein-Protein Interactions: Methods and Applications*, Methods in Molecular Biology, vol. 1278, DOI 10.1007/978-1-4939-2425-7_16, © Springer Science+Business Media New York 2015

sample preparation to obtain information. Techniques now exist to study complexes in solution as large as 80–90 kDa. A complete discussion of these techniques and pulse sequences is outside the scope of this chapter, but the reader is directed to many reviews that discuss these in greater detail [1–7]. The techniques discussed here are chemical shift perturbation (CSP), paramagnetic relaxation enhancement (PRE), and molecular docking.

NMR is based upon the ability of magnetically active nuclei, when placed in a strong magnetic field, to absorb electromagnetic radiation at distinct frequencies. These frequencies are determined by the chemical environment of each nucleus and are affected by (1) atoms bound to the nucleus detected, (2) atoms within 5 Å of the nucleus (through space), and (3) the dynamics of the environment (how fast and how drastically it changes). The diversity of chemical environments within the protein results in dispersion of the frequencies absorbed by the nuclei, giving rise to unique signals (chemical shifts) for each nucleus. A variety of pulse sequences exist to assign the nucleus associated with each signal [8–10]. Once the chemical shifts of each peak have been assigned, changes in either location or intensity of a specific peak can be associated to a change in the chemical environment of the corresponding nucleus. This is the principle behind chemical shift perturbation (CSP), where binding of a ligand to the observed protein can/may change the chemical environment for nuclei at the interface. Thus, changes in the chemical shift or intensity of a peak upon titration of the binding partner suggest that the observed nucleus is near the site of interaction. This does not always hold true, as some proteins undergo a conformational change associated with binding. Therefore changes in chemical shift or intensity may be the result of the conformational change at a greater distance, instead of direct contact with the ligand.

The specific changes observed during a CSP experiment depend upon the rate of exchange between the free and bound states. If the exchange rate is slow on the NMR time scale ($k < \delta\nu$, where k is the rate of exchange and $\delta\nu$ is the change in chemical shift for the nucleus between the two states), then as the ligand is titrated in, the signal for a specific nucleus will be split into two separate peaks, one for the free state and one for the bound state. The relative intensities of the two peaks provide information on the relative populations of the protein. However, if the exchange rate is fast on the NMR time scale ($k > \delta\nu$), only one peak will be observed. In this case, the chemical shift for the complex will be an average of the chemical shifts of the free and bound states, weighted by their respective populations. The third type of exchange is intermediate ($k \approx \delta\nu$). In this case, as the ligand is titrated, the signal broadens out and eventually disappears. These exchange regimes, slow, intermediate, and fast, generally tend to correspond with tight ($K_d < 1$ μM), intermediate (1 μM $< K_d$

<100 µM), or weak binding ($K_d > 100$ µM), assuming diffusion-limited association rates. Therefore, CSP is most useful when the binding interaction is either in slow or fast exchange.

PRE is another NMR technique that involves changing the chemical environment of nuclei involved in an interaction. Paramagnetic ions have a free, unpaired electron and therefore a net spin, which generates a magnetic field [11]. When the paramagnetic ion is allowed to rotate freely (isotropically), it exposes nearby nuclei to a constantly changing magnetic field. This increases the rate of relaxation of the signal from the nucleus and is observed as a decreased intensity of the signal and a significantly increased linewidth. This relaxation enhancement is related to the distance between the ion and the magnetically susceptible nucleus by a factor of $1/r^6$, where r is the distance between the two atoms. Thus, if a paramagnetic ion is attached to a ligand, then nuclei in the protein near the ligand-binding site should have increased relaxation rates. More complex experiments can be run to measure the rate of relaxation and determine the distance from the paramagnetic ion, but it is more common to use PRE to identify nuclei with decreased signal intensity and define them as being close to the paramagnetic ion when the ligand is bound [12].

The final method we discuss is NMR data-driven docking, which couples computational methods with limited NMR restraints. When NMR restraints are too limited for structural determination or a high-resolution structure is not critical, docking procedures can be used to develop a testable ligand-binding model. There are many available docking programs, each with its own distinct methods for performing docking and scoring its structures [13–18]. We used the program HADDOCK, which is designed to allow for straightforward input of restraints from NMR data, as well as other methods [13]. HADDOCK refers to these inputs as Ambiguous Interaction Restraints (AIRs) and relies on them to generate docked structures from the structures of the individual binding partners. The structures are then scored, with those scoring the highest being selected for semi-flexible simulated annealing. Next, a four-step, iterative annealing process that allows for the movement of amino acid residues at the interaction interface is performed. Finally, the structures are refined in explicit solvent (water), clustered into groups, and scored. The scoring function is based upon physical parameters and experimental restraints, and can be customized for each project.

In this chapter, we show how a combination of CSP, PRE, and docking were used to generate a model for an interaction between an EF-hand domain of α-actinin-2 and a short region of the protein palladin [19]. These two proteins are involved in regulation of the actin cytoskeleton and have been implicated in cancer. A structural model of the interaction between the two provides a starting point for understanding the functional role of the interaction. While other restraints were used for model generation (including residual dipolar couplings (RDCs) and mutagenesis), we are presenting these three techniques because they are some of the most common and straightforward NMR methods to study protein-protein interactions.

2 Materials

1. ^{15}N-enriched human α-actinin 2 EF-hand residues 823–894 (referred to as Act-EF34, numbered 1–73 herein).

2. ^{15}NH$_4$Cl.

3. Unlabeled palladin peptide (residues 235–252 of human palladin, numbered 1–19 herein).

4. Palladin peptide with an N-terminal cysteine labeled with S-(2,2,5,5-tetramethyl-2,5-dihydro-1H-pyrrol-3-yl)methyl methanesulfonothioate (MTSL).

5. Varian Inova 600 or 700 MHz NMR spectrometer.

6. 5.0 mm Wilmad standard NMR sample tube.

7. NMRPipe/NMRDraw (Delaglio, NIH).

8. NMRViewJ (One Moon Scientific, Inc.).

9. NMR Buffer: 20 mM 3-(N-morpholino)propanesulfonic acid, pH 6.6, 10 mM NaCl, 2 mM tris(2-carboxyethyl)phosphine, 0.01 % NaN$_3$, and 10 % D$_2$O.

10. 1 M ascorbic acid.

3 Methods

3.1 NMR Sample Preparation

Act-EF34 was expressed as a Z-tagged, histidine-tagged (Chapter 23) fusion protein and purified by affinity chromatography [20]. Uniform ^{15}N-enrichment was obtained by growing *Escherichia coli* in M9 media (*see* **Note 1**) with ^{15}NH$_4$Cl as the sole nitrogen source. NMR samples were prepared in NMR buffer with an Act-EF34 concentration of 0.22 mM and a variable peptide concentration described in Subheadings 3.3 and 3.4. The NMR sample volume was 600 μL in a 5 mm OD precision tube, standard wall (*see* **Note 2**).

3.2 Spectroscopy

NMR experiments were performed on a Varian Inova 700-MHz NMR spectrometer with a triple-resonance cryoprobe with z-axis pulsed-field gradients at 27 °C. All experiments described here are two-dimensional ^{1}H–^{15}N heteronuclear single quantum coherence spectroscopy (HSQC) experiments [21]. Data were acquired with 2,048 × 512 complex data points and a ^{1}H spectral width of

10,000 Hz and a ^{15}N spectral width of 2,000 Hz (*see* **Note 3**). Processing and analysis of data were done with NMRPipe, NMRDraw [22], and NMRViewJ.

3.3 Chemical Shift Perturbation

Proton and ^{15}N resonance assignments of apo Act-EF34 were obtained from the Biological Magnetic Resonance Bank (BMRB), entry 17627. Two-dimensional ^{1}H–^{15}N HSQC data were acquired on the Act-EF34:palladin peptide (*see* **Note 4**) complex at molar ratios of 1:0, 1:0.1, 1:0.5, 1:1, 1:2, 1:3, 1:4, and 1:5, with 220 μM Act-EF34. At a ratio of 1:5, complete saturation of binding was achieved as peaks no longer shifted. To prevent dilution effects, samples were made by mixing the 1:0 and 1:5 samples in the appropriate ratio. The K_d of the interaction, determined by isothermal titration calorimetry (Part II Chapter 11) to be ~16 μM, suggested that exchange between the free and bound state would likely be intermediate or fast on the NMR time scale (*see* **Note 5**). In this system, we see exchange on the fast time scale, which results in a single peak for each nucleus with a chemical shift that is the average of the weighted population of the free- and bound-state chemical shifts.

3.4 Paramagnetic Relaxation

An NMR sample with 220 μM Act-EF34 and 1.1 mM MTSL-labeled palladin peptide was created with a final volume of 600 μL. MTSL-labeled peptide was generated by incubation of MTSL and the palladin peptide with an N-terminal cysteine with an MTSL:peptide ratio of 4:1 in a nonreducing buffer [19]. This allows for formation of a disulfide bond between the MTSL label and the thiol group on the cysteine. Then excess, unbound MTSL label was removed by extensive dialysis into nonreducing NMR buffer. A 2D ^{1}H–^{15}N HSQC was acquired on the sample, using the same parameters as those given above. Then 2 μL of 1 M ascorbic acid was added to the sample to reduce the MTSL label, allowing it to diffuse into the solution, and a duplicate ^{1}H–^{15}N HSQC was acquired on this sample.

3.5 NMR Data Analysis

Both the chemical shift perturbation and paramagnetic relaxation studies relied on changes in the chemical shifts or peak intensities, respectively, as observed by 2D ^{1}H–^{15}N HSQC analyses. Additionally, in both experiments, ^{15}N-labeled Act-EF34 was used, meaning that the signals observed come only from Act-EF34. The 2D ^{1}H–^{15}N HSQC provides information on all ^{15}N atoms bound to a ^{1}H. This includes one NH peak for the backbone amide, which we expect for all residues except proline, giving a site-specific probe for each residue. Side-chain amides also provide peaks, though their assignment is more difficult. The signal from the proton is in the first dimension, while the signal from the ^{15}N nucleus is in the second. The peaks in the resulting spectrum appear as spots with contour lines to represent their height or intensity. Both the

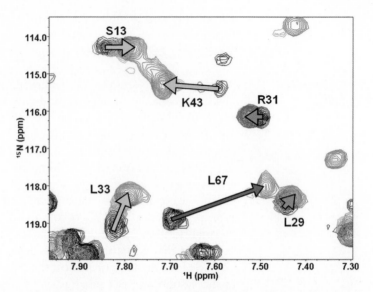

Fig. 1 Expanded region of a 2D ^{1}H–^{15}N HSQC spectral overlay showing NH peaks associated with 0.22 mM Act-EF34 in the presence of varied concentrations of WT palladin peptide. The Act-EF34 is uniformly enriched with ^{15}N, while the peptide is unlabeled; therefore peaks correspond to changes in chemical shift of Act-EF34 upon binding palladin peptide. Peptide concentrations are 0 (*black*), 1 (*purple*), 10 (*blue*), 50 (*green*), 100 (*yellow*), 200 (*orange*), and 500 (*red*) μM. Residues S13, L33, and K43 show significant changes in chemical shift upon titration with WT palladin peptide (*yellow arrows*), while residues L29 and R31 do not (*green arrows*). Residue L67 shows a very significant change in chemical shift (*red arrow*, greater than average CS + twice standard deviation)

amplitude and the chemical shift position of the signal provide information on the corresponding residue within Act-EF34. For the chemical shift perturbation titration, the peaks for the apo (or free) Act-EF34 had previously been assigned to their respective residues (BMRB ID:17626). The HSQC spectra collected during the titration were superimposed on each other using NMRViewJ (*see* Fig. 1). To select shifts that are significantly large for use in docking, a weighted chemical shift that incorporates both dimensions was calculated. The calculation was performed as follows:

$$CS_{residue} = \sqrt{\left(\frac{N_{free} - N_{bound}}{10}\right)^2 + (H_{free} - H_{bound})^2}$$

where N_{free} is the chemical shift of a peak in the nitrogen dimension in the free, or apo, state (*see* **Note 6**). The ^{15}N dimension is weighted at 1/10 of the proton dimension, because the absolute value for the gyromagnetic ratio for ^{15}N is approximately 1/10 that of ^{1}H. If $CS_{residue} > 0.173$ ppm (average CS + half the standard deviation), then the residue was deemed to have shifted significantly enough to be used as a docking restraint (*see* Fig. 2a).

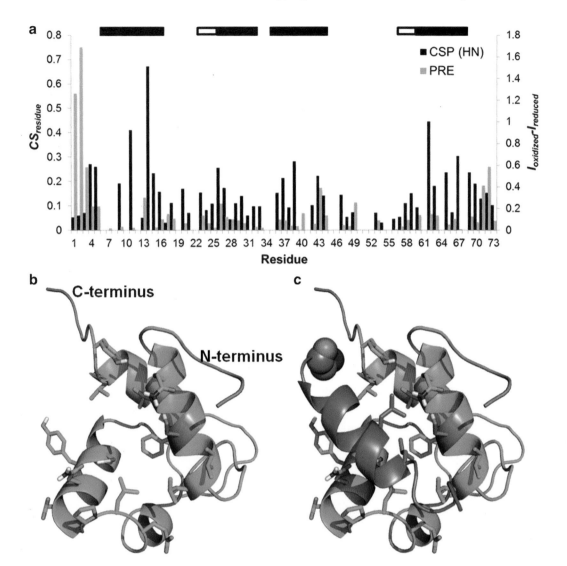

Fig. 2 Structural model of Act-EF34 complexed with WT palladin peptide. (**a**) CSP and PRE data mapped onto the primary structure of Act-EF34. CSP data are shown in *black*, with PRE data in *gray*. The secondary structure (from PDB 1H8B) of Act-EF34 is given by *black bars* (α-helices) and *white bars* (β-sheets). Figure modified from Beck et al. [19] (**b**) Model structure of Act-EF34 when bound to WT palladin peptide as created by PyMOL [28]. Sidechains shown as *sticks* belong to residues selected as AIRs for docking with HADDOCK. Residues colored in *pink* showed significant relaxation when bound to the MTSL-labeled palladin peptide. (**c**) Model of the ligand-bound complex. The color scheme is maintained, and the palladin peptide is shown in *blue*. The sidechains of L9 and L12 are represented as *sticks*. The N-terminus (where the MTSL label was added) is shown as *spheres*

Additional considerations for docking restraints were made on the basis of solvent exposure, chemical shift perturbations of the Cα and Cβ atoms, and absence of peaks upon palladin peptide titration. If a protein-protein interaction does not involve significant conformational rearrangement, then atoms involved in the interaction should be exposed to the solvent and available for mediating the interaction. Solvent exposure was determined by running NAC-CESS, a program that defines the solvent accessibility of atoms based on the known protein structure in a PDB file [23]. Residues chosen for AIRs had atoms with a solvent accessibility greater than 50 % [23]. Chemical shift perturbations for the Cα and Cβ atoms were determined during assignment of peaks for the palladin-bound Act-EF34; residues with a $CS_{residue} > 1.2$ ppm (average CS + half the standard deviation) were chosen as AIRs. As with the CSP for HN peaks described above, a $CS_{residue} > 1.2$ ppm suggested involvement in the binding interaction. Additionally, some HN peaks disappeared during titration, either due to line broadening (exchange on the intermediate time scale) or CSP that could not be easily tracked. Residues that show solvent accessibility and fit one or more of the above criteria are Q9, I11, S13, F14, I16, L17, E32, L33, D36, Q37, C41, Y68, and G69 (Fig. 2b). Additionally, many of the peaks that showed significant Act-EF23 CSP upon binding to the palladin fragment showed similar CSP upon binding of Act-EF34 to a titin peptide, suggesting that Act-EF34 bound titin and palladin in a similar fashion [19].

To analyze the paramagnetic relaxation data, the reduced (dia-magnetic probe) and oxidized (paramagnetic probe) HSQCs were superimposed. No changes in chemical shift were observed, which suggests that the presence of the MTSL label did not affect the binding of the peptide to Act-EF34. The intensities of the peaks were calculated using NMRViewJ, and a difference in intensities (heights) for each peak was determined:

$$PRE_{resdiue} = I_{residue,\ oxidized} - I_{residue,\ reduced}$$

The paramagnetic probe increases the rate of magnetic relaxation for nearby atoms, therefore peaks that exhibit a significantly lower R_I are likely close to the paramagnetic probe. Peaks corresponding to residues M1, A2, D3, T4, D5, S13, A26, K43, S71, D72, and L73 all exhibited a significant decrease in intensity in the presence of the paramagnetic probe, suggesting that these residues are near the N-terminus of the palladin peptide when it is bound to Act-EF34 (Fig. 2). These results were not used to generate AIRs for HADDOCK, but were used as an independent confirmation of the HADDOCK model generated (see below).

3.6 Docking

The docking program HADDOCK [13] was used to develop a model for the interaction between the palladin peptide and Act-EF34. HADDOCK starts with known structures of the two ligands and docks using experimental restraints and a traditional energy function; it does allow for small conformational changes, a common occurrence in protein-protein interactions. The starting structure for Act-EF34 was taken from the solution structure of

Act-EF34 bound to the seventh Z-repeat of titin (PDB 1H8B). This choice was justified by the similarity of CSP data in binding of the titin peptide and the palladin peptide [19]. Further validation for using this starting structure came from ^{1}H–^{15}N RDCs on the peptide-bound Act-EF34, which agreed very well with those calculated for the Act-EF34:titin structure (*see* **Note 7**). The starting structure for the palladin peptide was designed in MacPyMOL as an alpha helix. Circular dichroism data and the PSIPRED protein structure prediction server both suggested that the peptide was helical when bound [19]. Ambiguous interaction restraints (AIRs) for the docking interaction came from CSP and mutagenesis of leucines L9 and L12 in the palladin peptide, which resulted in loss of binding. Additional residues surrounding L9 and L12 on the palladin peptide were also defined by HADDOCK as AIRs. Docking was performed on the HADDOCK server website using the default parameters [24]. HADDOCK generated 1,000 initial structures, from which the 200 with the best score were selected. Scores are generated using a weighted sum of energy values corresponding to calculated van der Waals forces, electrostatic forces, solvation, and the AIRs. The highest-scoring models underwent an annealing stage with some flexibility and subsequent refinement in water. The final models were then clustered into groups of no fewer than 4 with clustering cutoff of 7.5 Å. The clustering cutoff ensures that two distinct clusters of models are grouped into one larger cluster. One cluster of 12 model structures had a significantly better score than all other clusters and a root mean square deviation in the Act-EF34 structure from PDB 1H8B (Act-EF34 bound to titin) of only 0.76 Å. This model was selected as the best model for the interaction between Act-EF34 and palladin (Fig. 2) and is further testable by mutagenesis. Model-driven mutagenesis can lead to refinement of the model and the development of specific tools (mutants or variants) to study the functional relevance of a protein-protein interaction in vivo.

4 Notes

1. Expression of proteins in M9 minimal medium can be much more difficult than expression in LB. The reader is encouraged to work test expression in M9 without isotopic enrichment before expressing with isotopic enrichment. Common techniques to improve growth and expression in M9 medium include using an overnight growth that is 50 % H_2O, 50 % LB, doubling the concentration of glucose in the M9 medium, growing in a baffled flask to increase aeration, and supplementing the medium with a vitamin mixture [25]. Other techniques that are also used to improve expression in LB, including expression at lower temperatures or use of a fermentor, may be useful.

2. A number of considerations should be made when preparing an NMR sample. As NMR is insensitive by nature, a large amount of soluble protein is required. Most NMR experiments use samples with protein concentrations between 0.1 and 2 mM, with higher concentrations generally preferred. To improve the ratio of signal-to-noise (S/N), NMR signals are frequently averaged multiple times. The S/N increases by the square root of the number of times the signal is averaged, but increases linearly with the protein concentration. A doubling in concentration therefore saves one four times the amount of acquisition time. Therefore, the NMR buffer should optimize protein solubility while maintaining low ionic strength, especially if using a magnet with a cryoprobe. Generally, a buffer with a pH between 5.5 and 7 is optimal. Many pulse sequences rely on the amide proton in the backbone as a starting point, and at higher pH, these protons exhibit increased exchange and become increasingly difficult to observe. Having sufficient sample volume is critical to obtaining quality spectra. An ideal volume is 500 μL for the Wilmad tube (or 350 μL for a Shigemi tube). Anything less makes shimming of the sample difficult and leads to poor water suppression and a decrease in S/N. Increasing sample volume is generally a poor idea, as the volume "seen" by the spectrometer is fixed.

3. For a 2D ^{1}H–^{15}N HSQC, determining the number of complex points and spectral width in the ^{15}N dimension is important. In the ^{1}H dimension, the signal is directly detected, so the number of complex points depends solely upon the number of times the signal is collected, and the spectral width is usually set sufficiently wide to capture all signals (backbone amide protons generally fall between 11 and 6 ppm). However, as the ^{15}N dimension is indirectly detected, these values should be optimized. A spectral width that is too large results in poorer resolution in the ^{15}N dimension, or requires more points (and time) to achieve the same resolution. When starting NMR experiments, one should first carry out an HSQC with a large spectral window to see where the data fall in the ^{15}N dimension. Subsequent HSQC experiments should be performed with the ^{15}N offset centered in the middle of the data and a smaller spectral width. If one or two peaks in the HSQC lie sufficiently far away from other peaks in the ^{15}N dimension, these peaks can be "folded" to further decrease the spectral width, though this is often reserved for 3D experiments. Once an optimal spectral width is determined, the minimum number of complex points required is equal to the spectral width divided by the linewidth of the peaks in the ^{15}N dimension. The linewidth is related to the size of the protein (larger proteins have larger linewidths) and can be determined from

an HSQC with a significantly large number of complex points in the ^{15}N dimension. A good estimate for the line width of a protein is approximately 0.6 Hz per kDa molecular weight.

4. With many proteins, expression and purification of the protein (or the domain of interest) at levels sufficient for NMR experiments is very difficult. Additionally, some protein-protein interactions are mediated primarily by a small part of one protein. In these cases, a small peptide can be used to study the interaction between proteins. Small peptides provide their own challenges, especially in determination of their concentration [26]. Solutions of peptide at high concentration often need pH adjustments.

5. Exchange between the bound and free states of a protein takes place in one of three NMR time scales: fast, intermediate, or slow. If the interaction is in intermediate exchange, the signal from the bound state will shrink in amplitude and broaden in linewidth, resulting in a disappearance of signal as the population shifts to the bound state. This is the least optimal exchange regime for studying protein-protein interactions by NMR, as it makes quantification of the relative populations difficult. Changing the temperature, buffer conditions, or the nature of the ligand can push the interaction into the fast or slow regimes.

6. The most commonly accepted formula for calculating weighted chemical shift differences is shown below, as reported by Mulder et al. [27].

$$CS_{residue} = \sqrt{\left(\frac{N_{free} - N_{bound}}{6.5}\right)^2 + (H_{free} - H_{bound})^2}$$

This formula uses a different weighting factor for ^{15}N chemical shift differences, which is based upon the disparity in the average dispersion of chemical shift values for ^{15}N and ^{1}H signals in a 2D ^{1}H–^{15}N HSQC.

7. A complete description of RDCs and their use in studying protein-protein interactions is outside the scope of this chapter. The reader is referred to reviews on the subject for a more detailed experimental and theoretical discussion [5, 6]. RDCs can be very useful for the study of protein-protein interactions. Simply put, they describe the relative orientation of bond vectors within a protein and can be used to compare structures. As described in Beck et al. [19], RDCs can be used to compare a known structure to an unknown one to determine a degree of similarity. Additionally, RDCs for the ligand-free and ligand-bound states of a protein can be compared to test for large conformational changes that sometimes occur upon binding.

References

1. Akke M (2002) NMR methods for characterizing microsecond to millisecond dynamics in recognition and catalysis. Curr Opin Struct Biol 12:642–647

2. Clore G, Gronenborn A (1991) Two-, three-, and four-dimensional NMR methods for obtaining larger and more precise three-dimensional structures of proteins in solution. Annu Rev Biophys Biophys Chem 20:29–63

3. Kay LE (1997) NMR methods for the study of protein structure and dynamics. Biochem Cell Biol 75:1–15

4. Vinogradova O, Qin J (2012) NMR as a unique tool in assessment and complex determination of weak protein–protein interactions. Top Curr Chem 326:35–45

5. O'Connell MR, Gamsjaeger R, Mackay JP (2009) The structural analysis of protein–protein interactions by NMR spectroscopy. Proteomics 9:5224–5232

6. Jensen MR, Ortega-Roldan JL, Salmon L et al (2011) Characterizing weak protein–protein complexes by NMR residual dipolar couplings. Eur Biophys J 40:1371–1381

7. Marintchev A, Frueh D, Wagner G (2007) NMR methods for studying protein–protein interactions involved in translation initiation. Methods Enzymol 430:283–331

8. Wittekind M, Mueller L (1993) Hncacb, a high-sensitivity 3d Nmr experiment to correlate amide-proton and nitrogen resonances with the alpha-carbon and beta-carbon resonances in proteins. J Magn Reson Ser B 101:201–205

9. Ikura M, Kay LE, Bax A (1990) A novel-approach for sequential assignment of H-1, C-13, and N-15 spectra of larger proteins – heteronuclear triple-resonance 3-dimensional Nmr-spectroscopy – application to calmodulin. Biochemistry 29:4659–4667

10. Grzesiek S, Bax A (1992) Correlating backbone amide and side-chain resonances in larger proteins by multiple relayed triple resonance NMR. J Am Chem Soc 114:6291–6293

11. Solomon I (1955) Relaxation processes in a system of two spins. Phys Rev 99:559–565

12. Jahnke W (2002) Spin labels as a tool to identify and characterize protein-ligand interactions by NMR spectroscopy. Chembiochem 3:167–173

13. Dominguez C, Boelens R, Bonvin AM (2003) HADDOCK: a protein–protein docking approach based on biochemical or biophysical information. J Am Chem Soc 125:1731–1737

14. Lyskov S, Gray JJ (2008) The RosettaDock server for local protein–protein docking. Nucleic Acids Res 36:W233–W238

15. Claussen H, Buning C, Rarey M et al (2001) FlexE: efficient molecular docking considering protein structure variations. J Mol Biol 308:377–395

16. Tovchigrechko A, Vakser IA (2006) GRAMM-X public web server for protein–protein docking. Nucleic Acids Res 34:W310–W314

17. Pierce BG, Hourai Y, Weng Z (2011) Accelerating protein docking in ZDOCK using an advanced 3D convolution library. PLoS One 6:e24657

18. Morris GM, Huey R, Lindstrom W et al (2009) AutoDock4 and AutoDockTools4: automated docking with selective receptor flexibility. J Comput Chem 30:2785–2791

19. Beck MR, Otey CA, Campbell SL (2011) Structural characterization of the interactions between palladin and alpha-actinin. J Mol Biol 413:712–725

20. Joseph C, Stier G, O'Brien R et al (2001) A structural characterization of the interactions between titin Z-repeats and the alpha-actinin C-terminal domain. Biochemistry 40:4957–4965

21. Bodenhausen G, Ruben DJ (1980) Natural abundance N-15 Nmr by enhanced heteronuclear spectroscopy. Chem Phys Lett 69:185–189

22. Delaglio F, Grzesiek S, Vuister GW et al (1995) NMRPipe: a multidimensional spectral processing system based on UNIX pipes. J Biomol NMR 6:277–293

23. Hubbard SJ, Thornton JM (1993) NACCESS Computer Program, Department of Biochemistry and Molecular Biology, University College London www.bionf.manchester.ac.uk/naccess/nac_intro.html

24. de Vries SJ, van Dijk M, Bonvin AM (2010) The HADDOCK web server for data-driven biomolecular docking. Nat Protoc 5:883–897

25. Venters RA, Calderone TL, Spicer LD et al (1991) Uniform 13C isotope labeling of proteins with sodium acetate for NMR studies: application to human carbonic anhydrase II. Biochemistry 30:4491–4494

26. Kuipers BJ, Gruppen H (2007) Prediction of molar extinction coefficients of proteins and peptides using UV absorption of the

constituent amino acids at 214 nm to enable quantitative reverse phase high-performance liquid chromatography-mass spectrometry analysis. J Agric Food Chem 55:5445–5451

27. Mulder FA, Schipper D, Bott R et al (1999) Altered flexibility in the substrate-binding site

of related native and engineered high-alkaline Bacillus subtilisins. J Mol Biol 292:111–123

28. DeLano WL (2006) MacPyMOL: The PyMOL Molecular Graphics System, Version 1.5.0.4 Schrödinger, LLC www.pymol.org

Chapter 17

Quantitative Protein Analysis by Mass Spectrometry

Vishwajeeth R. Pagala, Anthony A. High, Xusheng Wang, Haiyan Tan, Kiran Kodali, Ashutosh Mishra, Kanisha Kavdia, Yanji Xu, Zhiping Wu, and Junmin Peng

Abstract

Mass spectrometry is one of the most sensitive methods in analytical chemistry, and its application in proteomics has been rapidly expanded after sequencing the human genome. Mass spectrometry is now the mainstream approach for identification and quantification of proteins and posttranslational modifications, either in small scale or in the entire proteome. Shotgun proteomics can analyze up to 10,000 proteins in a comprehensive study, with detection sensitivity in the picogram range. In this chapter, we describe major experimental steps in a shotgun proteomics platform, including sample preparation in the context of studying protein-protein interaction, mass spectrometric data acquisition, and database search to identify proteins and posttranslational modification analysis. Proteome quantification strategies and bioinformatics analysis are also illustrated. Finally, we discuss the capabilities, limitations, and potential improvements of current platforms.

Key words Proteomics, Mass spectrometry, Posttranslational modifications, Spectral counting, Metabolic labeling, Isobaric labeling

1 Introduction

During the last two decades, mass spectrometry (MS)-based shotgun proteomics has emerged as the mainstream technology for protein characterization with unprecedented specificity, sensitivity, and throughput [1–3]. The technology is now capable of detecting all components in the yeast proteome and the vast majority of the human proteome [4]. Traditionally, protein-protein interactions are discovered by yeast two-hybrid analysis [5] and affinity purification followed by protein identification [6]. Rapid development in mass spectrometry markedly accelerates protein identification and expands the investigation of the protein interactome [7]. More recently, a large-scale analysis of human stable protein complexes has been carried out by chromatographic separation coupled with

Cheryl L. Meyerkord and Haian Fu (eds.), *Protein-Protein Interactions: Methods and Applications*, Methods in Molecular Biology, vol. 1278, DOI 10.1007/978-1-4939-2425-7_17, © Springer Science+Business Media New York 2015

quantitative mass spectrometry [8]. Here, we introduce the basics of current proteomics technology, as well as its advances and caveats.

A shotgun proteomics experiment typically consists of MS and MS/MS analysis of peptides derived from proteolytically digested proteins, followed by in silico matching of MS/MS spectra against a database of theoretical peptide spectra generated from protein sequences [9–11]. For complex protein mixtures, fractionation technologies are implemented to separate proteins (e.g., by gel electrophoresis) or peptides (e.g., by strong cation exchange or reverse phase columns) before MS analysis [12]. Because of solvent compatibility with ionization, reverse phase liquid chromatography is usually coupled with online tandem mass spectrometry, termed LC-MS/MS. Even with all the technological advances in the field, there are still challenges that must be addressed. Sensitivity, reproducibility, and comprehensiveness are interconnected factors that must be considered in order to design a successful MS-based experiment [13].

Current commercial mass spectrometers have limits of detection in the low femtomole or attomole range. This limit of detection appears to be sensitive enough to detect most proteins in their pure state but the true sensitivity of mass spectrometry is determined by the type of protein sample. [13] Biological samples have a wide range of protein abundance: more than six orders of magnitude in mammalian cells [4], namely, expression levels of proteins may range from a few copies per cell to over ten million copies per cell. Mass spectrometers have a difficult time dealing with this wide dynamic range. When complex protein/peptide mixtures are ionized together, ion suppression prevents the detection of species of low abundance. Sample fractionation is routinely used to improve the dynamic range.

During LC-MS/MS analysis, digestion of complex protein mixtures produces a massive number of peptides beyond the analytical capacity of mass spectrometers. The scan speed of MS/MS is not sufficient to analyze all peptides during the period of elution time. Only a small fraction of peptides are selected for MS/MS and the majority of peptides are ignored, often designated as a problem of "undersampling" [13]. In the case of total cell lysate, less than 20 % of detectable peptides are selected for MS/MS [14]. Therefore, lack of MS identification cannot be treated as evidence of absence, especially for proteins of low abundance. In general, only ~70 % of proteins in a complex mixture are repeatedly identified even in a replicate. This under-sampling issue presents a challenge for quantifying known proteins of interest in a reproducible manner. Undersampling can be partially corrected and reproducibility improved by repeating analysis or by targeted proteomics approaches [15, 16].

Selection of mass spectrometers is critical for the purpose of a proteomics study. A mass spectrometer consists of an ion source, ion optics, and a mass analyzer [17]. Mass analyzers are the defining analytical components that differentiate current MS platforms on the market. Ion trap, Orbitrap, and ion cyclotron resonance mass analyzers separate ions based on m/z resonance frequency, whereas quadrupoles use m/z stability as a method of ion separation. A time-of-flight (TOF) analyzer separates ions based on flight time. Each mass analyzer has multiple properties that must be considered before choosing the best fit for the experiment. Some of these properties include accuracy and resolution, mass range, scan speed, sensitivity, dynamic range, and the choices of MS/MS fragmentation methods (Table 1). Many current mass spectrometers on the market, known as hybrid instruments, combine more than one analyzer to achieve greater experimental flexibility.

The quantity of proteins in MS analysis can be evaluated by a number of approaches, including label-free quantification based on spectral counting (SC) or extracted ion current, as well as stable isotope labeling methods. The spectral counts are the total numbers of MS/MS spectra identifying a protein, which increase almost linearly with protein abundance after normalizing for protein size [18]. This method works reasonably well for proteins with high spectral counts, but the reliability decreases dramatically for proteins with low spectral counts [19]. Alternatively, the abundance of peptides in samples can be compared by extracted ion current of corresponding ions [20]. Because ionization efficiency of the peptides may vary among different LC runs, largely due to fluctuation of the LC system and ion suppression, considerable variations may be expected and need to be normalized in the label-free method.

To alleviate the problems associated with intrinsic LC-MS/MS variations, a gold standard in quantitative MS is to use stable isotope-labeled peptides as internal standards [1–3]. The internal standards and their counterparts are eluted and ionized simultaneously during the LC-MS/MS runs, so that relative quantification can be achieved by comparing peak areas of the peptide pairs. Stable isotopes can be introduced into samples by in vitro labeling, such as iTRAQ or TMT [21]. The iTRAQ/TMT reagents differentially label the amine group at the N-termini and Lys residues of peptides after protein digestion. Ten labeling reagents are now available to allow multiplex comparisons in a single experiment. Moreover, the strategy of stable isotope labeling with amino acids in cell culture (SILAC) [22] has been developed as a highly accurate, in vivo labeling method for large-scale proteomics. Given the simplicity of label-free quantification, high accuracy of SILAC, and multiplex comparison of iTRAQ/TMT, all of these methods are commonly used by researchers. More recently, the Gygi group has combined the SILAC and TMT strategies to analyze as many as 18 samples in a single experiment [23].

Table 1
Common tandem mass spectrometer specifications (adapted from http://masspec.scripps.edu/mshistory/whatisms_details.php)

	Triple quadrupole	Linear ion trap	Q-TOF	LTQ orbitrap (ELITE)	Q exactive
Accuracy	0.01 % (100 ppm)	0.01 % (100 ppm)	0.001 % (10 ppm)	<2 ppm	<2 ppm
Resolution	4,000	4,000	10,000	240,000	140,000
m/z Range	4,000	4,000	10,000	6,000	4,000
Scan speed (s)	~1	~1	~1	~1	~0.5
Comments	Good accuracy, good resolution, low-energy collisions, low cost	Good accuracy, good resolution, low-energy collisions, low cost	Excellent accuracy, good resolution, low-energy collisions, high sensitivity	Excellent accuracy, resolution, and sensitivity; flexible fragmentation (e.g., CID, HCD, and ETD)	Excellent accuracy, resolution, and sensitivity; only HCD available; bench top and easy to use

Here we describe a number of detailed protocols for routinely analyzing proteins in our proteomics facility, including sample preparation and protein digestion, protein identification by LC-MS/MS, dissection of protein posttranslational modifications, label-free protein quantification, protein profiling by SILAC or iTRAQ/TMT, and bioinformatics data processing.

2   Materials

2.1   Sample Digestion and LC-MS/MS Analysis for Protein Identification

1. 20 mg/ml bovine serum albumin as an internal standard.

2. Acetonitrile.

3. 5 mM dithiothreitol in 100 mM ammonium bicarbonate (in HPLC grade water).

4. 10 mM iodoacetamide in 100 mM ammonium bicarbonate (*see* **Note 1**).

5. 10 % formic acid.

6. 5 % formic acid in 50 % acetonitrile.

7. Trypsin (sequencing grade, 20 μg/50 μl aliquots in 50 mM acetic acid): 2 μg/5 μl aliquots are prepared and stored at −80 °C. An aliquot is thawed and diluted to 200 μl with 25 mM ammonium bicarbonate (working concentration is 0.01 μg/μl, pH ~8.0, *see* **Note 2**).

8. Solid urea (*see* **Note 3**).

9. 100 mM calcium chloride.

10. Lys-C (5 μg of lyophilized protein with salt in each vial, reconstituted in 50 μl of water, 0.1 μg/μl in 50 mM Tris, 10 mM EDTA, pH 8.0).

11. C18 Zip tip (loading capacity: 10 μg peptide per μl resin, *see* **Note 4**).

12. Trifluroacetic acid (TFA).

13. Equilibration buffer: 5 % acetonitrile and 0.1%TFA.

14. Elution buffer: 70 % acetonitrile and 0.1%TFA.

15. Standard peptide digest: tryptic BSA.

16. C18 column (e.g., 75 μm × 10–15 cm, 15 μm tip orifice, 2.7 μm HALO beads, New Objective, (*see* **Note 5**) on column selection).

17. LC loading buffer: 5 % formic acid and 0.1 % TFA.

18. Buffer A: 0.2 % formic acid.

19. Buffer B: 0.2 % formic acid, 70 % acetonitrile.

20. HPLC system (e.g., Waters nanoACQUITY UPLC or Thermo EASY-nLC 1000).

21. Tandem MS instrument (e.g., Thermo Q Exactive or LTQ Orbitrap Elite).

Additional reagents for phosphopeptide enrichment:

22. TiO_2 beads.

23. Glutamic acid.

24. 15% ammonium hydroxide and 40 % acetonitrile.

2.2 Protein Quantification Strategies

1. A cell line for expressing a protein of interest.

2. Base SILAC medium free of Arg and Lys amino acids.

3. Dialyzed fetal calf serum (containing no free amino acids).

4. L-arginine and L-lysine.

5. Heavy stable isotope labeled L-type amino acids: $[^{13}C_6{}^{15}N_4]$ arginine (+10.0083 Da) and $[^{13}C_6{}^{15}N_2]$ lysine (+8.0142 Da).

6. SILAC light medium: mixing the base SILAC medium, dialyzed fetal calf serum, and regular L-arginine and L-lysine.

7. SILAC heavy medium: similar to the light medium except equal molar concentration of the heavy stable isotope labeled L-arginine and L-lysine (*see* **Note 6**).

8. TMT 6-plex Isobaric Mass Tagging Kit.

9. 50 mM HEPES, pH 8.5.

10. 5 % Hydroxylamine.

2.3 Bioinformatics Data Processing

1. Database downloaded from NCBI or UniProtKB/Swiss-Prot websites.

2. Database search engines, such as Sequest [24], and Mascot [25].

3. In-house software suite for summarizing protein identification and quantification.

4. A computer cluster for data processing.

3 Methods

3.1 Sample Digestion and LC-MS/MS Analysis for Protein Identification

3.1.1 Protein In-Gel Digestion

Protein samples are usually separated by sodium dodecyl sulfate (SDS) polyacrylamide gel electrophoresis followed by protein staining with Coomassie blue or SYPRO Ruby (*see* **Note 7**). The SDS gel not only resolves proteins based on protein size, but also removes contaminants that interfere with mass spectrometry (e.g., salt and detergents, Fig. 1). Total protein level of the samples may be also evaluated according to staining intensity if known amount of bovine serum albumin is titrated on the same gel (e.g., 10, 100, 1,000 ng). As most protein quantification kits need microgram

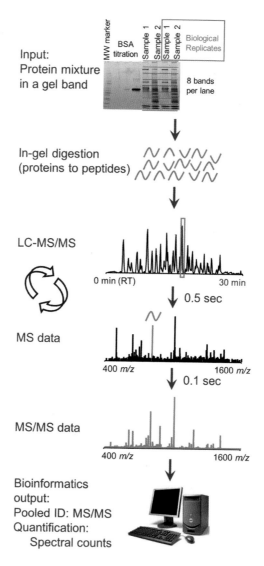

Input:
Protein mixture
in a gel band

In-gel digestion
(proteins to peptides)

LC-MS/MS

0 min (RT)   30 min

0.5 sec

MS data

400 m/z   1600 m/z

0.1 sec

MS/MS data

400 m/z   1600 m/z

Bioinformatics
output:
Pooled ID: MS/MS
Quantification:
  Spectral counts

Fig. 1 Major steps in spectral counting based protein quantification. Protein samples with biological replicates are resolved on a gel, followed by in-gel digestion, LC-MS/MS analysis, and computational data processing

levels of protein for accurate measurement, this gel-based assay provides a simple approach for evaluating protein concentration [26]. For gel excision, we determine the number of gel bands per lane based on the complexity of samples and the total protein level (e.g., at least 100 ng per gel band, Fig. 1). The in-gel digestion protocol is modified from a previously reported version [27] as listed below.

1. Cut protein bands from the gel (e.g., 50 μl of volume); chop each gel band into pieces of ~1 cubic millimeter with a scalpel, and transfer to a 0.5 ml microcentrifuge tube (*see* **Note 8**).

2. Wash the gel pieces twice with 400 μl of 50 % acetonitrile. Vortex the tubes for 20 s, wait for 5 min, and then remove the solution.

3. Reduce proteins with 150 μl of 5 mM DTT in 100 mM ammonium bicarbonate. Vortex the tubes for 20 s, incubate for 30 min at 37 °C, and then remove the solution.

4. Alkylate proteins with 150 μl of 10 mM iodoacetamide in 100 mM ammonium bicarbonate. Vortex the tubes for 20 s, incubate for 30 min in dark at room temperature, and then remove the solution.

5. Quench the reaction with 150 μl of 5 mM DTT in 100 mM ammonium bicarbonate. Vortex the tubes for 20 s, incubate for 30 min at 37 °C, and then remove the solution.

6. Wash twice with 400 μl of 50 % acetonitrile.

7. Wash once with 400 μl of 100 % acetonitrile.

8. Dry the gel pieces completely by speedvac for 10 min at low temperature. The dehydrated gel pieces will have a paper-like appearance.

9. Rehydrate the gel pieces with 40 μl of diluted trypsin in 25 mM ammonium bicarbonate (0.01 μg/μl, pH ~ 8.0) on ice for 20 min. Top with 25 mM ammonium bicarbonate if needed to cover the gel pieces.

10. Incubate at 37 °C overnight (*see* **Note 9**).

11. Spin the tubes to collect any condensation during the incubation, add 20 μl of 10 % formic acid, vortex, and incubate for 20 min.

12. Spin, remove, and save the supernatant.

13. Extract peptides with 50 μl of 5 % formic acid in 50 % acetonitrile.

14. Add 100 μl of 100 % acetonitrile to fully shrink the gel pieces to recover peptides.

15. Pool all peptide-containing supernatant together (~200 μl).

16. Spin at 20,000 × g for 10 min (*see* **Note 10**).

17. Transfer the supernatant to a 0.5 ml microcentrifuge tube, dry in speedvac, and store at −20 °C.

18. Dissolve the peptides right before LC-MS/MS analysis.

3.1.2 Protein In-Solution Digestion

If protein samples contain no detergents, the samples may be directly digested in solution to avoid the lengthy in-gel digestion protocol. In-solution digestion efficiency might be lower than in-gel digestion, which may be due to residual protein folding that reduces trypsin accessibility. Thus, a two-step digestion protocol is commonly used: (1) to cleave proteins under highly denaturing conditions (e.g., 8 M urea) with Lys-C, and (2) to dilute urea to 2 M and further digest the proteins with trypsin. After in-solution digestion, peptides are generally desalted and then analyzed by LC-MS/MS.

1. Evaluate protein complexity and concentration in the samples by running a SDS gel.

2. Add solid urea to the protein samples to final concentration of 8 M.

3. Add Lys-C at an enzyme-to-substrate ratio of 1: 100 (w/w), and add $CaCl_2$ to 10 mM.

4. Incubate at room temperature for 3 h.

5. Dilute the reaction fourfold with 25 mM ammonium bicarbonate to reduce the urea concentration to 2 M.

6. Add trypsin at an enzyme-to-substrate ratio of 1: 50 (w/w).

7. Incubate at room temperature overnight.

8. Add TFA to 0.5 % to acidify the reaction.

9. Spin at 20,000 × g for 10 min.

10. Desalt the supernatant by C18 Zip tip or another C18 column as follows.

11. Wet the Zip tip (~1 μl bed volume) and wash three times with 10 μl of 70 % acetonitrile and 0.1%TFA.

12. Equilibrate with 10 μl of 5 % acetonitrile and 0.1 % TFA three times.

13. Load the digest (<10 μg protein) on the Zip tip by aspirating and dispensing ten times to maximize peptide binding. Wash with 10 μl of 5 % acetonitrile and 0.1 % TFA three times.

14. Elute peptides twice with 10 μl of 70 % acetonitrile and 0.1% TFA.

15. Dry the eluent in a speedvac, and store at −20 °C.

3.1.3 Standard LC-MS/MS Analysis

Successful LC-MS/MS analysis relies on a robust and reproducible LC-MS/MS system, including a nanoscale HPLC system with autosampler, and an inline modern tandem mass spectrometer. Peptide samples are loaded on a reverse-phase C18 column by the HPLC system and eluted by a gradient. Eluted peptides are ionized and detected by the mass spectrometer (e.g., Thermo Q Exactive or LTQ Orbitrap Elite). MS spectra are collected first (in ~0.5 s), and the top 20 abundant ions are sequentially isolated for MS/MS analysis (each in ~0.1 s, totaling ~2 s). This process (~2.5 s) is cycled over the entire liquid chromatography gradient, and more than 14,000 MS/MS spectra are acquired during a 30-min elution (Fig. 1).

1. Calibrate and tune the MS instrument regularly according to the manufacturer's guidelines.

2. Thoroughly wash the C18 column (75 μm × 15 cm, bead volume of 0.7 μl) with 95 % buffer B + 5 % buffer A, and then equilibrate in 5 % buffer B + 95 % buffer A.

3. Run tryptic BSA (e.g., 0.2 μg) twice as a quality control step to evaluate the sensitivity and reproducibility of the LC-MS/MS system. Monitor signal intensity, peak width, and retention time of the selected BSA peptide ions, as well as run-to-run variability.

4. Dissolve dried samples in LC loading buffer before running samples. Extended storage of samples in low concentration results in significant peptide loss.

5. Load the sample on the LC-MS system (e.g., LTQ Orbitrap Elite) using an optimized platform [26]. Briefly, the peptides (~1 μg for the column above) are loaded, washed, and then eluted during a gradient of 8–30 % acetonitrile in 30 min (flow rate of 250 nl/min). The eluted peptides are detected by Orbitrap (400–1,600 m/z, 1,000,000 AGC target, 100 ms maximum ion time, resolution 240,000 FWHM) followed by 20 data-dependent MS/MS scans in LTQ (2 m/z isolation width, 35 % collision energy, 3,000 AGC target, 100 ms maximum ion time, 30 s dynamic exclusion, only +2 and +3 ions selected, *see* **Note 11**).

3.1.4 Analysis of Protein Posttranslational Modifications (PTM)

Approximately 300 forms of protein modifications are documented. In cells, the modifications range from the simple addition of small chemical groups (e.g., phosphorylation, methylation, and acetylation), to more complex moieties (e.g., glycosylation, lipidation, and ubiquitination). The modifications are reversible and catalyzed by an array of enzymes, leading to regulation of protein surfaces, structures, and interacting partners. Thus, protein modifications are central to signal transduction in cells.

Mapping modified residues by mass spectrometry is often challenging, which is highly different from protein identification analysis that requires one unique peptide (matched to a single MS/MS spectrum). For PTM mapping, approximately 100 % "sequence coverage" by MS/MS is needed. As some digested peptides (with extreme lengths or hydrophobicity) may not be compatible with the standard LC-MS/MS method, custom methods may be required, such as digestions with additional enzymes (e.g., Arg-C, Glu-C, Asp-N) to improve peptide coverage. In addition, the stoichiometry of PTM may be low and the modified sites may be heterogeneous in the samples, further limiting the amount of modified species. For instance, one protein can be modified by one or more modifications at a single site or multiple sites. Therefore, the assignment of modification sites requires a much higher starting sample amount (typically >100 ng of one purified protein excised in a gel band). During isolation of modified proteins, specific inhibitors are needed to prevent the loss of modifications, increasing the yield of the modified forms.

For a complex protein mixture, modified peptides are usually undetectable because of low abundance. Numerous strategies have been developed to enrich for modified peptides. For instance, phosphopeptides are isolated by immobilized metal-affinity chromatography (IMAC) enrichment incorporating metal oxides such as Fe^{3+} ion [28], TiO_2 [29], cation and anion exchange chromatography [30], antibody capture [31], chemical derivation [32], and a combination of these approaches. The TiO_2 method is widely used and the protocol for this method is described below:

1. Select the amount of TiO_2 beads based on the level of peptide digestion (0.4–0.6 mg beads for 0.1 mg digested protein, *see* **Note 12**).

2. Wash with 200 μl of 50 % acetonitrile saturated with glutamic acid.

3. Dissolve desalted and dried peptides with 100 μl of 50 % acetonitrile saturated with glutamic acid.

4. Incubate with TiO_2 for 1 h at room temperature.

5. Spin and remove supernatant.

6. Wash beads with 100 μl of 50 % acetonitrile with 0.1 % TFA three times.

7. Elute the phosphopeptides with 20 μl of 15 % ammonium hydroxide with 40 % acetonitrile, and acidify the samples (*see* **Note 13**).

8. Dry the peptides by speedvac and store at −80 °C until ready for analysis by LC-MS/MS.

3.2 Protein Quantification Strategies

3.2.1 Protein Profiling by Spectral Counting (SC)

SC-based quantification is straightforward and sensitive enough to identify large changes among protein samples. Using co-immunoprecipitation (IP) analysis as an example, the purpose of the analysis is to identify novel interacting proteins of the antigen. One expects that the antigen and associated proteins are abundant in the IPs but not in the controls. The LC-MS/MS procedure is basically the same as the standard protocol above, with the following exceptions (Fig. 1)

1. Biological replicates are highly recommended for quantitative comparison. Batch-to-batch variation during IP analysis may be large. It is critical to perform the analysis with at least one negative control and at least one biological replicate to evaluate the variation.

2. Examine the quality of IP samples. Make three aliquots (e.g., 10, 20, and 70 %) of the samples for western blotting, SDS PAGE, and MS analysis, respectively. The immunoblotting confirms the antigen captured by antibodies, whereas the SDS gel is stained with silver or SYPRO Ruby to assess the antigen level of and total protein in the IPs, assisted by BSA loading on the side of the gel (Fig. 1). Finally, the remaining 70 % of the samples are equally loaded on a SDS gel, stained by Coomassie blue or SYPRO Ruby, and subjected to in-gel digestion.

3. Extend the elution gradient from the standard 30 to 60 min to increase scan number (*see* **Note 14**).

4. Finally, evaluate the dataset by several critical parameters: (1) the antigen is expected to be one of the most abundant proteins in the IPs (e.g., among the top 10. If not, the IP conditions may need to be further refined to improve purity; (2) the SC of the antigen should be high (e.g., >50). Otherwise, the likelihood of finding its interacting proteins is low, because genuine interacting partners are usually present at substoichiometric level; (3) the variation between biological replicates should be small.

3.2.2 Protein Profiling by SILAC

The SILAC approach can accurately compare the relative intensities of peptides from protein samples. First, cells are grown in medium containing either light or heavy stable isotope labeled Lys and Arg. Then the differentially labeled cells are treated under different conditions, such as transfection with mock or a protein-expressing plasmid in a co-IP experiment. The cells are harvested and lysed for immunoprecipitation. Finally, the heavy and light IP samples are pooled together for SDS PAGE and LC-MS/MS (Fig. 2, *see* **Note 15**). We continue to use the co-IP experiment to explain details in a SILAC analysis.

1. Select appropriate cell lines and culture in light or heavy medium to ensure full labeling.

2. Examine for cell viability in SILAC medium, as certain cell lines are not able to thrive in SILAC medium.

3. Examine if the use of SILAC medium alters the biology of the cells (e.g., expression of protein of interest).

4. Test for complete labeling of proteins. Certain cell types may require extra passages to achieve full labeling of all proteins. It is necessary to achieve >95 % labeling for accurate quantification.

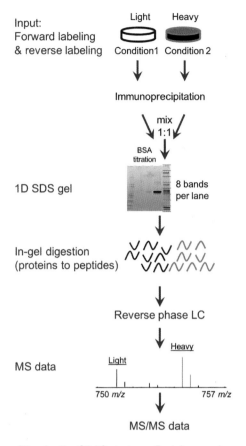

Input:
Forward labeling
& reverse labeling

Light   Heavy

Condition1 Condition 2

Immunoprecipitation

mix
1:1

BSA
titration

1D SDS gel

8 bands
per lane

In-gel digestion
(proteins to peptides)

Reverse phase LC

MS data

Heavy

Light

750 m/z   757 m/z

MS/MS data

Fig. 2 Protein profiling by the SILAC strategy. Protein samples are differentially labeled by light and heavy SILAC media. The two samples are harvested and subjected to immunoprecipitation. The IP samples are then pooled together, subjected to gel separation, in gel digestion and reverse phase LC-MS/MS analysis. Biological replicates can be performed using a forward and reverse labeling strategy

5. Treat the cells under different conditions.

6. Perform a biological replicate by swapping the labeling order.

7. Harvest and lyse the cells for immunoprecipitation.

8. Resolve pooled IP samples on an SDS gel, followed by in-gel digestion and LC-MS/MS analysis.

3.2.3 Protein Profiling
by iTRAQ/TMT

The development of multiplex isobaric labeling methods (e.g., iTRAQ and TMT) greatly improves the throughput of proteomics. In this protocol, we describe how to perform a 6-plex TMT analysis (Fig. 3). Every labeling reagent consists of three groups: an amine-specific active ester group, a balance group, and a reporter group. Six reagents can label six different samples by reacting with free

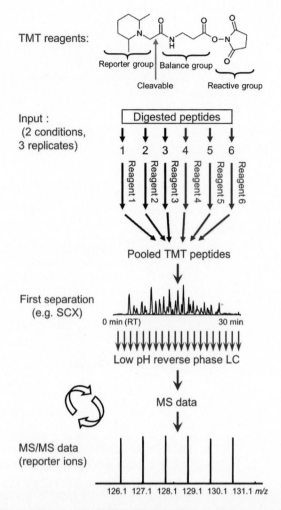

Fig. 3 Protein profiling by the TMT technology. Six samples can be differentially labeled by 6-plex TMT reagents for a quantitative analysis. The samples are first digested separately, labeled by TMT reagents, followed by pooling and LC-MS/MS analysis. If necessary, two dimensional LC may be implemented, including strong cation exchange (SCX) chromatography and reverse phase chromatography

alpha N-termini of peptides and the epsilon amino groups of lysine residues. The isobaric TMT tags, defined by different distribution of isotopes (e.g., ^{13}C and ^{15}N) between the reporter group and the balance group, allow labeled peptides to coelute during chromatography and ionize at the same mass-to-charge in MS scans. The labeled samples are equally mixed, further fractionated, and analyzed by LC-MS/MS. During ion fragmentation, the TMT tags are cleaved between the balance group and the reporter group, producing reporter ions with different mass for quantification.

1. Digest six protein samples (~100 µg) using an in-gel or in-solution protocol.

2. Desalt the peptides to remove salt and nonpeptide amine-containing contaminants (e.g., ammonium bicarbonate).

3. Resuspend each sample in 20 µl of 50 mM HEPES buffer.

4. Reconstitute each TMT reagent in 41 µl of anhydrous acetonitrile, vortex and spin to collect all of the solution.

5. Mix each pair of peptide sample and TMT reagent; incubate for 1 h at room temperature.

6. Quench the reaction by adding 8 µl of 5 % hydroxylamine and incubate for 15 min.

7. Make aliquots for the labeled samples and store at −80 °C until ready for analysis.

8. Perform a trial mix using a small aliquot of each sample (e.g., 5 µl), analyze by LC-MS/MS to obtain the global peptide ratio of all six samples.

9. Use the trial data to ensure equal mixing of the remaining samples.

10. Desalt pooled TMT peptides by C18 cartridge, during which the washing buffer is changed to 5 % acetonitrile with 0.2 % formic acid to completely remove quenched TMT reagents.

11. The desalted peptides are ready for subsequent LC-MS/MS analysis. If necessary, two dimensional LC separation may be implemented to improve the throughput (Fig. 3, *see* **Note 16**).

3.3 Bioinformatics Data Processing

The pipeline of MS data processing is outlined in Fig. 4. First, a database search is used to convert MS raw data into peptide/protein identification information. Then these proteins are quantified and subjected to statistical evaluation. Finally, proteins with altered expression are reported for users to derive subsequent hypotheses for targeted studies. However, any large-scale datasets inevitably contains false positives, and proteomics results also have certain pitfalls. Thus, it is important to validate MS results prior to significant investment. Currently, there is no consensus on the best software package in the proteomics field. We describe the pipeline used in our proteomics facility, which integrates in-house software and algorithms developed by colleagues.

3.3.1 Protein Identification

Numerous computation approaches have been developed to search databases for peptide identification [9, 10]. These approaches compare one acquired MS/MS spectrum to all theoretical fragmentation spectra in databases, and the best matched peptide is selected as putative identification. There are three fundamental steps during a database search, including preprocessing, database search and postprocessing. While preprocessing improves data quality and removes

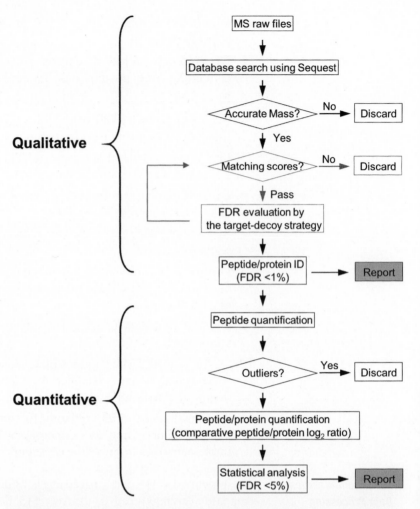

Fig. 4 The bioinformatics pipeline uses computer programs to match experimental MS/MS spectra to theoretical patterns in protein databases, resulting in the identification of peptides/proteins. The target-decoy strategy is used to filter the peptide matches by mass accuracy and matching scores until the protein false discovery rate is lower than 1 %. For protein quantification analysis, additional programs are used for peak extraction, sample comparison and normalization, and statistical inference

low quality spectra to save time and reduce risk of false identifications, post-processing filters database search results to maintain the false discovery rate at a tolerable level (e.g., 1 %).

Preprocessing

1. Extract MS/MS data from raw files. Mass spectrometers from different companies generate raw data in different formats. For instance, Thermo MS instruments produce files in raw format. The raw files are first converted to mzXML format with "ReAdW", then to dta files by "MzXML2Search". Both programs are available from the Institute for Systems Biology

(http://tools.proteomecenter.org). Each dta file contains mass information for one MS/MS scan, including precursor ion, charge state, product ions, and intensities.

2. Decharge and deisotope precursor ions in MS scans. Firstly, the precursor ion recorded in MS/MS scans is searched in the corresponding survey scan within a defined isolation window (e.g., 2 *m/z*). Secondly, charge state and monoisotopic mass of precursor ions are defined based on high resolution survey scan data.

3. Correct mass in dta files. Systematic mass shift in the survey scan is determined based on the polycyclodimethylsiloxane ion (445.120025 *m/z*). On the basis of the mass shift, monoisotopic precursor ion and charge state, the dta files are edited to increase accuracy.

4. Consolidate peaks in MS/MS scans to remove weak ions. The top 10 peaks within every 100 *m/z* window are selected to remove weak ions (*see* **Note 17** for TMT data).

Database Search

1. Construct a target-decoy concatenated database. The target database includes downloaded proteins and common contaminants, while the decoy database is created by reversing all target sequences. The two databases are concatenated together for MS/MS search. False discovery rate (FDR) is estimated by the number of decoy matches (nd) and the total number of assigned matches (nt). FDR=2*nd/nt, assuming that mismatches in the target database have equal possibility of being derived from the decoy database.

2. When using the Sequest search program (version 28), we use the following search parameters: partially tryptic restriction, five maximal missed cleavage sites, five maximal modification sites, and 50 ppm mass tolerance for precursor ion. Product ion mass tolerance is set to 0.02 Da for high resolution MS/MS spectra from Q Exactive MS, or 0.5 Da for low resolution data from LTQ Orbitrap Elite MS. Static modification on Cys is carbamidomethylation (+57.02146 Da), and dynamic modification includes Met oxidation (+15.99492 Da), and Ser/Thr/Tyr phosphorylation (+79.96633) if needed. The search result for each dta file is saved in an out file (*see* **Note 18**).

Post-processing

1. Prefilter the MS/MS matched peptides by minimal peptide length (e.g., 7 for human database), and minimal matching scores (e.g., XCorr and ΔCn in a Sequest search).

2. Filter the matched peptides by mass accuracy (e.g., 2 ppm).

3. Further filter the peptides by XCorr and ΔCn scores. The peptides are divided into groups according to trypticity and

charge states, and then filtered by the two matching scores to reduce peptide and protein FDR to approximately 1 %.

4. Finally, when peptides are shared among multiple members of a protein family, all matched protein members are clustered into a single group. On the basis of the parsimony principle, the group is represented by one protein with the highest number of assigned peptides. Other proteins in the same group are not accepted unless they are matched by additional unique peptide(s).

3.3.2 Protein Quantification

As described above, three protein quantification strategies, spectral counting, SILAC, and iTRAQ/TMT, require different sample preparation protocols. Their related data processing methods are also distinct. After obtaining raw quantitative data for multiple sample comparisons, statistical inference is implemented to evaluate p-values, and a fraction of proteins are claimed to be altered. Two commonly used methods are adjusted p-value (i.e., q-value) and FDR derived from null experiments. The q-value is similar to p-value, but it measures the proportion of false positives incurred when the particular test is called significant. The latter is considerably straightforward, simply by estimating false positives in null comparisons.

Protein Profiling by Spectral Counting (SC)

Spectral count of a given protein is the summed number of its matching MS/MS spectra, similar to the number of reads in RNA-seq analysis. Since peptide ions are selected for MS/MS based on the rank of ion intensity, the spectral count of an identified protein is correlated with its abundance. Some pitfalls are that the spectral counting method is influenced by LC-MS/MS run-to-run variation, and small SC number (e.g., less than 5) markedly decreases the reliability of quantitative comparison.

1. Acquire spectral counts for proteins assigned by database search.

2. Normalize spectral counts to assume that the average SC per protein should be the same in all samples. This assumption works only when protein composition and concentration among samples are highly similar. For IP samples, this assumption may not apply well if the control and IP samples are vastly different.

3. Use a G-test to determine statistical significance (p-value) of protein abundance difference (*see* **Note 19**)

4. Calculate FDR based on null experiments (e.g., comparison between biological replicates) and set up a threshold for filtering the dataset.

Protein Profiling by SILAC

The SILAC method offers many benefits compared to label-free approaches. It can detect relatively small changes in protein expression levels or posttranslational modifications during comparison. Whereas SILAC quantification may be carried out prior to protein identification, as in the MaxQuant program [33], we usually derive SILAC data after identifying proteins [34].

1. Extract ion currents in MS scans for identified peptides. Define peak intensity, area intensity, and single-to-noise (S/N) ratio.

2. Match light and heavy isotope labeled peaks using predicted m/z difference and a defined mass tolerance (e.g., ±6 ppm for high resolution MS data).

3. Compare peak intensity of matched peptides to obtain abundance ratios of the peptides/proteins, and convert to log2 (ratio).

4. Summarize peptide ratios into protein ratios. Remove outliers by Dixon's Q test. The S/N ratio of a protein is represented by that of the most abundant peptide ion.

5. Fit the data to Gaussian distribution to obtain mean and standard deviation (SD). If the vast majority of proteins are not expected to change, the mean can be used to correct for systematic bias (e.g., experimental loading bias).

6. Select proteins with significantly changes in expression by the cutoff of log2(ratio) that is outside of a 95 % confidence interval (~2 SD) of the Gaussian distribution. FDR is evaluated by null experiments of biological replicates. We also notice that increasing the protein S/N cutoff leads to less false positives.

Protein Profiling by iTRAQ/TMT

Current isobaric mass tags (e.g., iTRAQ and TMT) provide simultaneously analysis of 4, 6, 8, or 10 biological samples. Quantitative information is extracted from report ions detected in MS/MS spectra.

1. Acquire ion intensity of reporter ions in MS/MS scans for all matched peptides, and define the S/N ratio of the ions.

2. For every binary comparison, compare the peak intensity of reporter ions to obtain abundance ratios, and convert to log2 (ratio). The S/N ratio of a protein is represented by that of the most abundant peptide ion.

3. The remaining protocol is the same as that in the SILAC analysis.

3.3.3 Validation of MS Results

Theoretically, a computer program could match almost any MS/MS spectrum to a putative peptide sequence. Post-processing is crucial to reduce false positives. The widely used target-decoy strategy provides a general estimation of the FDR of an entire dataset, but the reliability of individual peptides/proteins is not well evaluated, especially for proteins matched by only one single peptide, often termed one-hit-wonders. The same problem may occur for PTM identification that is typically matched by one MS/MS spectra. Errors in protein quantification also exist, because of misidentification, mismatching of ion peaks, variation of weak peaks, and peptide sharing among proteins. Thus, it is highly recommended that extra validation strategies are used to affirm the MS data.

1. Manually examine raw files for proteins of interest. Although this is a tedious step, manual interpretation can eliminate most errors in protein identification and quantification, as no current computer programs consider all possibilities during data processing.

2. Obtain both high resolution MS and MS/MS scans to improve the fidelity of the dataset.

3. Perform additional targeted MS/MS analysis with different fragmentation methods (*see* **Note 20**) for database search.

4. Chemically synthesize the peptides of interest, and then use them as internal standards to confirm the identification of native peptides. Their MS/MS patterns and retention time during LC-MS/MS should be identical.

5. Verify protein changes by targeted MS strategy approaches [16].

6. Verify protein changes by antibody-based approaches (e.g., Western blotting and immunohistochemistry).

4   Notes

1. Iodoacetamide (IAA) is used as a Cys-alkylation reagent and should be freshly made. At high temperatures (e.g., heating in SDS gel loading buffer), IAA modifies a fraction of Lys residues. Additional modification (i.e., two IAA molecules reacting with one amino group) form a tag of 114.0429 Da, the same as the mass of a tag generated by tryptic digestion of ubiquitin [35]. At low temperature (i.e., room temperature or lower) and concentration, this side reaction is essentially eliminated [34].

2. Unmodified trypsin undergoes auto-proteolysis to generate peptides that may interfere with mass spectrometric analysis. Also auto-proteolysis of trypsin can produce psuedotrypsin that has chymotrypsin-like activity. Acetylation of the ε-amino groups of lysine residues prevents autolysis. It is therefore desirable to use modified trypsin for protein digestion.

3. Urea can be converted to ammonium cyanate in solution, which is highly active and causes carbamylation of amine groups in proteins. The side reaction is accelerated by heating, so we generally use fresh urea solutions and perform digestion at room temperature in the presence of urea. In addition, when making 8 M urea solution, the dissolved urea occupies a significant volume.

4. It is important to use an appropriate amount of C18 resin for desalting. Either an excessive or limited amount of resin may lead to low recovery of peptides.

5. The sample loading amount should vary based on the LC system. In general, we load peptides digested from 1 μg of protein onto a 10–15 cm × 75 μm ID column. Increasing the gradient elution time may increase the number of identified proteins, but a plateau can be reached at a certain point due to peak broadening. [26]. A higher loading level may be applied for long columns (up to 1 m) installed on ultrahigh pressure HPLC systems. The operation pressure of a LC system is determined by column internal diameter, length, resin particle size and porous feature, and viscosity of buffers.

6. To avoid heavy isotope-labeled Arg-Pro conversion, extra proline is often included in the SILAC medium. Alternatively, Arg may be omitted and only Lys labeling used. In this case, samples are digested with Lys-C to ensure quantification of identified peptides. In addition to culturing cells in SILAC medium, flies [36] and mice [37] can also be fully labeled by feeding SILAC food.

7. Coomassie Blue G-250 colloidal protein staining can detect proteins as low as 10 ng and has little effect on in-gel digestion. A barely visible protein band is sufficient for sensitive MS identification. Sypro ruby stain has higher sensitivity and also works for MS. An MS-compatible silver staining protocol is recommended but could work with reduced sensitivity.

8. Special care needs to be taken to prevent keratin contamination of samples. Wear gloves at all times, and rinse them occasionally as they readily accumulate static charge and attract dust and pieces of hair and wool. If possible, perform all operations under laminar flow hood. Always visually check flasks, tubes, and pipette tips for contaminating particles.

9. Keep a jar of water inside the incubator to avoid a temperature gradient between the bottom and the lid of the tube. This prevents condensation of water at the inner surface of the lid and, consequently, premature dehydration of the gel pieces.

10. Centrifugation removes any residual gel pieces or other particles in extracted peptide solutions. This step is important to

avoid clogging of the nanoscale LC system, in particular, if in-gel digested samples are directly injected onto nanoscale LC columns.

11. MS instrumentation settings are optimized to achieve maximal peptide identification during gradient elution. The LTQ Orbitrap Elite hybrid instrument is capable of acquiring data dependent MS/MS in the ion trap while acquiring a high resolution full scan MS spectrum in the Orbitrap without affecting the duty cycle. The Q Exactive instrument can scan ions in the Orbitrap and accumulate ions simultaneously. The duty cycle is mainly affected by scan AGC, maximum ion injection time, and resolution.

12. Selectivity of phosphopeptide enrichment methods is affected by the ratio between beads, total level of peptides, the percentage of phosphopeptides, and the use of competitive reagent to reduce background binding [38]. Numerous competitive reagents have been compared and glutamic acid appears to one of the best choices [39].

13. A high pH in the elution step tends to degrade phosphopeptides. Acidify the samples immediately after elution.

14. During SC-based quantitative analysis, we shorten the dynamic exclusion time from 30 to 8 s in MS to increase the SC of proteins on a LTQ Orbitrap Elite instrument. When dynamic exclusion is set to 30 s, once one peptide ion is selected for MS/MS, it will not be selected again in the following 30 s, as its elution peak width is ~30 s. This function is designed to avoid repetitive selection of highly abundant ions. Reducing the dynamic exclusion may slightly affect the total number of proteins identified, but greatly improves the accuracy of SC-based quantification.

15. In SILAC analysis, one may think that mixing cell lysates before IP would reduce experimental variation. However, after these experiments failed, we recognized that noncovalently associated proteins are exchangeable during IP, and the exchange rate is rapid [40]. Thus it is important to perform IP first and then mix the SILAC-labeled proteins for LC-MS analysis.

16. Isobaric labeling strategies suffer from quantification distortion, a problem caused by co-eluted peptides that give rise to the same reporter ions, raising noise level of reporter ions. For example, a tenfold change may be detected as a fourfold change. In most cases, the co-eluted peptides may be very low in intensity but could be present in a large number, which may explain the distortion that occurs even when the precursor ions look highly abundant with no obvious co-eluted peaks. The problem may be alleviated by reducing precursor isolation window or extensive fractionation, but neither

method solves the issue. Alternatively, the problem can be addressed by gas phase ion purification [41], isolation of MS2 ions for MS3 fragmentation, and measurement of C-terminal peptide ions [42, 43].

17. The MS/MS consolidation step is inactivated during TMT analysis, because this function may remove TMT reporter ions (e.g., 6 ions between 126 and 131 *m/z*).

18. Although the Orbitrap allows the acquisition of high resolution data with mass accuracy within a few ppm or even sub-ppm dependent on the setting and intensity of ion signal, a wide window (50 ppm) is used during the search and a much narrower mass window is used later during data filtering to remove false positives. However, the narrow mass window may be applied during the search step, and then cannot be used for filtering [26]. The same scenario can be applied to the settings of tryptic restriction, maximal missed cleavage sites, and maximal modification sites.

19. The *G*-value of each protein is calculated by the following equation [19]:

$$G = 2 \times \left(S_1 \times \ln\left(\frac{S_1}{2 \times (S_1 + S_2)}\right) + S_2 \times \ln\left(\frac{S_2}{2 \times (S_1 + S_2)}\right) \right)$$

Where S_1 and S_2 are the detected spectral counts of a given protein in any of the two samples for comparison, respectively. Neither S_1 nor S_2 can be zero in the equation. In practice, we use a relatively small number (e.g., 0.000001) to estimate *G*-value for zero spectral counts in either of the two samples.

20. Peptide ions can be fragmented by multiple technologies, such as collision-induced dissociation (CID), higher-energy collision dissociation (HCD), and electron transfer dissociation (ETD), generating different product ion MS/MS patterns. Combination of these MS/MS spectra can significantly improve the confidence of protein identification [44].

Acknowledgements

This work was partially supported by the National Institutes of Health (R21NS081571, R21AG039764, and P30CA021765), the American Cancer Society (RSG-09-181), and ALSAC (American Lebanese Syrian Associated Charities).

References

1. Cravatt BF, Simon GM, Yates JR 3rd (2007) The biological impact of mass-spectrometry-based proteomics. Nature 450:991–1000

2. Choudhary C, Mann M (2010) Decoding signalling networks by mass spectrometry-based proteomics. Nat Rev Mol Cell Biol 11:427–439

3. Gstaiger M, Aebersold R (2009) Applying mass spectrometry-based proteomics to genetics, genomics and network biology. Nat Rev Genet 10:617–627

4. Mann M, Kulak NA, Nagaraj N et al (2013) The coming age of complete, accurate, and ubiquitous proteomes. Mol Cell 49:583–590

5. Chien CT, Bartel PL, Sternglanz R et al (1991) The two-hybrid system: a method to identify and clone genes for proteins that interact with a protein of interest. Proc Natl Acad Sci U S A 88:9578–9582

6. Behrends C, Sowa ME, Gygi SP et al (2010) Network organization of the human autophagy system. Nature 466:68–76

7. Vidal M, Cusick ME, Barabasi AL (2011) Interactome networks and human disease. Cell 144:986–998

8. Havugimana PC, Hart GT, Nepusz T et al (2012) A census of human soluble protein complexes. Cell 150:1068–1081

9. Eng JK, Searle BC, Clauser KR et al (2011) A face in the crowd: recognizing peptides through database search. Mol Cell Proteomics 10(R111):009522

10. Nesvizhskii AI, Vitek O, Aebersold R (2007) Analysis and validation of proteomic data generated by tandem mass spectrometry. Nat Methods 4:787–797

11. Peng J, Gygi SP (2001) Proteomics: the move to mixtures. J Mass Spectrom 36:1083–1091

12. Xie F, Smith RD, Shen Y (2012) Advanced proteomic liquid chromatography. J Chromatogr A 1261:78–90

13. Duncan MW, Aebersold R, Caprioli RM (2010) The pros and cons of peptide-centric proteomics. Nat Biotechnol 28:659–664

14. Michalski A, Cox J, Mann M (2011) More than 100,000 detectable peptide species elute in single shotgun proteomics runs but the majority is inaccessible to data-dependent LC-MS/MS. J Proteome Res 10:1785–1793

15. Gerber SA, Rush J, Stemman O et al (2003) Absolute quantification of proteins and phosphoproteins from cell lysates by tandem MS. Proc Natl Acad Sci U S A 100:6940–6945

16. Picotti P, Aebersold R (2012) Selected reaction monitoring-based proteomics: workflows, potential, pitfalls and future directions. Nat Methods 9:555–566

17. Aebersold R, Mann M (2003) Mass spectrometry-based proteomics. Nature 422:198–207

18. Liu H, Sadygov RG, Yates JR 3rd (2004) A model for random sampling and estimation of relative protein abundance in shotgun proteomics. Anal Chem 76:4193–4201

19. Zhou JY, Afjehi-Sadat L, Asress S et al (2010) Galectin-3 is a candidate biomarker for amyotrophic lateral sclerosis: discovery by a proteomics approach. J Proteome Res 9:5133–5141

20. Wang W, Zhou H, Lin H et al (2003) Quantification of proteins and metabolites by mass spectrometry without isotopic labeling or spiked standards. Anal Chem 75:4818–4826

21. Ross PL, Huang YN, Marchese JN et al (2004) Multiplexed protein quantitation in Saccharomyces cerevisiae using amine-reactive isobaric tagging reagents. Mol Cell Proteomics 3:1154–1169

22. Ong SE, Blagoev B, Kratchmarova I et al (2002) Stable isotope labeling by amino acids in cell culture, SILAC, as a simple and accurate approach to expression proteomics. Mol Cell Proteomics 1:376–386

23. Dephoure N, Gygi SP (2012) Hyperplexing: a method for higher-order multiplexed quantitative proteomics provides a map of the dynamic response to rapamycin in yeast. Sci Signal 5:rs2

24. Eng J, McCormack AL, Yates JR 3rd (1994) An approach to correlate tandem mass spectral data of peptides with amino acid sequences in a protein database. J Am Soc Mass Spectrom 5:976–989

25. Perkins DN, Pappin DJ, Creasy DM et al (1999) Probability-based protein identification by searching sequence databases using mass spectrometry data. Electrophoresis 20:3551–3567

26. Xu P, Duong DM, Peng J (2009) Systematical optimization of reverse-phase chromatography for shotgun proteomics. J Proteome Res 8:3944–3950

27. Shevchenko A, Wilm M, Vorm O et al (1996) Mass spectrometric sequencing of proteins silver-stained polyacrylamide gels. Anal Chem 68:850–858

28. Ficarro SB, McCleland ML, Stukenberg PT et al (2002) Phosphoproteome analysis by mass spectrometry and its application to

Saccharomyces cerevisiae. Nat Biotechnol 20:301–305

29. Larsen MR, Thingholm TE, Jensen ON et al (2005) Highly selective enrichment of phosphorylated peptides from peptide mixtures using titanium dioxide microcolumns. Mol Cell Proteomics 4:873–886

30. Ballif BA, Villen J, Beausoleil SA et al (2004) Phosphoproteomic analysis of the developing mouse brain. Mol Cell Proteomics 3:1093–1101

31. Rush J, Moritz A, Lee KA et al (2005) Immunoaffinity profiling of tyrosine phosphorylation in cancer cells. Nat Biotechnol 23:94–101

32. Zhou H, Watts JD, Aebersold R (2001) A systematic approach to the analysis of protein phosphorylation. Nat Biotechnol 19:375–378

33. Cox J, Mann M (2008) MaxQuant enables high peptide identification rates, individualized p.p.b.-range mass accuracies and proteome-wide protein quantification. Nat Biotechnol 26:1367–1372

34. Xu P, Duong DM, Seyfried NT et al (2009) Quantitative proteomics reveals the function of unconventional ubiquitin chains in proteasomal degradation. Cell 137:133–145

35. Nielsen ML, Vermeulen M, Bonaldi T et al (2008) Iodoacetamide-induced artifact mimics ubiquitination in mass spectrometry. Nat Methods 5:459–460

36. Sury MD, Chen JX, Selbach M (2010) The SILAC fly allows for accurate protein quantification in vivo. Mol Cell Proteomics 9:2173–2183

37. Kruger M, Moser M, Ussar S et al (2008) SILAC mouse for quantitative proteomics uncovers kindlin-3 as an essential factor for red blood cell function. Cell 134:353–364

38. Li QR, Ning ZB, Tang JS et al (2009) Effect of peptide-to-TiO2 beads ratio on phosphopeptide enrichment selectivity. J Proteome Res 8:5375–5381

39. Kettenbach AN, Gerber SA (2011) Rapid and reproducible single-stage phosphopeptide enrichment of complex peptide mixtures: application to general and phosphotyrosine-specific phosphoproteomics experiments. Anal Chem 83:7635–7644

40. Wang X, Huang L (2008) Identifying dynamic interactors of protein complexes by quantitative mass spectrometry. Mol Cell Proteomics 7:46–57

41. Wenger CD, Lee MV, Hebert AS et al (2011) Gas-phase purification enables accurate, multiplexed proteome quantification with isobaric tagging. Nat Methods 8:933–935

42. Ting L, Rad R, Gygi SP et al (2011) MS3 eliminates ratio distortion in isobaric multiplexed quantitative proteomics. Nat Methods 8:937–940

43. McAlister GC, Huttlin EL, Haas W et al (2012) Increasing the multiplexing capacity of TMTs using reporter ion isotopologues with isobaric masses. Anal Chem 84:7469–7478

44. Guthals A, Bandeira N (2012) Peptide identification by tandem mass spectrometry with alternate fragmentation modes. Mol Cell Proteomics 11:550–557

Chapter 18

Using Peptide Arrays Created by the SPOT Method for Defining Protein-Protein Interactions

Yun Young Yim, Katherine Betke, and Heidi Hamm

Abstract

Evaluating sites of protein-protein interactions can be an arduous task involving extensive mutagenesis work and attempts to express and purify individual proteins in sufficient quantities. Peptide mapping is a useful alternative to traditional methods as it allows rapid detection of regions and/or individual residues important for binding, and it can be readily applied to numerous proteins at once. Here we describe the use of the ResPep SL SPOT method to evaluate protein–protein binding interactions such as that between G-protein βγ subunits and SNARE proteins, identifying both regions of interest and subsequently individual residues which can then be manipulated in further biochemical assays to confirm their validity.

Key words ResPep SL, Fmoc chemistry, Peptide, Protein-protein interaction, Alanine screening

1 Introduction

1.1 History/Theory

Automated peptide arrays such as that performed by the ResPep SL enable protein-protein interactions to be easily studied. Using the SPOT method, overlapping linear peptides are simultaneously synthesized on cellulose membranes and used to predict the biological activity of a protein. The SPOT method was first described by Ronald Frank in 1992 [1] as a method for in situ peptide synthesis. It creates short, immobilized peptides on a glass plate or cellulose membrane, which are easy to handle and retain some of the biological characteristics derived from the primary sequence [2]. Cellulose membranes have become the preferred medium for these arrays as the membranes are inexpensive, stable in aqueous solutions, nontoxic to biological samples, and able to withstand organic solvents and acids [3]. Further, for a given membrane, the SPOT method allows the parallel synthesis of up to 600 different peptides in a highly reliable manner. This is done using Fmoc-chemistry whereby the amino group of each amino acid is temporarily protected by a 9-fluorenylmethoxycarbonyl group (Fmoc) to prevent polymerization and allow the addition

Cheryl L. Meyerkord and Haian Fu (eds.), *Protein-Protein Interactions: Methods and Applications*, Methods in Molecular Biology, vol. 1278, DOI 10.1007/978-1-4939-2425-7_18, © Springer Science+Business Media New York 2015

of only one amino acid to a growing peptide chain at each step. It enables repetitive amino terminus deprotection by piperidine and allows a mild-acid labile peptide–resin linkage [4]. Other reactive functional side chain groups are also preserved by protection groups such as tert. butyl-oxy(OtBu) and trityl (Trt), preventing interaction between side chains and the amino terminal of activated amino acids. As the synthesis proceeds, the machine cycles through deprotection steps whereby the Fmoc groups are removed by piperidine, the membranes washed, and activated amino acids coupled to their respective partners according to the desired sequence until the full peptide is assembled [5]. Using the ResPep SL SPOT method, we have shown that such peptide arrays can be used to identify biologically active protein-protein interaction sites. Despite the fact that peptides only share some of the biological activities of their full-length counterparts, this method enables the generation of a large number of individual peptides at once to facilitate screening for biological activity with a diverse group of effectors.

1.2 The Applications of Peptide Arrays

Peptides arrays generated using the SPOT method have wide biological applications, including evaluating the actions of enzymes, cell adhesion, and the binding of metals [2] but are particularly useful for identifying sites of protein-protein interactions and the development of inhibitors or activators of enzymes [6]. To understand protein-protein interactions, investigators often use co-immunoprecipitation, pull-down, or two hybrid assays, as well as in vivo Förster resonance energy transfer (FRET) [7]. While such assays can identify the presence of protein-protein interactions, they cannot determine the interaction sites without conjunction to other methods such as X-ray crystallography. At present, X-ray crystallography is widely used to evaluate sites of protein-protein interaction; however, this process is both challenging and time consuming, particularly with proteins that are difficult to express, crystallize, or obtain diffraction. In such instances, peptide arrays offer an easy alternative as they can rapidly highlight regions of residues which may be involved in binding without the challenge of expressing and crystallizing full-length proteins. While caution must be taken as this method does not use full-length expressed proteins, it offers a method to determine regions of interaction which can then be followed up with alanine screening and other biochemical methods to test identified residues in the context of the full-length proteins for their functional role in the interaction. As an example, Wells et al. [8] recently described this method to characterize the interaction sites between G-protein βγ subunits and a member of the SNARE complex, SNAP25. Following initial peptide screens, they determined which specific amino acid residues were important using Ala screening, before demonstrating that those residues were important in the context of the whole protein using biochemical and functional assays. Comparatively, peptide arrays generated using the SPOT method can also be used in a high throughput manner to develop enzyme inhibitors, as well as active substrates for various enzymes. As Thiele et al. shows, by generating peptide libraries on a glass plate or cellulose membrane, the peptide-enzyme activity, such as phosphorylation of peptide, can be detected using autoradiography, chemiluminescence, or enzymatic color development [6]. By monitoring the phosphorylation of peptides by a particular enzyme, researchers can select lead peptides which have the potential to be developed as enzyme inhibitors or activators. This is exemplified in Mukhija et al. [9], where the authors successfully used this method to identify peptides inhibiting enzyme I (EI) of the bacterial phosphotransferase system (PTS). Using immobilized combinatorial peptide libraries and phosphorimaging, they identified heptapeptides and octapeptides which selectively inhibit EI in vitro. Similarly, Collet et al. [10] has screened potential substrates of proteases by synthesizing peptides on a fluorous-coated glass. For the first time, the authors describe the use of fluorous to probe peptide sequences in protease screening. With the change in material where the peptide is immobilized and the detection methods [5], the peptide array by SPOT method was used as a screening method for the development of inhibitors or activators of enzymes. Thus, peptide arrays generated by the SPOT method have wide biological implications including significantly enhancing protein-protein interaction studies, as well as aiding in the development of reagents such as inhibitors and active substrates of enzymes.

In this chapter, we demonstrate how to use peptide arrays created by the SPOT method to understand protein-protein interactions. We first describe how to prepare all the reagents and program software for the SPOT method. With appropriate reagents and correctly programmed software, peptides are synthesized on cellulose membrane. Then, we show how to de-protect side chain and use far-western technique to identify peptides involved in protein-protein interaction. Lastly, we describe how to identify amino acids involved in protein-protein interaction by alanine screen.

2 Materials

Prepare all solutions using Milli-Q water and analytical grade reagents. Reagents should be stored at room temperature unless otherwise noted and materials disposed of in the appropriate manner. Carry out all procedures at room temperature and inside a fumehood unless otherwise specified.

1. Auto SPOT robot, such as the ResPep SL from Intavis loaded with ResPep software.

2.1 SPOT Method Preparation

2. Spotter needle.

3. 1 ml ResPep glass syringe.

4. Protein sequence: From UniProt (www.uniprot.org), find the protein sequence of interest and save as a text file.

5. 1 l of anhydride dimethylformamide (DMF): Pour DMF into a glass 1 l bottle and load into the Auto SPOT robot in the dilutor 1 reservoir.

6. 4 l of anhydride dimethylformamide (DMF): Connect 4 l glass bottle of DMF to ResPep SL equipment as the dilutor 2 solvent 1 (*see* **Note 1**).

7. 2 l of ethanol: Connect to ResPep SL equipment as the dilutor 2 solvent 2 (*see* **Note 1**).

2.2 Peptide Array Synthesis

1. Amino acid derivatives: Remove amino acids derivatives from the −20 °C freezer and thaw until they reach room temperature (approximately 2 h). Weigh out the appropriate amount of each amino acid derivative into 2 ml microcentrifuge tubes and dissolve in *N*-methyl pyrrolidone (NMP) using an oscillating shaker (*see* **Note 2**).

2. 2 ml ResPep tubes with caps.

3. Membrane preparation: Pre-swell cellulose membrane from INTAVIS in 30 ml of DMF in a black container covered with a lid for 1 h. Agitate gently every 15 min.

4. 1.1 M hydroxybenzotriazole (HOBT•H_2O): Weigh out the appropriate amount of HOBT•H_2O into a 30 ml conical tube and dissolve in DMF to produce a 1.1 M solution (*see* **Note 3**).

5. 1.1 M diisopropylcarbodiimide (DIC): Using a glass pipette, dilute the appropriate volume of DIC with DMF in a 50 ml conical tube to produce a 1.1 M solution (*see* **Note 3**).

6. 20 % piperidine: In a 250 ml plastic bottle, add the appropriate volume of piperidine to DMF to produce a 20 % solution (*see* **Note 3**).

7. 5 % acetic anhydride: Using a glass pipette, transfer the appropriate volume of acetic anhydride to a 50 ml conical tube and mix with DMF to produce a 5 % solution (*see* **Note 3**).

2.3 Side Chain Deprotection

1. Deprotection solution (10 ml/membrane): Using a glass pipette, add 9.5 ml of trifluoroacetic acid (TFA) to a glass beaker. Add 300 μl of triisopropylsilane and 200 μl of ddH_2O to the beaker and mix thoroughly.

2. Dichloromethane (DCM): Using a glass pipette, transfer 80 ml of DCM into a glass beaker.

3. DMF: Using a glass pipette, transfer 80 ml of DMF into a glass beaker.

4. Ethanol: Using a glass pipette, transfer 40 ml of 100 % ethanol into a glass beaker.

2.4 Far-Western Development

1. Plastic container for cellulose membrane.

2. Ethanol.

3. Tris Buffered Saline (TBS): 50 mM Tris–HCl, pH 7.6, and 150 mM NaCl.

4. TBS containing 0.1 % Tween (TBST).

5. Blocking solution: Dissolve 5 % dried skim milk (w/v) in TBST (0.1 % Tween). Store at 4 °C until needed.

6. Primary antibody solution: Dissolve 5 % dried skim milk (w/v) in TBST (0.2 % Tween) and 0.1 % *n*-octylglucoside (OG). Add the primary antibody at the appropriate dilution and store at 4 °C until needed (*see* **Note 4**).

7. Secondary antibody solution: Dissolve 5 % dried skim milk (w/v) in TBST (0.2 % Tween) and 0.1 % *n*-octylglucoside (OG). Add the secondary antibody at the appropriate dilution and store at 4 °C until needed (*see* **Note 4**).

8. Wash buffer: TBST (0.1 % Tween) and 0.1 % OG (*see* **Note 4**).

9. TBS with *n*-octylglucoside (OG): 50 mM of Tris–HCl, pH 7.6, 150 mM NaCl, and 0.1 % OG (*see* **Note 4**).

10. Protein binding buffer: 20 mM HEPES-KOH, pH 7.5, 2 mM $MgCl_2$, 0.1 % *n*-octylglucoside, and 5 % glycerol (*see* **Note 5**).

11. Binding partners: Purified protein(s) of interest to examine interaction with immobilized peptides (*see* **Note 6**).

12. Antibodies: Primary and horseradish peroxidase (HRP) conjugated secondary antibodies (*see* **Note 7**).

13. Enhanced chemiluminescence (ECL) reagents.

2.5 Alanine Screen

1. This process requires the same materials listed in Subheadings 2.1–2.4.

3 Methods

Carry out all procedures at room temperature unless otherwise specified. Thoroughly read the manual of the auto spotter equipment such as the ResPep SL manual to understand how to use the equipment and programs.

3.1 SPOT Method Preparation (Fig. 1)

1. Turn on Auto SPOT robot such as ResPep SL and vacuum pump by pressing the green ON/OFF switch on the side of the AUTO spot robot.

2. Start the spotter software such as ResPep SL Spotter software on the computer (*see* **Note 8**).

3. Change waste bottles as needed.

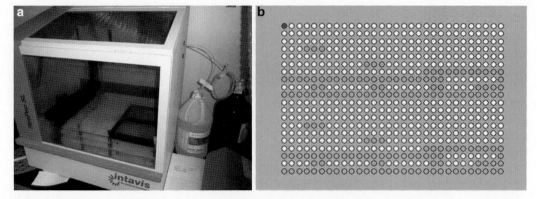

Fig. 1 ResPep SL and membrane design to study protein-protein interactions. (**a**) The ResPep SL is a peptide synthesizer that needs to be connected to a computer, vacuum pump, and various solutions to synthesize peptides on a membrane. The 4 L of anhydride Dimethylformamide and 2 L of ethanol are connected to the equipment from the outside as dilutor 2 solvent 1 and solvent 2, respectively (permission from Intravis AG). (**b**) Shown is a representative image of membrane design. *White dots* represent the location where peptides will be synthesized. *Grey dots* represent negative controls which are left without peptide synthesis. On a membrane, the positive control to the protein of interest and primary antibody are also synthesized. For example, the sequences for the SIRK peptide, QEHA peptide, βARK peptide, the Gβγ binding domain of the calcium channel CaV2.2, and the C-terminus of $G\beta_1$ are used to synthesize the positive control peptides to study the interaction of SNAP-25 and $G\beta_1\gamma_2$ (permission from Intravis AG)

4. Complete **steps 5–12** below for each protein to be synthesized on the membrane (*see* **Note 9**).

5. Under the "Edit sequence" tab, choose "open" and open text file.

6. Convert sequence to proper form that can be read by ResPep.

7. Type the name of the protein in the format ";<name of protein>" (e.g., human SNAP25).

8. On the next line, enter the length of the peptides to be synthesized and the extent of overlap between subsequent peptides (e.g., .seq,14,3 indicates peptides are to be 14 amino acids in length and overlap by three residues).

9. To add a single space between peptide spots, enter the command: ".space."

10. To clear the rest of the row after your last peptide for a particular protein, enter the command: ".newline."

11. To leave a line completely blank between two proteins, enter the command: ".newline"; Go to next line, type ".space" and again go to a new line and type ".newline."

12. Type ".end" where you want the machine to stop.

13. Once all sequences have been entered, press the "clean sequence" button under the "edit sequence" tab to ensure that all information is in the correct format (*see* **Note 10**).

14. Press the "SHOW PEPTIDES" tab and click any peptide to ensure it is located at the right position.

15. Load the method: Click the "EDIT METHODS" tab, press "file" and open the method called "SPOTpreactiv." (*see* **Note 11**).

16. Click the "REPORT" tab and check the amount of amino acid and reagents needed for the prep (*see* **Note 12**).

17. Mount and calibrate the spotter needle and 1 ml syringe (*see* **Note 13**).

18. Test the vacuum pump by clicking the ON and OFF buttons under the "MANUAL" tab in the ResPep software.

19. Fill the DMF reservoir bottle in the ResPep machine and ensure that the ethanol and DMF bottles on the outside are attached to the Auto SPOT robot as appropriate (*see* **Note 14**).

20. Prime dilutor 1 and 2 to remove air bubbles (*see* **Note 15**).

21. Wash the membrane three times with ethanol using the wash protocol in the method folder.

3.2 Peptide Array Synthesis (Fig. 2)

1. In the fumehood, transfer the dissolved amino acid derivatives into 2 ml ResPep tubes with caps (*see* **Note 16**).

2. Place the pre-swelled membrane onto the SPOT synthesis frame in fumehood (*see* **Note 17**).

3. Load the SPOT synthesis frame into the ResPep SL equipment and connect the vacuum tube to bottom of the SPOT synthesis frame.

4. Wash the membrane three times with ethanol using the wash protocol in methods (*see* **Note 18**).

5. Reload the "SPOTpreactiv" method for peptide generation.

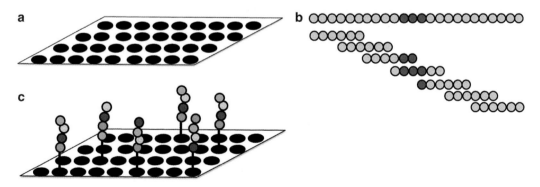

Fig. 2 Generating peptide arrays. (**a**) The cellulose membrane is pre-swelled by DMF. (**b**) Each peptide is generated as 14mers, and peptides overlap as shown. The *blue* residues are part of three different peptides. (**c**) Linear peptides are generated on a membrane as shown. Each *color dot* represents amino acids, which are synthesized into peptides

6. Remove the caps from reagents (1.1 M HOBt·H$_2$O, 1.1 M DIC, 20 % piperidine, and 5 % acetic anhydride solutions) and cover with tin foil. Create a hole in the tin foil to allow the spotter needle to pass into the reagent tubes unimpeded (*see* **Note 19**).

7. Load all reagents (1.1 M HOBt·H$_2$O, 1.1 M DIC, 20 % piperidine, and 5 % acetic anhydride solutions) into the specified positions on the reagent rack in the ResPep SL equipment.

8. Place the amino acid tubes into the appropriate locations on the amino acid rack as directed by the report file (*see* **Note 20**).

9. Remove the lids from the 0.6 ml microcentrifuge tubes and place into the small holes next to each amino acid tube (*see* **Note 21**).

10. Ensure all reagents are in place and close the cabinet door.

11. Verify the protein sequences and peptides to be generated.

12. Verify the loaded method.

13. Under "Run Synthesis" tab, hit the START button.

14. Monitor the reagents over the course of the run as reagents may need to be refilled (*see* **Note 22**).

15. Once the ResPep SL finishes the peptide synthesis, remove all remaining reagents from the rack and discard in appropriate waste containers (*see* **Note 23**).

3.3 Side Chain Deprotection

1. At the end of the run, use a pencil to mark the top and bottom corners of the membrane and any other location required for cutting the membrane (*see* **Note 24**).

2. In the fumehood, remove the membrane from the SPOT synthesis frame and place it in a container labeled TFA.

3. Add 10 ml of deprotection solution to the container and let react for 1 h in the fumehood (*see* **Note 25**). Following incubation, discard in a TFA waste container.

4. Wash the membrane four times in 20 ml dichloromethane (DCM) for 10 min per wash. Discard each wash in dichloro waste container.

5. Wash the membrane four times in 20 ml DMF for 2 min per wash. Discard waste in dichloro waste container.

6. Wash the membrane two times in 20 ml EtOH for 2 min per wash. Discard in dichloro waste container.

7. Allow the membrane to dry in the fumehood until completely dry. Wrap loosely in tin foil, and place in a labeled container in the refrigerator until ready to perform the far western experiment.

3.4 Far-Western Development (Fig. 3)

1. In a gel box, soak the membrane in ethanol for 5 min and wash twice with distilled water for 5 min on a shaker table at room temperature.

2. Block the membrane for 1 h on a shaker in 20 ml of 5 % milk in TBS/0.1 % Tween.

3. Wash the membrane five times for 5 min per wash in TBS/0.1 % Tween on shaker table.

4. Incubate the membrane overnight with an appropriate concentration of the protein of interest in protein binding buffer at 4 °C (*see* **Note 26**).

5. Wash the membrane three times for 5 min per wash with an appropriate wash solution on a shaker table

6. Incubate the membrane in 20 ml of primary antibody in primary antibody solution for 1 h (*see* **Note 7**).

7. Wash the membrane three times for 5 min/wash with an appropriate wash solution on shaker table.

8. Incubate the membrane in 20 ml of horseradish peroxidase (HRP) conjugated secondary antibody in secondary antibody solution for 1 h (*see* **Note 7**).

9. Wash the membrane two times for 5 min/wash in an appropriate wash solution on shaker.

10. Wash the membrane two times for 10 min/wash in TBS with OG.

11. Visualize using HRP: in a container, combine 3 ml of each enhanced chemiluminescence (ECL) reagent and mix well. Place membrane in the container and cover with tin foil. Swish container back and forth for 1 min. Transfer the membrane to a plastic sheet cover and use a pipette to smooth away bubbles. Set up the camera and place the membrane in the machine to focus as per instructions. Set number of pictures, exposures, etc. and take pictures (*see* **Note 27**).

12. Analyze data using a densitometry program (*see* **Note 28**).

3.5 Alanine Screen (Fig. 4)

For the alanine screen, use the methods listed in Subheadings 3.1–3.4. At Subheading 3.1, **step 5**, command ".replace, A" is needed for alanine screening. This will generate overlapping peptides as shown in Fig. 2b. The amount of amino acid derivatives and reagents needed for the experiment will change according to the peptide sequence, but the protocol is identical to that of general peptide array synthesis.

316   Yun Young Yim et al.

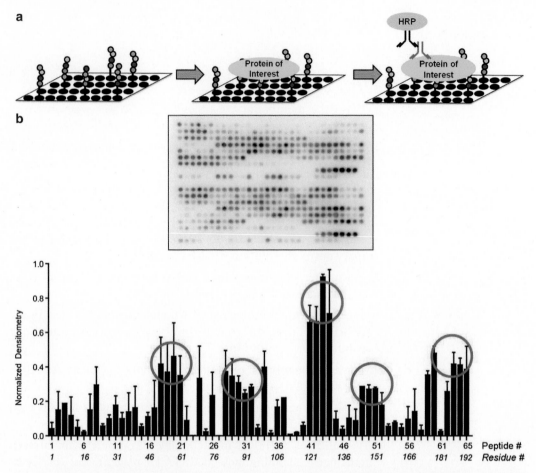

Fig. 3 Far-Western technique to screen for interactions with a protein of interest. Peptides are synthesized on a cellulose membrane, and binding interactions assessed using a far-western technique as stated in Subheading 3. (**a**) Peptides on the membrane are washed with ethanol and water, and then the blot is exposed to the protein of interest overnight. The next day, membranes are incubated with a primary antibody for the binding partner of interest, and a HRP-conjugated secondary antibody. (**b**) Using enhanced chemiluminescence (ECL) reagents that detect horseradish peroxidase (HRP) enzyme activity, peptides on membranes are screened for interaction with potential binding partners. Selective peptides will bind to their binding partners. Shown is a representative image of SNAP-25 peptides on a membrane exposed to $G\beta_1\gamma_2$. Using image J and Prism, densitometry analysis is used to quantify the intensity of each *dot*, and values are plotted via bar graph. Shown are representative densitometry results from the membrane examining SNAP-25 and $G\beta_1\gamma_2$ interactions. The *x*-axis reflects both the peptide number and residue number of SNAP-25. *Circled* regions, representing regions that interact with of the protein of interest, are used for subsequent alanine screening (reproduced from *Molecular Pharmacology* December 2012 82:1136–1149 with permission of American Society for Pharmacology and Experimental therapeutics)

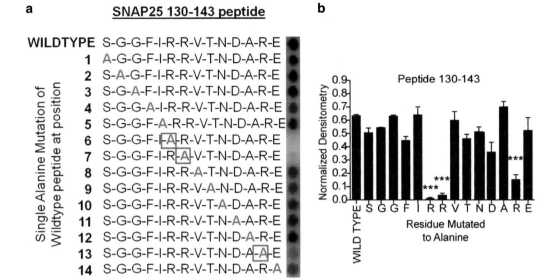

Fig. 4 Alanine Screening. Clusters of SNAP-25 peptides that are circled in red in Fig. 3b are tested via alanine screen. By monitoring the change in the intensity of each *dot*, the alanine screening suggests residues that are important for the protein-protein interaction. (**a**) Shown is a representative image of alanine screening for the SNAP-25 130–143 peptide that was synthesized on a membrane. The first spot is a wild type peptide, while the following 14 are mutated with a single alanine replacement of the residue at position 1–14 from wild type. *Boxes* highlight residues which are shown to interrupt the SNAP25-$G\beta_1\gamma_2$ interaction by alanine mutation. (**b**) The membrane is developed using the enhanced chemiluminescence (ECL) reagents. Densitometry is quantified and plotted into a graph as shown above. When arginine residues in the SNAP25 130–143 peptide are mutated to alanine, a significant decrease in the intensity of the dots is observed (*** $p < 0.001$) (reproduced from *Molecular Pharmacology* December 2012 82:1136–1149 with permission of American Society for Pharmacology and Experimental therapeutics)

4 Notes

1. ResPep SL equipment has two solvent connectors for dilutor 2. DMF is connected to dilutor 2 solvent 1, and ethanol is connected to dilutor2 solvent 2.

2. In the "REPORT" tab, the mass of each amino acid needed for the prep is calculated. Calculations are for a 0.5 M solution of each amino acid. Solutions are made according to the amount calculated by dissolving the amino acid derivatives in *N*-methyl pyrrolidone (NMP) overnight on a rotator in the fumehood.

3. In the "REPORT" tab, amounts of each reagent needed for the prep are calculated. Reagents are made fresh on the day of synthesis in DMF.

4. 0.1 % OG has to be added right before the solution is used.

5. Protein binding buffer depends on the protein you are using to study the interaction and must be determine empirically. This buffer cannot have high detergent levels as it will interfere with

the detection signal. Detergents have to be added right before the solution is used.

6. The purified protein should be concentrated so it can be diluted with the protein binding buffer when it is used in this experiment.

7. Primary and secondary antibodies are chosen based on the protein of interest. The amount of antibody needed for optimal signal detection must be optimized.

8. The system will start in simulation mode if a computer is not connected or the system is not switched ON. The following procedure in SPOT method preparation section is specific to ResPep SL from Intravis. If you are using different equipment, please follow the manual of your equipment.

9. You can view the spaces or lines that you've introduced by going to the "SHOW PEPTIDES" tab. This will show you the position of each peptide for every protein, as well as the number of spaces or lines that have been designated.

10. When pressing the "clean sequence" button, options such as single letter format, delete spaces, etc. will show up. Ignore these and press OK.

11. Peptide synthesis is a cyclic procedure with a number of steps carried out for each amino acid addition. Step 1 is Fmoc deprotection. Step 2 is DMF wash. Step 3 is ethanol wash. Step 4 is air drying of the membrane. Step 5 is spotting of activated amino acids and waiting for reaction time. Step 6 is capping (optional). Step 7 is DMF wash. These steps are repeated until the desired peptides have been assembled. After peptide assembly, the side chain protection groups will be removed using a side chain deprotection protocol.

12. Under the "DERIVATIVES" tab, the amount (in mg) of each amino acid that you will need to weigh out for your synthesis, the volume (in ml) of solvent needed to dissolve amino acids, and the final volume you will end up with will be shown. It will also have the amount of various reagents needed for the experiment.

13. Check the robot arm is capable of moving to various locations using the "go home," "go to vial," and "check vial positions" buttons under the manual tab. Ensure the needle contacts the membrane properly.

14. ResPep SL equipment is connected to dilutor 1 and dilutor 2. Dilutor 1 represents the DMF bottle inside the equipment. Dilutor 2 represents the DMF and ethanol bottles connected outside of the equipment. During the peptide synthesis, DMF is used to wash the membrane and ethanol is used to both wash the membrane and start the drying process.

15. For priming, allow depressing five times and emptying out waste reservoir by turning on the vacuum pump and completely draining into waste. For dilutor 2, there is solvent 1 (DMF) and solvent 2 (EtOH), both solutions need to be primed. By clicking the dilutor 2 solvent 1 prime button, you can prime the dilutor 2 solvent 1. Once it is finished, click the dilutor 2 solvent 2 prime button to prime the dilutor 2 solvent 2.

16. Only 2 ml ResPep tubes will fit onto the amino acid holder in the ResPep equipment. Remember to label each cap and tube appropriately.

17. Ensure that there are no bubbles or wrinkles on the membrane once mounted on the SPOT membrane frame.

18. If there are bubbles or the membrane dries unevenly, remove the membrane from the frame and repeat the DMF and wash steps.

19. Put a hole on foil for the needle to go through.

20. Before placing the amino acid tubes onto the rack, uncap all tubes.

21. These tubes will be used as mixing tubes for the activation of amino acid derivatives.

22. During refilling, synthesis is paused by clicking the PAUSE key. This will suspend the operation. Only do this when the needle is away from the reagent rack and NOT pressing down on the membrane or in an activation tube. A good time to pause is when the software is counting down between piperidine deprotection steps.

23. Waste should be collected following the hazardous waste rules at each institution. A carboy is connected to the ResPep SL equipment during peptide array synthesis. There should be separate waste containers for the deprotection solution and washes in side-deprotection steps.

24. Use a pencil to mark the top and bottom corners of the membrane. This will be important to identify the orientation of the membrane later.

25. Ensure the lid is on for the container. Every 15 min or so, swish solution around.

26. Protein and detergent concentrations have to be determined empirically. Individual proteins may need different concentrations to obtain optimal signals for analysis. The purity of expressed proteins may also affect the amount of protein needed for detection. For example, 0.4 μM Gβγ protein is needed to identify Gβγ-SNARE interactions. Similarly, for these experiments, the membrane was determined to tolerate 0.1 % OG.

27. Set number of pictures is 6; exposures are up to 60 s. Starting at 10 s, it takes a photo every 10 s. This depends on the strength of signals.

28. Analysis programs, which have a densitometry analysis function, can be used to analyze the data. Image J can be used to quantify the intensity of each dot. Each dot is normalized to the most intense dot on the membrane and plotted into a bar graph by prism.

References

1. Frank R (2002) The SPOT synthesis technique – synthetic peptide arrays on membrane supports – principles and applications. J Immunol Methods 267:13–26

2. Min DH, Mrksich M (2004) Peptide arrays: towards routine implementation. Curr Opin Chem Biol 8:554–558

3. Winkler DF, Hilpert K, Brandt O et al (2009) Synthesis of peptide arrays using SPOT-technology and the CelluSpots-method. Methods Mol Biol 570:157–174

4. Frank R, Doring R (1988) Simultaneous multiple peptide-synthesis under continuous-flow conditions on cellulose paper disks as segmental solid supports. Tetrahedron 44:6031–6040

5. Dikmans A, Beutling U, Schmeisser E et al (2006) SC2: a novel process for manufacturing multipurpose high-density chemical microarrays. Qsar Comb Sci 25:1069–1080

6. Thiele A, Zerweck J, Schutkowski M (2009) Peptide arrays for enzyme profiling. Methods Mol Biol 570:19–65

7. Phizicky EM, Fields S (1995) Protein–protein interactions – methods for detection and analysis. Microbiol Rev 59:94–123

8. Wells CA, Zurawski Z, Betke KM et al (2012) G beta gamma inhibits exocytosis via interaction with critical residues on soluble N-ethyl-maleimide-sensitive factor attachment protein-25. Mol Pharmacol 82:1136–1149

9. Mukhija S, Germeroth L, Schneider-Mergener J et al (1998) Identification of peptides inhibiting enzyme I of the bacterial phosphotransferase system using combinatorial cellulose-bound peptide libraries. Eur J Biochem 254:433–438

10. Collet BY, Nagashima T, Yu MS et al (2009) Fluorous-based peptide microarrays for protease screening. J Fluor Chem 130:1042–1048

Part III

Tag/Affinity-Based Methods

Chapter 19

Fluorescence Polarization Assay to Quantify Protein-Protein Interactions: An Update

Ronald T. Raines

Abstract

A fluorescence polarization assay can be used to evaluate the strength of a protein-protein interaction. A green fluorescent protein variant is fused to one of the protein partners. The formation of a complex is then deduced from an increase in fluorescence polarization, and the equilibrium dissociation constant of the complex is determined in a homogeneous aqueous environment. The assay is demonstrated by using the interaction of the S-protein and S-peptide fragments of ribonuclease A as a case study.

Key words Fluorescence anisotropy, Fluorescence polarization, Fusion protein, Green fluorescent protein, Protein-protein interaction

1 Introduction

Fluorescence polarization can be used to analyze macromolecular interactions in which one of the reactants is labeled with a fluorophore (*see* **Note 1**). In this assay, the formation of a complex is deduced from an increase in fluorescence polarization, and the equilibrium dissociation constant (K_d) of the complex is determined in a homogeneous aqueous environment (*see* **Note 2**). Most fluorescence polarization assays have used a small molecule such as fluorescein as a fluorophore [1–4].

Here, a variant (S65T) of green fluorescent protein (GFP) is used as the fluorophore in a polarization assay [5–7]. The advantages of using S65T GFP as the fluorophore are the ease with which a protein can be fused to GFP by using recombinant DNA techniques, the high integrity of the resulting chimera, and the broad chemical and physical stability of GFP compared to small-molecule fluorophores. To quantify the formation of a complex of two proteins (X and Y), a GFP fusion protein (GFP–X) is produced by using recombinant DNA technology (*see* **Note 3**).

Protein GFP–X is then titrated with protein Y, and the equilibrium dissociation constant is obtained from the increase in

Cheryl L. Meyerkord and Haian Fu (eds.), *Protein-Protein Interactions: Methods and Applications*, Methods in Molecular Biology, vol. 1278, DOI 10.1007/978-1-4939-2425-7_19, © Springer Science+Business Media New York 2015

fluorescence polarization that accompanies complex formation. Like a free fluorescein-labeled ligand, free GFP–X is likely to rotate more rapidly and therefore to have a lower rotational correlation time than does the GFP–X·Y complex. An increase in rotational correlation time upon binding results in an increase in fluorescence polarization, which can be used to assess complex formation [8].

In a fluorescence polarization assay, the interaction between the two proteins is quantified in a homogeneous solution. The fluorescence polarization assay thereby allows for the determination of accurate values of K_d in a wide range of solution conditions. GFP is particularly well suited to this application because its fluorophore is held rigidly within the protein, as revealed by the three-dimensional structure of wild-type GFP and the S65T variant [9, 10]. Such a rigid fluorophore minimizes local rotational motion, thereby ensuring that changes in polarization report on changes to the *global* rotational motion of GFP, as affected by a protein-protein interaction.

2 Materials

1. 20 mM Tris–HCl buffer (varying pH).

2. Solution of aqueous NaCl (varying concentration).

3. Purified GFP–X and purified protein Y.

4. Fluorometer equipped with polarization measurement capability.

5. Graphics software capable of nonlinear regression analysis (e.g., DeltaGraph or SigmaPlot).

3 Methods

3.1 Fluorescence Polarization Assay

1. Mix protein GFP–X (0.50–1.0 nM) with various concentrations of protein Y in 1.0 mL of 20 mM Tris–HCl buffer, pH 8.0, with or without NaCl at 20 °C (*see* **Note 4**). Conditions such as buffer, pH, temperature, and salt can be varied as desired.

2. After mixing, take five to seven polarization measurements at each concentration of protein Y (*see* **Notes 5** and **6**). For a blank measurement, use a mixture that contains the same components except for protein GFP–X.

3.2 Data Analysis

1. Fluorescence polarization (P) is defined as

$$P = \frac{I_{\parallel} - I_{\perp}}{I_{\parallel} - I_{\perp}} \tag{1}$$

where $I_{\parallel}$ is the intensity of the emission light parallel to the excitation light plane and $I_{\perp}$ is the intensity of the emission light perpendicular to the excitation light plane. P, the ratio of light intensities, is a dimensionless number with a maximum value of 0.5. Calculate values of K_d by fitting the data to the equation:

$$P = \frac{\Delta P \cdot F}{K_d + F} + P_{min} \qquad (2)$$

In Eq. 2, P is the measured polarization, $\Delta P (= P_{max} - P_{min})$ is the total change in polarization, and F is the concentration of free protein Y (*see* **Note 7**).

2. Calculate the fraction of bound protein (f_B) by using the equation

$$f_B = \frac{P - P_{min}}{\Delta P} = \frac{F}{K_d + F} \qquad (3)$$

Plot f_B versus F to show the binding isotherms.

3.3 Case Study

Fluorescence polarization was used to determine the effect of salt concentration on the formation of a complex between the S15 and S-protein fragments of ribonuclease A [11–15]. A GFP chimera of S-peptide [S15–GFP(S65T)–His$_6$] was produced from bacteria and titrated with free S-protein. The value of K_d increased fourfold when NaCl was added to a final concentration of 0.10 M (Fig. 1). A similar salt dependence for the dissociation of RNase S had been observed previously [16]. The added salt is likely to disturb the

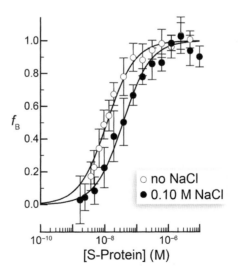

Fig. 1 Fluorescence polarization assay of a protein-protein interaction: S15–GFP (S65T)–His$_6$ with S-protein. S-Protein is added to 20 mM Tris–HCl buffer, pH 8.0, in a volume of 1.0 mL. Each data point is an average of 5–7 measurements. Curves are obtained by fitting the data to Eq. 3. The values of K_d in the presence of 0 and 0.10 M NaCl are 1.1×10^{-8} M and 4.2×10^{-8} M, respectively

water molecules hydrating the hydrophobic patch in the complex between S15 and S-protein, resulting in a decrease in the binding affinity [17]. Finally, the value of $K_d = 4.2\ \text{Å} \times 10^{-8}$ M observed in 20 mM Tris–HCl buffer (pH 8.0) containing NaCl (0.10 M) is similar (i.e., threefold lower) to that obtained by titration calorimetry in 50 mM sodium acetate buffer (pH 6.0) containing NaCl (0.10 mM) [18].

4 Notes

1. We used the term "polarization" instead of "anisotropy" herein. Fluorescence polarization (P) and fluorescence anisotropy (A) are related $[A = 2P/(3 - P)]$ and contain equivalent physical information with respect to monitoring macromolecular complex formation. Many instruments report on both polarization and anisotropy and either parameter can be used to evaluate K_d.

2. Polarization is proportional to the rotational correlation time, which is defined as

$$P \propto \tau = \frac{3\eta V}{RT} \qquad (4)$$

In Eq. 4, rotational correlation time (τ) is the time taken for a molecule to rotate 68.5° and is related to the solution viscosity (η), molecular volume (V), gas constant (R), and absolute temperature (T). Thus, under conditions of constant viscosity and temperature, polarization is directly proportional to the molecular volume, which increases upon complex formation.

3. Using the "superfolder" variant of GFP could facilitate the creation of a GFP–X fusion protein [19].

4. In the assay solution, [GFP–X] should be significantly lower than the value of K_d ([GFP–X] $< <$ K_d) but still be high enough to generate detectable fluorescence in the spectrometer. In the case study, [GFP–X] = 1 nM and $K_d > 10$ nM.

5. Data collection must be done at equilibrium. To estimate the time to reach equilibrium, a pilot experiment can be performed in which Y is added at [Y] = K_d, and the polarization is monitored until it reaches a stationary value.

6. At each [Y], the sample should be blanked with an identical mixture that lacks GFP–X.

7. The change in polarization (ΔP) upon complex formation must be detectable. For example, if the value of τ for GFP–X does not change significantly upon formation of the GFP–X·Y complex, then the value of ΔP is small and the data analysis is difficult.

References

1. Jameson DM, Ross JA (2010) Fluorescence polarization/anisotropy in diagnostics and imaging. Chem Rev 110:2685–2708

2. Smith DS, Eremin SA (2008) Fluorescence polarization immunoassays and related methods for simple, high-throughput screening of small molecules. Anal Bioanal Chem 391:1499–1507

3. Owicki JC (2000) Fluorescence polarization and anisotropy in high throughput screening: perspectives and primer. J Biomol Screen 5:297–306

4. Royer CA, Scarlata SF (2008) Fluorescence approaches to quantifying biomolecular interactions. Methods Enzymol 450:79–106

5. Park S-H, Raines RT (1997) Green fluorescent protein as a signal for protein–protein interactions. Protein Sci 6:2344–2349

6. Park SH, Raines RT (2000) Green fluorescent protein chimeras to probe protein–protein interactions. Methods Enzymol 328:251–261

7. Park SH, Raines RT (2004) Fluorescence polarization assay to quantify protein–protein interactions. Methods Mol Biol 261:161–166

8. Jameson DM, Sawyer WH (1995) Fluorescence anisotropy applied to biomolecular interactions. Methods Enzymol 246:283–300

9. Ormö M, Cubitt AB, Kallio K et al (1996) Crystal structure of the *Aequorea victoria* green fluorescent protein. Science 237:1392–1395

10. Yang F, Moss LG, Phillips GN Jr (1996) The molecular structure of green fluorescent protein. Nat Biotechnol 14:1246–1251

11. Richards FM, Vithayathil PJ (1959) The preparation of subtilisin modified ribonuclease and separation of the peptide and protein components. J Biol Chem 234:1459–1465

12. Watkins RW, Arnold U, Raines RT (2011) Ribonuclease S redux. Chem Commun 47:973–975

13. Richards FM (1958) On the enzymic activity of subtilisin-modified ribonuclease. Proc Natl Acad Sci U S A 44:162–166

14. Raines RT (1998) Ribonuclease A. Chem Rev 98:1045–1065

15. Kim J-S, Raines RT (1993) Ribonuclease S-peptide as a carrier in fusion proteins. Protein Sci 2:348–356

16. Schreier AA, Baldwin RL (1977) Mechanism of dissociation of S-peptide from ribonuclease S. Biochemistry 16:4203–4209

17. Baldwin RL (1996) How Hofmeister ion interactions affect protein stability. Biophys J 71:2056–2063

18. Connelly PR, Varadarajan R, Sturtevant JM et al (1990) Thermodynamics of protein–peptide interactions in the ribonuclease S system studied by titration calorimetry. Biochemistry 29:6108–6114

19. Pédelacq JD, Cabantous S, Tran T et al (2006) Engineering and characterization of a superfolder green fluorescent protein. Nat Biotechnol 24:79–88

Chapter 20

Förster Resonance Energy Transfer (FRET) Microscopy for Monitoring Biomolecular Interactions

Alexa L. Mattheyses and Adam I. Marcus

Abstract

Förster or Fluorescence resonance energy transfer (FRET) can be used to detect protein-protein interactions. When combined with microscopy, FRET has high temporal and spatial resolution, allowing the interaction dynamics of proteins within specific subcellular compartments to be detected in cells. FRET microscopy has become a powerful technique to assay the direct binding interaction of two proteins in vivo. Here, we describe a sensitized emission method to determine the presence and dynamics of protein-protein interactions in living cells.

Key words Fluorescence, FRET, Sensitized emission, Dynamics, Microscopy, Live cell imaging

1 Introduction

The spatial resolution of the light microscope is on the order of several hundred nanometers. Therefore, standard fluorescence microscopy techniques such as co-localization cannot establish whether two proteins are directly interacting, only whether proteins are in the same general area. Förster or Fluorescence resonance energy transfer (FRET) can overcome this limitation and directly detect dynamic protein-protein interactions in living systems. FRET only occurs over distances less than approximately 10 nm and when combined with microscopy provides the resolution necessary to observe protein-protein interactions with high spatial and temporal resolution.

FRET is a non-radiative transfer of energy between two molecules, in this case fluorophores [1]. Fluorophores are excited by absorption of photons at one wavelength and emit photons at a lower energy, longer wavelength. In FRET there are two fluorophores, a donor and an acceptor, where an excited donor fluorophore transfers energy to an acceptor fluorophore resulting in acceptor emission. This non-radiative energy transfer does not involve emission of a photon from the donor and results from dipole coupling.

Cheryl L. Meyerkord and Haian Fu (eds.), *Protein-Protein Interactions: Methods and Applications*, Methods in Molecular Biology, vol. 1278, DOI 10.1007/978-1-4939-2425-7_20, © Springer Science+Business Media New York 2015

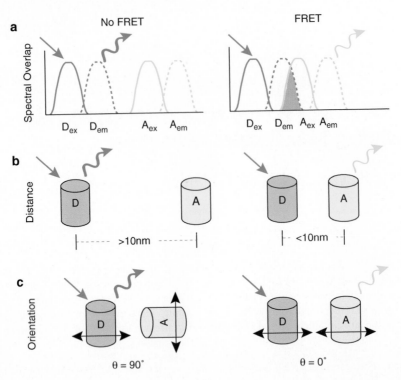

Fig. 1 Conditions for FRET include (**a**) A donor and acceptor pair where the emission spectra of the donor overlaps with the excitation spectra of the acceptor (spectral overlap; *grey shading*). (**b**) The distance between the donor and acceptor must be less than approximately 10 nm, at larger distances there is no FRET. (**c**) FRET efficiency is highest when the excitation and emission dipoles are oriented parallel and there is no FRET if they are perpendicular

There are three main conditions that need to be met for efficient FRET (Fig. 1). First, there must be "spectral overlap" of the donor's emission and the acceptors excitation spectra (Fig. 1a). Spectral overlap is a major consideration when the donor and acceptor fluorophores (termed the FRET pair) are selected. Second, the donor and acceptor fluorophores must be in close proximity to one another. If more than 10 nm separates the donor and acceptor, there is very little FRET; if they are within 10 nm, there is FRET (Fig. 1b). This distance dependence is the critical factor that enables detection of protein-protein interactions. Third, the dipoles of the donor and acceptor must be aligned. The efficiency of transfer is modulated by the angle between the excitation/emission dipoles, which is typically assumed to be random. Orientation is critical in systems where the fluorophores are not random but rigid with respect to one another: parallel dipoles have the potential for energy transfer while perpendicular dipoles do not (Fig. 1c).

The efficiency of energy transfer (E) is modulated by the distance between the two molecules (r):

$$E = \frac{1}{1 + (r/R_0)^6}$$

where R_0, also called the Förster radius, is the characteristic distance between molecules where the FRET efficiency is 50 %. The $1/r^6$ distance dependence allows FRET to be used as a molecular ruler for distances close to R_0 [2]. However, for most cell biology applications detecting interactions between two proteins, FRET is used as a binary readout: high FRET indicating the presence of an interaction and no FRET indicating the absence of interaction.

The Förster radius can be calculated for any pair of fluorescent molecules:

$$R_0 = \left(2.8 \times 10^{17} \times \kappa^2 \times Q_D \times E_A \times J(\lambda)\right)^6$$

where κ^2 is the orientation factor between the two fluorescent dipoles, Q_D is the quantum yield of the donor, E_A is the maximum extinction coefficient of the acceptor, and $J(\lambda)$ is the spectral overlap integral between the donor and acceptor. For the fluorescent protein FRET pairs R_0 is typically 5 ± 1 nm [3]. To assay protein-protein interactions in cells, fluorescent proteins are most commonly employed as the FRET donor and acceptor. Some common FRET pairs are listed in Table 1.

There are several basic approaches for measuring FRET in microscopy, including acceptor photobleaching, fluorescence lifetime imaging microscopy (FLIM), spectral imaging, fluorescence polarization, and sensitized emission [4–10]. This chapter

Table 1
Common FRET pairs

Donor–acceptor	Donor		Acceptor	
	Excitation	Emission	Excitation	Emission
ECFP-EYFP	433	475	514	527
ECFP-Venus	433	475	515	528
Cerulean-Venus	433	475	515	528
mTFP-Venus	462	492	515	528
EGFP-mCherry	488	507	587	610
Venus-mCherry	515	528	587	610
Clover-mRuby2	505	515	559	600

describes a sensitized emission method for measuring FRET that can be applied to dynamic systems. This basic technique can be modified, for example to measure intramolecular FRET, the stoichiometry of the FRET interaction, or protein conformation changes [9, 11].

The sensitized emission method, also called 3-cube FRET, is based on the detection of acceptor fluorescence after donor excitation. Sensitized emission measurements can be achieved using a standard confocal or widefield microscope with the appropriate excitation sources and emission filters. Three images are acquired for imaging donor, acceptor, and FRET channels (Fig. 2, Table 2). In the donor image, the sample is excited at the donor excitation wavelength and imaged with a donor emission filter. In the acceptor image, the sample is excited at the acceptor excitation wavelength and imaged with an acceptor emission filter. In the FRET image, the sample is excited at the donor excitation wavelength and imaged with an acceptor emission filter. Generally, it is not possible to image the sensitized emission directly due to contamination of the FRET image by both donor and acceptor. One source of contamination in the FRET image is that the donor excitation wavelength also excites the acceptor. Therefore, it is critical to select an acceptor that has minimal excitation at the donor excitation wavelength. Another source of contamination is donor fluorescence "bleed-through" that is detected in the acceptor emission channel. Therefore, the "FRET" signal measured when the sample is excited at the donor excitation wavelength and imaged at the acceptor emission wavelength consists of three components: (1) bleed-through from donor emission, (2) acceptor emission due to direct excitation of the acceptor, and (3) sensitized emission due to energy transfer. Thus, the measurement must be corrected for bleed-through and direct excitation. These corrections require calibration factors that are measured with reference samples containing either donor or acceptor molecules alone, as described in detail in Subheading 3.

The materials and methods are written to probe the interaction of protein-A and protein-B with FRET using fusion proteins with protein-A tagged with the donor CFP and protein-B tagged with the acceptor Venus. Protein-A and protein-B can be any two proteins of interest thought to interact based on biochemical or other experiments. Other Cyan and Yellow variants (i.e., Cerulean, Citrine) can be used in the constructs with no adjustments to the microscope. Other suitable FRET pairs can be substituted with appropriate adjustments to the constructs, filters, and lasers.

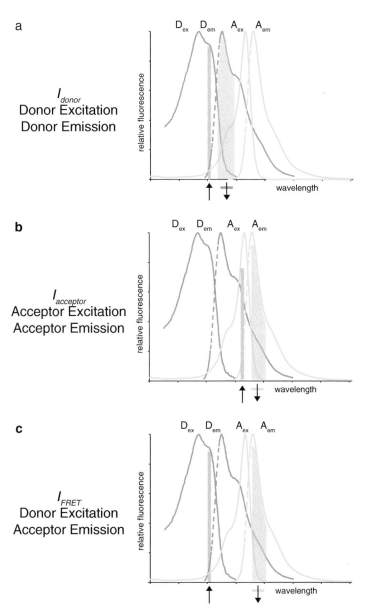

Fig. 2 Microscope setup for FRET. Two lasers, 458 nm for donor excitation and 514 nm for acceptor excitation, are selected with an AOTF and directed into the microscope by a polychroic mirror. After excitation of the sample, fluorescence is collected by the objective lens and the fluorescence split based on wavelength and imaged through a 480/20 filter (donor) or 535/30 filter (acceptor)

Table 2
Imaging parameters

Image	Excitation	Emission
I_{donor}	Donor; 458 nm	Donor; 480/20 nm
$I_{acceptor}$	Acceptor; 514 nm	Acceptor; 535/30 nm
I_{FRET}	Donor; 458 nm	Acceptor; 535/30 nm

2 Materials

2.1 Sample

There are three required and two optional control samples and one experimental sample. Before transfection with fluorescently tagged protein constructs cells should be plated in an appropriate live cell imaging dish, such as MatTek #1.5 coverslip bottom dish (MatTek Corporation, Ashland, MA). Fixed cell samples should be plated on #1.5 coverslips, fixed and mounted for imaging (*see* **Note 1**).

1. *Required* Control sample, donor only. Transfect cells with cytosolic CFP (or Protein-A-CFP, *see* **Note 2**).

2. *Required* Control sample, acceptor only. Transfect cells with cytosolic Venus (or Protein-B-Venus, *see* **Note 2**).

3. *Required* Control sample, medium only. Prepare a dish containing no cells with the imaging medium.

4. *Required* Experimental sample. Transfect cells with protein-A-CFP and protein-B-Venus. Ideally the expression level of the proteins should be similar (*see* **Note 3**).

5. *Suggested* Control sample, FRET positive. Transfect cells with CFP-Venus tandem protein.

6. *Suggested* Control sample, FRET negative. Transfect cells with cytoplasmic CFP and cytoplasmic Venus.

2.2 Microscope

This protocol describes data acquisition on a laser scanning confocal microscope (Fig. 3). It can be easily adapted for widefield microscopy (*see* **Note 4**).

1. Inverted laser scanning confocal fluorescence microscope with excitation lasers 458 nm (CFP) and 514 (Venus) and emission filters 480/40 (CFP) and 535/30 (Venus).

2. Objective lens: $60\times$ 1.4 NA oil immersion or similar high resolution objective lens.

3. Environmental control: heated stage and optional CO_2 to maintain live cells on the microscope (not required for fixed cell imaging).

2.3 Image Analysis

1. Image analysis software: Fiji (ImageJ pre-packaged with plugins) (free download: fiji.sc/Fiji) or equivalent commercial software.

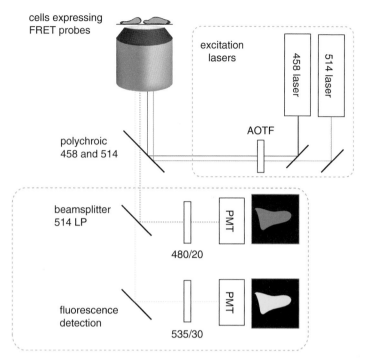

Fig. 3 The three acquisition conditions for sensitized emission FRET. (**a**) Excite at the donor wavelength and image at the donor wavelength (I_{donor}). (**b**) Excite at the acceptor wavelength and image at the acceptor wavelength ($I_{acceptor}$). (**c**) Excite at the donor wavelength and image at the acceptor wavelength (I_{FRET})

3 Methods

3.1 Optimize Imaging Conditions

The imaging parameters must be held constant for acquisition of all control and experimental images. First, you will optimize these parameters for all three imaging channels. Use the experimental sample expressing protein-A-CFP and protein-B-Venus.

1. Prepare the microscope. Select the 60× 1.4 NA oil high-resolution objective. Place a drop of immersion oil on the objective and focus on the doubly labeled sample. Set the temperature and CO_2 environmental control (*see* **Note 5**).

2. Set the image dimensions. The zoom/magnification and number of pixels should be selected to achieve optimum resolution as defined by Nyquist [12]. For our 60× 1.4 NA objective this is 90 nm/pixel. The ideal pixel size is determined by the optics and the NA of the objective (*see* **Note 6**).

3. Set the scan speed and line averaging. For live cell imaging the scan speed should be relatively fast and the line averaging low to minimize sample movement artifacts.

4. Sequentially for each channel, set the acquisition parameters for laser power, and detector gain and offset. The signal levels

for each channel should be roughly similar. There should be no saturation in the image, and there should be some "room" to accommodate an increase or decrease in fluorescence as the FRET signal dynamically changes. Check for and minimize photobleaching.

(a) Optimize the I_{donor} channel. Excite the sample at the donor excitation wavelength 458 nm, and select the donor emission filter 480/20.

(b) Optimize the $I_{acceptor}$ channel. Excite the sample at the acceptor excitation wavelength 514 nm, and select the acceptor emission filter 535/30.

(c) Optimize the I_{FRET} channel. Excite the sample at the donor excitation wavelength 458 nm, and select the acceptor emission filter 535/30.

3.2 Data Collection

All images are collected with the parameters established in Subheading 3.1.

1. Image the required control for donor bleed-through. Imaging the CFP only sample, collect I_{donor} and I_{FRET} for 10–20 individual cells.

2. Image the required control for acceptor bleed-through. Imaging the Venus only sample, collect $I_{acceptor}$ and I_{FRET} for 10–20 individual cells.

3. Image the required control for background. Imaging the medium only sample, collect I_{donor}, $I_{acceptor}$, and I_{FRET} from 20 independent positions. Ensure the microscope is focused in the medium (*see* **Note 7**).

4. Image the experimental sample. Imaging the protein-A-CFP and protein-B-YFP sample, collect I_{donor}, $I_{acceptor}$, and I_{FRET}. A time lapse should be set up to record dynamics.

Suggested controls

5. Image the recommended FRET positive control. From the CFP-Venus tandem protein sample collect I_{donor}, $I_{acceptor}$, and I_{FRET}.

6. Image the recommended FRET negative control. Imaging the cytosolic CFP and Venus sample, collect I_{donor}, $I_{acceptor}$, and I_{FRET}.

3.3 Image Processing and Analysis

All image processing steps can be performed with Fiji (*see* **Note 8**). The menu navigation in Fiji for the required functions is in Table 3. Images are referred to by the sample and the imaging condition, for example CFPI_{donor} is the CFP sample acquired with donor imaging conditions.

1. Calculate background from the medium only I_{donor}, $I_{acceptor}$, and I_{FRET} images. Background is a 20 frame average of the medium only sample:

Table 3
ImageJ/Fiji operations

Operation	Menu location
Adjust LUT	Image → Adjust → Brightness/contrast
Apply LUT	Image → Look up tables
Image math	Process → Image calculator
Math	Process → Math
Measure	Image → Measure
Open images	Plugins → LOCI → Bioformats importer
Save ROIs	Analyze → Tools → ROI manager
Stack average	Image → Stacks → Z project...; projection type = average intensity

$$\mathrm{BG_{donor}} = \mathrm{Average}(\mathrm{medium}I_{\mathrm{donor}})$$

$$\mathrm{BG_{acceptor}} = \mathrm{Average}(\mathrm{medium}I_{\mathrm{acceptor}})$$

$$\mathrm{BG_{FRET}} = \mathrm{Average}(\mathrm{medium}I_{\mathrm{FRET}})$$

2. Subtract $\mathrm{BG_{donor}}$ from all I_{donor} images, $\mathrm{BG_{acceptor}}$ from all I_{acceptor} images, and $\mathrm{BG_{FRET}}$ from all I_{FRET} images. All subsequent steps assume background subtracted images (*see* **Note 9**).

3. Calculate the contribution factor α of directly excited acceptor fluorescence in the FRET channel from the Venus only images. The result should be a 32-bit float image.

$$\alpha = \frac{\mathrm{Venus}I_{\mathrm{FRET}}}{\mathrm{Venus}I_{\mathrm{acceptor}}}$$

Select regions of the image where there was Venus fluorescence and measure the average α. Measure each field of view and average all α to obtain the final bleed-through correction value α (*see* **Note 10**).

4. Calculate the contribution factor β of donor fluorescence bleed-through in the FRET image from the CFP only images. The result should be a 32-bit float image.

$$\beta = \frac{\mathrm{CFP}I_{\mathrm{FRET}}}{\mathrm{CFP}I_{\mathrm{donor}}}$$

Select regions of the image where there was CFP fluorescence and measure the average β. Measure for each field of view and average all β to obtain the final bleed-through correction value β.

5. Calculate corrected FRET (N_{FRET}) for experimental and control data. The result should be a 32-bit float image.

$$N_{\mathrm{FRET}} = \frac{I_{\mathrm{FRET}} - \alpha I_{\mathrm{acceptor}} - \beta I_{\mathrm{donor}}}{\sqrt{I_{\mathrm{acceptor}} \times I_{\mathrm{donor}}}}$$

4 Notes

1. This protocol can be adapted for fixed cell samples that do not require labeling with antibodies. Transfect samples with donor and acceptor fusion proteins as described. Following adequate expression time, fix with 4 % paraformaldehyde and mount the cells on slides for microscopy.

2. The donor and acceptor only controls are needed to measure the bleed-through of direct excitation and emission into I_{FRET}. Cytosolic expression of CFP and Venus will simplify accurate measurements of these bleed-through contributions; however, protein-A-CFP and protein-B-Venus fusion proteins can be utilized.

3. The placement of the fluorescent tags in the fusion protein is critical for FRET. A negative "no FRET" result does not rule out protein interaction. If the distance between the donor and acceptor is too great (opposite sides of a complex) or if the orientation is not optimal, there may be no FRET even when protein-A and protein-B are in fact interacting. The labeling strategy should be carefully considered and different tag locations explored [13].

4. Sensitized emission FRET measurements can be conducted on a widefield or epi-fluorescence microscope. The widefield microscope should be equipped with a fluorescence light source (Hg lamp or LED), a scientific CCD or CMOS camera, and three filter cubes corresponding to the three imaging channels (Table 2). For CFP-Venus FRET the three filter cubes should be: Donor Cube: 430/20 ex 480/30 em; Acceptor Cube: 500/20 ex 535/30 em; and FRET Cube: 430/20 ex 535/30 em. The protocol can be followed substituting the appropriate filter cube for I_{donor}, I_{acceptor}, and I_{FRET}.

5. To reduce background, the cell imaging medium should be phenol-red free. When observed with widefield fluorescence, the sample is photobleaching. Therefore it is important to minimize the time the fluorescence lamp shutter is open and to use neutral density (ND) filters to attenuate the lamp illumination intensity.

6. The Nyquist sampling theorem is commonly utilized to determine the ideal pixel size for optimal high-resolution imaging.

7. To focus in the medium it is easiest to focus on some dust or particles on the coverslip and then up into the medium. Make sure the areas imaged do not contain any non-uniformities.

8. This analysis can be performed in many different imaging software. Macros can be developed in Fiji and there are plugins available for ImageJ/Fiji that streamline the analysis (i.e., pixFRET; free download: http://www.unil.ch/cig/page16989.html).

9. If the illumination intensity varies across the field of view, a shade correction, obtained by imaging a uniform fluorescence sample, should be applied.

10. The bleed-through of acceptor and donor (α and β) are often considered constants. This is generally true, however, intensity correlations have been reported. If the intensities of the sample varies significantly or if the intensity of the controls is significantly different than that of the sample this should be tested. If an intensity correlation exists, it can be corrected by fitting the bleed-through values.

Acknowledgements

The authors thank Dr. Claire E. Atkinson for critical reading of the manuscript. This work was supported by NIH grant 1RO1CA142858 awarded to A.I.M. and NIH grant 1R21AR066920 awarded to A.L.M.

References

1. Lakowicz JR (1988) Principles of frequency-domain fluorescence spectroscopy and applications to cell membranes. Subcell Biochem 13:89–126

2. Stryer L (1978) Fluorescence energy transfer as a spectroscopic ruler. Annu Rev Biochem 47:819–846

3. Lam AJ, St-Pierre F, Gong Y et al (2012) Improving FRET dynamic range with bright green and red fluorescent proteins. Nat Methods 9:1005–1012

4. Gautier I, Tramier M, Durieux C et al (2001) Homo-FRET microscopy in living cells to measure monomer-dimer transition of GFP-tagged proteins. Biophys J 80:3000–3008

5. Mattheyses AL, Hoppe AD, Axelrod D (2004) Polarized fluorescence resonance energy transfer microscopy. Biophys J 87:2787–2797

6. Padilla-Parra S, Tramier M (2012) FRET microscopy in the living cell: different approaches, strengths and weaknesses. Bioessays 34:369–376

7. Piston DW, Kremers GJ (2007) Fluorescent protein FRET: the good, the bad and the ugly. Trends Biochem Sci 32:407–414

8. Piston DW, Rizzo MA (2008) FRET by fluorescence polarization microscopy. Methods Cell Biol 85:415–430

9. Hoppe A, Christensen K, Swanson JA (2002) Fluorescence resonance energy transfer-based stoichiometry in living cells. Biophys J 83:3652–3664

10. Gordon GW, Berry G, Liang XH et al (1998) Quantitative fluorescence resonance energy transfer measurements using fluorescence microscopy. Biophys J 74:2702–2713

11. Goedhart J, Hink MA, Jalink K (2014) An introduction to fluorescence imaging techniques geared towards biosensor applications. Methods Mol Biol 1071:17–28

12. North AJ (2006) Seeing is believing? A beginners' guide to practical pitfalls in image acquisition. J Cell Biol 172:9–18

13. Miyawaki A, Tsien RY (2000) Monitoring protein conformations and interactions by fluorescence resonance energy transfer between mutants of green fluorescent protein. Methods Enzymol 327:472–500

Chapter 21

Utilizing ELISA to Monitor Protein-Protein Interaction

Zusen Weng and Qinjian Zhao

Abstract

Enzyme-linked immunosorbent assay (ELISA) is a commonly used method in analyzing biomolecular interactions. As a rapid, specific, and easy-to-operate method, ELISA has been used as a research tool as well as a widely adopted diagnostic method in clinical settings and for microbial testing in various industries. Inhibition ELISA is a one-site binding analysis method, which can monitor protein-protein interactions in solution as opposed to more commonly used sandwich ELISA in which the analyte capture step is required on a solid surface either through specific capture or through passive adsorption. Here, we introduce inhibition ELISA procedures, using a recombinant viral protein as an example, with emphasis on how inhibition ELISA could be used to probe subtle protein conformational changes in solution impacting protein–protein binding affinity. Inhibition ELISA is used to probe one binding site at a time for binding partners in solution with unrestricted conformation. The assay can be performed in a quantitative manner with a serially diluted analyte in solution for solution antigenicity or binding activity assessment.

Key words Inhibition ELISA, Monoclonal antibody, Antigen–antibody interaction, The half-maximal inhibitory concentration (IC_{50}), Conformational change, Disulfide bond, Binding affinity, Binding analysis in solution

1 Introduction

Enzyme-linked immunosorbent assay (ELISA) is a popular immunochemical assay [1] that uses a solid-phase enzyme immunoassay (EIA, or an earlier and more sensitive method, radiolabeled immunoassay commonly referred to as RIA) [2, 3] to detect the presence of a substance in a liquid sample. The specific binding was normally achieved through highly specific antigen–antibody interactions. ELISA has replaced RIA as the most popular method for analyzing a specific interaction due to the lack of need to use radiolabeled compounds. The ELISA was first reported in the 1960s [4] and now has been used as a diagnostic tool in medicine and plant pathology, as well as a quality-control assays in various industries and in quarantine laboratories at border controls and during pandemics [5, 6]. There are many ways one can perform an ELISA to probe a specific interaction in a qualitative or quantitative way.

Cheryl L. Meyerkord and Haian Fu (eds.), *Protein-Protein Interactions: Methods and Applications*, Methods in Molecular Biology, vol. 1278, DOI 10.1007/978-1-4939-2425-7_21, © Springer Science+Business Media New York 2015

In one example of ELISA, an antigen is affixed to a solid surface or onto synthetic beads through passive adsorption, and then a specific antibody or sample is applied over the surface so that it can bind to the antigen via a highly specific interaction. This detection antibody (or a secondary antibody) is covalently linked to an enzyme (the most common examples are horseradish peroxidase, HRP, and alkaline phosphatase, AP), and in the final step a substance is added that the enzyme can convert a substrate to a different form with detectable absorption or fluorescence signals [7–9].

There are several different types of ELISA, such as direct binding ELISA on Ag-coated plate, sandwich ELISA, competitive ELISA and so on. These methods are usually used to monitor antigen–antibody interaction through the detectable signals in the assay with an enzymatic reaction for amplification in the final step. Similarly, these methods can also be used to monitor two interacting proteins such as a protein and its receptor or an enzyme and its effector, and are not limited to an antigen–antibody pair. As a practical example, a competitive ELISA (or inhibition ELISA) was performed by Cuervo et al. to monitor the interaction between the Hepatitis B virus surface antigen (HBsAg) lots and polyclonal Ab for quality control on product consistency of a commercial vaccine [10]. The inhibition ELISA is a kind of competitive ELISA where a pair of binding partners is premixed prior to transferring to the assay plate to detect the residual labeled partner [11, 12]. The samples in solution to be detected in the inhibition ELISA compete for protein binding sites with the same protein coated on a solid surface, namely, the higher the binding activity in solution, the less the labeled binding partner left available for the surface immobilized partner to bind [13].

There are two unique characteristics for the inhibition ELISA we introduce here: it monitors a single epitope at a time as opposed two different epitopes in more commonly used sandwich ELISA [14]; and the test protein is interacting in solution with no restriction on its conformation as opposed to binding events on a surface-adsorbed protein as in the case for most ELISAs [15]. Therefore, it reflects more faithfully the biological interactions owing to fewer artifacts for this solution interaction based method and no wash cycles disrupting the interaction being probed [16]. Since there is no need for the analyte to be bound onto solid surface and to survive the wash cycles of most ELISAs [17], the inhibition ELISA, unlike dot blot, Western blot, direct binding ELISA and Biacore assay, can also be used on particulate proteins, such as a protein with the binding sites (i.e., epitopes) of interests on a cell surface or an antigen adsorbed onto particulate adjuvant in a vaccine formulation [18].

In this chapter, as an example of inhibition ELISA, we monitored the solution interaction of a pair of interacting proteins (recombinant HBsAg and a monoclonal antibody IgG1 5F11) in a quantitative manner. The binding strength can be quantitatively described by IC_{50} (the half-maximal inhibitory concentration) values of the analyte in solution [19]. IC_{50} is a measure of the effectiveness of a test sample (or a reference) in solution in binding to a biomolecule as a tracer in the assay. The larger an IC_{50} value (normally expressed in ng/mL) is, the weaker the interaction between an HBsAg sample and 5F11 in solution.

Since recombinant HBsAg is a cysteine-rich protein and the epitopes on HBsAg are highly dependent on the correct disulfide bind pairings [20], we tested the binding activity of 5F11 to HBsAg after the reduction of the disulfide bonds during pretreatment to probe the effect on the binding affinity. To reduce the disulfide bonds, the HBsAg are treated with 0.5 mM dithiothreitol (DTT). Subsequently, DTT was removed by dialysis to allow a certain degree of maturation (HBsAg would spontaneously undergo oxidative maturation after DTT removal) [20–22]. The epitope was found to mature back to some degree after the removal of the added reductant. We demonstrated that 0.5 mM DTT treated HBsAg and "Post DTT removal HBsAg with oxidative maturation" showed lower affinity (68- and 8-fold weaker, respectively) to 5F11 through their IC_{50} change relative to the untreated HBsAg control. The usefulness of this convenient solution-based inhibition ELISA method to monitor protein-protein interaction was demonstrated in a quantitative way.

2 Materials

2.1 Solutions

Prepare all solutions using double-distilled or deionized water (dd- or di-H_2O) and analytical grade reagents. Prepare and store all reagents at 4 °C (unless indicated otherwise). The reagents should be prepared as listed below.

1. Carbonate buffer (CB, 20×): 0.3 M Na_2CO_3, 0.7 M $NaHCO_3$. Weigh 31.8 g of Na_2CO_3 and transfer to a 1-L graduated cylinder or a glass beaker. Add 500 mL deionized water and mix with a glass bar until the Na_2CO_3 is completely dissolved. Then add 58.8 g of $NaHCO_3$ to the above solution and bring up to 1 L with dd-H_2O. Mix until no solid can be seen in the solution.

2. Saturated ammonium sulfate: Transfer 90 g of ammonium sulfate to a 100 mL graduated cylinder. Bring up to 100 mL with dd-H_2O. Incubate the solution at 80 °C in a water bath and dissolve the ammonium sulfate by stirring until the solid is completely dissolved. Incubate at 4 °C until the solution reaches 4 °C. Adjust the pH to 7.4 with 1.0 M NaOH (see **Note 1**).

3. Phosphate buffered saline (PBS) (10×): 20 mM KH_2PO_4, 0.1 M Na_2HPO_4, 1.37 M NaCl, 27 mM KCl. Weigh 2.7 g of

KH_2PO_4, 14.2 g of Na_2HPO_4, 80.0 g of NaCl, and 2.0 g of KCl and transfer to a 1-L graduated cylinder. Add dd-H_2O to a volume of 800 mL. Mix and adjust the pH to 7.4 with HCl. Bring up to 1.00 L with dd-H_2O.

4. Blocking buffer: PBS (2×), 0.5 % casein, 2 % gelatin, 0.1 % ProClin. Add 100 mL of PBS (10×) to a 1-L graduated cylinder. Weigh 5.0 g of casein and transfer to the cylinder. Incubate gelatin at 37–40 °C for 10 min to make the gelatin less sticky. Then add 20 mL of gelatin to the PBS (*see* **Note 2**). Add 1 mL of ProClin to the above solution and bring up to 1 L with dd-H_2O. Mix the solution with a magnetic stirring apparatus until all reagents are dissolved (*see* **Note 3**). Store the blocking buffer at 4 °C in a well-sealed bottle.

5. Assay diluent: 20 mM Tris–HCl (pH 8.0), newborn calf serum, 1 % casein, 10 % sucrose. Dissolve 2.42 g of Tris base to 100 mL dd-H_2O and adjust the pH to 8.0 with HCl. Then add 40 mL of newborn calf serum, 10 g of casein, and 100 g of sucrose to the solution. Bring up to 1 L with dd-H_2O (*see* **Note 4**). Mix the solution using a magnetic stirring apparatus until all reagents are dissolved.

6. PBS-T (wash buffer): PBS containing 0.05 % (w/v) Tween 20. Add 1 mL of Tween 20 to 2 L 1× PBS.

2.2 Antigens and Conjugates

1. Purified HBsAg: The HBsAg (NIDVD, Xiamen, China) was expressed in CHO (Chinese hamster ovary) cells and purified to >95 % purity. The concentration of HBsAg was 1.0 mg/mL.

2. Purified HBsAg conformational monoclonal antibody IgG1 5F11 (*see* **Note 5**).

3. Horseradish peroxidase (HRP), DTT, $NaIO_4$, glycol, and $NaBH_4$ were all bought from Sigma.

4. TMB (Tetramethylbenzidine) Reagent A and TMB Reagent B (INNOVAX, Beijing, China).

5. TMB Stop Solution: 2 M H_2SO_4 (Beijing Wantai Biopharmaceuticals, Beijing, China).

3 Methods

3.1 HRP (Horseradish Peroxidase) Conjugation to Antibody 5F11

1. Dialyze 4 mL of 5F11 solution (1 mg/mL) in 1 L of 50 mM CB (1×). Stir with a stirrer at 4 °C for 2 h. Change the CB to newly prepared CB twice. Each time stir for 2 h with a magnetic stirring apparatus.

2. Dissolve 0.2 g of HRP in 10 mL of dd-H_2O and dissolve 0.2 g of $NaIO_4$ in 10 mL dd-H_2O (the final concentration of both HRP and $NaIO_4$ is 20 mg/mL).

3. Mix 200 μL of 20 mg/mL HRP and 200 μL of 20 mg/mL NaIO$_4$ gently (*see* **Note 6**). Incubate the mixture at 4 °C for 30 min.

4. Add 4 μL of glycol to above mixture slowly with gently mixing (*see* **Note 7**). Incubate the solution in a dark place at room temperature for 30 min.

5. Add the solution in **step 4** to the dialyzed 5F11 solution (*see* **Note 8**) and dialyze the resulting mixture in 1 L of 50 mM CB three times, each time for 2 h at 4 °C.

6. Dissolve 0.2 g of NaBH$_4$ in 10 mL of dd-H$_2$O (*see* **Note 9**). Add 40 μL of NaBH$_4$ to the dialyzed mixture in **step 5**. Incubate the mixture in a dark place at 4 °C for 2 h. Mix every 30 min.

7. Slowly add an equal volume of saturated ammonium sulfate solution to the mixture in **step 6** while stirring (*see* **Note 10**). After sufficient mixing, incubate the mixture at 4 °C for 4 h.

8. Transfer the mixture to a 4 mL Eppendorf tube and centrifuge at 14,000 × *g* for 10 min. Remove the liquid carefully.

9. Add 200 μL of newborn calf serum to 1.8 mL of dd-H$_2$O and mix with 2 mL of glycerol. Dissolve the precipitate in **step 8** with the mixed solution. The final mixture should contain 1.0 mg/mL of 5F11-HRP (*see* **Note 11**). Store at −20 °C.

3.2 Determination of the Proper Working Concentration of 5F11-HRP

1. Dilute the HBsAg stock to 1.0 μg/mL with PBS for plate coating.

2. Add 100 μL to each well on 96-well plates. Seal the plates and incubate at 37 °C for 6 h.

3. Wash the 96-well plates with 300 μL of PBS-T per well. Tap the plates on bibulous paper when the wells point down until the wells are dry.

4. Add 200 μL of assay diluent to each well on 96-well plates (*see* **Note 12**). Seal the plates with plate sealers and incubate at 37 °C for 2 h.

5. Wash the 96-well plates with PBS-T and tap again. Seal the plates and store the plates at 4 °C.

6. Dilute 1 μL of 1.0 mg/mL 5F11-HRP solution in 1 mL of PBS. Then dilute the diluent at serial twofold dilutions (*see* **Note 13**).

7. Transfer 100 μL of the diluents in **step 6** to each well on the plates in **step 5**. Seal plates and incubate at 37 °C for 1 h.

8. Wash the 96-well plates with 300 μL of PBS-T per well five times. Then tap the plates until the wells are dry as mentioned before.

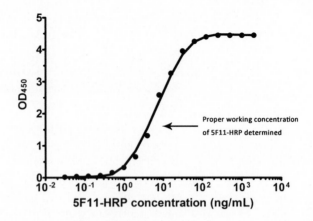

Fig. 1 The ELISA profile of different concentration of 5F11-HRP in a direct binding assay to determine the working concentration of the enzyme conjugate. The 5F11-HRP was diluted using a set of twofold serial dilutions in order to determine the appropriate working concentration/dilution in the assay. The initial concentration of 5F11-HRP is 2.0 ug/mL

9. Mix 6 mL of TMB Reagent A with 6 mL TMB Reagent B for each plate prior to use. Transfer 100 μL to each well. Incubate at 37 °C (5–20 min) for color development. To stop the color development, add 50 μL of TMB Stop Solution (*see* **Note 14**).

10. Read the optical density (OD) for each well with a microplate reader set to 450 nm. Analyze the results and determine the appropriate concentration of 5F11-HRP used (*see* **Note 15**) (Fig. 1).

3.3 Inhibition ELISA

The flowchart of solution inhibition ELISA is shown in Fig. 2. The protocol for inhibition ELISA below should correspond to the flowchart shown (Fig. 2).

1. Dilute the 5F11-HRP with assay diluent to 10 ng/mL.

2. Dilute the HBsAg samples used to detect twofold serial dilutions—starting with 4×10^5 ng/mL (*see* **Note 16**). Each sample should be more than 60 μL.

3. Mix 60 μL of diluted HBsAg sample with 60 μL of diluted 5F11-HRP in each well on plates (*see* **Note 17**). Seal the plates and incubate at 37 °C for 30 min. Then centrifuge the plates for 5 min at $1,500 \times g$.

4. Transfer 100 μL of the supernatant in **step 3** to each well of plates from **step 5** of Subheading 3.2 (*see* **Note 18**). Seal the plates and incubate at 37 °C for 1 h.

5. Wash the 96-well plates with 300 μL of PBS-T per well five times. Then tap the plates until no residual liquid remains, as mentioned before.

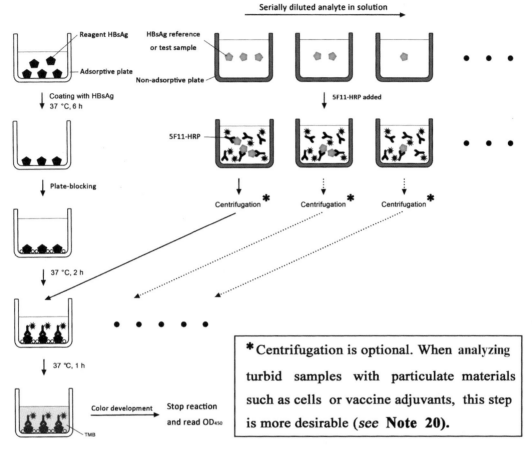

Fig. 2 Flowchart of a solution inhibition ELISA. The reagent HBsAg passively adsorbed on the surface in the microwells is shown *black*. The serially diluted analyte in solution, the HBsAg reference or test sample, are shown in *gray*. The IC$_{50}$ range of analyte to reference indicates binding affinity changes to antibody 5F11 (*see* **Note 20**)

6. Mix 6 mL of TMB Reagent A with 6 mL of TMB Reagent B for each plate immediately prior to use. Transfer 100 μL to each well. Incubate at 37 °C for color development as mentioned before. Add 50 μL of TMB Stop Solution to stop the color reaction.

7. Read the optical density (OD) for each well with a microplate reader set to 450 nm. Analyze the results and monitor the interaction between HBsAg and 5F11 through IC$_{50}$ shift (*see* **Note 19**) (Fig. 3).

4   Notes

1. The temperature of the water bath can range from 60 °C to 90 °C since the purpose of this step is to dissolve the ammonium sulfate at a high temperature. During incubation at 4 °C,

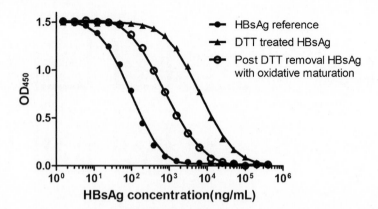

Fig. 3 Inhibition ELISA curves for different HBsAg samples (i.e., HBsAg reference, 0.5 mM DTT treated HBsAg, Post DTT removal HBsAg with oxidative maturation). Fitted curves and the IC_{50} values were obtained using "Prism 5." The IC_{50} values for HBsAg reference, 0.5 mM DTT treated HBsAg, Post DTT removal HBsAg with oxidative maturation are 97, 6,572, and 810 ng/mL, respectively (or relative IC_{50} of 1, 0.015, and 0.12, respectively). Nearly tenfold increase in solution antigenicity was observed as a results of HBsAg spontaneous maturation

the ammonium sulfate precipitates out. Ammonium sulfate solid can be seen at the bottom of the solution. 1.0 M NaOH can be used to adjust the solution pH. When making the NaOH solution, be aware of heat release during the dissolution of solid NaOH.

2. Gelatin is stored at 4 °C when it is solid. It can melt to liquid at a temperature of 37–40 °C. The measuring cylinder used to measure gelatin always contains residual gelatin on the walls. The residual gelatin should be washed with dd-H_2O and transferred to the solution to ensure the volume of gelatin is accurate.

3. The stirring time should be more than 1 h since it is difficult to dissolve casein powder. It is better that the rotor is set at high rotate speed.

4. If desired, a pH indicator such as phenol red can be added to the solution to indicate pH during storage or experiments.

5. The ELISA with conformationally sensitive monoclonal antibody can be used to monitor the integrity of protein binding sites [21], which is critical in quality control of pharmaceuticals or vaccines. When the binding affinity to a conformationally sensitive antibody weakens, the epitope integrity of the protein analyte may have been compromised.

6. This step is to oxidize the glycan components on HRP using $NaIO_4$. Mixing should be gentle. The weight of HRP and $NaIO_4$ in the mixture should be equal.

7. The step is to terminate the oxidation of HRP, volume of glycol (V_{glycol}, µL)–weight of HRP (M_{HRP}, mg) = 1:1.

8. The weight of antibody (M_{5F11}, mg)–the weight of HRP (M_{HRP}, mg) = 1:1. The experiment in this chapter is to monitor protein interactions through inhibition ELISA in solution with one binding partner labeled with HRP.

9. The NaBH$_4$ solution should be prepared immediately prior to use. NaBH$_4$ is a strong reductant, making the protein–protein conjugate into a more stable form.

10. Saturated ammonium sulfate should be added slowly with stirring at 4 °C. It is used to salt out the 5F11-HRP.

11. The concentration of 5F11-HRP is reported as the 5F11 concentration as it is hard to quantify the HRP conjugate. Loss of 5F11 is considered minimal in the process of HRP conjugation procedures.

12. This step is to block the nonspecific binding sites in the microwells. The blocking buffer should be added at a larger volume than the coating HBsAg solution so that blocking buffer can cover the walls that HBsAg-coated areas.

13. Serial dilution of 5F11-HRP should cover a wide enough range so that an appropriate concentration can be determined. A negative control should be used to monitor the ELISA background.

14. The color reagent solution turns blue as a result of surface bound HRP catalysis. The solution turns yellow after adding acidic TMB Stop Solution. The same color development time should be used throughout all the experiments.

15. In this experiment, the initial concentration of 5F11-HRP is 2,000 ng/mL; it can be larger when the proper initial concentration is unknown. The concentration of 5F11-HRP we determined is 5.0 ng/mL because the OD$_{450}$ that correlates to this concentration is about 1.5. An OD$_{450}$ that is too large leads to inaccurate measurement due to too little light transmitted. An antibody concentration that is too high could lead to incomplete inhibition, while too low of an antibody concentration would result in low signals.

16. The test HBsAg samples could be different production lots or different samples that underwent different pretreatments, such as 0.5 mM DTT-treated. Reference (HBsAg sample without any pretreatment) should be used as a control analyte for quantitatively analyzing the changes in IC$_{50}$ values. Duplicate wells are used to improve assay reproducibility.

17. The plates used here are non-adsorptive plates. Other plates mentioned in this chapter are adsorptive plates. If the test analyte were allowed to bind to the binding partner first in

solution (prior to being transferred to the assay plate), the inhibition ELISA can also be used to determine the true solution affinity (K_D) constant of biologically relevant binding partners such as antibody–antigen or receptor–ligand pairs when reaching equilibrium for all the mixtures is needed [23]. There is no need to reach equilibrium to determinate the IC_{50}.

18. The labeled binding partner and a serially diluted protein analyte can also be added directly to the plates without premixing. This is more commonly referred to as "competitive ELISA" as the labeled molecules bind to either the partner in solution or to the "competing" partner on the solid surface. It is an alternative method to monitor protein-protein interactions; however, different IC_{50} values will be obtained when compared to the method described in this chapter.

19. The binding curve (as indicated by OD_{450}) is shown in Fig. 3. The IC_{50} (the half-maximal inhibitory concentration) can be calculated by software such as "Prism 5". IC_{50} is a measure of the effectiveness of a sample in inhibiting biological or biochemical function. The larger an IC_{50} is, the lower affinity a HBsAg sample has to the same antibody. The HBsAg reference is an untreated sample. The HBsAg samples were also treated with 0.5 mM DTT and the DTT was later removed by dialysis. The results showed that the 0.5 mM DTT treated HBsAg has the largest IC_{50}, whereas the IC_{50} of HBsAg reference is the lowest (differing by 68-fold). This means that 0.5 mM DTT treatment made the interaction between HBsAg and 5F11 much weaker. After DTT removal, only a fraction of the 5F11 binding activity of HBsAg was observed in "Post DTT removal HBsAg with oxidative maturation." The relative IC_{50} (Reference IC_{50}/Sample IC_{50}) can also be used to show the IC_{50} range between sample and reference. The interaction between two interacting proteins, with one binding partner labeled with an enzyme or a fluorophore/chromophore [7], can be monitored in a similar way, particularly when the binding site could be changed subtly but with an impact on the binding affinity.

20. Since the immune complex of analyte and labeled binding partner forms in solution and there is no need for such a complex to bind onto a surface, depletion of the labeled binding partner is the measure of the analyte binding activity [18]. Therefore, inhibition ELISA can be employed in the study of active binding site in whole cells or particulate vaccine antigens on adjuvants. Serial dilution of the analyte enables the quantitative analysis of the binding activity of different lots or preparations of a given protein. With a sensitive probe, impact on the binding affinity due to some intentional manipulations could be monitored using the assay described here.

Acknowledgement

The work on chapter writing was enabled with the funding of Chinese Ministry of Science and Technology 863 Project (2012AA02A408) and National Science Foundation of China (81471934) to Q.Z.

References

1. Leng SX, McElhaney JE, Walston JD et al (2008) ELISA and multiplex technologies for cytokine measurement in inflammation and aging research. J Gerontol A Biol Sci Med Sci 63:879–884

2. Catt K, Tregear G (1967) Solid-phase radioimmunoassay in antibody-coated tubes. Science 158:1570

3. Aviñó A, Gómara M, Malakoutikhah M et al (2012) Oligonucleotide-peptide conjugates: solid-phase synthesis under acidic conditions and use in ELISA assays. Molecules 17:13825–13843

4. Yalow RS, Berson SA (1996) Immunoassay of endogenous plasma insulin in man. Obes Res 4:583–600

5. Smitsaart EN, Fernandez E, Maradei E et al (1994) Validation of an inhibition ELISA using a monoclonal antibody for foot-and-mouth disease (FMD) primary diagnosis. Zentralbl Veterinarmed B 41:313–319

6. Aleshukina A, Iagovkin É, Bondarenko V (2011) Change of proinflammatory cytokine profile in human intestine in dysbacteriosis caused by the antibiotics therapy. Zh Mikrobiol Epidemiol Immunobiol: 81–85

7. Meng Y, High K, Antonello J et al (2005) Enhanced sensitivity and precision in an enzyme-linked immunosorbent assay with fluorogenic substrates compared with commonly used chromogenic substrates. Anal Biochem 345:227–236

8. Dähnrich C, Komorowski L, Probst C et al (2013) Development of a standardized ELISA for the determination of autoantibodies against human M-type phospholipase A2 receptor in primary membranous nephropathy. Clin Chim Acta 421:213–218

9. Crowther JR (2000) The ELISA guidebook, vol 149. Humana press, Totowa, New Jersey

10. Cuervo MLC, de Castro Yanes AF (2004) Comparison between in vitro potency tests for Cuban Hepatitis B vaccine: contribution to the standardization process. Biologicals 32:171–176

11. Li H, Shen DT, Jessup DA et al (1996) Prevalence of antibody to malignant catarrhal fever virus in wild and domestic ruminants by competitive-inhibition ELISA. J Wildl Dis 32:437–443

12. Tai HC, Campanile N, Ezzelarab M et al (2006) Measurement of anti-CD154 monoclonal antibody in primate sera by competitive inhibition ELISA. Xenotransplantation 13:566–570

13. Raj GD, Rajanathan TM, Kumar CS et al (2008) Detection of peste des petits ruminants virus antigen using immunofiltration and antigen-competition ELISA methods. Vet Microbiol 129:246–251

14. Dessy FJ, Giannini SL, Bougelet CA et al (2008) Correlation between direct ELISA, single epitope-based inhibition ELISA and pseudovirion-based neutralization assay for measuring anti-HPV-16 and anti-HPV-18 antibody response after vaccination with the AS04-adjuvanted HPV-16/18 cervical cancer vaccine. Hum Vaccin 4:425–434

15. Giffroy D, Mazy C, Duchene M (2006) Validation of a new ELISA method for in vitro potency assay of hepatitis B-containing vaccines. Pharmeuropa Bio 2006:7

16. Chovel Cuervo ML, Sterling AL, Nicot AI et al (2008) Validation of a new alternative for determining in vitro potency in vaccines containing Hepatitis B from two different manufacturers. Biologicals 36:375–382

17. Friguet B, Chaffotte AF, Djavadi-Ohaniance L et al (1985) Measurements of the true affinity constant in solution of antigen-antibody complexes by enzyme-linked immunosorbent assay. J Immunol Methods 77:305–319

18. Zhao Q, Modis Y, High K et al (2012) Disassembly and reassembly of human papillomavirus virus-like particles produces more virion-like antibody reactivity. Virol J 9:52

19. Harrison RO, Goodrow MH, Hammock BD (1991) Competitive inhibition ELISA for the s-triazine herbicides: assay optimization and antibody characterization. J Agric Food Chem 39:122–128

20. Zhao Q, Wang Y, Abraham D et al (2011) Real time monitoring of antigenicity development of HBsAg virus-like particles (VLPs) during heat- and redox-treatment. Biochem Biophys Res Commun 408:447–453

21. Zhao Q, Towne V, Brown M et al (2011) In-depth process understanding of RECOMBIVAX HB® maturation and potential epitope improvements with redox treatment: Multifaceted biochemical and immunochemical characterization. Vaccine 29:7936–7941

22. Mulder AM, Carragher B, Towne V et al (2012) Toolbox for non-intrusive structural and functional analysis of recombinant VLP based vaccines: a case study with hepatitis B vaccine. PLoS One 7:e33235

23. High K, Meng Y, Washabaugh MW et al (2005) Determination of picomolar equilibrium dissociation constants in solution by enzyme-linked immunosorbent assay with fluorescence detection. Anal Biochem 347:159–161

Chapter 22

Glutathione-*S*-Transferase (GST)-Fusion Based Assays for Studying Protein-Protein Interactions

Haris G. Vikis and Kun-Liang Guan

Abstract

Glutathione-*S*-transferase (GST)-fusion proteins have become an effective reagent to use in the study of protein-protein interactions. GST-fusion proteins can be produced in bacterial and mammalian cells in large quantities and purified rapidly. GST can be coupled to a glutathione matrix, which permits its use as an effective affinity column to study interactions in vitro or to purify protein complexes in cells expressing the GST-fusion protein. Here, we provide a technical description of the utilization of GST-fusion proteins as both a tool to study protein-protein interactions and also as a means to purify interacting proteins.

Key words GST, Affinity chromatography, Protein-protein interactions, Ras, Raf

1 Introduction

A convenient method for the analysis of protein-protein interactions is through the use of chimeric proteins comprising glutathione-*S*-transferase (GST) linked in frame to a protein of interest. Commonly referred to as GST-fusion proteins, these chimeras can typically be expressed in *Escherichia coli* (*E. coli*) and rapidly affinity-purified under non-denaturing conditions. Fusion of GST to a particular protein commonly enhances the protein's solubility and because GST has virtually no toxicity in cells, these proteins can generally be expressed in large amounts. Numerous convenient expression vectors now exist for production of GST-fusions in both bacterial and mammalian cells.

GST-fusion proteins are commonly used to study protein-protein interactions mainly because GST has a strong affinity for glutathione and thus can be coupled at high concentration to an immobilized glutathione matrix. Furthermore, the interaction of GST with glutathione is particularly robust and resistant to stringent buffer conditions. The GST moiety of the fusion protein generally does not interfere with the accessibility of the fused protein because a long flexible linker region exists which permits

Cheryl L. Meyerkord and Haian Fu (eds.), *Protein-Protein Interactions: Methods and Applications*, Methods in Molecular Biology, vol. 1278, DOI 10.1007/978-1-4939-2425-7_22, © Springer Science+Business Media New York 2015

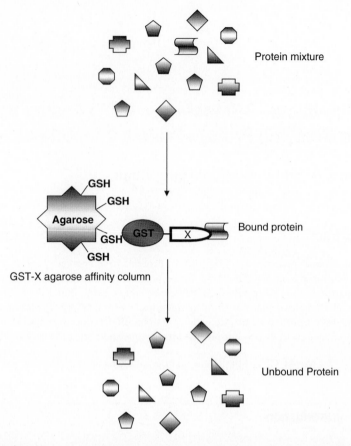

Fig. 1 A GST-fusion protein affinity column. Cell lysates comprising numerous proteins are incubated with GST-X fusion protein bound to a glutathione (GSH)–agarose matrix. After unbound proteins are washed away, only proteins that interact with the GST-X fusion protein can be retrieved from the mixture

each half of the fusion to behave as separate domains. All of these attributes make GST-fusion proteins bound to a glutathione matrix simple and effective affinity columns which many investigators have used as a tool to isolate novel interacting proteins (Fig. 1). In this chapter we use GST-Ras expressed in *E. coli* and GST-B-Raf expressed in mammalian cells to illustrate how to express and purify GST-fusion proteins and use them in the analysis of protein-protein interactions.

2   Materials

1. GEX-KG expression vector (or a similar GST-fusion expression vector).

2. (DH5α, BL21(DE3)-RIL) *E. coli* strains.

3. LB Media (agar, liquid).

4. Antibiotic.

5. 37 °C, 30 °C, and room temperature incubators.

6. IPTG.

7. Spectrophotometer.

8. Lysis buffer A: 50 mM Tris–HCl, pH 7.5, 100 mM NaCl, 1 mM DTT, 0.2 mM PMSF, EDTA-free protease inhibitor cocktail.

9. French press or sonicator.

10. Triton X-100 (TX100) detergent.

11. Refrigerated centrifuge.

12. Elution buffer: 50 mM Tris–HCl, pH 8, 100 mM NaCl, 10 mM reduced glutathione, 1 mM DTT.

13. SDS-PAGE equipment.

14. Glutathione (GSH)–agarose.

15. Reduced glutathione.

16. Liquid N_2, −80 °C storage.

17. Rocking/rotating platform.

18. Transfection reagent (e.g., Lipofectamine, Invitrogen).

19. Lysis buffer B: 20 mM Tris–HCl, pH 7.5, 100 mM NaCl, 1 % NP-40, 5 mM $MgCl_2$, 1 mM DTT, 0.2 mM PMSF, Complete EDTA-free protease inhibitor tablet.

20. Antibodies: α-C-Raf, α-HA, α-GST.

3 Methods

Here we outline the methods involved in the general cloning, expression, and purification of GST-fusion proteins and then describe the specific use of bacterially expressed GST-Ras mutants to analyze their interaction with C-Raf. We also describe the expression of a mammalian GST-fusion protein, GST-B-Raf, and analyze its interaction with AKT.

3.1 Expression Plasmids

3.1.1 pGEX-KG E. coli Expression Vector

Vectors that direct expression of GST fusion proteins in *E. coli* are widely available commercially. Furthermore, a large collection of empty vectors and vectors with cDNA inserts are also available through resources such as Addgene (addgene.org) and PlasmID (plasmid.med.harvard.edu/PLASMID). We commonly use the pGEX-KG expression vector for production of GST-fusion proteins in *E. coli* [1] (Fig. 2). This vector is based on a set of vectors initially constructed by Pharmacia Biotech. It is typical of many prokaryotic expression vectors in that it contains an origin of replication (pBR322 ori), an ampicillin resistance gene (Amp^r), and a strong

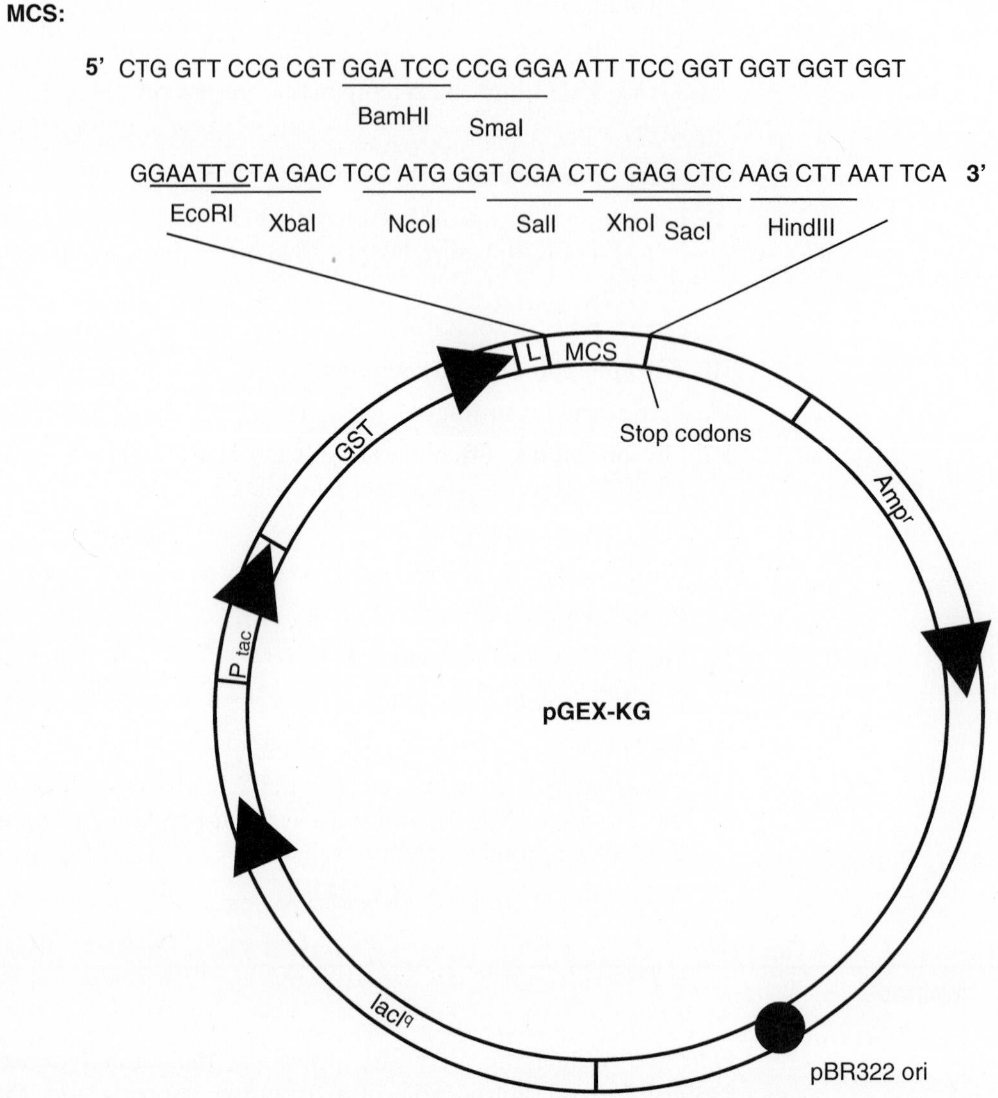
MCS:
5' CTG GTT CCG CGT GGA TCC CCG GGA ATT TCC GGT GGT GGT GGT
BamHI
SmaI
GGAATT CTA GAC TCC ATG GGT CGA CTC GAG CTC AAG CTT AAT TCA 3'
EcoRI
XbaI
NcoI
SalI
XhoI
SacI
HindIII
L
MCS
Stop codons
GST
Ampr
P tac
pGEX-KG
lacIq
pBR322 ori

Fig. 2 pGEX-KG *E. coli* expression vector. pGEX-KG vector shown with essential components: GST gene, L (linker region), MCS (multiple cloning site), Amp^r (ampicillin resistance gene), pBR322 ori (origin of replication), $lacI^q$ (lac repressor), and P_{tac} (promoter). The MCS is shown with restriction sites

promoter (P_{tac}) which is kept in the "off" state by the lac repressor ($lacI^q$). Addition of the non-hydrolyzable lactose analog, isopropyl β-D-1-thiogalactopyranoside (IPTG) inactivates the lac repressor and permits activation of the promoter. The GST gene lies downstream of the promoter and is followed by a linker region which is a stretch of amino acids predicted to be very flexible and with no apparent structure. This is then followed by the multiple cloning site (MCS) which contains a number of commonly used restriction sites for insertion of the cDNA of interest.

<table>
<tr><td>

3.1.2 Cloning Strategy

</td><td>

Insertion of the cDNA of interest into the pGEX-KG expression vector MCS is performed by conventional molecular biology cloning manipulations. It is worth noting that the cDNA need not contain an initiator methionine codon; however, it must be inserted in the same frame as the GST gene. DNA sequencing may be a necessary diagnostic method to confirm the construct.

</td></tr>
</table>

3.2 Production of GST-Fusion Proteins in E. coli

Following construction and verification of the expression construct, the construct must then be transformed into the appropriate bacterial host strain. We typically use two *E. coli* strains for protein expression: (1) DH5α and (2) BL21(DE3)-RIL (*see* **Note 1**).

3.2.1 E. coli Transformation and Induction of Protein Expression

1. Transform plasmid expression vector into *E. coli* strain using standard heat shock methods and plate on an LB/Amp agar plate overnight at 37 °C.

2. Pick an individual colony the next day (after 12–18 h of growth), inoculate 10 mL of LB/Amp (50–100 μg/mL) liquid media, and grow overnight in a 37 °C shaker (*see* **Note 2**).

3. The next morning pour the 10 mL culture of *E. coli* into 1 L of LB/Amp and grow for 2–5 h until an OD_{600} of 0.5–0.8 has been reached (log phase growth).

4. Add IPTG to 0.1 mM and shake cells for 3–4 h at 37 °C or overnight at room temperature (*see* **Note 3**).

3.2.2 Protein Extraction

1. Centrifuge cells at $5,000 \times g$ for 10 min in polypropylene bottles.

2. Decant supernatant. If desired, cells can be stored as a pellet at −20 °C.

3. Resuspend cells in 15–30 mL of cold lysis buffer (*see* **Note 4**). The use of an aluminum beaker to perform the following lysis step will ensure the cells stay cold.

4. Cell lysis: there are two methods commonly used; (a) pass through a French press twice or (b) sonicate (4× 20–30 s bursts). The French press is more cumbersome than the sonicator, but it is usually preferred when purifying a protein whose activity can be destroyed by heat generated by the metal tip of the sonicator. Therefore, we recommend keeping the sample as cold as possible while sonicating (e.g., keep on ice with constant mixing) and performing four 20–30 s sonication bursts allowing 1–2 min intervals between bursts.

5. Add Triton X-100 (TX100) detergent to 1 %. This is added after the lysis step to help solubilize proteins and to avoid any frothing that may occur during sonication.

6. Centrifuge lysate at $30,000 \times g$ for 20 min at 4 °C.

7. At this point NaCl can be added up to 1 M to prevent co-purifying nonspecific proteins. Addition of NaCl will not affect the affinity of GST for the glutathione matrix.

8. Add 1 mL of GSH–agarose or sepharose (50 % slurry, pre-washed in lysis buffer) to the *E. coli* supernatant and mix for at least 1 h at 4 °C. The binding capacity of glutathione–agarose is approximately 10 mg of GST protein per mL of 50 % slurry.

9. After incubation, the glutathione–agarose beads must be washed (*see* **Note 5**). Wash steps: (a) 2× lysis buffer + 1 M NaCl; (b) 2× lysis buffer; (c) 3× lysis buffer (- TX100).

   At this point, the protein can be stored bound to the glutathione beads (*see* Subheading 3.3) or be eluted off the beads.

3.2.3 Elution and Dialysis of GST-Fusion Protein

1. Elute the protein twice with 0.5 mL of elution buffer (10 min each).

2. To remove the glutathione from the buffer, the eluted protein can be dialyzed against a buffer of choice at 4 °C. We typically dialyze 1 mL of eluted protein against 1 L of elution buffer (without glutathione) and replace 1 mM DTT with 0.1 % β-Me. However, we recommend choosing whichever dialysis buffer system you deem most appropriate for your protein.

3. Quantify the amount of protein by running an SDS-PAGE gel against known protein quantities (e.g., BSA). Electrophoresis/ Coomassie staining is preferable to using the Bradford assay because of the ability to assess protein purity as well (Fig. 3). Performing a Western blot with α-GST antibody will also allow you to assess purity and the presence of degradation products.

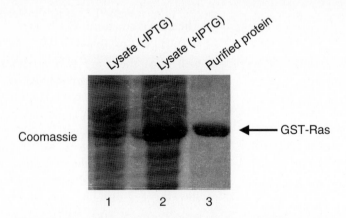

Fig. 3 IPTG induction and purity of GST-Ras protein. Samples were run on a 12.5 % SDS-PAGE gel and stained with Coomassie blue. *Lane 1*, *E. coli* lysate prior to IPTG induction; *lane 2*, *E. coli* lysate after 3 h of IPTG induction; *lane 3*, purified GST-Ras protein. The *arrow* indicates the location of the GST-Ras protein in the induced lysate and after purification by glutathione–agarose

3.2.4 Storage of
GST-Fusion Proteins

Various methods of storage are used and each should be tested to see whether it affects protein stability, activity, etc. GST-fusions can be stored on beads or in solution. Common storage methods include:

1. Addition of glycerol to 50 % and storage at −20 °C (prevents sample from freezing).

2. Addition of glycerol to 5–10 % and storage at −80 °C.

3. Addition of glycerol to 5–10 % and snap-freeze in liquid N_2, prior to storage at −80 °C.

After thawing and prior to use in an experiment, the beads should be washed to remove glycerol.

3.3 Using GST-Ras to Analyze the Interaction with Raf

To illustrate the use of a GST-fusion protein in a binding reaction, we analyzed the interaction of the H-Ras GTPase with the Raf kinase. We constructed pGEX-KG-H-RasV12 and -H-RasN17 expression vectors which express two mutants of the human H-Ras protein. The mutation G12V locks Ras in the GTP nucleotide bound conformation, while the T17N mutation locks the protein in the GDP nucleotide bound conformation [2, 3]. GTP bound Ras is in the "on" conformation and is able to interact with numerous effector molecules such as the protein kinase, Raf, through the Raf N-terminal domain. GDP bound Ras is in the "off" conformation and unable to interact with Raf.

We purified GST-RasV12 and GST-RasN17 using the protocol described in Subheading 3.2. It is important to note that these small GTPases are required to be purified in the presence of 5 mM $MgCl_2$ to prevent the loss of bound nucleotide. The following procedure describes the analysis of the interaction between *E. coli*-expressed GST-Ras mutants and mammalian expressed N-terminal domain (amino acids 1–269) of Raf. Raf 1–269 expression was directed from the plasmid pCDNA3-cRaf-1-269, which was transfected into mammalian HEK293 cells according to standard lipofection protocols.

1. Transfect a 10 cm dish of mammalian HEK293 cells (using Lipofectamine (Invitrogen)) with 10 μg of the pCDNA3-cRaf (1–269) plasmid and grow for a further 48 h.

2. Wash cells once with 10 mL of phosphate buffered saline (PBS) and lyse with 1 mL of lysis buffer B.

3. Place cells on ice for 10 min, scrape and collect into an eppendorf tube.

4. Centrifuge the lysate at $15,000 \times g$ for 15 min at 4 °C and collect the supernatant.

5. Add 10 μg GST, GST-RasV12, and GST-RasN17 into three separate tubes and to these tubes add the 293 cell lysate collected in **step 4**. Mix by incubating at 4 °C for 1–2 h on a rocking/rotating platform.

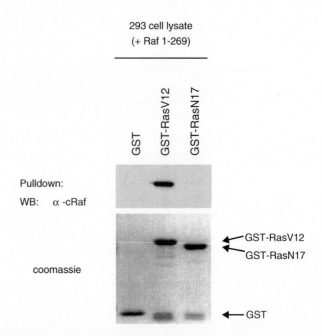

Fig. 4 Interaction of Raf and Ras. *Top panel*, Raf (1–269) binds GST-RasV12 as detected by α-Raf Western blot. *Bottom panel*, Coomassie blue stained gel of the GST-fusions used in the assay

6. Add 10–15 μL GSH–agarose to each tube and rock for a further 0.5–1 h (*see* **Note 6**). Wash the GSH–agarose beads four times with 1 mL of lysis buffer B.

7. Elute bound GST-Ras with the elution buffer as explained in Subheading 3.2.3, or with 1× SDS sample buffer (*see* **Note 7**).

8. Run samples on an SDS-PAGE gel and subject to a Western blot using α-c-Raf antibody (Fig. 4).

3.4 Production of GST-Fusion Proteins in Mammalian Cell Culture

One benefit of expression in mammalian cells is that many eukaryotic proteins undergo modifications that do not occur in *E. coli*. For example, the Ras protein is C-terminal prenylated when expressed in mammalian cells, but does not undergo this modification in *E. coli* because the appropriate modification enzymes are not present in bacteria. Furthermore, mammalian proteins that typically cannot be expressed in *E. coli* or are easily degraded in *E. coli* can often be expressed in mammalian cells where conditions (tRNA, folding machinery, etc.) are more suitable. However, the fact that expression in mammalian cells results in less protein compared to *E. coli* makes it less economical.

Using glutathione–agarose to purify GST-fusion proteins from mammalian cells bypasses the need for using antibodies and thus immunoprecipitation methods. This is particularly useful because immunopurified proteins cannot easily be eluted from the antibody and must be eluted by boiling which releases the antibody into the mixture. Since GST-fusions can easily be eluted with glutathione, no antibody is present. This makes identification of co-purified proteins by protein sequencing or mass spectrometry much easier because of the lack of large amounts of contaminating antibody present.

Note that the purification protocol for mammalian GST-fusion proteins is the same as in Subheading 3.2.2 except during the lysis step detergents such as 1 % TX-100 or 1 % NP-40 are used to lyse the cells instead of a French press or sonicator.

3.5 Detecting the Interaction Between B-Raf and AKT Using Mammalian GST-Fusion Proteins

There are a number of commercially available cytomegalovirus (CMV) promoter driven GST-fusion expression vectors for use in mammalian cells. The vector pEBG-3X drives the expression of GST-fusion proteins from a very strong E1Fα promoter. We have constructed a modified version of pEBG-3X vector where more restriction sites have been added to the MCS and have named it pEBG-3X-HV (Fig. 5). Transfection of a 10 cm plate of HEK293 cells with pEBG-3X-HV can produce up to 5–10 μg of GST protein which is roughly an order of magnitude more protein than what most CMV promoter based vectors can direct.

1. Transfect HEK 293 cells (in a 6-well plate) with the following combination of plasmids:

   (a) pEBG-3X-HV (0.5 μg) + pCDNA3-HA-B-Raf (0.5 μg)

   (b) pEBG-3X-HV-AKT   (0.5   μg) + pCDNA3-HA-B-Raf (0.5 μg)

   (c) pEBG-3X-HV (0.5 μg) + pDNA3-HA-AKT (0.5 μg)

   (d) pEBG-3X-HV-B-Raf   (0.5   μg) + pDNA3-HA-AKT (0.5 μg)

2. Forty-eight hours post-transfection, lyse the cells in 250 μL of lysis buffer B (without $MgCl_2$).

3. Centrifuge the lysate at 15,000 × *g* for 10 min at 4 °C.

4. Collect the supernatant and add 15 μL of GSH–agarose (50 % slurry).

pEBG-3X-HV multiple cloning site:

		BamHI	NdeI	EcoRV	NheI
5' ATC GAA GGT CGT GGG ATC		GGA TCC	CAT ATG	GAT ATC	GCT AGC

XmaI	SalI	SpeI	ClaI	NotI		
CCC GGG	GTC GAC	ACT AGT	ATC GAT	GCG GCC	GCT GAA TAG	**3'**

Fig. 5 pGEX-3X-HV vector multiple cloning site

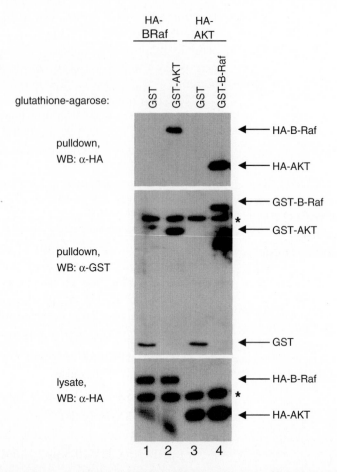

Fig. 6 Interaction of B-Raf with AKT. HEK293 cells were transfected with the plasmid combinations indicated and grown for 48 h. Cells were lysed and GST-fusion proteins were purified by addition of glutathione–agarose. To detect the presence of co-purified proteins, SDS-PAGE was performed followed by an α-HA Western blot. *Asterisk* nonspecific band

5. Incubate for 1–2 h on a rocker/rotating platform at 4 °C.

6. Wash GSH–agarose beads four times with lysis buffer.

7. Elute using glutathione elution buffer (Subheading 3.2.3) or boil in 1× SDS sample buffer.

8. Perform SDS-PAGE followed by Western blot analysis with α-HA (Fig. 6).

3.6   Passing Radiolabelled Lysates over GST-Fusion Columns

A common method to identify proteins that interact with a GST-fusion protein is to metabolically label cells with ^{35}S-Met/Cys and pass the cell lysates over a GST-fusion affinity column. This provides a very sensitive method of identifying novel interacting proteins. However, there are several procedural changes to be aware of to make this an effective method.

1. When lysing radiolabelled cells in mammalian culture, layer the lysis buffer onto the radiolabelled cells so as to not detach the cells from the cell culture plate. Furthermore, do not scrape the cells after the lysis. Simply tilt the dish and remove as much of the soluble lysate as possible. This will help reduce any insoluble fragments that might not be completely removed in the centrifugation step.

2. Often nonspecific proteins bind the glutathione–agarose matrix during the binding step. One way to reduce this is to block the beads in lysis buffer with the addition of 1 % BSA. In addition, preclearing the lysate with glutathione–agarose prior to incubating with glutathione–agarose/GST-fusion protein will help remove matrix-interacting proteins. This step typically requires about 100 µL of glutathione agarose per 300 µL of lysate.

3.7 Different Lysis Buffers to Use

Reagents in the binding buffer may influence protein-protein interactions. It is often useful to try different buffer systems when analyzing binding. Three commonly used buffer systems are provided in order of increasing stringency:

1. NP-40 buffer: 20 mM Tris–HCl, pH 7.5, 100 mM NaCl, 1 % Nonidet P-40

2. Triton buffer: 20 mM Tris–HCl, pH 7.5, 100 mM NaCl, 1 % TX-100

3. RIPA buffer: 0.1 % SDS, 1 % Triton X-100, 0.5 % deoxycholate, 50 mM Tris–HCl, pH 7.5, 150 mM NaCl.

4 Notes

1. BL21(DE3)-RIL cells contain some human tRNAs that are underrepresented in *E. coli* and hence enhance expression of certain human proteins in *E. coli*. Production of soluble mammalian proteins can be aided by cloning shorter cDNAs and/or encoding for less contiguous hydrophobic amino acids [4].

2. We recommend starting from a fresh colony for maximum expression. This is particularly important in the case of the RIL cells because often we see no protein expression if using colonies more than 1 day old.

3. We find 0.1 mM IPTG is adequate and higher concentrations do not seem to increase expression. To optimize expression, one might try different conditions such as length of induction and temperature of induction on small scale cultures first. For example, if solubility is a problem, this can often be improved if cells are induced overnight at 30 °C.

4. We have found that the extent of protein degradation in *E. coli* depends more on the expression conditions and less on the concentration of protease inhibitors during the purification. Hence it is important to vary the conditions as explained in **Note 3** to minimize protein degradation.

5. For washing of 1 mL of glutathione–agarose matrix, we typically spin down beads for 20 s at $5,000 \times g$ and wash in Eppendorf tubes using multiple 1 mL washes. The washes need not contain protease inhibitors.

   Since proteins expressed in *E. coli* are produced at a rapid rate, they sometimes do not fold efficiently. When this happens the *E. coli* chaperone Hsp70 binds these misfolded proteins. Therefore, it is common to find Hsp70 (at 70 kDa) associated with purified GST-fusion proteins. One method to remove Hsp70 from the GST-fusion is to wash the beads twice (prior to elution) with 1 mL of 500 mM triethanolamine–HCl (pH 7.5), 20 mM $MgCl_2$, 50 mM KCl, 5 mM ATP, 2 mM DTT for 10 min at room temperature.

6. Another method is to use GST-Ras which is still bound to GSH–agarose. Either way is acceptable as we have not observed any differences between the two methods.

7. Eluting the protein is more specific, but will release less GST-fusion protein than boiling for 3 min in SDS sample buffer.

Acknowledgments

The authors would like to thank Huira Chong and Jennifer Aurandt for critical review of the manuscript. This work was supported by grants from National Institutes of Health and Walther Cancer Institute (K.L.G.). K.L.G. is a MacArthur Fellow. H.V. is supported by a Rackham Predoctoral Fellowship.

References

1. Guan KL, Dixon JE (1991) Eukaryotic proteins expressed in Escherichia coli: an improved thrombin cleavage and purification procedure of fusion proteins with glutathione S-transferase. Anal Biochem 192:262–267

2. Katz ME, McCormick F (1997) Signal transduction from multiple Ras effectors. Curr Opin Genet Dev 7:75–79

3. Bourne HR, Sanders DA, McCormick F (1991) The GTPase superfamily: conserved structure and molecular mechanism. Nature 349:117–127

4. Dyson MR, Shadbolt SP, Vincent KJ, Perera RL, McCafferty J (2004) Production of soluble mammalian proteins in Escherichia coli: identification of protein features that correlate with successful expression. BMC Biotechnol 4:32

Chapter 23

Hexahistidine (6xHis) Fusion-Based Assays for Protein-Protein Interactions

Mary C. Puckett

Abstract

Fusion-protein tags provide a useful method to study protein-protein interactions. One widely used fusion tag is hexahistidine (6xHis). This tag has unique advantages over others due to its small size and the relatively low abundance of naturally occurring consecutive histidine repeats. 6xHis tags can interact with immobilized metal cations to provide for the capture of proteins and protein complexes of interest. In this chapter, a description of the benefits and uses of 6xHis-fusion proteins as well as a detailed method for performing a 6xHis-pulldown assay are described.

Key words Hexahistidine tag, 6xHis pull down, Affinity chromatography, Protein-protein interactions

1 Introduction

Protein-protein interactions play vital roles in the regulation of cellular functions, both physiological and pathophysiological. These interactions have been traditionally studied by immunoprecipitation assays where one protein is isolated through the use of a specific antibody and bound protein partners are analyzed. Immunoprecipitation, however, is extremely dependent on the characteristics of the antibody used, leading to significant disadvantages if an antibody is of poor quality or shows a high level of nonspecific binding [1]. Furthermore, the light and heavy chains of the antibody itself can frequently be detected by Western blot, which has the potential to obscure results obtained by this method.

To eliminate some of the disadvantages of immunoprecipitations in protein-protein interaction studies, many fusion-protein tags have been developed to specifically isolate proteins of interest. A variety of protein tags have been developed, including the commonly used glutathione S-transferase (GST), FLAG, HA, and hexahistidine (6xHis) tags [2]. While all of these tags have advantages

Cheryl L. Meyerkord and Haian Fu (eds.), *Protein-Protein Interactions: Methods and Applications*, Methods in Molecular Biology, vol. 1278, DOI 10.1007/978-1-4939-2425-7_23, © Springer Science+Business Media New York 2015

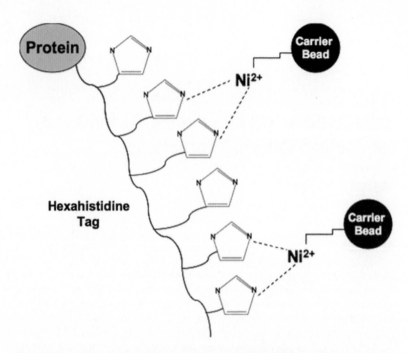

Fig. 1 Schematic of immobilized metal ions with histidine residues. Imidazole rings form coordination bonds with transition metal cations

and disadvantages, 6xHis tags are particularly useful for a variety of reasons and are discussed further in this chapter.

Polyhistidine tags were first used to aid in protein purification [3], and have since come to have many useful applications in protein-protein interaction studies. 6xHis tags are particularly useful due to their strong interactions with immobilized metal cations. The imidazole rings found in histidine residues can form coordination bonds with transition metal ions through the imidazole's Nitrogen atoms (Fig. 1), which serve as electron donors [4]. While Ni^{2+} is commonly used, other transition metals such as Co^{2+}, Cu^{2+}, and Zn^{2+} can also interact with polyhistidine tags [2]. The low abundance of naturally occurring histidine chains and the minimal interaction of other amino acid residues with these metals allows for highly specific capture of the tagged protein. Additionally, a low concentration of imidazole in the lysis buffer is sufficient to compete with native histidine residues on untagged proteins and the metal–histidine interactions can easily be competed by additional free imidazole or altered pH to recover the protein or protein complexes of interest.

In addition to their unique ability to bind immobilized metal cations, 6xHis tags have advantages due to their small size. Their small size minimizes tag interference in native protein folding and protein-protein interactions, which is sometimes noted with larger fusion tags. Similarly, 6xHis tags can be added to either the N- or

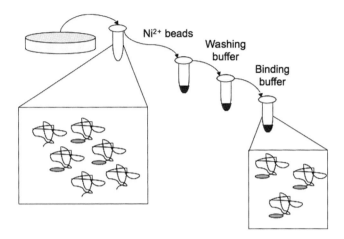

Fig. 2 Schematic of a 6xHis Pulldown assay. Lysates containing 6xHis tagged proteins and other proteins of interest are prepared. The lysates are then incubated with Ni^{2+} beads, which bind the 6xHis tag. Ni^{2+} beads are then washed with washing and binding buffers to decrease nonspecific interactions, leaving 6xHis-tagged protein complexes bound. The resulting complexes can then be recovered and analyzed

C-terminus of a protein, allowing for greater flexibility. Vectors also exist for the production of a cleavable His tag, allowing for non-tagged recombinant proteins to be recovered.

Due to these advantages, many applications have been developed utilizing 6xHis tags. Purification of proteins from bacteria is often accomplished by fusing the protein to a 6xHis tag, which can then be used to isolate the protein quickly and efficiently and can be cleaved, leaving an untagged protein available if desired [5]. Furthermore, with the development of fluorophore-conjugated antibodies, 6xHis-tagged fusion proteins can be used to analyze protein-protein interactions via fluorescence resonance energy transfer (FRET) as an alternative to traditional methods. The most common use of 6xHis fusion proteins, however, is in the His Pulldown assay. In this assay, a protein of interest is fused with a 6xHis tag and overexpressed in mammalian cells. The 6xHis fusion protein can then be isolated, and complexed proteins analyzed (Fig. 2).

2   Materials

1. 6xHis mammalian expression vector (e.g., pDEST26, Invitrogen).

2. 6xHis-tag affinity resin.

3. Chromatography column.

4. Lysis buffer: 1 % NP-40, 137 mM NaCl, 40 mM Tris–HCl pH 8.0, 60 mM imidazole, 5 mM $Na_4P_2O_7$, 5 mM NaF, 2 mM Na_3VO_4, 1 mM PMSF, 10 mg/L aprotinin, 10 mg/L leupeptin.

5. 1× PBS.

6. Washing buffer: 1 % NP-40, 500 mM NaCl, 20 mM Tris–HCl pH 8.0, 60 mM imidazole.

7. Binding buffer: 500 mM NaCl, 20 mM Tris–HCl pH 8.0, 5 mM imidazole.

8. Charging buffer: 50 mM $NiSO_4$.

9. 6× SDS sample buffer: 7 mL 4× Tris–HCl/SDS (6.05 g Tris base, 0.4 g SDS, 40 mL H_2O, pH 6.8), 3.0 mL glycerol, 1.0 g SDS, 0.93 g DTT, 1.2 mg bromophenol blue, H_2O to 10 mL total volume (aliquot and store at −20 °C).

10. Rotator.

11. Antibodies for Western blotting: His-probe (Santa cruz, sc-803), His-tag (Cell signaling, #2365).

3 Methods

3.1 Expression Plasmids

Many commercial plasmids are available for the production of 6xHis fusion proteins. Two commonly used vectors are pDEST17 for bacterial expression and pDEST26 for mammalian expression (Invitrogen). These plasmids can be used within the gateway cloning system to easily shuttle genes of interest into various vectors. However, other 6xHis vectors are commercially available.

3.2 Expression in Mammalian Cell Culture and Lysate Preparation

The following procedure is based on mammalian expression of 6xHis-tagged proteins grown in six well plastic tissue culture plates. Figure 3 shows interaction of 6xHis-14-3-3γ and HA-ASK1 as an example.

1. Transfect mammalian cells, such as HEK293T or COS7, with 6xHis-fusion protein expression plasmids using an appropriate transfection reagent (e.g., Fugene HD, Xtremegene HP, or Lipofectamine) according to the manufacturer's protocol.

2. Forty-eight hours after transfection, aspirate the medium off cells and wash each well with 1 mL of 1× PBS. After washing, add 200 μL of lysis buffer (containing imidazole) to each well (see Notes 1 and 2).

3. Scrap cells into lysis buffer and transfer to microcentrifuge tubes. Incubate samples on ice for 10–30 min.

4. Spin samples down in a microcentrifuge for 10 min at maximum speed at 4 °C to clarify lysates.

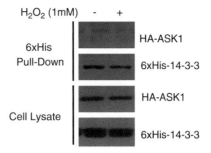

Fig. 3 Example of a 6xHis pulldown assay. Cells were transfected with HA-ASK1 and 6xHis-14-3-3γ. 48 h after transfection, cells were treated with H_2O_2 and harvested. Lysate samples were analyzed by performing a 6xHis pulldown assay, SDS-PAGE, and Western blot

5. Reserve 20 μL of clarified lysate supernatant to serve as the input or whole cell lysate control. Add 6 μL of 6× SDS sample buffer and boil for 5 min. Store samples at −20 °C or −80 °C until ready for SDS-PAGE/Western blot analysis.

3.3 Performing a His Pulldown

1. Add 160 μL of the lysate to 25 μL of charged his resin beads (*see* Subheading 3.4) in a microcentrifuge tube.

2. Rotate samples slowly at 4 °C for 2 h (*see* **Note 3**).

3. Quickly centrifuge beads and discard the supernatant.

4. Wash the beads twice by adding 200 μL of washing buffer and once with 200 μL of binding buffer. For each wash, gently agitate the beads by turning the tube up and down four to five times, quickly centrifuge the beads, and discard the supernatant (*see* **Note 4**).

5. Recover bound proteins from the beads by boiling in 20 μL of 2× SDS sample loading buffer for 5 min.

6. Analyze proteins by SDS-PAGE and Western blotting. Anti-6xHis antibodies are available from commercial sources (*see* **Note 5**). A couple commonly used antibodies are listed in Subheading 2.

3.4 Charging 6xHis Resin Beads

1. Add 3 mL his resin to a chromatography column, such as the Bio-Rad Poly-Prep column. (Volumes can be scaled up or down depending on need.)

2. Add 10 mL of nanopure water to the column, and allow to flow through by gravity. Repeat a total of three times. Allow the water to flow through completely before each repetition.

3. Add 15 mL of charging buffer to the column.

4. Add 20 mL of binding buffer to the column.

5. Add 3 mL of binding buffer to the column to make a 50 % slurry of charged beads.

6. Store at 4 °C for in capped columns or microcentrifuge tubes for up to 6 months.

4 Notes

1. Detergent and NaCl concentrations can be adjusted in the lysis buffer to alter the stringency of the assay. Increased detergent and salt concentrations will increase the stringency of the assay and decrease nonspecific binding.

2. The length of time between cell transfection and lysis can be altered to optimize protein expression.

3. The time of incubation at 4 °C can be lengthened to increase capture of 6xHis fusion proteins on the surface of the immobilized metal beads.

4. The number of washes following a 6xHis pulldown can be increased to decrease nonspecific binding and weak interactions.

5. Before performing Western blot analysis, confirm that the 6xHis antibody chosen can recognize the fusion tag utilized in the assay. 6xHis tags may be fused to either the N- or C-terminus of a protein, and the antibody used should be able to recognize the desired orientation. Similarly, anti-His antibodies may be raised against epitopes with varying numbers of histidine residues, and the ability of the antibody to recognize the 6xHis tag used should be confirmed by checking the manufacturer's datasheet.

References

1. Bjerrum OJ (1977) Immunochemical investigation of membrane proteins. A methodological survey with emphasis placed on immunoprecipitation in gels. Biochim Biophys Acta 472:135–195

2. Terpe K (2003) Overview of tag protein fusions: from molecular and biochemical fundamentals to commercial systems. Appl Microbiol Biotechnol 60:523–533

3. Hochuli E, Bannwarth W, Dobeli H, Gentz R, Stuber D (1988) Genetic approach to facilitate purification of recombinant proteins with novel metal chelate adsorbent. Nat Biotechnol 6:1321–1325

4. Kuo WH, Chase HA (2011) Exploiting the interactions between poly-histidine fusion tags and immobilized metal ions. Biotechnol Lett 33:1075–1084

5. Bornhorst JA, Falke JJ (2000) Purification of proteins using polyhistidine affinity tags. Methods Enzymol 326:245–254

Chapter 24

Studying Protein-Protein Interactions via Blot Overlay/Far Western Blot

Randy A. Hall

Abstract

Blot overlay is a useful method for studying protein-protein interactions. This technique involves fractionating proteins on SDS-PAGE, blotting to nitrocellulose or PVDF membrane, and then incubating with a probe of interest. The probe is typically a protein that is radiolabeled, biotinylated, or simply visualized with a specific antibody. When the probe is visualized via antibody detection, this technique is often referred to as "Far Western blot." Many different kinds of protein-protein interactions can be studied via blot overlay, and the method is applicable to screens for unknown protein-protein interactions as well as to the detailed characterization of known interactions.

Key words Protein-protein interactions, Blot overlay, Far Western blot, Protein, Receptor, Association, Nitrocellulose, SDS-PAGE, Binding

1 Introduction

During preparation for SDS-PAGE, proteins are typically reduced and denatured via treatment with Laemmli sample buffer [1]. Since many protein-protein interactions rely upon aspects of secondary and tertiary protein structure that are disrupted under reducing and denaturing conditions, it might seem likely that few if any protein-protein interactions could survive treatment with SDS-PAGE sample buffer. Nonetheless, it is well-known that many types of protein-protein interaction do in fact still occur even after one of the partners has been reduced, denatured, run on SDS-PAGE, and Western blotted. Blot overlays are a standard and very useful method for studying interactions between proteins.

In principle, a blot overlay is similar to a Western blot. For both procedures, samples are run on SDS-PAGE gels, transferred to nitrocellulose or PVDF, and then overlaid with a soluble protein that may bind to one or more immobilized proteins on the blot. In the case of a Western blot, the overlaid protein is antibody. In the case of a blot overlay, the overlaid protein is a probe of interest,

Cheryl L. Meyerkord and Haian Fu (eds.), *Protein-Protein Interactions: Methods and Applications*, Methods in Molecular Biology, vol. 1278, DOI 10.1007/978-1-4939-2425-7_24, © Springer Science+Business Media New York 2015

often a fusion protein that is easy to detect. The overlaid probe can be detected either via incubation with an antibody (this method is often referred to as a "Far Western blot"), via incubation with streptavidin (if the probe is biotinylated), or via autoradiography if the overlaid probe is radiolabeled with ^{32}P. The specific method that will be described here is a Far Western blot overlay that was used to detect the binding of blotted hexahistidine-tagged PDZ domain fusion proteins to soluble GST fusion proteins corresponding to adrenergic receptor carboxyl-termini [2]. However, this method may be adapted to a wide variety of applications.

2 Materials

1. SDS-PAGE mini-gel apparatus.

2. SDS-PAGE 4–20 % mini gels.

3. Western blot transfer apparatus.

4. Power supply.

5. Nitrocellulose.

6. SDS-PAGE pre-stained molecular weight markers.

7. SDS-PAGE sample buffer: 20 mM Tris–HCl, pH 7.4, 2 % SDS, 2 % β-mercaptoethanol, 5 % glycerol, 1 mg/ml bromophenol blue.

8. SDS-PAGE running buffer: 25 mM Tris–HCl, pH 7.4, 200 mM glycine, 0.1 % SDS.

9. SDS-PAGE transfer buffer: 10 mM Tris–HCl, pH 7.4, 100 mM glycine, 20 % methanol.

10. Purified hexahistidine-tagged fusion proteins.

11. Purified GST-tagged fusion proteins.

12. Anti-GST monoclonal antibody.

13. Goat anti-mouse HRP-coupled secondary antibody.

14. Phosphate-buffered saline: 137 mM NaCl, 2.7 mM KCl, 10 mM Na_2HPO_4, 2 mM KH_2PO_4, pH 7.4.

15. Blocking buffer: 2 % nonfat powdered milk, 0.1 % Tween-20 in PBS.

16. Enhanced chemiluminescence (ECL) kit.

17. Blot trays.

18. Autoradiography cassette.

19. Clear plastic sheet protector.

20. Film.

3   Methods

3.1   SDS-PAGE and Blotting

The purpose of this step is to immobilize the samples of interest on nitrocellulose or an equivalent matrix, such as PVDF. It is very important to keep the blot clean during the handling steps involved in the transfer procedure, since contaminants can contribute to increased background problems later on during detection of the overlaid probe.

1. Place gel in SDS-PAGE apparatus and fill chamber with running buffer.

2. Mix purified hexahistidine-tagged fusion proteins with SDS-PAGE sample buffer to a final concentration of approximately 0.1 μg/μl of fusion protein (*see* **Note 1**).

3. Load 20 μl of fusion protein (2 μg total) in each lane of the gel. If there are more lanes than samples, load 20 μl of sample buffer in the extra lanes (*see* **Note 2**).

4. In at least one lane of the gel, load 20 μl of SDS-PAGE molecular weight markers.

5. Run gel for approximately 1 h at 150 V using the power supply.

6. Stop gel, turn off the power supply, remove the gel from its protective casing, and place in transfer buffer.

7. Place pre-cut nitrocellulose in transfer buffer to wet it.

8. Put nitrocellulose and gel together in transfer apparatus, and transfer proteins from gel to nitrocellulose for 90 min at 200 V using a power supply.

3.2   Overlay

During the overlay step, the probe is incubated with the blot and unbound probe is then washed away. The potential success of the overlay depends heavily on the purity of the overlaid probe. GST and hexahistidine-tagged fusion proteins should be purified as extensively as possible. If the probe has many contaminants, this may contribute to increasing the background during the detection step, making visualization of the specifically bound probe more difficult.

1. Block blot in blocking buffer for at least 30 min (*see* **Note 3**).

2. Add GST fusion proteins to a concentration of 25 nM in 10 ml of blocking buffer.

3. Incubate GST fusion proteins with blot for 1 h at room temperature while rocking slowly.

4. Discard GST fusion protein solution and wash blot three times for 5 min each with 10 ml of blocking buffer while rocking slowly.

5. Add anti-GST antibody at 1:1,000 dilution (approximately 200 ng/ml final) to 10 ml of blocking buffer.

6. Incubate anti-GST antibody with blot for 1 h while rocking slowly.

7. Discard anti-GST antibody solution and wash blot three times for 5 min each with 10 ml of blocking buffer while rocking slowly.

8. Add goat anti-mouse HRP-coupled secondary antibody at 1:2,000 dilution to 10 ml of blocking buffer.

9. Incubate secondary antibody with blot for 1 h while rocking slowly.

10. Discard secondary antibody solution and wash blot three times for 5 min each with 10 ml of blocking buffer while rocking slowly (*see* **Note 4**).

11. Wash blot one time for 5 min with phosphate-buffered saline, pH 7.4.

3.3 Detection of Overlaid Proteins

The final step of the overlay is to detect the probe that is bound specifically to proteins immobilized on the blot. In viewing different exposures of the visualized probe, an effort should be made to obtain the best possible signal-to-noise ratio. Nonspecific background binding will increase linearly with time of exposure. Thus, shorter exposures may have more favorable signal-to-noise ratios.

1. Incubate blot with enhanced chemiluminescence solution for 60 s (*see* **Note 5**).

2. Remove excess ECL solution from blot and place blot in a clear plastic sheet protector.

3. Tape sheet protector into an autoradiography cassette.

4. Move to the darkroom and place one sheet of film into the autoradiography cassette with the blot.

5. Expose film for 5–2,000 s, depending on the intensity of the signal.

6. Develop the film in standard film developer.

4 Notes

1. The protocol described here is intended for the in-depth study of a protein-protein interaction that is already known. However, blot overlays can also be utilized in preliminary screening studies to detect novel protein-protein interactions. For this application, tissue lysates would typically be loaded onto the SDS-PAGE gel instead of purified fusion protein samples. The blotted tissue lysates would then be overlaid with the

probe of interest. The advantages of this technique are (1) many tissue samples can be screened in a single blot and (2) the molecular weight and tissue distribution of probe-interacting proteins can be immediately determined. The disadvantages of this method are (1) due to the multiple washing steps involved in the procedure, a fairly high-affinity interaction is required for the interaction to be detected, (2) detection of probe-interacting proteins is dependent upon their level of expression in native tissues, and (3) interactions requiring native conformations of both proteins will not be detected. Tissue lysate overlays have been utilized as screening tools to detect not only the interaction of the β_1-adrenergic receptor with MAGI-2 described here [2] (Fig. 1) but also the interaction of the β_2-adrenergic receptor with NHERF-1 [3] and the interactions of a number of different proteins with actin [4–6], calmodulin [7, 8], and the cyclic AMP-dependent protein kinase RII regulatory subunit [9–12]. Tissue lysate overlay approaches have also been effectively utilized to detect phosphorylation-specific interactions between SH2 domains and various phosphoproteins [13–16]. Associations involving modular protein domains such as SH2 or PDZ domains are often detected extremely well via blot overlay approaches, since such modular domains typically interact with short motifs on

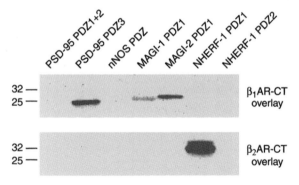

Fig. 1 Overlay of GST-tagged adrenergic receptor carboxyl-termini onto hexahistidine-tagged PDZ domains. Equal amounts (2 μg) of purified His-tagged fusion proteins corresponding to PDZ domains from PSD-95, nNOS, MAGI-1, MAGI-2, and NHERF-1 were immobilized on nitrocellulose. Overlays with the carboxyl-terminus of the β_1-adrenergic receptor expressed as a GST fusion protein (β_1AR-CT-GST) (25 nM) revealed strong binding to PSD-95 PDZ3 and MAGI-2 PDZ1, moderate binding to MAGI-1 PDZ1, and no detectable binding to the first two PDZ domains of PSD-95 or to the PDZ domains of nNOS or NHERF-1. In contrast, overlays with the β_2-adrenergic receptor expressed as a GST fusion protein (β_2AR-CT-GST) (25 nM) revealed strong binding to NHERF-1 PDZ1 but no detectable binding to any of the other PDZ domains examined. These data demonstrate that selective and specific binding can be obtained in overlay assays

their binding partners in a manner that is not disrupted by denaturation of the binding partners on SDS-PAGE gels. Screens for novel protein-protein interactions via blot overlay have also been performed by probing panels of purified fusion proteins instead of tissue lysates; this method has been successfully utilized to identify a number of novel PDZ domain-mediated interactions [17–26].

2. Since some probes can exhibit extensive nonspecific binding to blotted proteins, it is important in overlay assays to have negative controls for probe binding. When the blotted proteins are GST fusion proteins, GST by itself is a good negative control (as illustrated in Fig. 2). When the blotted proteins are His-tagged fusion proteins, as illustrated in Fig. 1, it is helpful to have one or more His-tagged fusion proteins on the same blot that will not bind to the probe. In this way, it is possible to demonstrate the specificity of binding and to rule out the possibility that the observed interaction is due to the tag.

3. The blocking of the blot is a very important step in every overlay assay. The idea is to block potential nonspecific sites of protein attachment to the blot, so that nonspecific binding of the probe will be minimized. When a high amount of

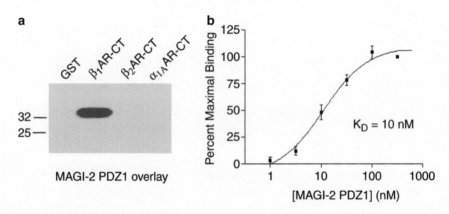

Fig. 2 Overlay of hexahistidine-tagged MAGI-2 PDZ1 onto GST-tagged adrenergic receptor carboxyl-termini. (**a**) In the reverse of the overlay experiments illustrated in Fig. 1, equal amounts (2 μg) of purified GST fusion proteins corresponding to the carboxyl-termini of various adrenergic receptor subtypes were immobilized on nitrocellulose. Overlay with His/S-tagged MAGI-2 PDZ1 (20 nM) revealed strong binding to β_1AR-CT-GST but no detectable binding to control GST, β_2AR-CT-GST or α_{1A}AR-CT-GST. These data demonstrate that the interaction between the β_1AR-CT and MAGI-2 PDZ1 can be visualized via overlay in either direction. (**b**) Estimate of the affinity of the interaction between β_1AR-CT and MAGI-2 PDZ1. Nitrocellulose strips containing 2 μg β_1AR-CT-GST (equivalent to lane 2 in the preceding panel) were incubated with His/S-tagged MAGI-2 PDZ1 at six concentrations between 1 and 300 nM. Specific binding of MAGI-2 PDZ1 did not increase between 100 and 300 nM, and thus the binding observed at 300 nM was defined as "maximal" binding. The binding observed at the other concentrations was expressed as a percentage of maximal binding within each experiment. The bars and error bars shown on this graph indicate mean ± SEM ($n = 3$). The K_D for MAGI-2 PDZ1 binding to β_1AR-CT was estimated at 10 nM (*see* **Note 6**)

nonspecific background binding is observed, it is often helpful to block for a longer time or with a higher concentration of milk. Some investigators favor bovine serum albumin or other proteins in place of milk for blocking blots prior to overlay.

4. The washing of the blot is of critical importance. If the washes are not rigorous enough, the nonspecific background binding of the probe will be undesirably high. Conversely, if the washes are too rigorous, specific binding of the probe may be lost and the protein-protein interaction of interest may be difficult to detect. Thus, if a large amount of nonspecific background binding is observed, one should consider increasing the rigor of the washes, while conversely if the background is low but little or no specific binding is observed, one should consider decreasing the rigor of the washes. The rigor of the washes is dependent upon (1) time, (2) volume, (3) speed, and (4) detergent concentration. To make washes more rigorous, one should wash for a longer time, wash in a larger volume, increase the rate at which the gels are rocked during the washes, and/or increase the detergent concentration in the buffer used for washing.

5. There are a number of ways to visualize bound probe in an overlay assay. The method described here depends upon detection of the probe with an antibody, which is often referred to as a "Far Western blot." One alternative approach is to biotinylate the probe and then detect it with a streptavidin/enzyme conjugate [5–7]. The appeal of this approach is that it can be quite sensitive, since the streptavidin-biotin interaction is one of the highest affinity interactions known. The main drawback of this approach is that biotinylation of the probe may alter its properties, such that it may lose the ability to interact with partners it normally binds to. An additional approach to probe detection is phosphorylation of the probe using ^{32}P-ATP to make the probe radiolabeled [10–12]. A primary advantage of this method is that once the probe is overlaid onto the blot, no further detection steps are necessary (i.e., no incubations with antibody or streptavidin are required). This cuts down on the number of washing steps and may aid in the detection of protein-protein interactions that are of somewhat lower affinity. The main disadvantages of the phosphorylation approach are (1) radioactive samples require special handling, and (2) as with biotinylation, phosphorylation of the probe may alter its properties, such that certain protein-protein interactions may be disrupted.

6. As is illustrated in Figs. 1 and 2, detection of the interactions between adrenergic receptor carboxyl-termini and their PDZ domain-containing binding partners is completely reversible. Either partner can be immobilized on the blot and overlaid

with the other. Many other protein-protein interactions can similarly be detected in a reversible manner, but some interactions can only be detected in one direction due to a requirement for the native conformation of one of the partners. As is also illustrated in Fig. 2, the affinity of a given protein-protein interaction may be estimated via blot overlay saturation binding curves. This method involves increasing the concentration of overlaid probe until a maximal amount of specific binding is obtained. An estimate for the affinity constant (K_D) of the interaction can then be determined from the slope of the binding curve. Estimates such as these must be evaluated with the caveat that they are derived under artificial conditions involving many hours of incubation time, washing, and detection. Nonetheless, affinity constant estimates derived via this method are useful in comparing affinities between proteins examined under the same conditions and overlaid with the same probe.

Acknowledgments

R.A.H. is supported by grants from the National Institutes of Health.

References

1. Laemmli UK (1970) Cleavage of structural proteins during the assembly of the head of bacteriophage T4. Nature 227:680–685

2. Xu J, Paquet M, Lau AG et al (2001) beta 1-adrenergic receptor association with the synaptic scaffolding protein membrane-associated guanylate kinase inverted-2 (MAGI-2). Differential regulation of receptor internalization by MAGI-2 and PSD-95. J Biol Chem 276:41310–41317

3. Hall RA, Premont RT, Chow CW et al (1998) The beta2-adrenergic receptor interacts with the Na+/H+−exchanger regulatory factor to control Na+/H+ exchange. Nature 392:626–630

4. Luna EJ (1998) F-actin blot overlays. Methods Enzymol 298:32–42

5. Li Y, Hua F, Carraway KL et al (1999) The p185(neu)-containing glycoprotein complex of a microfilament-associated signal transduction particle. Purification, reconstitution, and molecular associations with p58(gag) and actin. J Biol Chem 274:25651–25658

6. Holliday LS, Lu M, Lee BS et al (2000) The amino-terminal domain of the B subunit of vacuolar H+−ATPase contains a filamentous actin binding site. J Biol Chem 275:32331–32337

7. Pennypacker KR, Kyritsis A, Chader GJ et al (1988) Calmodulin-binding proteins in human Y-79 retinoblastoma and HTB-14 glioma cell lines. J Neurochem 50:1648–1654

8. Murray G, Marshall MJ, Trumble W et al (2001) Calmodulin-binding protein detection using a non-radiolabeled calmodulin fusion protein. Biotechniques 30:1036–1042

9. Lohmann SM, DeCamilli P, Einig I et al (1984) High-affinity binding of the regulatory subunit (RII) of cAMP-dependent protein kinase to microtubule-associated and other cellular proteins. Proc Natl Acad Sci U S A 81:6723–6727

10. Bregman DB, Bhattacharyya N, Rubin CS (1989) High affinity binding protein for the regulatory subunit of cAMP-dependent protein kinase II-B. Cloning, characterization, and expression of cDNAs for rat brain P150. J Biol Chem 264:4648–4656

11. Carr DW, Hausken ZE, Fraser ID et al (1992) Association of the type II cAMP-dependent protein kinase with a human thyroid RII-anchoring protein. Cloning and

characterization of the RII-binding domain. J Biol Chem 267:13376–13382

12. Hausken ZE, Coghlan VM, Scott JD (1998) Overlay, ligand blotting, and band-shift techniques to study kinase anchoring. Methods Mol Biol 88:47–64

13. Machida K, Thompson CM, Dierck K, Jablonowski K, Karkkainen S, Liu B, Zhang H, Nash PD, Newman DK, Nollau P, Pawson T, Renkema GH, Saksela K, Schiller MR, Shin DG, Mayer BJ (2007) High-throughput phosphotyrosine profiling using SH2 domains. Mol Cell 26:899–915

14. Machida K, Mayer BJ (2009) Detection of protein-protein interactions by far-western blotting. Methods Mol Biol 536:313–329

15. Evans JV, Ammer AG, Jett JE, Bolcato CA, Breaux JC, Martin KH, Culp MV, Gannett PM, Weed SA (2012) Src binds cortactin through an SH2 domain cystine-mediated linkage. J Cell Sci 125:6185–6197

16. Gao Z, Poon HY, Li L, Li X, Palmesino E, Glubrecht DD, Colwill K, Dutta I, Kania A, Pawson T, Godbout R (2012) Splice-mediated motif switching regulates disabled-1 phosphorylation and SH2 domain interactions. Mol Cell Biol 32:2794–2808

17. Fam SR, Paquet M, Castleberry AM, Oller H, Lee CJ, Traynelis SF, Smith Y, Yun CC, Hall RA (2005) P2Y$_1$ purinergic receptor signaling is controlled by interaction with the PDZ scaffold NHERF-2. Proc Natl Acad Sci U S A 102:8042–8047

18. He J, Bellini M, Inuzuka H, Xu J, Xiong Y, Wang X, Castleberry AM, Hall RA (2006) Proteomic analysis of β$_1$-adrenergic receptor interactions with PDZ scaffold proteins. J Biol Chem 281:2820–2827

19. Paquet M, Asay MJ, Fam SR, Inuzuka H, Castleberry AM, Oller H, Smith Y, Yun CC, Traynelis SF, Hall RA (2006) The PDZ scaffold NHERF-2 interacts with mGluR5 and regulates receptor activity. J Biol Chem 281:29949–29961

20. Balasubramanian S, Fam SR, Hall RA (2007) GABA$_B$ receptor association with the PDZ scaffold Mupp1 alters receptor stability and function. J Biol Chem 282:4162–4171

21. Lee SF, Kelly M, McAlister A, Luck SN, Garcia EL, Hall RA, Robins-Browne RM, Frankel G, Hartland EL (2008) A C-terminal class I PDZ binding motif of EspI/NleA modulates the virulence of attaching and effacing *Escherichia coli* and *Citrobacter rodentium*. Cell Microbiol 10:499–513

22. Kunkel MT, Garcia EL, Kajimoto T, Hall RA, Newton AC (2009) The protein scaffold NHERF-1 interacts with PKD and controls the amplitude and duration of localized PKD activity. J Biol Chem 284:24653–24661

23. Stalker TJ, Wu J, Morgans A, Traxler EA, Wang L, Chatterjee MS, Lee D, Quertermous T, Hall RA, Hammer DA, Diamond SL, Brass LF (2009) Endothelial cell specific adhesion molecule (ESAM) localizes to platelet-platelet contacts and regulates thrombus formation *in vivo*. J Thromb Haemost 7:1886–1896

24. Ritter SL, Asay MJ, Paquet M, Paavola KP, Reiff RE, Yun CC, Hall RA (2011) GLAST stability and activity are regulated by interaction with the PDZ scaffold NHERF-2. Neurosci Lett 487:3–7

25. O'Neill A, Gallegos L, Justilien V, Garcia EL, Leitges M, Fields A, Hall RA, Newton AC (2011) PKCα promotes cell migration through a PDZ-dependent interaction with its novel substrate discs large homolog (DLG) 1. J Biol Chem 286:43559–44368

26. Matsumoto M, Fujikawa A, Suzuki R, Shimizu H, Kuboyama K, Hiyama TY, Hall RA, Noda M (2012) SAP97 promotes the stability of Na$_x$ channels at the plasma membrane. FEBS Lett 586:3805–3812

Chapter 25

Co-immunoprecipitation from Transfected Cells

Yoshinori Takahashi

Abstract

Co-immunoprecipitation (Co-IP) is one of the most widely used methods to identify novel proteins that associate with a protein of interest or to determine complex formation between known proteins. For this technique, a protein of interest is captured using a specific antibody. The antibody-bound protein, as well as any proteins bound to the protein of interest, is then precipitated using a resin (immunoprecipitation, IP). Proteins that are not bound to the protein of interest are then removed from the sample with a series of washes. The resulting immunocomplexes are then analyzed by immunoblot. As the requirements for protein-protein interactions vary, optimal experimental conditions for examining the interacting partners of different proteins of interest must be determined empirically. Once appropriate experimental conditions have been established, the IP/Co-IP procedure is simple and straightforward. In this chapter, a standard protocol for IP/co-IP, with several key factors for the success of IP/co-IP analyses, is discussed.

Key words Immunoprecipitation, Co-immunoprecipitation, Protein-protein interaction, Antibody, Transfection

1 Introduction

Protein-protein interactions regulate many biological processes both within and between cells [1, 2]. Co-immunoprecipitation (Co-IP) is a simple, yet effective method to determine complex formation between proteins of interest [3–5]. In this assay, proteins that directly or indirectly interact with a protein of interest can be co-purified using specific antibodies (Fig. 1) (*see* **Note 1**). The results of Co-IP are highly reproducible, and the assay is relatively inexpensive.

The first step of the co-IP procedure is to prepare lysate from cells or tissue samples that express (either endogenously or exogenously) the protein of interest. In order to ensure that protein-protein interactions will remain intact, cells need to be lysed using a proper lysis buffer. In general, non-ionic detergents (e.g., NP-40, Triton X-100) disrupt lipid–protein but not protein-protein interactions, while ionic detergents (e.g., Chaps) affect both. Therefore, lysis buffers containing non-ionic detergents

Cheryl L. Meyerkord and Haian Fu (eds.), *Protein-Protein Interactions: Methods and Applications*, Methods in Molecular Biology, vol. 1278, DOI 10.1007/978-1-4939-2425-7_25, © Springer Science+Business Media New York 2015

**1. Transfect expression vectors
and/or treat cells**

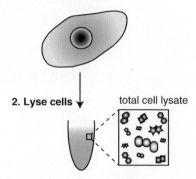

2. Lyse cells   total cell lysate

3. Add antibody-immobilized sepharose/agarose resin

4. Precipitate immune-complexes & wash the precipitates

immune-complex

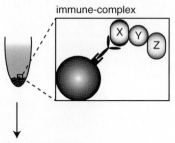

5. Analyze by immunoblotting

Fig. 1 Schematic of the co-immunoprecipitation procedure. Cells may be transfected with expression vectors that encode proteins of interest and/or treated to induce protein complex formation (step 1). The cells are then lysed in an optimal lysis buffer (step 2). Total cell lysate (TCL) is then incubated with resin-immobilized antibodies, which can specifically recognize the protein of interest, to form immune-complexes (step 3). The resultant immune-complexes are then precipitated by centrifugation, washed to remove unbound proteins, and analyzed by immunoblotting (steps 4 and 5)

tend to preserve protein-protein interactions more than those containing ionic detergents (*see* **Note 2**). Proteins of interest are then captured by incubating total cell/tissue lysate with specific antibodies. The resultant immunocomplexes (composed of antibody, protein of interest (antigen), and antigen-associated proteins) can be precipitated using a resin (e.g., agarose, sepharose, or magnetic beads) that is conjugated with IgG-binding Protein A/G (*see* **Note 3**). Following a series of washes to remove irrelevant, non-binding proteins, antigens and any proteins that are bound are eluted by boiling the precipitated resins in denaturing Laemmli buffer or by incubating with large amounts of peptides containing the epitope of the immunoprecipitation antibody. The eluted proteins are then analyzed by SDS-PAGE/immunoblotting and/or mass spectrometry.

As mentioned above, proper experimental conditions must be determined for each protein-protein interaction. Selection of an optimal lysis buffer and immunoprecipitation antibody are the two most important aspects for the success of a co-IP experiment. However, the requirement of antibodies that specifically recognize the antigen without interfering with its interaction with binding partners limits the application of co-IP for detection of complexes between many endogenously expressed proteins. To overcome this hurdle, the protein of interest (and in some cases associating proteins of interest) is frequently fused with a small peptide sequence (~15aa), known as an epitope tag (e.g., flag, myc, HA, his, V5), or a fluorescence protein (e.g., GFP, DsRed) at the amino- or carboxy-terminus, and ectopically expressed in cells (*see* **Note 4**). Antibodies for these "tags" that are compatible with IP/co-IP have been well developed and are commercially available from multiple manufacturers. In this chapter, a protocol for transient transfection of 293 T cells using the calcium-phosphate method, the most frequently used and least expensive strategy to express proteins of interest, in addition to standard procedures for co-IP, is described.

2   Materials

1. Cell line: HEK293, 293 T, or a variant such as 293 T/F17 (all available from ATCC).

2. Growth medium: Dulbecco's Modified Eagle's Medium (DMEM) supplemented with 10 % fetal bovine serum and 1 % penicillin/streptomycin.

3. Expression plasmids.

4. Transfection reagents: 2.5 M $CaCl_2$; 2 × HEPES-buffered saline (HeBS): 140 mM NaCl, 1.5 mM Na_2HPO_4, 50 mM HEPES, pH 7.05 (*see* **Note 5**).

5. Cell scraper/lifter (for adherent cells).

6. Phosphate buffered saline (PBS): 137 mM NaCl, 2.7 mM KCl, 10 mM Na_2HPO_4, 2 mM KH_2PO_4, pH 7.4.

7. Lysis buffer: prepare stock solution (e.g., 137 mM NaCl, 20 mM Tris–HCl, pH 8.0, 10 % glycerol, 1 % NP-40) and add protease and phosphatase (optional) inhibitor cocktails prior to use (*see* **Note 6**).

8. Commercially available kit for measuring protein concentration (e.g., BCA kit, Thermo Scientific).

9. Antibody that is specific for the protein of interest, and is indicated for use in immunoprecipitation experiments.

10. Protein A or G-immobilized resin (e.g., sepharose, agarose) (see **Note 3**).

11. 2 × Laemmli sample buffer: prepare stock solution (125 mM Tris–HCl, pH 6.8, 4 % SDS, 0.01 % bromophenol blue, 20 % glycerol) and add ~5–10 % of β-mercaptoethanol prior to use.

12. Equipment: Basic lab equipment including a tube shaker/rotator and cold room/chromatography refrigerator.

13. Reagents and equipment for immunoblotting.

3   Methods

3.1   Transfection of Plasmids Expressing the Gene/Proteins of Interest (See Note 7)

Day 1

1. Trypsinize 293 T/F17 cells grown under standard culture conditions and pellet 4.5×10^6 cells per sample.

2. Resuspend the cells in 2 ml of medium/sample. Seed the cells in 10 cm culture dishes containing 8 ml of growth medium. Place the dishes in a 37 °C incubator supplemented with 5 % CO_2 and allow the cells to attach and grow overnight.

Day 2 (The cells should be approximately 50–60 % confluent prior to transfection)

3. Mix the following reagents in a 5 ml tube for each sample:

Expression plasmids (5 µg each) + sterile deionized water	450 µl
2.5 M CaCl$_2$	50 µl

4. Gently add 500 µl of 2 × HeBS in a drop-wise manner. Mix by vortexing (2 s × 5) and incubate at room temperature for 15 min.

5. Gently add the transfection mixture in a drop-wise manner to the cells.

6. Change the medium 6–20 h after transfection.

7. Treat the cells if necessary (see **Note 8**).

3.2   Preparation of Total Cell Lysate

1. Twenty-four to 48 h after transfection, harvest the cells using a cell scraper/lifter and pellet the cells by centrifugation at $1{,}500 \times g$ for 5 min at 4 °C.

2. Wash the cells by resuspending the pellet in 1 ml of ice-cold PBS, transfer the samples to a new 1.7 ml tube, centrifuge at $1{,}500 \times g$ for 5 min at 4 °C, and discard as much of the supernatant as possible (see **Note 9**).

3. Add ice-cold lysis buffer (~3 × –5 × the pellet volume) and lyse the pellet by gently pipetting until all clumps disappear (~5–10 times).

4. Incubate the samples on ice for 30 min.

5. Centrifuge at 15,000× g for 10 min at 4 °C and transfer the supernatant to a new 1.7 ml tube (*see* **Note 9**).

6. Measure protein concentration.

7. Check the expression of proteins of interest by SDS-PAGE/immunoblot analyses (this step is optional, but highly recommended) (*see* **Note 10**).

3.3 Preclearing of Cell Lysate (Optional) (See Note 11)

1. Using a large-orifice tip, transfer the appropriate IgG-cross-linked resins (15–20 μl × (sample number + 1)) (*see* **Notes 12** and **13**) to a microcentrifuge tube containing 1 ml of PBS. Pellet the resin by centrifugation at 6,000 × g for 30 s and aspirate the supernatant.

2. Resuspend the resin in 1 ml lysis buffer (without protease/phosphatase inhibitors), pellet the resin by centrifugation at 6,000 × g for 30 s, and aspirate the supernatant.

3. Repeat **step 2** three times for a total of four washes.

4. Resuspend the resin in ice-cold lysis buffer (~30–35 μl × (sample number + 1)).

5. Transfer 750 μg of total cell lysate to a new tube and adjust the total volume to 200 μl with ice-cold lysis buffer.

6. Add 50 μl of the resin slurry prepared in Subheading 3.3 **step 4** using a large-orifice tip. Incubate the tubes on a tube shaker/rotator at a slow speed for 1 h at 4 °C.

7. Pellet the resin by centrifugation at 6,000 × g for 30 s at 4 °C and transfer 200 μl of the supernatant to a new tube. Save the pelleted resins to use as negative controls for the experiment (*see* **Note 14**).

3.4 Immunoprecipitation

1. In a new tube, dilute the appropriate amount of immunoprecipitation antibodies (1–10 μg of affinity purified antibodies, 1–5 μl of immune serum, or 10–100 μl of hybridoma supernatant,) with lysis buffer (containing the appropriate inhibitors) to 100 μl/sample (*see* **Note 15**).

2. Add 100 μl of the antibody diluent to a precleared lysate-containing tube prepared in **step 7** in Subheading 3.3 and incubate the tubes on a tube shaker/rotator for 1 h (to overnight) at 4 °C to allow the formation of immune complexes.

3. During the incubation, prepare Protein A or G-immobilized resin slurry (as described in **steps 1–4** in Subheading 3.3).

4. Add 50 µl of the resin slurry to each reaction tube in **step 2** and incubate the tubes on a tube shaker/rotator for 1 h at 4 °C to allow the antibody to bind to the protein complexes.

5. Pellet the resin to precipitate the immune complexes by centrifugation at 6,000 × *g* for 30 s at 4 °C. Transfer 100 µl of supernatant to a new tube and aspirate the remaining supernatant using an 18G needle (*see* **Note 16**).

6. Resuspend the pelleted resin in 500 µl of ice-cold lysis buffer, centrifuge at 6,000 × *g* for 30 s at 4 °C, and aspirate the remaining supernatant to remove non-binding proteins.

7. Repeat the wash step two to three more times.

8. Resuspend the resin-bound immune complexes in 20 µl of 2 × Laemmli buffer, boil for 5 min, and analyze by SDS-PAGE/immunoblot analysis. If detecting proteins of interest by the indirect method, avoid using antibodies developed in the same species as those used for immunoprecipitation (*see* **Notes 17, 18**, and **19**).

4   Notes

1. A protein detected in immune-complexes does not necessarily mean that it can directly bind to the antigen. For example, while Beclin 1 (Protein Z in Fig. 1), a component of the class III PI3-kinase that directly interacts with UVRAG (Protein Y), can be detected in an immune-complex isolated using anti-Bif-1 antibodies (immunoprecipitation antibodies), knockdown of *UVRAG* or inhibition of UVRAG-Beclin 1 interaction results in failure to detect Beclin 1in the anti-Bif-1 immune-complex [6]. This indicates that Bif-1 (Protein X) indirectly associates with Beclin 1 through UVRAG. In addition, physiologically irrelevant interactions may be artificially induced during the preparation of whole cell lysates (e.g., association of a nuclear protein with a cytoplasmic protein). Therefore, it is important to validate protein-protein interactions found during co-IP experiments by other methods such as GST-pull down with purified proteins (for the detection of direct protein-protein interactions in vitro), FRET and/or BiFC (for spatial and temporal analyses of protein-protein interactions) (*see* Chapters 20, 22, 31–33; 35–37). Colocalization analysis by confocal microscopy or deconvolution microscopy can also demonstrate that two (or more) proteins localize in the same compartment of a cell (*see* Chapter 34).

2. It is important to keep in mind that this is not always the case and optimal detergents vary depending on proteins of interest. For example, non-ionic detergents have been shown to induce

the conformational change of the inactive form of proapoptotic Bax that artificially leads to the heterodimerization of Bax with antiapoptotic Bcl-xL [7].

3. Protein A and G are immunoglobulin-binding proteins derived from bacteria. The affinity of each protein binding to immunoglobulins differs among species and the subclass of IgG. Typically, Protein A and Protein G are used for immunoprecipitation of antibodies developed in rabbit and mouse systems, respectively. For further information regarding which immunoglobulin should be used for a specific antibody, see the manufacture's product data sheet.

4. Fusion with an epitope tag/fluorescent protein may affect the intracellular localization or function of a protein of interest. It is therefore important to confirm that a tagged-protein is functional in cells (e.g., by expressing it in cells that are deficient for the protein of interest and comparing the results to wild-type cells).

5. pH is a critical factor for the formation of DNA precipitates. It is important to precisely adjust the pH to 7.05. If the efficiency of transfection is too low, recalibrate the pH meter and make solutions with several different pHs (e.g., $\pm$ 0.3) to optimize an ideal pH condition for your particular experiment.

6. Selection of an appropriate lysis buffer is one of the most important steps in a co-immunoprecipitation assay. A suitable lysis buffer varies depending on the proteins of interest (*see* Subheading 1). Addition of glycerol (5–10 %, final) in the lysis buffer may increase the stability or formation of protein complexes.

7. If a specific IP antibody for the protein of interest is available, choose an optimal cell line/tissue lysate in which your proteins of interest are known to be functionally expressed, treat the cells if necessary (*see* **Note 8**) and start from Subheading 3.2 **step 2**.

8. Many protein-protein interactions are regulated by posttranscriptional modifications (e.g., phosphorylation) that are induced upon exposure to certain circumstances. However, in some cases, overexpressing proteins of interest can induce the formation of protein complexes that would normally require a certain stimulation to be detected at the endogenous expression level (e.g., protein complex formation that is regulated by post-transcriptional modifications of proteins of interest).

9. If you wish to stop and continue the following steps later, the samples can be stored at −80 °C. However, multiple freeze–thaw cycles should be avoided to prevent/minimize protein denaturation and degradation.

10. If the proteins of interest are routinely expressed in your lab, the expression levels can simply be quantified by dot blotting. The dot blotting procedure is as follow: (1) Spot 5 μl of sequentially diluted lysates (e.g., 0, 2, 4, and 8 μg/μl) onto a strip of nitrocellulose membrane and mark each spot with a pencil. (2) Dry the membrane strip. (3) Soak the membrane strip in Tris-buffered saline with Tween 20 (TBS-T: 0.05 % Tween 20, 150 mM NaCl, 20 mM Tris–HCl, pH 7.5) and follow the procedure for immunoblotting starting at the blocking step.

11. Preclearing lysates will remove proteins that nonspecifically bind to resins and/or immunoglobulins and thus reduce background signals.

12. Resins are generally stored as slurry in a buffer containing ethanol. Calculate the amount of resin in the slurry and use a volume of 15–20 μl of resin per reaction (e.g., if the resin you are using is stored as a 50 % slurry, use 30–40 μl of slurry/sample).

13. To preclear lysates, use IgG derived from the same species as the antibodies being used for immunoprecipitation. Alternatively, Protein A or G-immobilized resin without immunoprecipitation antibodies can be used.

14. Avoid taking the pelleted resin. After transferring the supernatant, aspirate the remaining supernatant using a 27 G needle (avoid aspirating resin), wash the resin 3–4 times with lysis buffer, resuspend in 20 μl of 2 × Laemmli buffer, and boil for 5 min along with the co-IP samples. The resins used to preclear the lysate can be used as a negative control for the experiment.

15. The appropriate amount of immunoprecipitation antibody will vary by their affinity to the antigen. Contact the antibody source to obtain detailed information on appropriate dilutions.

16. The post-IP supernatant can be used to determine whether the amount of antibodies (and Protein A or G-immobilized resin) used for the experiment was sufficient for immunoprecipitating all, or the majority of, the protein of interest, and how much of the protein of interest can be associated with the antibody. If a substantial amount of the protein of interest remains in the supernatant, increase the amount of immunoprecipitation antibody (or decrease the amount of total cell lysate).

17. In the indirect detection system, proteins of interest are detected using HRP-conjugated secondary antibodies that recognize the IgG of the primary antibody. Boiling of the precipitated resin in Laemmli buffer results in the elution of not only the antigens and antigen-associated proteins but also the immunoprecipitation antibodies that are composed of heavy (~50 kDa) and light (~25 kDa) chains. Therefore, immunoblotting by the indirect

method with antibodies developed in the same species as those used for IP results in the detection of very intense signals from heavy and light chain proteins that frequently mask the signals from proteins of interest. If selection of antibodies developed from different species is not optional, it is strongly recommended to crosslink the immunoprecipitation antibodies to resins or to use the direct detection method for immunoblotting. Alternatively, using detection antibodies that preferentially recognize the native disulfide forms of the IP antibodies (e.g., TrueBlot reagents, Rockland Immunochemicals Inc.), the appearance of heavy and light chains can be minimized from blots.

18. Once a protein (prey) has successfully been co-precipitated with a protein of interest (bait), it is important to switch the "bait" and "prey" and confirm that the two proteins can still be co-precipitated in order to exclude false positives from nonspecific binding.

19. Factors that may affect co-IP results: (1) lysis buffer; (2) immunoprecipitation antibodies; (3) the location where an epitope tag is fused; (4) expression levels of proteins of interest; (5) culture conditions (including certain treatments); and (6) the amount of lysate, antibodies, and/or resin used for IP.

References

1. Fu H (2004) Protein-Protein Interactions. Methods and Applications. Methods in Molecular Biology 261.

2. Braun P, Gingras AC (2012) History of protein-protein interactions: from egg-white to complex networks. Proteomics 12:1478–1498

3. Dwane S, Kiely PA (2011) Tools used to study how protein complexes are assembled in signaling cascades. Bioeng Bugs 2:247–259

4. Markham K, Bai Y, Schmitt-Ulms G (2007) Co-immunoprecipitations revisited: an update on experimental concepts and their implementation for sensitive interactome investigations of endogenous proteins. Anal Bioanal Chem 389:461–473

5. Berggard T, Linse S, James P (2007) Methods for the detection and analysis of protein-protein interactions. Proteomics 7:2833–2842

6. Takahashi Y, Coppola D, Matsushita N et al (2007) Bif-1 interacts with Beclin 1 through UVRAG and regulates autophagy and tumorigenesis. Nat Cell Biol 9:1142–1151

7. Hsu YT, Youle RJ (1997) Nonionic detergents induce dimerization among members of the Bcl-2 family. J Biol Chem 272:13829–13834

Chapter 26

In Vivo Protein Cross-Linking

Fabrice Agou and Michel Véron

Abstract

In the cell, homo- and hetero-associations of polypeptide chains evolve and take place within subcellular compartments that are crowded with many other cellular macromolecules. In vivo chemical cross-linking of proteins is a powerful method to examine changes in protein oligomerization and protein-protein interactions upon cellular events such as signal transduction. This chapter is intended to provide a guide for the selection of cell membrane permeable cross-linkers, the optimization of in vivo cross-linking conditions, and the identification of specific cross-links in a cellular context where the frequency of random collisions is high. By combining the chemoselectivity of the homo-bifunctional cross-linker and the length of its spacer arm with knowledge on the protein structure, we show that selective cross-links can be introduced specifically on either the dimer or the hexamer form of the same polypeptide in vitro as well as in vivo, using the human type B nucleoside diphosphate kinase as a protein model.

Key words In vivo cross-linking, Oligomerization, NDPK-B, Nucleoside nucleotide metabolism, Multifunctional enzyme, Cysteine-mediated cross-linking, Histidine kinase, DNA binding protein, Metastasis

1 Introduction

Chemical cross-linking is a powerful technique that has long been used to characterize protein-protein interactions [1, 2]. Over the past decade, in vitro cross-linking combined with mass spectrometry has further augmented the technique's utility. By harnessing these two techniques, investigators have studied protein subunit composition and architectural organization of large protein complexes. They have also detailed structural aspects of protein-protein interfaces and identified unknown protein partners within proteome-wide protein interaction networks [3, 4]. In vitro cross-linking has also been successfully employed as a complementary method for determining the crystal structure of large multi-subunit protein assemblies when classical structural biology approaches failed [5]. However, all these studies take place outside of a cellular context, and only a few examples of cross-linking performed in living cells currently exist in the literature. This is mainly due to

Cheryl L. Meyerkord and Haian Fu (eds.), *Protein-Protein Interactions: Methods and Applications*, Methods in Molecular Biology, vol. 1278, DOI 10.1007/978-1-4939-2425-7_26, © Springer Science+Business Media New York 2015

the difficulty of targeting cross-links in specific proteins in a cellular environment like those of *E. coli* and human cells, in which the total concentrations of protein are in the range of 200 and 50–100 g/l, respectively [6]. Such a situation increases the frequency of random collisions, which may lead to nonspecific intermolecular cross-linking. For instance, tetrameric hemoglobin was formerly shown to form octamers when intact erythrocyte cells were treated with a membrane permeable cross-linker [7]. Moreover, cross-linker application to cells usually produces less in vivo cross-linked sites than those observed in vitro [8]. Therefore, it requires more sophisticated cross-linking reagents and advanced technology and informatics, especially when cross-linking experiments are coupled with mass spectrometry analysis.

Using structural information provided by X-ray crystallography or NMR, it is possible to overcome some of the problems related to cross-linking in living cells. Indeed, more selective cross-links can be introduced in a protein taking into account the spatial arrangement of its nucleophilic residues.

This chapter presents a brief general survey of in vivo cross-linking approaches based on structural design to probe the quaternary structure of the human type B nucleoside diphosphate kinase (NDPK-B, also known as NME2/NM23-H2), an enzyme which is involved in the maintenance of the cellular nucleoside triphosphate (NTPs) pool [9, 10]. This enzyme has also been identified as a DNA binding protein and is involved in transcriptional activation of the *c-myc* gene [11, 12]. Only the hexameric enzyme is active in NTPs synthesis, whereas the dimeric protein is not [13, 14]. Conversely, the dimer binds to DNA with a higher affinity than the hexamer [15], suggesting the possibility that the dimer may act as a more efficient transcriptional activator than the hexamer. As this dual activity depends on hexameric and dimeric structures, different oligomeric states of the protein were investigated in cells. In the first attempt, S100 extracts from different tumor cells were analyzed by gel filtration method. However the resolution of the gel was too low to separate the different oligomeric forms due to the interconversion of different species during the protein elution. Thus, we developed an in vivo cross-linking method to monitor the quaternary structure of NDPK-B.

We first showed by in vitro cross-linking that specific cross-links on cysteine residues could be selectively introduced on the NDPK dimer or hexamer, depending on the length of the spacer arm of the homo-bifunctional cross-linker. We used this difference in the protein reactivity with long and short cross-linkers to establish an in vitro direct correlation between the oligomeric state of protein and the amount of cross-linked dimeric subunits. Differential cross-linking was then performed with long and short membrane permeable cross-linkers in HeLa cells after overexpressing either the dimer or the hexamer. A positive correlation could be established between the amount of overexpressed dimeric mutant protein and the amount of cross-links quantified by immunoblotting. This shows that our in vivo protein cross-linking method based on structural design is highly selective and can be used to probe the quaternary structure of proteins in intact cells.

2   Materials

2.1   Pure Recombinant Proteins

The recombinant wild type hexamer was purified as described in ref. [16]. The recombinant dimeric mutant P96S-A146 stop was constructed by the overlap extension method [17]. The point mutation P96S and the removal of the 7 C-terminal amino acids in the crystallographic hexamer interface has been shown to greatly affect hexamerization of NDPK [14]. The double mutant was purified according to a procedure similar to the one used for the wild type protein except that all purification buffers contained 0.05 mM of dodecyl-β-D-maltoside (DDM) detergent. Proteins were stored at −20 °C in 20 mM potassium phosphate pH 7.0 buffer containing 1 mM DTE and 50 % (v/v) glycerol.

2.2   SDS-PAGE

1. Laemmli SDS-PAGE equipment and buffers.

2. Coomassie R250 dye and Coomassie staining buffers.

2.3   Western Blotting

1. Western blotting equipment (nitrocellulose membrane, transfer chamber, filter paper, X-ray film).

2. Western blotting buffers: Tris–HCl/glycine transfer buffer (pH 8.3), blocking and washing buffers.

3. Specific antibodies against the protein of interest and anti-Ig-peroxidase (secondary antibody). In our experiments we routinely used polyclonal anti-NDPK-B antibodies at a final concentration of 10 μg/ml.

4. Enhanced chemiluminescence Plus reagents (ECL Plus) with a laser scanner system such as Storm (GE Healthcare life sciences) for detection and quantification.

2.4   Protein Cross-Linking via Cysteine Residues

1. Membrane permeable sulfydryl-reactive homo-bifunctional cross-linker: Bis-maleimidoethane (BMOE); bis-maleimidohexane (BMH) freshly dissolved in DMSO.

2. Membrane impermeable and water-soluble sulfhydryl-reactive homo-bifunctional cross-linker: 1,8-bis-maleimidotriethylene-glycol (BM[PEO$_3$]) freshly dissolved in water according to the manufacturer's instructions.

3. Quenching and reaction buffers for in vitro cross-linking: 2 × quenching buffer, 2 × Laemmli loading buffer containing

100 mM DTE; 2 × reaction buffer, 100 mM potassium phosphate pH 7.2, and 300 mM potassium chloride.

4. 6 × quenching buffer for in vivo cross-linking: 180 mM DTE in water.

2.5 Cell Culture

1. Cell culture equipment (plastic culture vessels, laminar flow hood, CO_2 incubator, water bath).

2. Complete culture medium (DMEM, fetal calf serum (FCS), antibiotics).

3. HeLa cells (ATCC CCL-2) in serum supplemented DMEM.

4. Ca^{2+} and Mg^{2+}-free phosphate-buffered saline (CMF-PBS).

5. Trypsin–EDTA solution in CMF-PBS.

2.6 DNA Transfections

1. Mammalian expression vectors bearing the cytomegalovirus promoter (e.g., pcDNA3, Invitrogen).

2. Cationic liposomes (e.g., FuGENE 6, Roche).

2.7 Whole Cell Extracts

Urea lysis buffer (SBLU): 62.5 mM Tris–HCl, pH 7.8, 2 % SDS, 6 M urea, 100 mM DTE, and 0.05 % bromophenol blue.

3 Methods

3.1 General Considerations: In Vivo Selectivity in Cross-Linking Based on Structural Design

Cross-links can be more efficiently introduced in a protein when its three-dimensional structure is known. The criteria to be considered for optimal selectivity in in vivo protein cross-linking (hetero- or homo-bifunctional reagent; type of reactive group; length and rigidity of the linker structure) are almost the same as those previously described in vitro [1, 18] except for two:

1. The cross-linker must diffuse across the membrane, and therefore its linker structure should be mostly hydrophobic.

2. Individual amino acids likely to react with cross-linker should be limited in the 3D space to minimize nonspecific intermolecular bridges in vivo (see **Note 1**). The use of a low cells–cross-linker ratio is also advised to limit the extent of protein modifications, thereby enhancing the relative differences in chemical reactivity between the amino acids.

We used a non-cleavable, membrane-permeable, homo-bifunctional reagent containing maleimide groups at both ends for introducing selective cysteine cross-links in vivo. In addition to the advantages of cysteine-based cross-linking, the extent of protein modification through cysteine residues is usually reduced due to the low content of cysteine in protein, minimizing nonspecific intermolecular cross-linking in vivo.

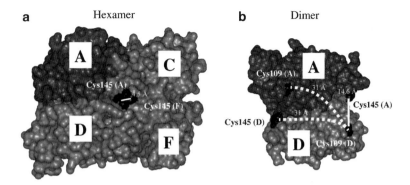

Fig. 1 Protein cross-linking based on structural design to probe quaternary structure. (**a**) Side view of the hexameric structure of NDPK-B. Out of the six identical subunits which compose the NDPK-B hexamer, only four subunits are clearly visible in this Connelly surface representation and are referred as A, D, C, and F subunits. (**b**) Connelly surface representation of the dimer rotated by 180° around the threefold axis (hexameric interface view). Each subunit contains two cysteines C109 and C145 (colored in *black*) which are differently exposed to solvent in dimeric and hexameric states. In the hexamer, only the C145 cysteines are solvent-accessible and close in the 3D space (*solid line* 4.5 Å, hexameric contact between A chain and F chain), whereas in the dimer, C145 (A) of one monomer and C109 (D) of the second monomer are solvent-accessible and are distant from each other by 14.6 Å (*solid line*). All other cysteine pairs in the dimer are separated by more than 31 Å and are indicated in *dotted lines*. Given the different spatial arrangement of cysteine pairs in the dimeric and hexameric states, selective cross-linking can be introduced on the dimer and the hexamer using a homo-bifunctional reagent varying only in length and flexibility of the spacer arm which connects both specific cysteine reactive groups

We chose Human NDPK-B as a model protein because (1) its crystal structure has already been determined [19] and (2) the protein only contains two cysteine residues (C145 and C109) which are differently exposed to solvent in dimeric and hexameric states (*see* Fig. 1 for details).

Whatever the protein studied, those who wish to use commercially available reagents may consult the "Crosslinking Technical Handbook," which can be freely downloaded from Thermo Scientific Pierce (http://www.piercenet.com). The website also provides an online interactive cross-linker selection guide, which is a practical starting point for those who do not know which reagents and protocols to use to cross-link their protein specifically in vivo, taking also into account all the criteria we have listed above.

3.2 In Vitro Protein Cross-Linking of NDPK-B via Cysteine Residues Using Variable Spacer Arms

BM[PEO]₃ and BMOE are homo-bifunctional uncleavable bis-maleimide cross-linkers that selectively react with thiols to form stable covalent thioether linkages at pH 6.5–7.5. BMOE and BM[PEO]₃ differ only in the length of the spacer arm and in their solubility in water. The two maleimide groups are distant from 14.7 Å and 8 Å in BM[PEO]₃ and BMOE, respectively. In the following section, BM[PEO]₃ will therefore be referred to as the long cross-linker and BMOE as the short cross-linker.

3.2.1 In Vitro Cross-Linking of the NDPK-B Hexamer (See Fig. 2)

1. Dialyze extensively or desalt the protein with a small G25 column into 50 mM sodium phosphate pH 7.2 reaction buffer containing 150 mM sodium chloride to remove any trace of reducing agents.

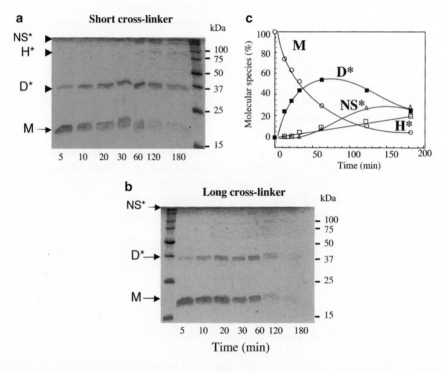

Fig. 2 In vitro cross-linking of the NDPK-B hexamer with short and long cysteine specific cross-linkers. NDPK-B, 0.02 mg/ml (1.2 μM, subunit concentration), was reacted with either the short (BMOE) (**a**) or the long cross-linker (BM[PEO]₃) (**b**) using a concentration of 0.1 mM following the experimental procedure described in Subheading 3.2.1. At the time indicated, the reaction was quenched by addition of SDS-PAGE loading buffer and the samples were subjected to 15 % SDS-PAGE followed by Coomassie staining. M, D*, H*, and NS* refer to the positions of monomers, cross-linked dimers, cross-linked hexamers, and nonspecific higher order cross-linked species of NDPK-B subunits, respectively. (**c**): Scans of the SDS-PAGE gel shown in (**a**) in order to quantify cross-linked species. The hexameric cross-linked species was very low as compared to the cross-linked dimer even though the protein at 0.02 mg/ml was a hexamer as judged by the equilibrium sedimentation [16]. This reflects the very low efficiency of these short and long cross-linkers to cross-bridge all six subunits together. Patterns were similar using a higher concentration of cross-linker or using different protein concentrations (0.002 mg/ml or 0.2 mg/ml) confirming the low cross-linking efficiency in the formation of the hexameric cross-linked species

2. Dissolve the long cross-linker in water according to the manufacturer's instructions and the short cross-linker in DMSO, both at 1 mM.

3. Prepare several eppendorf tubes each containing 25 μl of 2 × reaction buffer, 1 or 10 μg of the protein of interest and complete with water to a 50 μl final volume. Add also 10 μl of 2 × quenching buffer in one extra tube to control the quenching efficiency.

4. Start the reaction by adding 5 μl of the cross-linker solution (1/10 volume) and mix well by vortexing.

5. Stop the reaction at different times by adding 10 μl of 2 × quenching buffer and stir vigorously by vortexing.

6. Heat samples for 3–5 min at 100 °C.

7. Load samples on a Laemmli gel with a desired percentage of acrylamide. Generally, use 5 % gels for SDS-denatured proteins of 60–200 kDa, 10 % gels for 16–70 kDa, and 15 % gels for 12–45 kDa. Visualize different cross-linked species with Coomassie blue staining according to the standard protocol (*see* **Note 2**).

3.2.2 In Vitro Cross-Linking of the NDPK-B Dimeric Mutant (P96S-A146stop) (See Fig. 3)

Gel filtration analyses showed that the purified P96S-A146stop mutant formed a stable dimer in the protein concentration range of 0.5–50 μM (subunit concentration). Experimental procedures to cross-link the NDPK-B dimeric mutant with the long and short cross-linkers were similar to those used for the wild type hexamer except that the reaction buffer contained 0.05 mM of dodecyl-β-D-maltoside (DDM) detergent, to stabilize the dimer.

Obviously the in vitro characterization of protein cross-linking with pure components (as illustrated in Fig. 4) is not a prerequisite step to carry out in vivo cross-linking. This was shown here for a direct comparison of the selectivity of protein cross-linking in vitro and in a cellular context.

3.3 Cross-Linking of NDPK-B in Living Cells (See Note 3)

Since the long cross-linker BM[PEO]$_3$ is membrane-impermeable, an analogous reagent such as BMH which crosses the cell membrane was used. Like BM[PEO]$_3$, BMH is a thiol-specific, non-cleavable, homo-bifunctional cross-linker that connects both maleimide groups via a spacer arm whose length is close to that of BM[PEO]$_3$ (16.1 Å). In the following protocol of in vivo protein cross-linking, the long and short cross-linkers are then referred to as BMH and BMOE, respectively. To check whether overexpressing the stable dimeric mutant P96S-C145stop in HeLa cells induced a shift towards a hexameric state, GFP fusion proteins with the hexamer wild type or the dimeric mutant were made. After transient transfections of HeLa cells, analysis by fluorescence imaging showed different protein localization patterns. Whereas the wild type hexamer-GFP fusion protein was mainly located in the cytosol,

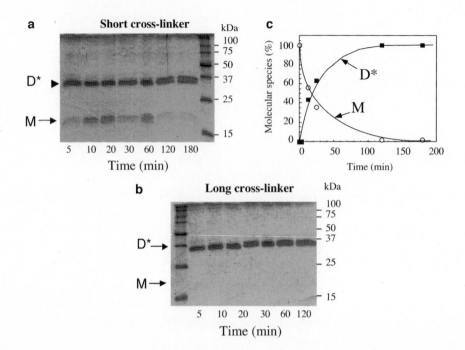

Fig. 3 In vitro cross-linking of the dimeric mutant of NDPK-B with short and long cysteine specific cross-linkers. Pure dimeric mutant, 0.02 mg/ml 1.2 µM (subunit concentration) was treated with either the short (BMOE) (**a**), or the long cross-linker (BM[PEO]₃) (**b**), according to the instructions described in Subheading 3.2.2 using 0.1 mM of the cross-linker. The reaction was quenched by addition of SDS-PAGE loading buffer at the time indicated. Cross-linked species were then analyzed by SDS-PAGE and Coomassie blue staining. M and D* denote the monomer and the cross-linked dimer. (**c**) Scans of the SDS-PAGE gel shown in (**a**). The graph represents the percentage of each species with time and each curve was fitted with a monoexponential (*solid line*) (*see* also **Note 3**). Scans of the SDS-PAGE gel shown in (**b**) are not represented under these experimental conditions because the reaction was too fast

the dimeric mutant-GFP fusion protein was present in both compartments (nucleus and cytosol, data not shown). These results suggested that the dimeric mutant was not totally converted into a hexamer when highly expressed in HeLa cells.

3.3.1 Transient Transfections of HeLa Cells to Overexpress the Dimeric Mutant and the Wild Type Hexamer

1. The day before the transfections, trypsinize and count the HeLa cells, plate them at the appropriate plating density in a complete medium so that they are 50–60 % confluent on the day of transfection. For cationic lipid-mediated transfections a 6-well plate was used, with each well containing 0.2×10^6 HeLa cells in a total complete media volume of 2 ml (DMEM, 10 % FCS).

2. On the day of transfection, add the requisite volume of serum-free medium as diluent to a total volume of 100 µl in a sterile polystyrene or polypropylene tube, and then the cationic lipid transfection reagent such as FuGENE 6 (Roche). Tap gently to mix.

DIMER   HEXAMER

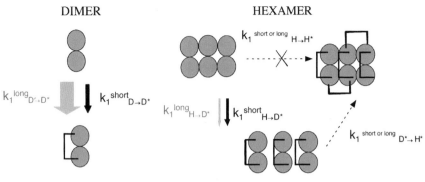

Cross-linked dimeric intermediates

Fig. 4 Diagram of in vitro cross-linking efficiencies of NDPK-B with short and long cysteine-specific cross-linkers as a function of the oligomeric states. Bimolecular rate constants relative to short (BMOE) and long (BM[PEO]$_3$) cross-linkers with the dimer and the hexamer were determined using three independent cross-linker concentrations as described in Subheading 3.2. (*see* **Note 3**). The *arrow* surfaces relative to cross-linking reactions with short (in *black*) and long cross-linkers (in *gray*) are proportional to the magnitude of each rate constant. The reaction D → D* is the formation of the dimeric mutant into the cross-linked dimer; H → D* is the formation of the hexamer into two cross-linked subunits; D* → H* is the formation of the cross-linked dimer into the cross-linked hexamer and H → H* is the direct conversion of the wild type into the cross-linked hexamer. The rate constants k_1^{long} D → D* and k_1^{short} D → D* were 900 ± 50 M^{-1}.s^{-1} and 9.6 ± 0.5 M^{-1}.s^{-1}, respectively, meaning that the cross-linking efficiency on the dimer is enhanced about 100-fold with the long cross-linker compared to the short cross-linker. The cross-linking efficiency corresponding to the formation of two cross-linked subunits within the hexamer is lower with both reagents. Bimolecular rate constants, k_1^{long} H → D* and k_1^{short} H → D*, were 0.9 M^{-1}.s^{-1} and 3 M^{-1}.s^{-1}, respectively, indicating that the short cross-linker is slightly more effective as compared to the long cross-linker in hexamer cross-linking. Non-measurable rates of the cross-linked hexamer or the intermediate cross-linked species (trimer, tetramer, or pentamer) are shown as *dotted arrows*

3. Add the DNA solution at a final concentration of 0.1 μg/μl to the prediluted transfection reagent. In our experiments a 3:1 (μl/μg) ratio of FuGENE 6 reagent to DNA was used for optimal transfection efficiency into HeLa cells. Mix gently and let the DNA–lipid complex form for 45 min at room temperature.

4. While DNA–lipid complexes are forming, replace medium on cells with the appropriate volume of fresh complete medium.

5. Add DNA-lipid complexes directly to each well with the desired range of DNA. In our experiments 0, 2, 5, 10, and 20 μl of transfection medium containing 0.1 μg/ml of DNA were added directly on the cells. Distribute the DNA-lipid mixture around each well and swirl the 6-well plate to ensure even dispersal. Incubate for 24 h at 37 °C in 5 % CO$_2$.

3.3.2 In Vivo Protein Cross-Linking (See Fig. 5)

On the day of the treatment with cross-linkers, HeLa cells should be ≈80 % confluent.

1. Dissolve the long and short cross-linkers in DMSO to a final concentration of 20 mM.

2. Trypsinize each well with 100 μl of Trypsin–EDTA solution and incubate for 5 min at 37 °C. Stop the reaction by adding 400 μl of complete medium and count the cells.

3. Immediately after, transfer the cells still in suspension into a sterile polystyrene or polypropylene tube. Wash the cells twice with 500 μl of complete medium. Incubate cells for 30 min on ice.

4. Distribute HeLa cells in sterile polystyrene or polypropylene tubes with a cell density of ≈0.8 × 10^6 cells in 20 μl of complete medium.

5. Add the cross-linker directly to the cell suspension at a final concentration of 1 mM (1 μl of 20 mM stock in DMSO in a total medium volume of 20 μl). Immediately mix gently. Take one half of the cells without cross-linker treatment and add the same volume of DMSO as a control. To check the quenching efficiency, add the cross-linker in a tube already containing 30 mM of DTE. Incubate samples for 1 h at 37 °C in 5 % CO_2. Swirl the suspension occasionally during the 37 °C incubation (*see* **Notes 4** and **5**).

6. Stop the cross-linking reaction by adding 4 μl of 6 × quenching buffer (for a final concentration of 30 mM). Let the cells incubate for 10 min at 37 °C to allow the quencher to fully diffuse into the cells.

7. Visualize and compare cells which were treated and mock treated by light microscopy, in order to check for cellular death.

8. Pellet cells by centrifugation at 4 °C and wash twice with cold CMF-PBS buffer at 4 °C.

3.3.3 Preparation of Whole Cell Extracts

1. Prepare protein extracts by directly adding 40 μl of SBLU 1 × buffer, vortex well, and heat cells at 100 °C until lysis is complete.

2. Let cell lysates return to room temperature before centrifuging at 16,000 × *g* for 10 min at 4 °C. Transfer the soluble protein fraction to a new microcentrifuge tube without touching the pellet (*see* **Note 6**).

3. Samples are then analyzed by SDS-PAGE after appropriate dilution with SBLU buffer (in general 1/10) and cross-linked proteins are identified by Western blotting according to standard protocols.

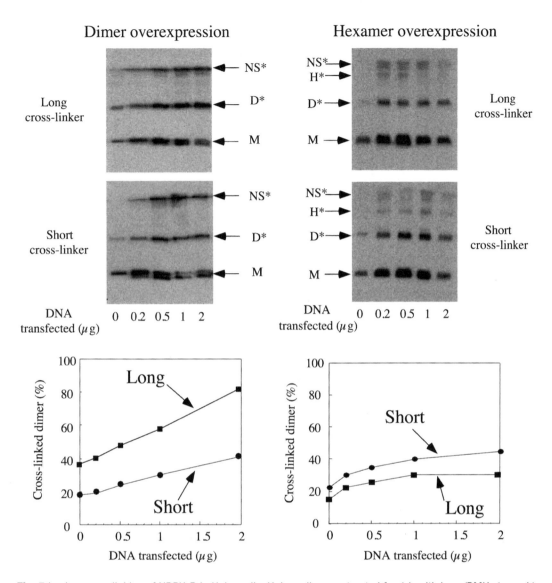

Fig. 5 In vivo cross-linking of NDPK-B in HeLa cells. HeLa cells were treated for 1 h with long (BMH, *top gels*) or short (BMOE, *bottom gels*) membrane permeable cross-linkers after transient transfections with a variable amount of DNA encoding either the dimeric mutant or the wild type hexamer as indicated. After quenching the reaction with DTE, samples were taken up in SDS-urea gel-solubilizing extraction buffer (SBLU) and were analyzed by SDS-PAGE and Western blotting using a polyclonal antibody against NDPK-B as described in Subheading 3.3.3. M, D*, H*, and NS* refer to the electrophoretic migrations of monomers, cross-linked dimers, cross-linked hexamers, and higher-order nonspecific multimers of NDPK-B subunits, respectively. Dimeric cross-links were quantified using a Storm system and the percentages of cross-linked dimers related to either dimer (*left*) or hexamer (*right*) overexpression were represented versus the amount of DNA transfected into HeLa cells. Note the double position of monomers especially when HeLa cells overexpressing the dimer were treated with the short cross-linker. This was due to unmodified and one-point modified monomers (*see* **Note 2**). Note also the considerable increase of dimeric cross-linked species with dimer overexpression in HeLa cells treated by the long cross-linker. This increase was not as significant when the same transfected cells were treated with the short cross-linker. In cells that overexpressed the hexamer there were only small increases of dimer cross-links and the cross-linking was slightly more efficient with the short cross-linker. Taken together, these results were very similar to those obtained in vitro (Fig. 3), indicating the high selectivity of our protein cross-linking in a cellular context

3.3.4 Quantification of Cross-Linked Species on Western Blots

The technique we currently use for immunodetection of cross-linked species is based on a chemiluminescent substrate system for horseradish peroxidase (HRP) which can be visualized both on film and blot imaging system (ECL Plus, Amersham Biosciences). Immunodetection is carried out according to the supplier's instructions and the quantification is performed using Storm (Amersham Biosciences), which is better than film detection.

3.4 Concluding Comments

In vivo protein cross-linking based on structural design can greatly facilitate the introduction of selective cross-links on proteins to study protein oligomerization or protein-protein interactions. The combination of the chemoselectivity and regioselectivity of the reagent with the protein structure can even allow examination of changes in protein oligomerization upon cellular events and specific protein-protein interactions upon ligand binding. By considering the specificity of the maleimide group for cysteine residues and the spatial arrangement of these residues, selective cross-linking can be introduced specifically on either the dimer or the hexamer in vitro as well as in intracellular environments that are crowded with many other cellular macromolecules. The method described here allows investigation into the oligomerization of the NDPK-B protein in metastatic, tumor, and normal cells (unpublished results). This approach can be extended to signaling proteins, such as the NEMO protein, an essential modulator involved in the NF-κB pathway, which are usually less abundant in cells. We have successfully used these methods to show that NEMO forms a protein complex in cells, termed IκB kinase (IKK) complex, with two IKKα/β kinases [20]. Strikingly, the molecular mass of the cross-linked IKK complex perfectly matched the subunit composition of the IKK complex, which was accurately determined in vitro more than one decade after the in vivo cross-linking experiments [21]. Given the high sensitivity of this method to detect unstable homo- and hetero-associations of proteins, the experimenter will also have to address the question of whether the cross-links correspond to a mature protein association or to an assembly intermediate. This is particularly true when in vivo cross-linking results are compared in different cells at variable stages of proliferation.

4 Notes

1. Precautions should be taken when interpreting in vivo chemical cross-linking results. If cross-links are observed in vivo, it may be specific (reflecting a relevant structure) or it may be nonspecific. Cross-linking on living cells increases the probability of detecting unstable oligomers, but it also increases the probability of nonspecifically cross-linking proteins together. When cross-links

are observed, the assay should be repeated with reduced concentrations of cross-linker, at a reduced temperature or for shorter periods of time to detect preferential cross-links. If identical results are obtained over a broad concentration range of cross-linkers or with several reagents, the in vivo cross-links are likely to be specific.

2. Often a monomer containing an internal cross-link forms a complex with SDS that has a smaller Stokes radius than does the uncross-linked monomer-SDS complex. This cross-linked complex will then be observed on SDS-PAGE as a band migrating with lower apparent molecular weight than does the uncross-linked monomer. This slight change of the electrophoretic migration can also be observed for the cross-linked species of dimer, trimer, etc., on Laemmli SDS-PAGE.

3. General equation for in vivo protein cross-linking

   The formation of a specific cross-linking with a homo- or hetero- cross-linker can be written as follows:

   $$[P] \xrightarrow{k_1 = k_0[CL]} [P^*] \xrightarrow{k_2} [P^{**}]$$

   where $[P]$ is the unmodified protein, $[CL]$ the cross-linker, $[P^*]$ the protein cross-linked to one end of the bifunctional reagent, and $[P^{**}]$ the protein cross-linked to both ends of the reagent. Usually for protein cross-linking $[CL]_{Total} >> [P]_{Total}$ we can then approximate $[CL]$ by $[CL]_T$ and we can consider the first reaction as pseudo-first-order. The general equation for the variation of $[P^{**}]$ with time is:

   Equation 1: $[P^{**}] = \left\{ 1 + \dfrac{k_1 \exp(-k_2 t) - k_2 \exp(-k_1 t)}{k_2 - k_1} \right\} [P_0]$

   where t is the time, k_1 the pseudo-first-order rate constant, and P_0 the protein concentration at time 0. The first reaction is usually the rate-determining step of the reaction with a bimolecular reactant, meaning that whenever a P^* protein is formed, it is rapidly converted into P^{**} ($k_2 >> k_1$). Equation 1 then becomes a more simple expression:

   $$[P^{**}] = \{1 - \exp(-k_1 t)\}[P_0]$$

4. Because many proteins contained in the serum can interfere with in vivo cross-linking, it is recommended to carry out the reaction in PBS buffer, if cells can endure such a treatment up to 2 h. There is no general rule to calculate the effective concentration of cross-linker in serum-containing medium. In our experiments with NDPK-B the effective concentration of both cross-linkers in PBS is tenfold less compared to that in medium containing 10 % serum.

5. It is possible to incubate the cross-linker for longer periods. We performed a range of experiments with different times of incubation, 10 , 60 and 120 min. The cross-linking efficiency was optimal at 60 and 120 min, whereas the reaction was not complete at 10 min. This obviously depends on the chemoselectivity of the cross-linker as well as the type of functional groups that are accessible on the protein. It also depends mostly on the subcellular localization of the protein. The in vivo cross-linking of the NEMO protein, a cytosolic signaling protein that is located near the membrane, was optimal after an incubation time of only 10 min [20].

6. It is important to note that the use of a high cross-linker concentration can reduce the total amount of extracted proteins. This is due to the in vivo formation of nonspecific intermolecular cross-links with a high molecular mass that are removed by centrifugation. It is therefore crucial for any in vivo protein cross-linking to determine the minimal effective concentration of the cross-linker in order to reduce the probability of random collisions and to get a satisfactory yield of protein after extraction. To examine the extent of global protein cross-linking, it is also advised to analyze the crude extracts from the mock-treated and the reagent-treated cells by SDS-PAGE followed by Coomassie or silver staining. If only a few proteins were entirely cross-linked into species of high molecular mass, the cross-linking is likely specific.

Acknowledgments

The authors are very grateful to Samuel Levy, MD for his critical reading of the manuscript. This work was supported, in whole or in part, by the Fondation ARC pour la recherche sur le cancer and La Ligue contre le cancer.

References

1. Ji TH (1983) Bifunctional reagents. Methods Enzymol 91:580–609

2. Kluger R, Alagic A (2004) Chemical cross-linking and protein-protein interactions-a review with illustrative protocols. Bioorg Chem 32:451–472

3. Petrotchenko EV, Borchers CH (2010) Cross-linking combined with mass spectrometry for structural proteomics. Mass Spectrom Rev 29:862–876

4. Stengel F, Aebersold R, Robinson CV (2012) Joining forces: integrating proteomics and cross-linking with the mass spectrometry of intact complexes. Mol Cell Proteomics 11 (R111):014027

5. Pornillos O, Ganser-Pornillos BK, Kelly BN et al (2009) X-ray structures of the hexameric building block of the HIV capsid. Cell 137:1282–1292

6. Mika JT, Poolman B (2011) Macromolecule diffusion and confinement in prokaryotic cells. Curr Opin Biotechnol 22:117–126

7. Wang K, Richards FM (1975) Reaction of dimethyl-3,3'-dithiobispropionimidate with intact human erythrocytes. Cross-linking of membrane proteins and hemoglobin. J Biol Chem 250:6622–6626

8. Bruce JE (2012) In vivo protein complex topologies: sights through a cross-linking lens. Proteomics 12:1565–1575

9. Boissan M, Lacombe ML (2011) Learning about the functions of NME/NM23: lessons from knockout mice to silencing strategies. Naunyn Schmiedebergs Arch Pharmacol 384:421–431

10. Lascu I, Gonin P (2000) The catalytic mechanism of nucleoside diphosphate kinases. J Bioenerg Biomembr 32:237–246

11. Dexheimer TS, Carey SS, Zuohe S et al (2009) NM23-H2 may play an indirect role in transcriptional activation of c-myc gene expression but does not cleave the nuclease hypersensitive element III(1). Mol Cancer Ther 8:1363–1377

12. Postel EH, Berberich SJ, Flint SJ et al (1993) Human c-myc transcription factor PuF identified as nm23-H2 nucleoside diphosphate kinase, a candidate suppressor of tumor metastasis. Science 261:478–480

13. Mesnildrey S, Agou F, Karlsson A et al (1998) Coupling between catalysis and oligomeric structure in NDP kinase. J Biol Chem 273:4436–4442

14. Karlsson A, Mesnildrey S, Xu Y et al (1996) Nucleoside diphosphate kinase. Investigation of the intersubunit contacts by site-directed mutagenesis and crystallography. J Biol Chem 271:19928–19934

15. Mesnildrey S, Agou F, Veron M (1997) The in vitro DNA-binding properties of NDP kinase are related to its oligomeric state. FEBS Lett 418:53–57

16. Agou F, Raveh S, Mesnildrey S et al (1999) Single strand DNA specificity analysis of human nucleoside diphosphate kinase B. J Biol Chem 274:19630–19638

17. Pogulis RJ, Vallejo AN, Pease LR (1996) In vitro recombination and mutagenesis by overlap extension PCR. Methods Mol Biol 57:167–176

18. Kluger R (1997) Chemical cross-linking and protein function. In: Creighton TE (ed) Protein function a practical approach. IRL Press, Oxford, p 185

19. Morera S, Lacombe M-L, Xu Y et al (1995) X-Ray structure of nm23 Human Nucleoside Diphophate Kinase B complexed with GDP at 2A resolution. Structure 3:1307–1314

20. Agou F, Ye F, Goffinont S et al (2002) NEMO trimerizes through its coiled-coil C-terminal domain. J Biol Chem 277: 17464–17475

21. Napetschnig J, Wu H (2013) Molecular Basis of NF-kappaB Signaling. Annu Rev Biophys 42:443–468

Part IV

Cell-Based Bimolecular Interaction Reporter Assays

Chapter 27

Identification of Protein-Protein Interactions by Standard Gal4p-Based Yeast Two-Hybrid Screening

Jeroen Wagemans and Rob Lavigne

Abstract

Yeast two-hybrid (Y2H) screening permits identification of completely new protein interaction partners for a protein of interest, in addition to confirming binary protein-protein interactions. After discussing the general advantages and drawbacks of Y2H and existing alternatives, this chapter provides a detailed protocol for traditional Gal4p-based Y2H library screens in *Saccharomyces cerevisiae* AH109. This includes bait transformation, bait auto-activation testing, prey library transformation, Y2H evaluation, and subsequent identification of the prey plasmids. Moreover, a one-on-one mating protocol to confirm interactions between suspected partners is given. Finally, a quantitative α-galactosidase assay protocol to compare interaction strengths is provided.

Key words Protein-protein interactions, Gal4p, Yeast two-hybrid, α-galactosidase assay

1 Introduction

The development of the yeast two-hybrid (Y2H) system by Stanley Fields in 1989 [1] represented a major milestone in the study of protein-protein interactions, which play a crucial role in almost all biological processes. Although originally used to detect binary protein-protein interactions between known interaction partners, it is now clear that Y2H facilitates the identification of completely new protein interaction partners for a protein of interest.

1.1 Gal4p-Based Yeast Two-Hybrid

The standard Y2H system exploits the modular nature of eukaryotic transcription factors like Gal4p of *Saccharomyces cerevisiae*. This transcriptional activator has two functional domains: an amino terminal DNA binding domain and a carboxy terminal activation domain. While the former binds to the upstream activating sequence (UAS) of the galactose metabolism genes in *S. cerevisiae*, the latter activates their transcription by recruiting the RNA polymerase to the promoter region. Moreover, Keegan et al. [2] demonstrated that Gal4p remains functional even if the two

Cheryl L. Meyerkord and Haian Fu (eds.), *Protein-Protein Interactions: Methods and Applications*, Methods in Molecular Biology, vol. 1278, DOI 10.1007/978-1-4939-2425-7_27, © Springer Science+Business Media New York 2015

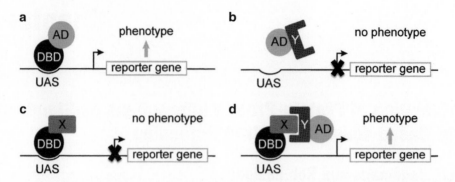

Fig. 1 The Y2H system. (**a**) Transcriptional activator Gal4p of *S. cerevisiae* is composed of two functional domains: the DNA binding domain (DBD) that binds to the upstream activating sequence (UAS) and the activation domain (AD) that recruits the RNA polymerase and activates transcription of a reporter gene. Both components are required to produce a selectable phenotype. Y2H uses the modularity of Gal4p. (**b**) The bait protein X is fused to the Gal4p DBD, which cannot activate transcription on its own. (**c**) The prey protein Y is fused to the Gal4p AD. Since this fusion protein is not recruited to the promoter, it is also unable to activate transcription on its own. (**d**) Only when X and Y interact, the DBD and AD are both present at the promoter site, restoring Gal4p function resulting in a detectable phenotype

domains are not covalently attached as long as they are both present at the promoter site. Fields [1] picked up this idea to fuse interacting proteins to the two Gal4p domains, which physically brings the two domains together and restores Gal4p function.

In Gal4p-based Y2H, the protein fused to the C-terminus of the Gal4p DNA binding domain is known as the bait, while the other interaction partner, fused to the C-terminus of the Gal4p activation domain, is called the prey. These constructs are both transformed into the same yeast cell. Finally, a third component of the Y2H system (Fig. 1) is the reporter construct in this bait- and prey-expressing yeast cell. This reporter gene is located downstream from a promoter with a UAS recognized by the Gal4p DNA binding domain. If bait and prey interact, the Gal4p activation domain is tethered to the promoter region and transcription of the reporter gene is activated. This results in a specific, selectable phenotype.

While Fields originally designed the system to investigate known interactions, Chien et al. [3] applied the technique 2 years later to unravel completely new interactions for a particular protein of interest. The known protein, fused to the Gal4p DNA binding domain, was used as a bait to screen for interaction partners in a whole library of potential interaction partners, fused to the activation domain. After Y2H interaction analysis, the prey plasmid was harvested from those yeast colonies able to activate all reporter genes. Finally, by sequencing the prey plasmid inserts, the interaction partner of the protein of interest was identified.

1.2 Drawbacks of the Yeast Two-Hybrid System and Possible Solutions

As for any other protein-protein interaction analysis technique, one should realize it is almost impossible to detect all interacting proteins using Y2H. In contrast, Y2H can also identify interactions that actually do not occur in vivo. These are called false negative and false positive results, respectively. These often arise due to the use of artificial fusion proteins in combination with an artificial reporter construct, the removal of the proteins from their natural biological context, their translocation to the yeast nucleus, or the production of the hybrid protein at a different level than its natural concentration in the cell [4].

One potential problem is the bait protein, which could independently lead to reporter gene activation (e.g., if the protein has transcription activation activity or interacts nonspecifically with the Gal4p activation domain). This is called auto-activation or self-activation and can lead to false positive results. While this activity can be related to the biological function of the protein, it can also be a consequence of the fusion with the DNA binding domain or the removal of the protein from its natural context [4]. Because of this phenomenon, auto-activation of the bait protein should always be checked before performing the YH2 screen. If this auto-activation test is positive, the bait as such is not suited for the screen. To resolve this problem, the transcription activation domain could be removed and the screen performed with the other domain, or another Y2H system could be used.

While only the *lacZ* reporter gene, encoding β-galactosidase, was originally used in a classical blue/white screening assay [1], current Y2H reporter yeast strains have several secondary auxotrophic reporter genes, which permit selection for interaction by monitoring yeast cell growth on selective media. For instance, Y2H strain *S. cerevisiae* AH109 [5] has the *HIS3* and *ADE2* reporter genes, encoding imidazole glycerol phosphate dehydratase and phosphoribosylaminoimidazole carboxylase, respectively. These enzymes are crucial in the biosynthesis of histidine and adenine. By plating on medium lacking histidine and/or adenine, interacting baits and preys are selected. In addition, *S. cerevisiae* AH109 contains the *MEL1* reporter gene, encoding α-galactosidase, a secreted enzyme enabling direct blue/white screening on X-α-gal indicator plates. The use of more than one reporter gene reduces the number of false positive results [5].

Moreover, when using *HIS3* as a reporter gene, as is the case with strain *S. cerevisiae* AH109, 3-amino-1,2,4-triazole (3-AT) should be added to the medium. *HIS3* is the most leaky construct of the three reporter genes in AH109, so there is always a low background transcription activation of this gene. 3-AT is a competitive inhibitor of the *HIS3* product and will reduce false positive results due to the background activation [6]. At the same time, the 3-AT concentration should not be too high to avoid losing identification of weaker interactions. The optimal concentration that should be used in the Y2H screen is also determined in the auto-activation test.

Although Y2H tests for interactions in vivo, the use of yeast as a host can be a potential disadvantage. It is possible that the fusion proteins will not be stable in the yeast cell or will require specific posttranslational modifications. Moreover, some proteins might be toxic to yeast. Bacterial two-hybrid [7] and mammalian two-hybrid [8] assays are possible alternatives to overcome these problems.

Targeting of non-nuclear proteins to the yeast nucleus can also be inefficient or could cause nonspecific interactions, which is likely for proteins with hydrophobic domains. For instance, membrane proteins forced into the nucleus tend to bind nonspecifically to other proteins [4]. To address this issue, the membrane yeast two-hybrid system, based on the split-ubiquitin assay [9], was developed.

Furthermore, interactions depending on more than two partners are not detected using Y2H. The yeast three-hybrid assay [10] provides a possible solution.

A last but common class of false positive results occurs because of spontaneous mutations in the yeast strain, bait or prey constructs causing reporter gene activation [11]. Hence, it is important to confirm the potential interacting proteins in an independent Y2H experiment using fresh yeast cells.

In addition to false positive results, the use of artificial fusion proteins can also lead to false negative results (e.g., if the expression levels are too low or when the fusion alters the actual conformation of the bait or prey protein). This could result in a reduced activity or the exposure of artificial surfaces that nonspecifically interact with other proteins. Moreover, the domain fused to the protein could occlude the natural interaction domain or cause steric hindrance [4]. To reduce the number of false negative results and to maximize the number of identified interactions, it is advisable to perform screens in more than just one Y2H vector system. Although the original system uses the modularity of Gal4p, other DNA binding domains like LexA [12], the λ repressor protein cI [13], or the human estrogen receptor protein [14] could be used. Similarly, different transcription activation domains like the herpes simplex virus protein VP16, or its alternative B42, have been applied [12].

Application of different Gal4p-based vector systems is also a possibility. As the expression level of the bait and prey fusion constructs influences the outcome of the Y2H analysis, vectors with varying promoters and copy numbers due to a different origin of replication (*ori*) could be exploited [15, 16]. For Y2H, our lab uses a combination of the pGBKT7g-pGADT7g (Gateway variants of the Clontech Matchmaker 3 vectors) and the pDEST32-pDEST22 (Life Technologies) plasmids that lead to different levels of expression (*see* Table 1). Hence, their results are complementary. Although one can easily use classical restriction enzyme cloning to construct bait plasmids and prey libraries, it is preferable to use the

Table 1
Summary of applied yeast two-hybrid vectors

Vector	Characteristics			Selection		Reference
	Promoter	Ori		E. coli	Yeast	
pGBKT7g (bait)	Truncated p_{ADH1}	2 μ		Kanamycin	*TRP1*	[17]
pGADT7g (prey)	Full-length p_{ADH1}	2 μ		Ampicillin	*LEU2*	[17]
pDEST32 (bait)	Full-length p_{ADH1}	CEN		Gentamicin	*LEU2*	Life Technologies
pDEST22 (prey)	Full-length p_{ADH1}	CEN		Ampicillin	*TRP1*	Life Technologies

Gateway system from Life Technologies to facilitate subcloning into different Y2H systems. By first constructing a prey library in an entry vector (e.g., pENTR1A, Life Technologies), one can easily shuttle the library to different prey vectors and use these different libraries in subsequent screens.

1.3 Advantages of the Yeast Two-Hybrid Approach

Despite its drawbacks, the Y2H system has some clear advantages over other methods. Firstly, it is clear that an in vivo assay more closely resembles cellular conditions. Furthermore, Y2H is able to detect weak and transient interactions since a continuous activation of the reporter genes is not necessary to see a signal. The fact that Y2H only detects binary interactions can also be an advantage. In contrast to methods such as affinity purification, which detect whole complexes, Y2H directly identifies the two interaction partners. Moreover, there is no need for recombinant proteins, which is a huge advantage. Finally, although it is labor intensive, Y2H has a relatively low investment cost. Only minimal laboratory equipment and low-cost microbial growth media are needed [4].

This chapter provides a detailed protocol for Gal4p-based Y2H screens using one protein as bait against a prey library. Assuming a bait and prey library has been constructed, this protocol will first describe small-scale yeast transformation of the bait plasmid. In addition, this chapter includes a bait auto-activation test, large-scale yeast transformation of the prey library, evaluation of the Y2H results, yeast plasmid isolation, and identification of the inserts by DNA sequencing. Since Y2H is prone to false positive results, it is important to confirm the potential interactions. A one-on-one protocol using mating between the two different *S. cerevisiae* strains AH109 (mating type a) [5] and Y187 (mating type α) [18] will also be discussed. This protocol can also be applied to directly confirm interaction between two suspected interaction partners. Finally, a quantitative α-galactosidase assay to compare different interaction strengths will be provided.

2   Materials

2.1   Bait Plasmid Transformation

1. *Yeast strain*: *S. cerevisiae* AH109 with genotype *MATa, trp1-901, leu2-3,112, ura3-52, his3-200, gal4Δ, gal80Δ, LYS2:: GAL1$_{UAS}$-GAL1$_{TATA}$-HIS3,GAL2$_{UAS}$-GAL2$_{TATA}$-ADE2, URA3::MEL1$_{UAS}$-MEL1$_{TATA}$-lacZ* (Clontech) [5].

2. *Bait plasmids*: Gene of interest cloned in bait vector (e.g., pGBKT7g [17] or pDEST32 (Life Technologies)).

3. *YPDA liquid medium*: 10 g yeast extract, 20 g peptone. Bring up to 880 ml with deionized water and autoclave. After sterilization, let the medium cool to room temperature. Add 100 ml of 20 % glucose stock solution and 20 ml of 2 mg/ml adenine stock solution for a final volume of 1 L. Store at room temperature.

4. *YPDA solid medium*: 10 g yeast extract, 20 g peptone, 20 g agar. Bring up to 880 ml with deionized water and autoclave. After sterilization, let the medium cool to 60 °C. Add 100 ml of 20 % glucose stock solution and 20 ml of 2 mg/ml adenine stock solution for a final volume of 1 L. Pour plates in a sterile hood and let the medium solidify. Store plates at 4 °C for 2–3 weeks.

5. *2 mg/ml Adenine stock solution*: 200 mg adenine hemisulfate salt. Bring up to 100 ml with deionized water and sterilize through a 0.22 μm filter.

6. *20 % Glucose stock solution*: Put 800 ml of ultrapure water in a beaker and stir. Weigh 200 g α-D (+)-glucose (anhydrous) and add it to the ultrapure water in small amounts to ensure that it is completely dissolved. Autoclave.

7. *Sterile water*: autoclave ultrapure water.

8. *1 M LiOAc*: 5.1 g lithium acetate dihydrate. Bring up to 50 ml with ultrapure water and autoclave. Store at room temperature.

9. *100 mM LiOAc*: 0.51 g lithium acetate dihydrate. Bring up to 50 ml with ultrapure water and autoclave.

10. *50 % (w/v) PEG3350*: Add 50 g polyethylene glycol (average molecular weight 3,350) to 50 ml of ultrapure water. Stir until completely dissolved. Filter sterilize and store at room temperature, securely capped to prevent evaporation. Water loss will increase the PEG concentration and severely reduce the yield of transformants. To avoid this, make fresh PEG3350 solution every few months.

11. *ssDNA*: 10 mg/ml single-stranded fish sperm DNA, MB grade. Store at −20 °C in aliquots.

12. *Selective liquid yeast medium*: 6.9 g yeast nitrogen base without amino acids. Bring up to 700 ml with ultrapure water and

autoclave. Let the medium cool to room temperature and add 100 ml of 10× amino acid stock solution, 100 ml of 20 % glucose stock solution, and 100 ml of 10× amino acid dropout solution for a final volume of 1 L.

13. *Selective liquid yeast medium ("synthetic defined" or "SD medium")*: 6.9 g yeast nitrogen base without amino acids, 20 g bacto agar. Bring up to 700 ml with ultrapure water and autoclave. Let the medium cool to 60 °C and add 100 ml of 10× amino acid stock solution, 100 ml of 20 % glucose stock solution, and 100 ml of 10× amino acid dropout solution for a final volume of 1 L. If needed, add 3-AT and/or X-α-gal and pour the plates in a sterile culture hood. Store at 4 °C for up to 2–3 weeks.

14. *10× amino acid stock solution*: 500 mg L-arginine, 800 mg L-aspartic acid, 500 mg L-isoleucine, 500 mg L-lysine, 200 mg L-methionine, 500 mg L-phenylalanine, 1,000 mg L-threonine, 500 mg L-tyrosine, 1,400 mg L-valine, and 200 mg uracil. Bring up to 1 L with ultrapure water and sterilize with a 0.22 μm filter. Store at 4 °C for up to 1 month.

15. *10× amino acid dropout (DO) stock solutions*: Depending on the selective medium, add the remaining components not provided by the 10× amino acid stock solution.

    SD-Trp medium (medium without tryptophan): add *10× DO-Trp*: 40 mg adenine hemisulfate salt, 20 mg L-histidine, 100 mg L-leucine dissolved in 100 ml of ultrapure water (filter sterilize).

    SD-Leu medium (medium without leucine): add *10× DO-Leu*: 40 mg adenine hemisulfate salt, 20 mg L-histidine, 50 mg L-tryptophan in 100 ml of ultrapure water (filter sterilize).

    SD-Trp-Leu medium (medium lacking tryptophan and leucine): add *10× DO-Trp-Leu*: 40 mg adenine hemisulfate salt, 20 mg L-histidine in 100 ml of ultrapure water (filter sterilize).

    SD-Trp-Leu-His medium (medium lacking tryptophan, leucine, and histidine): add *10× DO-Trp-Leu-His*: 40 mg adenine hemisulfate salt in 100 ml of ultrapure water (filter sterilize).

    SD-Trp-Leu-His-Ade medium: add *10× DO-Trp-Leu-His-Ade*, which is the same as sterile ultrapure water.

16. *1 M 3-AT stock solution*: 0.84 g 3-amino-1,2,4-triazole in 10 ml of ultrapure water. Filter sterilize. Store at 4 °C.

17. *X-α-gal stock solution*: 200 mg 5-bromo-4-chloro-3-indolyl-α-D-galactopyranoside dissolved in 10 ml of *N,N*-dimethylformamide. Store at −20 °C in the dark (e.g., a conical tube wrapped

in aluminum foil). Add 2 ml of stock solution to 1 L of SD medium. As X-α-gal is a heat-sensitive compound, let the medium first cool to 60 °C before adding.

2.2 Bait Auto-activation Test

1. *Yeast strain*: *S. cerevisiae* AH109 containing the bait plasmid.
2. *Prey plasmids*: empty prey vectors or prey vectors with an independent protein inserted (e.g., if the bait is constructed in pGBKT7g, use the empty pGADT7g vector as a control; if the bait is constructed in pDEST32, use the empty pDEST22 as a control).
3. *Solid media needed* (25 ml medium/90 mm petri dish):

   2 SD-Trp-Leu plates

   1 SD-Trp-Leu-His plate

   1 SD-Trp-Leu-His +1 mM 3-AT plate (25 µl 1 M 3-AT/25 ml medium)

   1 SD-Trp-Leu-His +3 mM 3-AT plate (75 µl 1 M 3-AT/25 ml medium)

   1 SD-Trp-Leu-His +5 mM 3-AT plate (125 µl 1 M 3-AT/25 ml medium)

   1 SD-Trp-Leu-His +10 mM 3-AT plate (250 µl 1 M 3-AT/25 ml medium)

   1 SD-Trp-Leu-His-Ade plate

2.3 Prey Library Transformation

1. *Yeast strain*: *S. cerevisiae* AH109 containing the bait plasmid.
2. *Prey library*: library cloned in prey vector (e.g., pGADT7g [17] or pDEST22 (Life Technologies)).
3. *2× YPDA liquid medium*: 20 g yeast extract, 40 g peptone. Bring up to 760 ml with deionized water and autoclave. After sterilization, let the medium cool down. Add 200 ml of 20 % glucose stock solution and 40 ml of 2 mg/ml adenine stock solution for a final volume of 1 L.
4. *Solid media needed* (25 ml medium/90 mm petri dish; 80 ml medium/150 mm petri dish):

   12 SD-Trp-Leu 90 mm plates

   30 SD-Trp-Leu-His 150 mm plates + x mM 3-AT (concentration determined from auto-activation test)

2.4 Evaluation of Library Transformation

1. *Solid media needed* (40 ml medium/Nunc OmniTray dish or similar microtiter plate):

   1 OmniTray SD-Trp-Leu

   1 OmniTray SD-Trp-Leu-His + x mM 3-AT (concentration determined from auto-activation test)

   1 OmniTray SD-Trp-Leu-His-Ade + 40 mg/L X-α-gal

2.5 Yeast Plasmid Isolation	1. *0.67 mM K₂HPO₄ pH 7.5*: 0.0306 g K₂HPO₄. Dissolve in 150 ml of deionized water, adjust pH to 7.5, bring up to 200 ml with deionized water and autoclave.

2.5 Yeast Plasmid Isolation

1. *0.67 mM K_2HPO_4 pH 7.5*: 0.0306 g K_2HPO_4. Dissolve in 150 ml of deionized water, adjust pH to 7.5, bring up to 200 ml with deionized water and autoclave.

2. *Zymolyase solution*: 12.5 mg zymolyase 20T (250 U) in 50 µl of 0.67 mM K_2HPO_4, pH 7.5, for each yeast colony. Prepare fresh before each use.

2.6 Identification of Prey Inserts

1. *Competent Escherichia coli cells*: such as One Shot TOP10 chemically competent *E. coli* (Life Technologies).

2. *LB medium*: 10 g tryptone, 5 g yeast extract, 10 g NaCl (15 g agar). Bring up to 1 L with deionized water and autoclave. Let cool before adding antibiotics.

3. *Resuspension buffer:* 6.06 g tris base, 3.72 g ethylenediamine-tetraacetic acid disodium salt dihydrate ($Na_2EDTA \cdot 2H_2O$). Add approximately 800 ml of deionized water and adjust pH to 8. Bring up to 1 L with deionized water and autoclave. After autoclaving, add DNase- and proteinase-free RNase A to a final concentration of 100 µg/ml and store at 4 °C.

4. *Lysis buffer*: 8 g NaOH, 10 g sodium dodecyl sulfate (SDS). Bring up to 1 L with demineralized water. Filter sterilize and store in a plastic bottle at room temperature. When stored too cold, a white precipitate will appear. The precipitated SDS can be dissolved again by microwaving.

5. *Neutralization buffer*: 294.5 g potassium acetate. Dissolve in 500 ml of deionized water and adjust pH to 5.5 with glacial acetic acid (more than 100 ml will be needed). Bring up to 1 L and autoclave. Store at room temperature.

6. *70 % ice cold ethanol*: 70 ml ethanol, 30 ml deionized water. Store at −20 °C.

7. *Sequencing primers*: specific for the prey plasmids. E.g:

pDEST22_F: 5′-TATAACGCGTTTGGAATCACT-3′

pDEST22_R: 5′-AGCCGACAACCTTGATTGGAGAC-3′

pGADT7g_F: 5′-CTATTCGATGATGAAGATACCCCAC-CAAACCC-3′

pGADT7g_R: 5′-GTGAACTTGCGGGGTTTTTCAGTATC-TACGATT-3′

2.7 One-on-One Y2H Confirmation Test

1. *Yeast strains*:

S. cerevisiae AH109 containing the bait plasmid (generated in Subheading 3.1).

S. cerevisiae AH109 containing an empty bait vector or a bait vector with an independent gene inserted (this strain can be generated with the same protocol as in Subheading 3.1).

S. cerevisiae Y187 with genotype *MATα, ura3-52, his3-200, ade2-101, trp1-901, leu2-3,112, gal4Δ, met−, gal80Δ, URA3::GAL1_{UAS}-GAL1_{TATA}-lacZ* (Clontech) [18].

2. *Solid media needed* (40 ml medium/Nunc OmniTray dish)

1 OmniTray YPDA

1 OmniTray SD-Trp-Leu

1 OmniTray SD-Trp-Leu-His + *x* mM 3-AT

1 OmniTray SD-Trp-Leu-His-Ade + 40 mg/L X-α-gal

2.8 Quantitative α-Galactosidase Assay

1. *Assay buffer*: 2 volumes of 1× NaOAc buffer combined with 1 volume of 100 mM PNP-α-Gal solution. Mix well. Prepare fresh before each use.

2. *100 mM PNP-α-Gal solution*: 30.13 mg p-nitrophenyl-α-D-galactopyranoside in 1 ml of deionized water. Filter sterilize. Prepare fresh before each use.

3. *1X NaOAc*: 2.05 g sodium acetate in 50 ml of deionized water (0.5 M NaOAc). Adjust pH to 4.5.

4. *10X Stop solution*: 10.6 g Na_2CO_3 in 100 ml of deionized water (1 M Na_2CO_3).

3 Methods

3.1 Bait Plasmid Transformation

This protocol is adapted from the lithium acetate/single-stranded carrier DNA/polyethylene glycol method by Gietz and Woods [19], which permits high-efficiency transformation of yeast. This method generally yields 10^5 transformants per μg of DNA.

1. Streak *S. cerevisiae* strain AH109 on a fresh YPDA agar plate starting from a 25 % glycerol cell stock (stored at −80 °C). Incubate for 2–3 nights at 30 °C (*see* **Note 1**).

2. Inoculate two 5 ml YPDA cultures (in 50 ml flasks) with one big (3 mm) or two to four small (1–2 mm) colonies. Inoculate different numbers of colonies for the two cultures to ensure different cell densities the next day. Incubate overnight in a shaking incubator at 30 °C. The next morning, measure the optical density at 600 nm (OD_{600}). A preculture with a value below 1 is preferred as these cells are still not stationary and will have a shorter lag phase. Use this overnight culture to inoculate a fresh yeast culture.

3. Inoculate 50 ml of YPDA (in a 500 ml flask) until an OD_{600} value of approximately 0.120–0.140 is achieved (*see* **Note 2**). Incubate at 30 °C until the OD_{600} reaches ±0.480–0.560 (this usually takes 4–5 h). At least two cell divisions are needed for a good transformation efficiency. This efficiency remains constant during the next three to four cell divisions.

4. From this point forward, all steps are performed at room temperature. Centrifuge the cells for 5 min at 3,500 × g.

5. Discard the medium and resuspend the cells in a half volume of sterile water (25 ml). Centrifuge again for 5 min at 3,500 × g to wash the cells.

6. Discard the water and resuspend the cell pellet in 1 ml of 100 mM LiOAc (*see* **Note 3**). Transfer to a 1.5 ml microcentrifuge tube.

7. Pellet the cells (14,000 × g, 15 s) and remove the LiOAc with a micropipette. Resuspend the cells in 400 μl of 100 mM LiOAc. The final volume is now approximately 500 μl. This is sufficient for ten transformations (*see* **Note 4**). Divide the cells into 50 μl aliquots.

8. Centrifuge (14,000 × g, 15 s) and remove the LiOAc. The cells are now ready for transformation. The transformation mixture is added in the following order (*see* **Note 5**):

    240 μl PEG3350 (50 % w/v)

    36 μl 1 M LiOAc

    10 μl ssDNA (10 mg/ml)

    x μl (1 μg) plasmid DNA

    74 − x μl sterile water

    Total 360 μl

9. Vortex for 1 min and incubate the transformation mixture in a water bath set at 42 °C for 30 min. Invert tubes every 5–10 min (*see* **Note 6**).

10. Centrifuge the mixture for 15 s at 3,500 × g and remove the supernatant with a micropipette. Resuspend the cells in 1 ml of sterile water.

11. After resuspension, plate the cells on selective medium (10 μl, 100 μl, remaining volume) (e.g., SD-Trp for pGBKT7g and SD-Leu for pDEST32) and incubate at 30 °C for 3 nights (*see* **Note 7**).

3.2 Bait Auto-activation Test

This protocol also uses the LiOAc/ssDNA/PEG method. The important differences are highlighted in bold.

1. Streak *S. cerevisiae* AH109 containing the bait plasmid (obtained in Subheading 3.1) on a fresh *selective agar* plate. Incubate for 2–3 nights at 30 °C (*see* **Note 8**).

2. Inoculate two 5 ml *SD medium* cultures (in 50 ml flasks) with one big (3 mm) or two to four small (1–2 mm) colonies. Incubate overnight in a shaking incubator at 30 °C. The next morning, determine the OD_{600} and use the overnight culture to inoculate a fresh yeast culture.

3. **Steps 3–10** are exactly the same (*see* **Note 9**). In **step 8**, use 1 µg of empty prey vector or an independent prey construct to transform the *S. cerevisiae* AH109 containing the bait plasmid.

4. After resuspension in 1 ml of sterile water, 10 µl and 100 µl are plated on SD-Trp-Leu to check the transformation efficiency. The remaining volume (890 µl) is divided between SD-Trp-Leu-His +0, 1, 3, 5 and 10 mM 3-AT (100 µl on every plate) and SD-Trp-Leu-His-Ade (the remaining volume) plates. All plates are incubated at 30 °C for 5 days.

5. Evaluation: a background of very small colonies, due to leaky expression of the *HIS3* reporter, can usually be observed. Only large colonies on the SD-Trp-Leu-His plates (or maybe even on the SD-Trp-Leu-His-Ade plate) are an indication of auto-activation. Out of the different concentrations of 3-AT, choose one concentration that will be used in the Y2H screen. This concentration should not be too low to reduce the number of false positive "interactions" but also not too high, to reduce the chance of missing weaker interactions (*see* **Note 10**).

3.3 Prey Library Transformation

1. Streak *S. cerevisiae* AH109 containing the bait plasmid (obtained in Subheading 3.1) on a fresh selective agar plate. Incubate for 2–3 nights at 30 °C (*see* **Note 11**).

2. *Day 1*: Inoculate the freshly streaked strain in 5 ml of selective SD medium in a 50 ml flask (use several small colonies). Incubate overnight at 30 °C.

3. *Day 2*: Inoculate two 100 ml cultures of selective SD medium in 1 L baffled flasks with the overnight culture (use two different volumes: 1 and 2 ml) (*see* **Note 12**). The next day, one of the 100 ml cultures should have an OD_{600} below 1 (still exponentially growing, but preferably also close to 1 to have enough inoculum for a 600 ml culture). Prewarm 600 ml 2× YPDA in a 2 L baffled flask at 30 °C and autoclave two additional empty 2 L baffled flasks and a 250 ml graduated cylinder.

4. *Day 3*: Measure the OD_{600}. Centrifuge the culture with the value the closest to, but still below, 1 (divide between two 50 mL tubes) ($3,500 \times g$, 5 min). Discard the supernatant and resuspend the cell pellets in 1 ml of 2× YPDA. Use this concentrated cell solution to inoculate the 600 ml 2× YPDA culture to an OD_{600} of approximately 0.120–0.140. Divide the 600 ml culture between the three 2 L baffled flasks using the sterile graduated cylinder. Incubate the three cultures at 30 °C in a shaking incubator until the OD_{600} reaches ±0.480–0.560. This will take 4–5 h.

5. Divide the culture in twelve 50 ml sterile tubes. Centrifuge ($3,500 \times g$, 5 min).

6. Discard the supernatant and wash the cells in a half volume (300 ml) of sterile water (50 ml/2 cell pellets) (*see* **Note 13**). Centrifuge again.

7. Discard the supernatant and wash the cells a second time with 300 ml of sterile water. Collect the cells by centrifugation. Resuspend all cell pellets in 4 ml of sterile water.

8. Prepare the 120× transformation mixture (*see* **Note 14**):

   28.8 ml PEG3350 (50 % w/v)

   4.32 ml 1 M LiOAc

   1.2 ml ssDNA (10 mg/ml)

   x ml (120 µg) prey library DNA

   4.88 − x ml sterile water

9. Vortex the transformation mixture vigorously and add it to the competent cells which were resuspended in 4 ml of water. Vortex the entire mixture for 10 min (*see* **Note 15**).

10. Divide the solution into six 15 ml conical tubes and incubate for 30 min at 30 °C followed by 45 min at 42 °C in a water bath (*see* **Note 16**). Invert the tubes every 5–10 min.

11. Collect all cells in one 50 ml tube and centrifuge for 5 min at 3,500 × g. Discard the transformation solution and resuspend the cells in 12 ml of sterile water.

12. Plate three dilution series on SD-Trp-Leu to calculate the transformation efficiency. For each series, 10 µl of cells are diluted in 990 µl of sterile water (10^{-2}). 100 µl from this dilution is further diluted in 900 µl water (10^{-3}). Repeat for a 10^{-4} and 10^{-5} dilution. Mix every dilution again right before plating 100 µl on SD-Trp-Leu.

13. Divide the remaining volume (±400 µl per plate) over 30 SD-Trp-Leu-His + x mM 3-AT 150 mm plates (use the appropriate concentration of 3-AT determined from the auto-activation test) (*see* **Note 17**).

14. After 3 nights at 30 °C, the dilution series can be counted. Calculate the average. The transformant yield is the average × 10 (100 µl plated) × dilution factor × 12 (total volume). The minimum number of transformants needed depends on the library used (*see* **Note 18**). Incubate the 150 mm SD-Trp-Leu-His plates at 30 °C for 5–7 nights.

3.4 Evaluation of Library Transformation

1. Check all 30 plates and select a maximum of 96 yeast colonies to analyze (*see* **Note 19**).

2. Pick the selected colonies with a toothpick and transfer them to a microtiter plate containing 100 µl of sterile water.

3. Shake for 15 min on a shaker to resuspend all cells.

4. Spot in parallel 2 μl on an SD-Trp-Leu, SD-Trp-Leu-His + x mM 3-AT, and SD-Trp-Leu-His-Ade + 40 mg/L X-α-gal OmniTray plate (*see* **Note 20**).

5. Incubate at 30 °C for 5–7 nights.

6. Add 100 μl of sterile water +30 % glycerol to the remaining volume of resuspended cells. This backup cell stock can be stored at −80 °C.

7. Evaluation: everything should grow on the SD-Trp-Leu and the SD-Trp-Leu-His + x mM 3-AT plates. Interactions should also be strong enough to activate not only the *HIS3* reporter but also the *ADE2* and *MEL1* reporter genes, so colonies of interest are those that can grow on the SD-Trp-Leu-His-Ade + X-α-gal plate and have a blue halo around them caused by the α-galactosidase activity. A maximum of 24 blue colonies should be selected for further analysis (*see* **Note 21**).

3.5 Yeast Plasmid Isolation

1. Inoculate the colonies in 5 ml of selective SD medium and incubate for 2 nights at 30 °C (*see* **Note 22**). It is only necessary to select for the prey plasmid.

2. Centrifuge for 5 min at 3,500 × g and discard the supernatant. Resuspend the cell pellet in 250 μl of 0.67 mM K_2HPO_4, pH 7.5. Transfer to a 1.5 ml tube.

3. Add 50 μl of zymolyase solution and incubate in a water bath at 35 °C. Invert regularly.

4. Collect cells by centrifugation for 5 min at 3,500 × g. Discard the supernatant.

5. The zymolyase will degrade the yeast cell wall and create spheroplasts; from now on a standard bacterial miniprep kit can be used to isolate the yeast plasmid.

3.6 Identification of Prey Inserts

Because of its low yield and impurity, the isolated yeast prey plasmid cannot be directly used as a template in DNA sequencing analysis. To identify the prey inserts, the plasmid should first be transformed back to *E. coli*. A direct PCR is not advisable, as there is still some genomic yeast DNA present that may cause nonspecific products and there can be more than one type of prey plasmid present.

1. For each blue colony, transform 5 μl of yeast plasmid DNA to competent *E. coli* cells. Plate cells on LB medium selecting for the prey plasmid.

2. For each transformation, four colonies (derived from one yeast colony) are selected for further analysis (*see* **Note 23**). Grow the yeast overnight in a 4 ml selective LB culture and isolate the prey plasmid using a standard miniprep kit. In the case of a larger number of yeast colonies, it is better to use a microtiter plate scaled plasmid isolation method. Pick four colonies per

transformation in a microtiter plate with 100 μl selective LB and incubate for 2 h at 37 °C without shaking.

3. First inoculate 15 μl of the cells in a microtiter plate containing 150 μl of selective LB. Close the lid and seal the plate with parafilm around the borders. Grow overnight at 37 °C on a microtiter plate (MTP) shaker. To the remaining culture, add 100 μl of LB with 40 % glycerol. This backup glycerol cell stock can be stored at −20 °C.

4. The next day, collect the cells by centrifugation at 4,000 × *g* for 10 min. Discard the supernatant (*see* **Note 24**).

5. Add 35 μl of resuspension buffer, shake for approximately 5 min on an MTP shaker until everything is resuspended.

6. Add 35 μl of lysis buffer, shake for about 2 min until the solution is clear again. Do not allow lysis to proceed for longer than 5 min.

7. Add 49 μl of neutralization buffer, shake for another 5 min. A white precipitate of cell debris and genomic DNA will form.

8. Centrifuge for 20 min at 4,000 × *g*. Transfer the supernatant containing the plasmid DNA to a new round-bottomed microtiter plate (approximately 110 μl).

9. Add 60 μl of isopropanol and mix by pipetting up and down. Incubate for 15 min without shaking at room temperature.

10. Centrifuge 30 min at 4,000 × *g* to precipitate the plasmid DNA. Remove the supernatant by quickly inverting the plate above a sink.

11. Add 70 μl of 70 % ice cold ethanol to wash the pellet. Centrifuge 10 min at 4,000 × *g* at 4 °C. Discard supernatant by quickly inverting the plate above the sink and tapping the plate on paper tissues (*see* **Note 25**). Let the DNA dry for about 15 min at room temperature.

12. Resuspend the DNA pellets in 20 μl of ultrapure water by pipetting up and down.

13. Digest the plasmid DNA with a restriction enzyme cutting out the insert (for miniprep on a column: use 2 μl of plasmid DNA as template; for miniprep in MTP format: use 6 μl of plasmid DNA as template) and visualize using DNA agarose gel electrophoresis.

14. Evaluation: this digest checks whether all prey plasmids originated from one blue yeast cell contain the same insert (based on length alone). As yeast can contain more than one prey plasmid, in contrast to *E. coli* that can only harbor one, it is possible to see different insert lengths. If this is the case, both types of inserts have to be analyzed. One construct per insert length

should be chosen for DNA sequencing analysis using prey plasmid-specific primers to identify the potential interaction partners for the protein of interest.

3.7 One-on-One Y2H Confirmation Test

All the observed interactions could be false positive results and should be confirmed using an independent Y2H experiment. Choose one prey construct (the plasmid isolated from *E. coli* in Subheading 3.6) for each different potential interaction partner. These prey constructs will be transformed to the α-mating type yeast strain Y187 [18] and will be mated with an AH109 strain containing the bait construct and another strain containing an empty bait vector as a negative control. An independent protein inserted in the bait vector could also be used. If there are not too many different prey plasmids, the same protocol provided in Subheading 3.1 can easily be followed to transform the plasmids into *S. cerevisiae* Y187. After transformation, one can proceed with **step 15** from this protocol. In the case of a higher number of different preys or different Y2H screens confirmation tests combined in one experiment, the protocol given in this section, which is adapted for up to 96 yeast transformations in parallel, can be followed.

1. Streak *S. cerevisiae* Y187 on a fresh YPDA agar plate starting from a 25 % glycerol cell stock (stored at −80 °C). Incubate for 2–3 nights at 30 °C.

2. Inoculate two tubes of 4 ml of YPDA with several small colonies. Shake overnight at 30 °C.

3. Dilute the overnight culture with an OD_{600} value below 1 in 100 ml of YPDA (in a 1 L baffled flask) until an OD_{600} of ±0.120–0.140. Incubate at 30 °C until the OD_{600} reaches ±0.480–0.560.

4. Divide the culture between two conical tubes and centrifuge ($3,500 \times g$, 5 min).

5. Discard the medium and resuspend the cells in 2×25 ml sterile water. Centrifuge again.

6. Discard the supernatant and resuspend the cells in 2×1 ml of 100 mM LiOAc. Transfer to 1.5 ml microcentrifuge tubes.

7. Pellet the cells ($14,000 \times g$, 15 s) and remove the LiOAc with a micropipette. Resuspend each cell pellet in 400 μl of 100 mM LiOAc. The final volume is now around 2×500 μl. This is enough for 96 small-scale transformations. Centrifuge ($14,000 \times g$, 15 s) and remove the LiOAc.

8. Resuspend the two pellets in 1,480 μl of sterile water and transfer to a 15 ml conical tube. The transformation mixture is now added in the following order to the cells:

4,800 μl PEG3350 (50 % w/v)

720 µl 1 M LiOAc

200 µl ssDNA (10 mg/ml)

9. Vortex for 10 min. Prepare a PCR plate by adding to each well 100 ng of plasmid DNA (a different plasmid per well).

10. Divide the transformation mixture in the PCR plate by adding 25 µl to each well. Pipette up and down a few times to mix. Place a lid on the plate during incubation (*see* **Note 26**).

11. Incubate for 10 min at 30 °C followed by 30 min at 42 °C to heat shock the cells. This step can be done on a PCR block.

12. Centrifuge the plate at 3,500 × *g* for 1 min and discard the supernatant by pipetting. Resuspend the cells in 100 µl of sterile water.

13. Plate the entire cell mixture on selective SD medium and incubate at 30 °C for 3 nights (*see* **Note 27**).

14. On the same day as **step 13**, freshly streak the different AH109 strains on selective agar plates. Incubate for 3 nights at 30 °C.

15. Fill a microtiter plate with 50 µl of sterile water and resuspend the different *S. cerevisiae* Y187 and AH109 strains in different wells (pick several colonies).

16. Mating: spot 2 µl a-yeast on a YPDA agar plate. Let dry. Spot 2 µl α-yeast on the same position as the a-yeast. Do this for all different combinations (e.g., each prey in *S. cerevisiae* Y187 is mated with *S. cerevisiae* AH109 containing the bait construct and *S. cerevisiae* AH109 containing the empty bait vector). There will be two positions for each prey which has to be tested. Incubate overnight at 30 °C.

17. The next day, mating has occurred. To select for diploid cells, pick some cell material from every position in 50 µl of sterile water and spot the different wells (2 µl) on SD-Trp-Leu. Incubate for two nights at 30 °C.

18. Resuspend the diploid cells again in 50 µl of sterile water and spot 2 µl on SD-Trp-Leu-His + x mM 3-AT. Incubate for 5–7 nights at 30 °C.

19. Pick the different cells in 50 µl of sterile water and spot 2 µl on SD-Trp-Leu-His-Ade +40 mg/L X-α-gal. Incubate for 5–7 nights at 30 °C.

20. Evaluation: those prey plasmids that only generate colonies with a halo for the bait-prey combination but not when combined with the empty bait vector are confirmed interactions.

3.8 Quantitative α-Galactosidase Assay

To compare different interaction strengths, a quantitative assay to measure the α-galactosidase activity (encoded by the *MEL1* reporter gene) should be performed. The stronger the interaction between bait and prey, the more transcription of *MEL1*, and consequently the more α-galactosidase is produced. This enzyme is secreted in the medium, so there is no need to lyse the yeast cells in this protocol (adapted from ref. 23).

1. Inoculate 3 ml of liquid SD-Trp-Leu-His medium with a bait- and prey-expressing yeast cell (*see* **Note 28**). Do this in tripli- cate for every type of yeast colony that needs to be analyzed. Incubate overnight at 30 °C in a shaking incubator.

2. Vortex each culture for 30 s to disperse cell clumps, then transfer 1 ml to a cuvette to measure the OD_{600} (should be between 0.5 and 1; if higher, dilute the cell suspension).

3. Transfer 1 ml of the vortexed cell suspension to a 1.5 ml microcentrifuge tube. Centrifuge (14,000 × *g*, 2 min). Imme- diately proceed to **step 4** to reduce loss of enzyme activity.

4. Prepare enough assay buffer for all samples and controls, and let it equilibrate to room temperature. For each assay 48 μl of buffer is needed (*see* **Note 29**).

5. Transfer 16 μl of the supernatant (of **step 3**) into a well of a clear 96-well flat-bottom microtiter plate. Also include one well containing sterile SD-Trp-Leu-His. This blank will be used to zero the spectrophotometer in **step 8**.

6. Add 48 μl of assay buffer to each well and close the plate. Incubate at 30 °C for 1 h.

7. Stop the reaction by adding 136 μl of 10× stop solution to each well.

8. Measure the OD at 410 nm (OD_{410}) relative to the blank sample using a spectrophotometer that can read microtiter plates.

9. Calculate the α-galactosidase units. One unit is defined as the amount of enzyme that hydrolyzes 1 μmol of *p*-nitrophenyl-α- D-galactoside to *p*-nitrophenol and D-galactose in 1 min at 30 °C in acetate buffer, pH 4.5 [20].

   In the following formula, the OD_{600} of the overnight culture is used to normalize the OD_{410} to the amount of cells. Remember to take into account the dilution factor from **step 2**, if necessary.

   α-galactosidase in milliunits per ml per cell
   $$= (OD_{410} \times V_f \times 1,000)/(\varepsilon \times b \times t \times V_i \times OD_{600})$$

   t = elapsed time of incubation (in min) (60 min)

   V_f = final volume of the assay (200 μl)

   V_i = volume of supernatant added (16 μl)

   ε = *p*-nitrophenol molar absorptivity at 410 nm

   b = the light path (in cm)

   $\varepsilon \times b$ = 10.5 ml/μmol for the microtiter plate format

4   Notes

1. Always use freshly streaked cells for your transformations. This will give the best transformation efficiency. The plate used to inoculate the overnight culture can be up to 2 weeks old for routine transformations, but for the prey library transformation, it is definitely advisable to use fresh colonies.

2. The fresh culture in the morning should have 5×10^6 cells/ml because at the start of transformation 2×10^7 cells/ml are needed. For strain AH109, this start density corresponds to an OD_{600} of ± 0.130. To reach this, first vortex the overnight culture to disperse all cell clumps, add the cells in smaller steps, and measure the OD_{600} in between until an OD_{600} of 0.130 is reached. It is critical that all cell clumps are dispersed to obtain a healthy liquid culture. If the culture does not reach the final OD_{600} in the expected 4–5 h, it is best to start over with an overnight culture and do the transformation the following day, especially when the efficiency should be high (e.g., for the library transformation).

3. It is not required to prepare a separate 100 mM stock of LiOAc, a 1 M stock suffices. In this case, first resuspend the cells in sterile water and then add the required LiOAc. For instance, if the cells need to be resuspended in 400 μl of 100 mM LiOAc, first resuspend them in 360 μl of sterile water and then add 40 μl of 1 M LiOAc.

4. When less than ten transformations are needed, all washing volumes can be adapted. Five milliliters of yeast culture is needed for one transformation. For every 5 ml of exponentially growing cells, the washing volumes are subsequently 2.5 ml of water, 100 μl of 100 mM LiOAc, and 40 μl of 100 mM LiOAc.

5. Adding the PEG3350 before the LiOAc is important. LiOAc is harmful for the cells if the concentration is too high. The PEG will first make a shield around the cells. When transforming a higher number of different plasmids, it can be useful to resuspend the cells in batches in the transformation mixture without the plasmids and then divide the cells over different tubes each containing a different plasmid. This does not affect the transformation efficiency. For instance, for ten transformations, stop in **step 7** when the amount of cells is 500 μl. Centrifuge and remove the LiOAc. Then resuspend the cells in $740 - y$ μl of sterile water (where y is the total amount of plasmid DNA) and transfer to a small conical tube. Add 2.4 ml of PEG3350 and vortex. Add 360 μl of 1 M LiOAc and 100 μl of ssDNA and vortex again for 2 min. Add $360 - x$ μl of transformation mixture to each tube with x μl of plasmid DNA.

6. The vortexing step is crucial for a good transformation efficiency. Cells have to be fully resuspended and surrounded by all the compounds to take up the plasmid DNA. Cells that are not vortexed sufficiently will settle faster to the bottom of the tube and the transformant yield will be lower.

7. Usually the 100 μl plate will have a sufficient number of single colonies. Once you have become familiar with the protocol, only plate 100 μl and keep the rest of the transformation mixture at 4 °C. Check the plate after 2 days and when more colonies are needed, the remaining cells can be plated.

8. It is possible to directly use the colonies obtained in Subheading 3.1 if you directly proceed. Then there is no need to streak them again before performing the auto-activation test.

9. Also use YPDA for this culture. Growth in SD medium selecting for the bait would take too long.

10. Most of the time, 1 or 3 mM 3-AT is sufficient.

11. For the library transformation, one should always use fresh cells to obtain a high transformant yield.

12. One overnight culture would not give enough cells to inoculate 600 ml of culture, which is needed for the library transformation. Hence, two steps are needed. Inoculation of the second overnight culture (100 ml culture) early in the afternoon (2–3 PM) is usually successful to get an OD close to 1 the next morning.

13. First add ±2 ml of water per pellet to resuspend them by pipetting up and down. Then adjust to the final volume by pouring.

14. The transformation mixture can be made and vortexed during the washing steps. As the PEG3350 is viscous, pipette it slowly to obtain the correct volume.

15. Vortexing for 10 min can easily be done by taping the tubes on the vortex machine.

16. After dividing the mixture in the small conical tubes, make sure the level is below the water level. If this is not the case, divide over more tubes.

17. It is recommended to prepare the selective plates only 2–3 days before your planned experiment, especially for the library transformations. In that case, they can be kept at room temperature, so the plates will be dried, allowing the transformation mixtures to dry much faster after plating.

18. The number of transformants needed depends on the library. For random genomic libraries, this can be calculated using the Clarke and Carbon formula [21] that indicates that the actual number of clones required equals $\ln(1 - P)/\ln(1 - (f/G))$

where P is the probability, f is the mean fragment size of the plasmid inserts, and G is the genome size (e.g., for *E. coli* K-12 [22] with a genome size of 4,639,221 bp and a mean fragment size of 600 bp, 35,605 transformants are needed to cover the whole genome once with a probability of 0.99). Knowing that in Y2H fusion proteins the fragment orientation and reading frame has to be correct, the actual total number of yeast transformants needed is 6×35605 or 2.14×10^5.

19. The total number of colonies growing on the SD-Trp-Leu-His plates depends on the bait, the 3-AT concentration, and the efficiency. The higher the transformation efficiency, the more colonies will be growing. In this case the library is covered multiple times, so there will be colonies that contain the same prey plasmid. Because it is too labor intensive to analyze all of the colonies in this case, it is important to make a selection. Limit the selection to 96 colonies, which is still an easy format to work with. Careful inspection of all plates and numbering and sorting the colonies in groups of different sizes is one way to remove duplicates (e.g., 1, 2, 3, >3 mm). Although size might reflect interaction strength, there is no reason to assume that weak interactions have less biological significance, so always select different sizes of colonies from all different plates. When picking the colonies to array them in a microtiter plate, check the consistency of the colony. Based on our experience, granular colonies that fall apart after picking generally contain false positive interactions. So it is preferable to only pick smooth and creamy colonies.

20. Other single well microtiter plates or a small petri dish in the case of a smaller number of colonies can be used.

21. Limit this selection to 24 colonies again by carefully inspecting and sorting the colonies in groups of different colors (e.g., white, pale blue, dark blue, dark blue with a blue halo).

22. It is easier to use glass culture tubes than flasks. Yet, when using glass tubes, the cells will settle more easily to the bottom of the tube. To grow a culture sufficiently dense for plasmid yield, vortex after inoculation, shake the tubes vigorously (250 rpm), and vortex them again after 1 day of incubation.

23. Different *E. coli* colonies derived from the same yeast colony are selected because yeast can contain more than one prey plasmid. As *E. coli* can only contain one, this step will show the different inserts.

24. Supernatant removal can be performed using a vacuum pump. The microtiter plate method to recover plasmids requires some practice but is an inexpensive and efficient solution when screening large numbers of colonies.

25. To remove the supernatant, invert the plate very quickly to avoid well-to-well contamination.

26. Use a microtiter plate lid on top of the PCR plate. When using strips to close the plate, it is too difficult to remove the strips after incubation which will cause well-to-well contamination.

27. To reduce the amount of medium and to make the plating more convenient, QTray bioassay trays from Genetix can be used for plating. These contain a 48-position grid, so only two plates are needed to plate 96 transformation mixtures.

28. Inoculate the cells starting from an SD-Trp-Leu or SD-Trp-Leu-His plate. Use fresh cells (maximum 2 weeks old).

29. When the number of samples is smaller or a spectrophotometer to read microtiter plates is not available, a 1 ml assay can also be performed. Refer to [23]) for a detailed protocol.

Acknowledgements

Jeroen Wagemans holds a predoctoral fellowship of the *"Agentschap voor Innovatie door Wetenschap en Technologie in Vlaanderen"* (IWT, Belgium).

References

1. Fields S, Song OK (1989) A novel genetic system to detect protein protein interactions. Nature 340:245–246

2. Keegan L, Gill G, Ptashne M (1986) Separation of DNA binding from the transcription-activating function of a eukaryotic regulatory protein. Science 231:699–704

3. Chien CT, Bartel PL, Sternglanz R et al (1991) The two-hybrid system: a method to identify and clone genes for proteins that interact with a protein of interest. Proc Natl Acad Sci U S A 88:9578–9582

4. Van Criekinge W, Beyaert R (1999) Yeast two-hybrid: state of the art. Biol Proced Online 2:1–38

5. James P, Halladay J, Craig EA (1996) Genomic libraries and a host strain designed for highly efficient two-hybrid selection in yeast. Genetics 144:1425–1436

6. Durfee T, Becherer K, Chen PL et al (1993) The retinoblastoma protein associates with the protein phosphatase type 1 catalytic subunit. Genes Dev 7:555–569

7. Hu JC, Kornacker MG, Hochschild A (2000) *Escherichia coli* one- and two-hybrid systems for the analysis and identification of protein-protein interactions. Methods 20:80–94

8. Buchert M, Schneider S, Adams MT et al (1997) Useful vectors for the two-hybrid system in mammalian cells. Biotechniques 23:396–398, 400, 402

9. Johnsson N, Varshavsky A (1994) Split ubiquitin as a sensor of protein interactions in vivo. Proc Natl Acad Sci U S A 91:10340–10344

10. Brachmann RK, Boeke JD (1997) Tag games in yeast: the two-hybrid system and beyond. Curr Opin Biotechnol 8:561–568

11. Uetz P (2002) Two-hybrid arrays. Curr Opin Chem Biol 6:57–62

12. Golemis EA, Khazak V (1997) Alternative yeast two-hybrid systems. The interaction trap and interaction mating. Methods Mol Biol 63:197–218

13. Serebriiskii I, Khazak V, Golemis EA (1999) A two-hybrid dual bait system to discriminate specificity of protein interactions. J Biol Chem 274:17080–17087

14. Le Douarin B, Pierrat B, vom Baur E et al (1995) A new version of the two-hybrid assay for detection of protein-protein interactions. Nucleic Acids Res 23:876–878

15. Caufield JH, Sakhawalkar N, Uetz P (2012) A comparison and optimization of yeast two-hybrid systems. Methods 58:317–324

16. Chen YC, Rajagopala SV, Stellberger T et al (2010) Exhaustive benchmarking of the yeast two-hybrid system. Nat Methods 7:667–668, author reply 668

17. Uetz P, Dong YA, Zeretzke C et al (2006) Herpesviral protein networks and their interaction with the human proteome. Science 311:239–242

18. Harper JW, Adami GR, Wei N et al (1993) The p21 Cdk-interacting protein Cip1 is a potent inhibitor of G1 cyclin-dependent kinases. Cell 75:805–816

19. Gietz RD, Woods RA (2002) Transformation of yeast by lithium acetate/single-stranded carrier DNA/polyethylene glycol method. Methods Enzymol 350:87–96

20. Lazo PS, Ochoa AG, Gascon S (1978) alpha-Galactosidase (melibiase) from *Saccharomyces carlsbergensis*: structural and kinetic properties. Arch Biochem Biophys 191:316–324

21. Clarke L, Carbon J (1976) A colony bank containing synthetic Col El hybrid plasmids representative of the entire *E. coli* genome. Cell 9:91–99

22. Blattner FR, Plunkett G 3rd, Bloch CA et al (1997) The complete genome sequence of *Escherichia coli* K-12. Science 277:1453–1462

23. Clontech Yeast Protocols Handbook. Clontech Laboratories, Inc. www.clontech.com. Protocol No. PT3024-1. 2. Version No. PR973283

Chapter 28

Reverse Two-Hybrid Techniques in the Yeast *Saccharomyces cerevisiae*

Matthew A. Bennett, Jack F. Shern, and Richard A. Kahn

Abstract

Use of the yeast two-hybrid system has provided definition to many previously uncharacterized pathways through the identification and characterization of novel protein-protein interactions. The two-hybrid system uses the bifunctional nature of transcription factors, such as the yeast enhancer Gal4, to allow protein-protein interactions to be monitored through changes in transcription of reporter genes. Once a positive interaction has been identified, either of the interacting proteins can be mutated by site-specific or randomly introduced changes, to produce proteins with a decreased ability to interact. Mutants generated using this strategy are very powerful reagents in tests of the biological significance of the interaction and in defining the residues involved in the interaction. Such techniques are termed reverse two-hybrid methods. We describe a reverse two-hybrid method that generates loss-of-interaction mutations of the catalytic subunit of the *Escherichia coli* heat-labile toxin (LTA1) with decreased binding to the active (GTP-bound) form of human ARF3, its protein cofactor. While newer methods are emerging for performing interaction screens in mammalian cells, instead of yeast, the use of reverse two-hybrid in yeast remains a robust and powerful means of identifying loss-of-interaction point mutants and compensating changes that remain among the most powerful tools of testing the biological significance of a protein-protein interaction.

Key words Reverse two-hybrid, Two-hybrid, Protein interaction, Loss-of-interaction mutation

1 Introduction

Protein interactions define key biochemical and regulatory networks in the cell. The identification and characterization of protein-protein interactions through the use of two-hybrid assays have provided definition to many previously uncharacterized pathways. Yeast two-hybrid systems use the bifunctional nature of transcription factors, e.g., Gal4 or LexA, to assay protein interactions in live cells [1, 2]. The DNA binding and transcriptional activation domains of the transcription factor are each expressed as fusion proteins and promote transcription only when joined by the binding of the other portion of the fusion proteins (*see* Fig. 1 and **Note 1**). Screens of two-hybrid libraries have led to the

Cheryl L. Meyerkord and Haian Fu (eds.), *Protein-Protein Interactions: Methods and Applications*, Methods in Molecular Biology, vol. 1278, DOI 10.1007/978-1-4939-2425-7_28, © Springer Science+Business Media New York 2015

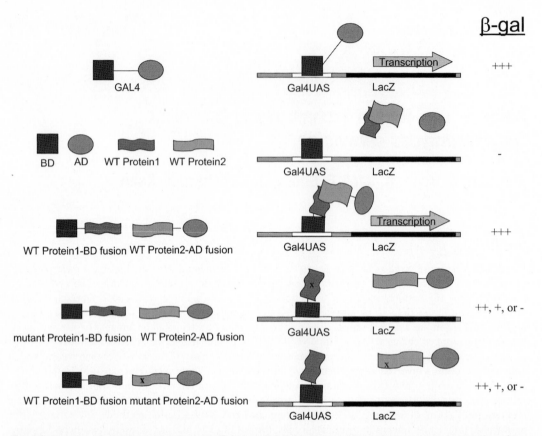

Fig. 1 Expression of reporter genes (e.g., LacZ) in two-hybrid systems is dependent upon the interaction of two fusion proteins that can be lost as a result of a point mutation. The Gal4 protein, two interacting proteins of interest, and resulting fusion proteins are shown pictorially on the left. An "x" in a protein represents a point mutation. Interaction between the two fusion proteins generates a transactivator capable of promoting transcription of reporter genes (LacZ) downstream of the Gal4 upstream activating sequence (UAS). Point mutations in either fusion protein that decrease binding result in decreases in or elimination of reporter gene expression. The level of expression is shown on the right as the amount of β-galactosidase (β-gal) activity

identification of a large number of protein binding partners (e.g., refs. 3, 4) and proteome based screens (e.g., refs. 5–8) have increased the number of putative interactions markedly. Both the methods and the level of complexity or detail in the interactomes continue to improve [7, 9–17].

After a two-hybrid interaction is identified, either of the interacting proteins can be mutated to decrease the binding affinity. The generation and use of such mutants is collectively referred to as "reverse two-hybrid" methods. Such loss-of-interaction mutants can become essential reagents to test the functional importance of specific protein interactions in live cells, to obtain a low-resolution map of the binding interface, or to generate pairs of mutants that interact independently of endogenous proteins (*see* also Chappell and Gray [18]). A number of variants and uses for reverse two-hybrid have emerged over the years that offer greater versatility, depending on needs [19–21]. These approaches are particularly attractive when one of the proteins under study binds multiple protein partners as mutagenesis and two-hybrid screens can identify specific loss of binding mutants that retain binding to other partners. In this chapter, we describe a reverse two-hybrid method for generating mutations of the catalytic subunit of the *Escherichia coli* heat-labile toxin (LTA1) with reduced binding to the activated (GTP bound) form of human ARF3, [Q71L]ARF3 [22, 23]. As seen here and in other studies (e.g., Das et al. [24]), the use of reverse two-hybrid is an attractive means of testing the relevance of protein interactions in pathogen biology.

The main features of this approach include PCR-based random mutagenesis over a defined region of any coding region, and identification of full-length mutants, expressed to the same levels as wild-type protein by immunoblot analysis using an epitope tag on the fusion proteins. The conditions used here have yielded approximately one change per 250 bp amplified. PCR allows the extent of mutagenesis to be controlled by modifying the concentration of nucleotides or manganese in the PCR reaction [25]. Although the use of selectable markers fused to the C-terminus of the mutated protein allows for genetic selection of full-length proteins [26–28], which is important to high-throughput screens and can speed up this rate limiting step in reverse two-hybrid screens, we prefer the use of immunoblots to allow estimates of protein expression as well as confirmation that the protein is full-length. Using the methods described below we have been able to generate hundreds of colonies with reduced β-galactosidase activity in less than a week and typically 25–50 % of those were found to result from point mutations in full-length proteins, suitable for further analysis. Thus, tens of mutants can be obtained, their interactions with multiple effectors characterized, and sequences determined in less than 1 month.

The loss of an interaction is limited in interpretations due to the negative nature of the data. However, loss of one binding partner with retention of others provides a powerful probe of activities in live cells. In addition, after the generation of loss-of-interaction mutants in one protein it is even easier to then mutate the other binding partner and screen for mutants that regain the interaction. Such pairs of mutants can provide a formal proof of specific functionalities in live cells. Such mutants may have the added benefit of not binding the wild-type protein and can allow cell studies to be carried out independently of interference or complications presented by the presence of endogenous proteins. As with any technique, the data and conclusions from the use of reverse two-hybrid techniques are strengthened by confirmation using independent methods.

2 Materials

1. pACT2 vector [29] (Genbank accession #U29899), a 2 μ plasmid carrying the selectable *LEU2* gene and the *ADH1* promoter driving expression of the activation domain (AD) of Gal4p (Gal4 Region II [1], residues 768–881), followed by the HA epitope, sites for insertion of the open reading frame of your protein, and the *ADH1* terminator.

2. pBG4D vector [30], a 2 μ plasmid carrying the selectable *TRP3* marker and the Gal4 DNA Binding Domain (BD) followed by the HA epitope, designed to add the Gal4-BD to the 3′ end of any inserted open reading frame, with expression of the resulting fusion protein driven by the *ADH1* promoter. pBG4D was made by Robert M. Brasas by subcloning the SacI/HindIII fragment from the pAS1-CYH plasmid into the plasmid D133.

3. Yeast strain Y190 (*MAT a gal4 gal80 his3 trp1-901 ade2-101 ura3-52 leu2-3,112 + URA3::GAL→lacZ, LYS2::GAL(UAS)-HIS3 cyh^R*) [29] kindly provided by Steve Elledge.

4. Whatman filter paper no. 1.

5. Nitrocellulose filters (Schleicher & Schuell Protran, 82 mm).

6. Monoclonal HA antibody 12CA5 (Boehringer), diluted 1:1,000 (final = 1 μg/mL) for immunoblots.

7. X-gal (5-bromo-4-chloro-3-indolyl-β-D-galactopyranoside), 100 mg/ml stock solution in *N,N*-dimethylformamide (DMF).

8. Low dATP nucleotide mix (2.5 mM dCTP, 2.5 mM dGTP, 2.5 mM dTTP, and 0.625 mM dATP in water).

9. Taq Polymerase.

10. Synthetic (SD) minimal, complete, and dropout media. Minimal SD medium: 1.5 g yeast nitrogen base lacking amino acids, 20 g glucose, 5 g $(NH_4)_2SO_4$ per liter. Complete SD medium: 20 mg adenine, 20 mg histidine, 20 mg methionine, 20 mg tryptophan, 20 mg uracil, 30 mg leucine, 30 mg lysine per liter, each is omitted as needed to generate the dropout media, e.g., SD leu− trp−.

11. SD Plates, made with the same components as SD media plus 20 g agar per liter.

12. SD leu−/cycloheximide, SD leu− containing 2.5 μg/mL cycloheximide.

13. Agarose gel and DNA sequencing equipment.

14. SDS-PAGE (sodium dodecylsulfate polyacrylamide gel electrophoresis) and protein electrotransfer equipment for immunoblotting.

15. 1× SDS sample buffer (50 mM Tris–HCl, pH 6.8, 10 % glycerol, 1 % SDS, 1 % (v:v) β-mercaptoethanol, bromophenol blue).

16. Z-Buffer (60 mM Na_2HPO_4, 40 mM NaH_2PO_4, 10 mM KCl, 2 mM $MgSO_4$, pH 7.0).

17. Oligonucleotide primers for PCR mutagenesis:

    #827 (sense primer): GCGTATAACGCGTTTGGAATC, binds ≈ 70 bp 5′ of the pACT2 MCS.

    #828 (anti-sense primer): GAGATGGTGCACGATGCACAG, binds ≈ 70 bp 3′ of the pACT2 MCS.

18. Sequencing Oligonucleotides:

    #1042 (sense primer): GCTTACCCATACGATGTTCCA—binds within the pACT2 MCS at the 5′ end.

    #1043 (anti-sense primer): TGAACTTGCGGGGTTTTT-CAG—binds within the pACT2 MCS at the 3′ end.

19. DNA purification kits (e.g., Qiagen mini prep, QiaQuick gel extraction, Wizard PCR purification kit).

20. pAB157, a plasmid encoding [Q71L]ARF3 with the Gal4-BD at the C-terminus by insertion into the BamHI site of pBG4D, to direct expression of [Q71L]ARF3-BD.

21. pXJ12, a plasmid carrying the open reading frame of the catalytic (A1) subunit of the *E. coli* heat labile toxin (LTA1) cloned into the NcoI and BamHI sites of pACT2, putting it in frame with the Gal4-AD, to direct expression of AD-LTA1 (*see* Fig. 2).

22. YAB457, the yeast strain, derived from Y190, carrying the pAB157 plasmid and thus expressing [Q71L]ARF3-BD.

23. SOS: 3.33 g/L yeast extract, 6.66 g/L dextrose, 6.66 g/L peptone, 6.5 mM $CaCl_2$.

24. TE: 10 mM Tris–HCl, pH 8.0 and 1 mM EDTA, pH 8.0.

25. 0.1 M lithium acetate in TE buffer.

26. LiOAc/PEG solution: 40 % (w:v) PEG 3350 suspended in 0.1 M lithium acetate in TE buffer.

27. Sonicated herring sperm DNA, 10 mg/mL.

28. STES: 500 mM NaCl, 200 mM Tris–HCl, pH 7.6, 10 mM EDTA, 1 % SDS.

29. Phenol–chloroform: 50 % phenol, 50 % chloroform.

30. Chloroform

31. Polymerase chain reaction (PCR) buffer (1×): 10 mM Tris–HCl, pH 8.3, 50 mM KCl, 0.0001 % gelatin, which is made from a 10× solution.

438   Matthew A. Bennett et al.

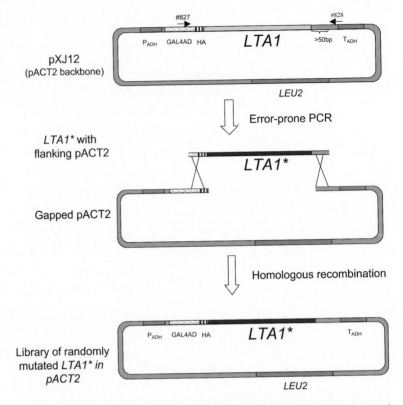

Fig. 2 Reverse two-hybrid screens use random PCR mutagenesis and gap repair in yeast to generate a large library of mutants, ready for screening. Amplification of the coding region of *LTA1* in pXJ12 is performed with primers that anneal ≥50 bp outside of the region targeted for mutagenesis to yield a product with ends homologous to those in the gapped plasmid, shown below. The error rate of the polymerase is increased by carrying out the PCR under conditions of reduced stringency. Note that any region of DNA can be targeted for mutagenesis. Co-transformation of the mutated PCR product with gapped pACT2 plasmid, prepared by restriction digestion, into yeast allows for repair of the plasmid by homologous recombination. When transformed into a yeast strain expressing the [Q71L]ARF3-BD fusion proteins, the resulting transformants can be assayed directly for loss of interaction using the colony β-galactosidase assay

3   Methods

The methods listed in this section describe: (1) PCR-introduced, random mutagenesis of LTA1, (2) co-transfection of mutated LTA1 and gapped pACT2 plasmids into YAB457 to yield yeast colonies expressing mutant LTA-AD and [Q71L]ARF3 fusion proteins, (3) assaying for loss of interaction via a colony β-galactosidase assay, (4) rescuing the plasmids expressing mutant LTA1, and (5) characterizing the loss-of-interaction LTA1 mutants. These methods begin after a two-hybrid interaction has been identified; e.g., between the protein products of the plasmids pXJ12 (AD-LTA1) and pAB157 ([Q71L]ARF3-BD).

3.1 Generation of Mutant Substrates for Plasmids

This method exploits the lack of proofreading by Taq polymerase and ability to replicate DNA with low fidelity and generate random changes into any region of DNA, in this case the LTA1-encoding portion of the AD-LTA1 plasmid (*see* Fig. 2 and **Note 2**).

3.1.1 Mutagenesis of LTA1

Mutagenesis of LTA1 is accomplished through PCR amplification of pXJ12 with primers #827 and #828 under conditions that promote polymerase-induced errors (*see* Fig. 2). These error-prone conditions include the addition of $MnCl_2$ and lowering the concentration of dATP, relative to other nucleotides, as described previously [31, 32]. The 50 μL PCR reaction consists of the following: 1.5 mM $MgCl_2$, 0.05 mM $MnCl_2$, 0.25 mM each of dCTP, dGTP, and TTP, 0.0625 mM dATP, 2 μM Primer #827, 2 μM Primer #828, 1 μL Taq Polymerase (0.05 U/μL), 25 ng pXJ12 in 1× PCR Buffer.

PCR amplification/mutagenesis is performed in 25 cycles using an annealing temperature of 55 °C (15 s) and an extension temperature of 72 °C (1 min for products <1 kb), with a denaturation step (94 °C, 15 s) before each cycle. Purify the product using a commercially available PCR cleanup kit according to manufacturer's directions. The resulting purified PCR product will contain randomly introduced mutations in the LTA1 coding region and approximately 80 bp of flanking pACT2-derived DNA upstream and downstream of the open reading frame. The yield from PCR under these conditions will be lower than normal but should produce plenty of product for later steps.

3.1.2 Gapped Plasmid (pACT2) Production

Create the gapped plasmid by digesting pXJ12 with NcoI and XhoI (*see* **Note 3**). Purify the resulting linearized, 8 kb vector backbone from a 1 % agarose gel using a commercially available gel extraction kit.

3.1.3 Estimating Vector and PCR Product Concentration

Estimate the concentration of the purified, gapped (pACT2-derived) plasmid and PCR products by comparing the ethidium bromide fluorescence intensity from each sample in a 1 % agarose gel to that of a known amount of a λ HindIII DNA marker of similar size.

3.2 Co-transformation of Gapped Plasmid and PCR Product

The insertion of the mutagenized open reading frame into the gapped plasmid can be accomplished by taking advantage of the high level of homologous recombination in yeast cells. Co-transformation of yeast with gapped plasmid and PCR products that contain regions of identity of ≥ 50 bp at each end is approximately 50 % as efficient as transformation with the circular plasmid alone (*see* Fig. 2 and **Note 4**).

1. Inoculate 50 mL of SD trp− with YAB457 and grow at 30 °C to OD_{600} ~ 0.6.

2. Harvest the cells by centrifugation at $1,000 \times g$ for 5 min.

3. Resuspend the cells in TE buffer and collect cells again by centrifugation.

4. Wash the cells in 0.1 M lithium acetate in TE and collect by centrifugation.

5. Resuspend the cells 0.5 mL of 0.1 M lithium acetate/TE.

6. Aliquot 100 µL of cell suspension into each of five sterile microfuge tubes.

7. Add 5 µL of freshly boiled sonicated herring sperm DNA and 0.7 mL 40 % PEG/LiAc/TE to each tube (single stranded carrier DNA improves transformation frequency).

8. Add the transforming DNA (gapped plasmid and PCR products) to each tube. Use equal molar amounts of gapped plasmid and PCR products. The amount of DNA will vary with the experimentally determined transformation frequency, designed to yield 400–600 transformants per 100 mm plate, but will probably be in the range of 0.1–1 µg per transformation (*see* **Notes 4** and **5**).

9. Incubate each transformation reaction at 30 °C for 30 min.

10. Transfer tubes to a 42 °C water bath and heat shock for 15 min.

11. Add 600 µL of SOS to each transformation tube.

12. Plate an appropriate volume (100 µL) of the transformed cell suspension on SD trp− leu− plates and incubate at 28 °C until small, distinct colonies became visible (approximately 3 days).

3.3 Assaying Transformants for Loss-of-Interaction

The X-gal filter assay is used to measure levels of β-galactosidase activity in yeast colonies as an indicator of protein interactions [33]. The two fusion proteins (often referred to as bait and prey) interact, thereby activating transcription of the lacZ gene and production of β-galactosidase, which catalyzes the hydrolysis of X-gal to yield a blue product. Both the time of incubation and the darkness of the blue product formed on the filter should be monitored and compared to the parent plasmid to allow initial determination of loss-of-interaction mutations.

1. Place a nitrocellulose filter on each plate for 30 s (or until wet) to replica the yeast colonies onto the filter.

2. Drop the filter into liquid nitrogen and carefully remove after 20 s. Avoid breaking the filters that will be brittle while frozen.

3. Transfer the frozen nitrocellulose filter, with the yeast side facing up, to the lid of a 100 mm dish on which has been placed Whatman filter paper wetted with X-gal in Z buffer (1.5 mL of Z-buffer with 15 µL of 100 mg/mL X-gal solution to yield a final X-gal concentration of 1 mg/mL).

4. Cover with the empty dish to prevent dehydration and incubate the filter at 30 °C. Compare the blueness of colonies that develop on the co-transformation plates with that of the control plates at different times. The time will vary with the bait–prey combination. Severe changes (white colonies) are often later found to result from the introduction of premature stop codons but can also produce a mutant with the most dramatic difference in binding. It is a good idea to initially select groups of colonies with small, intermediate, and large differences in activity.

5. Streak colonies from the experimental plates, which turned less blue or remained white in comparison to the (parental) positive control plate, onto SD trp− leu− plates to select for retention of both plasmids. After colonies appear, they may be re-screened using histidine auxotrophy for selection. We have found this to be a more sensitive assay than the X-gal filter assay.

3.4 Recovering Loss-of-Interaction Mutant Plasmids and Retesting

Recover plasmids from yeast using a modification of the standard glass bead method [34], also known as a "smash and grab" preparation. The modification simply adds the use of commercial plasmid purification columns to further remove contaminants that inhibit transformation. The use of plasmid purification columns is not required or used in all laboratories but we have found it increases success rates, sometimes markedly.

1. Streak colonies with diminished β-galactosidase activity onto SD leu− cycloheximide plates to select for loss of the [Q71L]ARF3-BD expressing plasmid. This step is not required but often facilitates the isolation of the desired (AD-LTA1 mutant) plasmid from strains carrying two different plasmids (*see* **Note 6**).

2. Incubate the plates for ≥3 days at 28 °C.

3. Inoculate 20 mL of SD leu− media with colonies from the SD leu− cycloheximide plates and grow each culture to an OD_{600} nm ~ 1.

4. Collect cells by centrifugation at $1{,}000 \times g$ for 10 min.

5. Wash each pellet in 1 mL of STES buffer and collect again.

6. Resuspend in 100 μL of STES and add an equal volume of glass beads (*see* **Note 7**).

7. Vortex each tube for 2 min.

8. Add 100 μL of STES and 200 μL phenol–chloroform and vortex for 5–30 min. Times will vary with the vortex used.

9. Pellet cells and debris and separate phases in a microfuge at maximal speed (approximately $14{,}000 \times g$) for 10 min, and transfer the aqueous phase to a new tube.

10. Extract with 200 μL of phenol–chloroform, and then with chloroform alone.

11. Further purify the plasmid in the aqueous phase using a commercial mini-prep plasmid prep, according to manufacturer's directions. Then use the product to transform competent bacterial (DH5α) cells.

12. Purify plasmids from transformed DH5α using a commercial plasmid mini-prep kit, according to manufacturer's instructions. Check by restriction digestion analysis to confirm the correct plasmid has been obtained.

13. Transform purified plasmids into YAB457 and assay transformants for interaction with [Q71L]ARF3 using the X-gal filter assay. The color developed at different times (15, 30, 60, 180 min) should be compared to that of the YAB457-derived strain expressing wild-type LTA1. Score activities visually on a three plus scale, with three plus equal to that of wild-type LTA1 (*see* **Note 8**).

3.5 Characterizing the Noninteracting Proteins

Multiple factors can be responsible for a loss of interaction, some of which are far less informative than a point mutation in a critical amino acid residue. Two types of uninteresting changes, premature stop codons and decreased levels of protein expression, can be screened for by immunoblotting cell lysates with an HA antibody to determine the presence and relative abundance of the epitope tag in the LTA1-AD fusion proteins. The levels of protein expressed in each strain expressing mutant LTA1 should be compared to wild-type LTA1 (*see* **Note 9**). Protein preparations from total yeast cell lysates can be obtained using a modification of Horvath et al. [35].

1. Grow five OD of each yeast strain in selective liquid medium.

2. Collect cells by centrifugation at $1,000 \times g$ for 5 min.

3. Wash cells with water and collect again at $1,000 \times g$ for 5 min.

4. Resuspend each pellet in 20 μL of 1× SDS sample buffer and boil at 95 °C for 5 min.

5. Add one cell volume of acid-washed glass beads and vortex 3 min at maximum speed.

6. Add 120 μL sample buffer and boil for 5 min.

7. Pellet cells and debris by centrifugation in a microfuge for 15 min at maximal setting. Transfer supernatant to another tube and boil for 5 min.

8. Load 7.5 μL of the sample on a 12 % SDS gel and resolve proteins using standard methods.

9. Transfer the resolved proteins to nitrocellulose at 60 V for 2 h and immunoblot with the 12CA5 (HA) antibody using standard methods.

If the mutant protein library was transfected into an ARF3-expressing strain, then you should detect two bands in immunoblots, at sizes corresponding to the AD-LTA1 and ARF3-BD fusion proteins. Y190-derived strains also express a protein of approximately 50 kDa that is bound by the 12CA5 antibody Y190 should be used as a control, at least initially. A shift to a smaller size or the absence of (or clearly reduced) immunoreactivity is suggestive of premature termination or alteration in the level of expression, respectively. Plasmids encoding full-length LTA1 mutants that were expressed at approximately the same level as wild type LTA1 should be sequenced with pACT2 sequencing primers #1259 and #1260 to determine any changes [23].

4   Notes

1. The two-hybrid system described here was that developed in the Elledge lab and uses a Gal4-based reporter system [29]. Y190 is a strain of *S. cerevisiae* engineered to express the *HIS3* and *lacZ* gene products as reporters, in response to the presence of a functional Gal4 transactivator. Y190 is able to detect protein-protein interactions by making two Gal4 hybrid proteins involving two different domains of Gal4: one protein is fused with the Gal4 Activation Domain to generate a Gal4-AD fusion protein, while the other protein is fused with the Gal4 DNA Binding Domain to yield a Gal4-BD fusion protein. If, when expressed in Y190, the two fusion proteins fail to interact, the yeast fail to activate transcription at the *GAL4* promoter and hence of the two reporters. However, if the two proteins do interact, a functional Gal4 transactivator is created. Thus, interaction can be observed by selecting and assaying for histidine prototrophy and β-galactosidase activity.

   Other two-hybrid systems have been described: that use other Gal4-based plasmids [36], that use more than two reporters [37], and that use other transcriptional activators (e.g., LexA; for reviews *see* refs. 38, 39). With only slight alterations, the steps outlined in this chapter can be applied to other such systems to perform a reverse two-hybrid screen.

2. The lack of proofreading by Taq polymerase facilitates the introduction of mutations, most often A-T and A-G mutations [25, 40]. Alternate DNA polymerases (e.g., Stratagene's Mutazyme) designed to introduce a broader variety of mutations can also be used.

3. Gapped plasmid was generated from pXJ12 (as opposed to empty pACT2) to ensure that any uncut plasmid present in the gapped plasmid isolation would encode a protein that would retain interaction with [Q71L]ARF3 and remain blue in the X-gal filter lift assay.

4. Transformation efficiency was determined by transforming YAB457 with undigested pXJ12. We estimated the efficiency for gapped plasmid repair in the co-transformation reaction to be half that of intact plasmid when transforming with an equimolar amount of gapped plasmid and PCR product. The volume of cell suspension resulting from the co-transformation reaction plated on each SD trp− leu− plate was adjusted to yield between 400 and 600 transformants per 100 mm plate.

5. Up to four control transformation reactions can be performed; including transformation of (1) uncut pXJ12 plasmid to determine general transformation efficiency, (2) no DNA, (3) the gapped plasmid alone or (4) the PCR product alone provide estimates of the background levels of transformations that will appear on the experimental co-transformation plates. Although not required, these controls are strongly encouraged the first time reverse two-hybrid techniques are used in a lab.

6. pAS1-CYH and pBG4D both contain the Cyh^S gene that causes sensitivity to cycloheximide. By growing cells containing pACT2 and pBG4D-based plasmids on cycloheximide-containing medium, cells not expressing pBG4D-based plasmids are selected and allow for easier isolation of pACT2-LTA1 plasmids.

7. This and the next step are critical and are the ones most likely to cause low recovery of plasmids. Too few beads and the cells simply rotate in the tube and too many beads and a slurry forms that also decreases shearing forces that produce cell lysis.

8. As an alternative to the filter lift assay, a liquid culture β-galactosidase activity assay can be used to generate a more quantifiable measure of activity [41]. Briefly, duplicate strains of yeast can be grown in 5 ml of selective medium overnight at 30 °C. The next day, 5 ml of fresh selective medium is added to 20–50 μL of the overnight culture and grown to mid log phase ($OD_{600} = 0.3–0.7$). Cells are collected by centrifugation and resuspended in 1 ml Z-Buffer and put on ice. Cells (50–100 μL) are then mixed with Z buffer to a volume of 1 ml, and a drop of 0.1 % SDS and two drops of chloroform are added to the sample with a Pasteur pipette. The mixtures are vortexed for 15 s and incubated for 15 min at 30 °C. O-nitrophenol α-D-galactopyranoside (ONPG, 0.8 mg) is added, the solution vortexed for 5 s, and incubated at 30 °C until a medium yellow color develops in the positive controls. The reaction is stopped by adding 0.5 ml of 1 M Na_2CO_3, and the time is recorded. OD_{420} and OD_{550} are determined after the cell debris is removed by centrifugation. Units of activity are determined using the equation:

$$U = 1,000(OD_{420} - OD_{550})/(T \times V \times OD_{600}),$$

where V is the volume of the culture used in the assay (μL), T is the time of the reaction (min), OD_{600} is the cell density at the start of the assay, OD_{420} represents the combination of absorbance by ONPG and the light scattering by cell debris, and OD_{550} represents the light scattering by cell debris.

9. In addition to providing the Gal4 activation domain, pACT2 also adds an HA-epitope after the Gal4-AD and before the protein of interest. Other plasmids encoding the Gal4 DNA binding domain, such as pBG4D or pAS1-CYH, also include the HA tag to their fusion protein products and therefore will also appear when blotting with an HA antibody. When comparing levels of protein expression by immunoblotting, it is necessary to determine the protein concentration of each sample and load equivalent protein onto the gel as differences in cell lysis and protein recovery can occur. Note that not all proteins are expressed in yeast to a level that can be detected by immunoblotting, but these proteins can still be used in two-hybrid and reverse two-hybrid methods.

References

1. Fields S, Song O (1989) A novel genetic system to detect protein-protein interactions. Nature 340:245–246

2. Chien CT, Bartel PL, Sternglanz R et al (1991) The two-hybrid system: a method to identify and clone genes for proteins that interact with a protein of interest. Proc Natl Acad Sci U S A 88:9578–9582

3. Boman AL, Zhang C, Zhu X et al (2000) A family of ADP-ribosylation factor effectors that can alter membrane transport through the trans-Golgi. Mol Biol Cell 11:1241–1255

4. Van Valkenburgh H, Shern JF, Sharer JD et al (2001) ADP-ribosylation factors (ARFs) and ARF-like 1 (ARL1) have both specific and shared effectors: characterizing ARL1-binding proteins. J Biol Chem 276:22826–22837

5. Ito T, Chiba T, Ozawa R et al (2001) A comprehensive two-hybrid analysis to explore the yeast protein interactome. Proc Natl Acad Sci U S A 98:4569–4574

6. Ito T, Tashiro K, Muta S et al (2000) Toward a protein-protein interaction map of the budding yeast: a comprehensive system to examine two-hybrid interactions in all possible combinations between the yeast proteins. Proc Natl Acad Sci U S A 97:1143–1147

7. Uetz P (2002) Two-hybrid arrays. Curr Opin Chem Biol 6:57–62

8. Uetz P, Giot L, Cagney G et al (2000) A comprehensive analysis of protein-protein interactions in *Saccharomyces cerevisiae* [see comments]. Nature 403:623–627

9. Maier RH, Maier CJ, Hintner H et al (2012) Quantitative real-time PCR as a sensitive protein-protein interaction quantification method and a partial solution for non-accessible autoactivator and false-negative molecule analysis in the yeast two-hybrid system. Methods 58:376–384

10. Pellet J, Meyniel L, Vidalain PO et al (2009) pISTil: a pipeline for yeast two-hybrid Interaction Sequence Tags identification and analysis. BMC Res Notes 2:220

11. Rajagopala SV, Uetz P (2009) Analysis of protein-protein interactions using array-based yeast two-hybrid screens. Methods Mol Biol 548:223–245

12. Ratushny V, Golemis E (2008) Resolving the network of cell signaling pathways using the evolving yeast two-hybrid system. Biotechniques 44:655–662

13. Cagney G, Uetz P (2001) High-throughput screening for protein-protein interactions using yeast two-hybrid arrays. Curr Protoc Protein Sci Chapter 19:Unit 19 16.

14. Parrish JR, Gulyas KD, Finley RL Jr (2006) Yeast two-hybrid contributions to interactome mapping. Curr Opin Biotechnol 17:387–393

15. Vidalain PO, Boxem M, Ge H et al (2004) Increasing specificity in high-throughput yeast two-hybrid experiments. Methods 32:363–370

16. Legrain P, Selig L (2000) Genome-wide protein interaction maps using two-hybrid systems. FEBS Lett 480:32–36

17. Walhout AJ, Boulton SJ, Vidal M (2000) Yeast two-hybrid systems and protein interaction mapping projects for yeast and worm. Yeast 17:88–94

18. Chappell TG, Gray PN (2008) Protein interactions: analysis using allele libraries. Adv Biochem Eng Biotechnol 110:47–66

19. Endoh H, Vincent S, Jacob Y et al (2002) Integrated version of reverse two-hybrid system for the postproteomic era. Methods Enzymol 350:525–545

20. Endoh H, Walhout AJ, Vidal M (2000) A green fluorescent protein-based reverse two-hybrid system: application to the characterization of large numbers of potential protein-protein interactions. Methods Enzymol 328:74–88

21. Vidal M, Endoh H (1999) Prospects for drug screening using the reverse two-hybrid system. Trends Biotechnol 17:374–381

22. Zhu X, Kahn RA (2001) The *Escherichia coli* heat labile toxin binds to Golgi membranes and alters Golgi and cell morphologies using ADP-ribosylation factor-dependent processes. J Biol Chem 276:25014–25021

23. Zhu X, Kim E, Boman AL et al (2001) ARF binds the C-terminal region of the Escherichia coli heat-labile toxin (LTA1) and competes for the binding of LTA2. Biochemistry 40:4560–4568

24. Das S, Kalpana GV (2009) Reverse two-hybrid screening to analyze protein-protein interaction of HIV-1 viral and cellular proteins. Methods Mol Biol 485:271–293

25. Cadwell RC, Joyce GF (1992) Randomization of genes by PCR mutagenesis. PCR Methods Appl 2:28–33

26. Leanna CA, Hannink M (1996) The reverse two-hybrid system: a genetic scheme for selection against specific protein/protein interactions. Nucleic Acids Res 24:3341–3347

27. Puthalakath H, Strasser A, Huang DC (2001) Rapid selection against truncation mutants in yeast reverse two-hybrid screens. Biotechniques 30:984–988

28. Vidal M, Brachmann RK, Fattaey A et al (1996) Reverse two-hybrid and one-hybrid systems to detect dissociation of protein-protein and DNA-protein interactions. Proc Natl Acad Sci U S A 93:10315–10320

29. Durfee T, Becherer K, Chen PL et al (1993) The retinoblastoma protein associates with the protein phosphatase type 1 catalytic subunit. Genes Dev 7:555–569

30. Boman AL, Kuai J, Zhu X et al (1999) Arf proteins bind to mitotic kinesin-like protein 1 (MKLP1) in a GTP-dependent fashion. Cell Motil Cytoskeleton 44:119–132

31. Kuai J, Kahn RA (2000) Residues forming a hydrophobic pocket in ARF3 are determinants of GDP dissociation and effector interactions. FEBS Lett 487:252–256

32. Muhlrad D, Hunter R, Parker R (1992) A rapid method for localized mutagenesis of yeast genes. Yeast 8:79–82

33. Bai C, Elledge SJ (1996) Gene identification using the yeast two-hybrid system. Methods Enzymol 273:331–347

34. Rose MD, Winston F, Hieter P (1990) In: Rose MD, Winston F, Hieter P (eds) Laboratory course manual for methods in yeast genetics. Cold Spring Harbor Laboratory, Cold Spring Harbor, NY

35. Horvath A, Riezman H (1994) Rapid protein extraction from *Saccharomyces cerevisiae*. Yeast 10:1305–1310

36. Bartel P, Chien CT, Sternglanz R et al (1993) Elimination of false positives that arise in using the two-hybrid system. Biotechniques 14:920–924

37. James P, Halladay J, Craig EA (1996) Genomic libraries and a host strain designed for highly efficient two-hybrid selection in yeast. Genetics 144:1425–1436

38. Brent R, Finley RL Jr (1997) Understanding gene and allele function with two-hybrid methods. Annu Rev Genet 31:663–704

39. Vidal M, Legrain P (1999) Yeast forward and reverse 'n'-hybrid systems. Nucleic Acids Res 27:919–929

40. Shafikhani S, Siegel RA, Ferrari E et al (1997) Generation of large libraries of random mutants in *Bacillus subtilis* by PCR-based plasmid multimerization. Biotechniques 23:304–310

41. Guarente L (1983) Yeast promoters and lacZ fusions designed to study expression of cloned genes in yeast. Methods Enzymol 101:181–191

Chapter 29

MAPPIT, a Mammalian Two-Hybrid Method for In-Cell Detection of Protein-Protein Interactions

Irma Lemmens, Sam Lievens, and Jan Tavernier

Abstract

MAPPIT (MAmmalian Protein-Protein Interaction Trap) is a two-hybrid technology that facilitates the detection and analysis of interactions between proteins in living mammalian cells. The system is based on type 1 cytokine receptor signaling. The bait protein of interest is fused to a chimeric signaling-deficient cytokine receptor, the signaling competence of which is restored upon recruitment of a prey protein that is coupled to a functional cytokine receptor domain. MAPPIT exhibits an excellent signal-to-noise ratio, detects a wide variety of protein-protein interactions (PPIs) including transient and indirect interactions, and has been shown to be highly complementary to other two-hybrid methods with respect to the interactions it can detect. Variants of the method were developed to allow large-scale PPI screening, mapping of protein interaction interfaces, PPI inhibitor screening and drug profiling. This chapter describes a basic 4-day MAPPIT protocol for the analysis of interaction between two designated proteins.

Key words MAPPIT, Two-hybrid, Protein-protein interaction, Cytokine receptor, Interactomics, Mammalian cells

1 Introduction

MAPPIT (MAmmalian Protein-Protein Interaction Trap) is a mammalian two-hybrid method that is based on type 1 cytokine receptor signaling. Type 1 cytokine receptors, such as the erythropoietin receptor (EpoR) and leptin receptor (LepR), lack intrinsic kinase activity but make use of constitutively associated Janus Kinases (JAK) to relay signals intracellularly. When the cytokine receptor is stimulated by its ligand, the associated JAKs are activated by cross-phosphorylation and subsequently phosphorylate tyrosine residues in the cytoplasmic tail of the receptor. These phosphorylated tyrosine motifs recruit STAT (Signal Transducer and Activator of Transcription) proteins that are in their turn phosphorylated by the activated JAKs, upon which the STAT molecules migrate to the nucleus and activate transcription of a selected set of target genes.

Cheryl L. Meyerkord and Haian Fu (eds.), *Protein-Protein Interactions: Methods and Applications*, Methods in Molecular Biology, vol. 1278, DOI 10.1007/978-1-4939-2425-7_29, © Springer Science+Business Media New York 2015

In the MAPPIT assay, a signaling-deficient chimeric type 1 cytokine receptor is used. Its extracellular domain is typically derived from the homodimeric EpoR and is fused to the transmembrane domain and cytoplasmic tail of the LepR. This chimeric receptor is made signaling-deficient by mutating the 3 tyrosines present in the cytoplasmic tail to phenylalanine. Upon ligand administration the associated JAKs of this chimeric receptor are activated, but since there are no tyrosines left in the receptor tail that can be phosphorylated, no STAT recruitment sites will be created and no signal is transmitted to the nucleus. The protein of interest that will serve as bait is C-terminally fused to this signaling-deficient receptor. The protein which one wants to test for interaction with the bait protein, referred to as the prey, is coupled to a portion of another cytokine receptor, the glycoprotein 130 receptor (gp130). The gp130 domain that is used contains tyrosine motifs that after phosphorylation by the JAKs serve as STAT3 recruitment sites. If the two proteins of interest (bait and prey) interact, the JAKs will be able to phosphorylate these tyrosines upon cytokine ligand activation. This generates functional STAT3 binding sites, and recruited STATs, upon activation by the JAKs will migrate to the nucleus to activate transcription of a luciferase reporter gene (Fig. 1) [1].

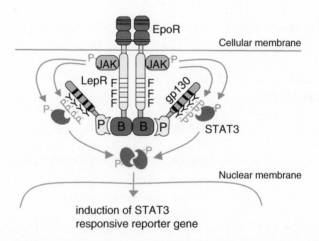

Fig. 1 Schematic representation of the MAPPIT principle. Interaction between bait and prey protein results in the recruitment of a gp130 fragment containing STAT3 recruitment sites, thereby complementing the signaling-deficient chimeric receptor bait. Activation of STAT3 can be monitored using the STAT3-responsive rat *Pap* reporter gene fused to luciferase (for details see text). EpoR, erythropoietin receptor; LepR, leptin receptor; JAK, Janus kinase; F, a tyrosine mutated to a phenylalanine; gp130, glycoprotein 130; STAT3, signal transducer and activator of transcription 3; B, bait; P, prey. Phosphorylation events are indicated by an *arrow* and *a light-colored P*

One of the main assets of MAPPIT is that the interaction occurs in intact mammalian cells, representing a near-physiological environment to study human proteins. The interaction (cytoplasm) and detector (nucleus) site are physically separated which reduces background problems (e.g., those caused by transcriptional auto-activators in the yeast two-hybrid system). Moreover, the fact that the readout is dependent on ligand administration provides an additional background control. MAPPIT has been successfully used for in-depth analyses of the protein interaction profiles of different type 1 cytokine receptors, such as the EpoR, LepR and growth hormone receptor, as well as for toll-like receptor signaling and for the detection of interactions with viral proteins [2–10]. These analyses have shown that the approach can detect transient as well as indirect interactions. Extensive benchmarking against other PPI methods indicated that MAPPIT is complementary to other two-hybrid methods with respect to the interactions it can detect, and is a valuable tool for large-scale protein-protein interactomics studies [11]. Downscaling the assay from a 96-well to a 384-well format using liquid handling robotics increases the throughput, allowing the parallel analysis of several thousands of interactions in order to validate the quality of various protein-protein interactomes, generated by a high-throughput yeast two-hybrid method [12–16]. This robotized platform is also applied for detailed characterization of protein-protein interaction interfaces in an unbiased manner, utilizing a random mutagenesis based strategy [17–19]. To use MAPPIT as a primary screening tool for the detection of novel interactions, ArrayMAPPIT was developed, which allows a bait protein to be screened against a large collection of preys (currently over 12.500) in only 4 days [20, 21]. Alternatively, a FACS-based cDNA library screening variant can also be used [22]. Owing to its robustness, the MAPPIT assay has also proven useful for high-throughput screening for small molecule disruptors of PPIs, which represent an emerging drug target class [23]. Finally, compound-protein interaction profiling can be performed using the MASPIT variant where a methotrexate fusion of a compound of interest is linked to a dihydrofolate reductase (DHFR) bait receptor [24]. For a more detailed overview of the different applications of the MAPPIT technology platform we refer to Lievens et al. [25].

Here we describe a basic MAPPIT protocol that enables detection of an interaction between two proteins of interest. In brief, this procedure involves the proteins of interest to be cloned in the appropriate MAPPIT bait and prey plasmids, co-transfecting these together with a luciferase reporter in mammalian (typically Hek293T) cells, stimulating the cells with the cytokine ligand, and lastly measuring luciferase reporter activity.

2 Materials

2.1 Plasmids

1. The pSEL(+2L) plasmid. The backbone of this plasmid is derived from pSV-SPORT, which carries an early SV40 promoter for low-level expression in mammalian cells and encodes the signaling-deficient chimeric receptor to which the bait should be C-terminally fused after a glycine–glycine–serine linker. Alternative plasmids can be used for cloning the bait (*see* **Note 1**, [1, 26]).

2. The pMG1 construct. The prey should be cloned C-terminally, after a glycine–glycine–serine hinge, in this plasmid that encodes an N-terminally FLAG-tagged part of the gp130 receptor (amino acids 760–918) using the strong SRα promoter of the pMET7 vector [1].

3. The reporter construct pXP2d2-rPAP1-luci. This reporter contains a STAT3-dependent promoter fragment derived from the rat Pancreatitis Associated Protein 1 (Pap) gene fused to luciferase [1].

2.2 Cell Culture

1. Human embryonic kidney (Hek) 293 T cell line.

2. Growth medium: Dulbecco's modified Eagle's medium (DMEM) supplemented with 10 % Fetal Calf Serum and antibiotics like gentamycin or penicillin/streptomycin.

3. Human Epo.

2.3 Transient $Ca_3(PO_4)_2$ Transfection of Hek293T Cells

1. 2.5 M $CaCl_2$. Prepare in distilled water. Filter-sterilize by passage through a 0.45 µM nitrocellulose membrane and store at −20 °C.

2. 2× Hepes-Buffered Saline (HeBS): 280 mM NaCl, 1.5 mM Na_2HPO_4, 50 mM HEPES. Adjust pH to 7.05 with NaOH. Filter-sterilize by passage through a 0.45 µM nitrocellulose membrane and store at −20 °C.

2.4 Luciferase Assay

1. Luciferase lysis buffer: 25 mM Tris-phosphate, pH 7.8, 2 mM DTT, 2 mM 2,2 diaminocyclohexane-N,N,N',N'-tetra-acetate (DCTA), 10 % glycerol, 1 % Triton X-100. Store at −20 °C.

2. Luciferase substrate buffer: 40 mM tricine, 2.14 mM $(MgCO_3)_4$ $Mg(OH)_2 \cdot 5H_2O$, 5.34 mM $MgSO_4$, 66.6 mM DTT, 0.2 mM EDTA, Coenzyme A, 734 µM Adenosine 5′ triphosphate, 940 µM D-luciferin. Store at −20 °C. The reagent is light sensitive.

3. Luminescence counter.

3 Methods

3.1 Cloning Bait and Prey of Interest

1. Design primers flanking your gene of interest that incorporate relevant restriction sites to clone the desired bait in the pSEL(+2L) plasmid and the prey in the pMG1 construct. Suitable restriction sites to clone the bait in the pSEL(+2L) vector are SalI or SacI in combination with NotI or XbaI. The prey can be cloned in the pMG1 vector using EcoRI combined with NotI or XbaI (*see* **Notes 1** and **2**). Both plasmids also exist in a gateway recombinatorial cloning compatible version, allowing the transfer of bait and prey using an LR clonase reaction (Invitrogen).

2. Perform a Polymerase Chain Reaction using a polymerase with proofreading activity (e.g., Pfu DNA polymerase).

3. Use traditional cloning methods or gateway recombination to insert the genes of interest into the appropriate vectors (Fig. 2).

4. Transform the resulting plasmids into a suitable bacterial line.

5. Prepare plasmid DNA of transfection-suitable quality, and validate the plasmids by restriction digest or sequencing.

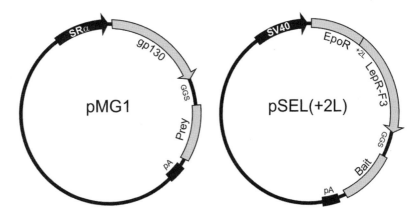

Fig. 2 Overview of the MAPPIT bait and prey plasmids. Plasmid pSEL(+2L) is used for cloning the bait; plasmid pMG1 to fuse the prey C-terminally to the gp130 moiety. A hinge (amino acids Gly–Gly–Ser) is placed between the fusions for extra flexibility. EpoR, erythropoietin receptor; +2L, two extra leucines were added in the transmembrane domain to optimize the fusion between the extracellular domain of the EpoR and the cytoplasmic tail of the LepR. LepRF3, leptin receptor variant with 3 tyrosines mutated to phenylalanines that lacks a functional STAT3 recruitment site; gp130, glycoprotein 130; SV40, simian virus 40 early promoter; SRα, promoter comprising SV40 and the R-U5 segment of human T-cell leukemia virus type 1 long terminal repeat; GGS, glycine–glycine–serine; pA, SV40 polyadenylation signal

3.2 Seeding Cells

1. For each bait–prey combination to be tested seed 1×10^4 subconfluent Hek293T cells in six wells of a black 96-well plate in 100 μl growth medium per well. Cell lines other than Hek293T cells can be used (*see* **Notes 3** and **4**).

    Besides testing the bait–prey interaction the following controls should be included: a bait–irrelevant prey and irrelevant bait–prey combination to control for a possible background caused by the bait or prey, respectively. Additionally, a prey (like SH2beta) that can interact with the bait receptor independent of the bait can serve as an additional control to check the signaling capacity of the bait receptor.

2. Grow overnight in a humidified atmosphere at 37 °C and 5–8 % CO_2.

3.3 Transient $Ca_3(PO4)_2$ Transfection

1. For each bait–prey combination, make a DNA and $CaCl_2$ mixture containing 250 ng of the bait plasmid, 500 ng of the prey plasmid, and 50 ng of the reporter (pXP2d2-rPAP1-luci) with 5 μl of sterile 2.5 M $CaCl_2$ in a total volume of 50 μl of distilled water in a 96-well plate (*see* **Notes 5** and **6**).

2. Add 50 μl of 2×HeBS to each well containing the DNA and $CaCl_2$ mixtures.

3. Shake the plate for 1 min at 800 rpm.

4. Resuspend the DNA, $CaCl_2$ and 2×HeBS mixture.

5. Add 10 μl to the cells in the 96-well plate in sextuple for every bait–prey combination.

6. Incubate the cells overnight in a humidified atmosphere at 37 °C and 5–8 % CO_2.

    The precipitate can be checked under a microscope: the particles should look like small speckles to obtain optimal transfection efficiencies (*see* **Note 7**). Alternative commercially available transfection reagents like lipofectamine can be used as well.

3.4 Stimulation of the Cells

1. For each bait–prey combination, stimulate three of the six wells with 50 μl Epo at a final concentration of 5 ng/μl. To the remaining three wells 50 μl of growth medium (without cytokine) should be added.

2. Incubate the cells overnight in a humidified atmosphere at 37 °C and 5–8 % CO_2.

3.5 Measurement of the Luciferase Activity

1. Remove growth medium from the black 96-well plates.

2. Add 50 μl of luciferase lysis buffer to each well and incubate for 10 min at room temperature.

3. Add 35 μl of luciferase substrate buffer to each well and immediately measure the luminescent signals using a chemiluminescence reader suitable for 96-well format.

3.6 Data Analysis

1. Calculate the fold induction value for each bait–prey combination by dividing the average of the values of the stimulated wells by the average of the values of the non-stimulated wells.

2. The bait–prey interaction is scored positive if the fold induction value obtained by the bait–prey combination is at least three times higher than both the fold induction value of the bait–irrelevant prey and irrelevant bait–prey interactions.

4 Notes

1. Alternative plasmids for cloning the bait can be used, such as the pCLL vector wherein the extracellular domain of the homodimeric EpoR receptor is replaced by the oligomeric LepR. Additionally, the cytoplasmic tail of the LepR after the JAK binding site can be replaced by glycine–glycine–serine repeats. For a more detailed description of these different plasmids, we refer to Lemmens et al. [26]. An alternative prey plasmid, named pMG2, can be used as well. In this plasmid, a part of the gp130 moiety (amino acids 905–918) is duplicated thereby adding two additional tyrosines [27]. Although this can increase the obtained signal, the background is often elevated as well.

2. When deciding which one of the protein pair of interest to be tested should be cloned as bait or prey, consider if it involves a protein that is targeted to a particular cellular organelle (like the nucleus or mitochondria). It is better to clone these genes as bait since detection of the interaction in MAPPIT takes place in the submembranal space of the cytoplasm. For a functional prey fusion, the targeting signals (such as a nuclear localization signal) should be omitted. Likewise, full-length integral transmembrane proteins will be problematic to study using MAPPIT.

3. Cells other than Hek293T cells (e.g., the hematopoietic TF-1 or neuronal N38 cell lines [2, 28]) can be used as long as they are transfectable and contain a sufficient amount of endogenous STAT3 molecules to transmit the signal to the nucleus. An alternative MAPPIT method wherein the gp130 moiety is replaced with a part of the beta common receptor chain, which carries STAT5 instead of STAT3 recruitment sites, can be more suitable for certain cell lines like the hematopoietic Ba/F3 cells which express little endogenous STAT3 but sufficient STAT5 [29].

4. It is important to use cells that are in a logarithmic growth phase (subconfluent), overgrown cultures won't transfect efficiently.

5. Different amounts of bait and prey DNA may yield better results in some settings.

6. If background caused by the prey is observed when testing the irrelevant bait–prey interaction, lowering the amount of prey DNA can reduce or diminish the background.

7. If no precipitation can be observed, check the pH of the 2×HeBS buffer, as this is critical for the formation of the precipitate.

Acknowledgements

pXP2d2 was a gift from Dr. S. Nordeen, Colorado Health Sciences Center, Department of Pathology, Denver, CO, 80262, USA. This work was supported by a grant from the Fund for Scientific Research-Flanders (FWO-V grant to I.L.).

References

1. Eyckerman S, Verhee A, Van der Heyden J et al (2001) Design and application of a cytokine-receptor-based interaction trap. Nat Cell Biol 3:1114–1119

2. Montoye T, Lemmens I, Catteeuw D et al (2005) A systematic scan of interactions with tyrosine motifs in the erythropoietin receptor using a mammalian 2-hybrid approach. Blood 105:4264–4271

3. Uyttendaele I, Lemmens I, Verhee A et al (2007) Mammalian protein-protein interaction trap (MAPPIT) analysis of STAT5, CIS, and SOCS2 interactions with the growth hormone receptor. Mol Endocrinol 21:2821–2831

4. Lavens D, Montoye T, Piessevaux J et al (2006) A complex interaction pattern of CIS and SOCS2 with the leptin receptor. J Cell Sci 119:2214–2224

5. Lavens D, Ulrichts P, Catteeuw D et al (2007) The C-terminus of CIS defines its interaction pattern. Biochem J 401:257–267

6. Piessevaux J, Lavens D, Montoye T et al (2006) Functional cross-modulation between SOCS proteins can stimulate cytokine signaling. J Biol Chem 281:32953–32966

7. Piessevaux J, De Ceuninck L, Catteeuw D et al (2008) Elongin B/C recruitment regulates substrate binding by CIS. J Biol Chem 283:21334–21346

8. Ulrichts P, Tavernier J (2008) MAPPIT analysis of early Toll-like receptor signalling events. Immunol Lett 116:141–148

9. Pattyn E, Lavens D, Van der Heyden J et al (2008) MAPPIT (MAmmalian Protein-Protein Interaction Trap) as a tool to study HIV reverse transcriptase dimerization in intact human cells. J Virol Methods 153:7–15

10. Ulrichts P, Peelman F, Beyaert R et al (2007) MAPPIT analysis of TLR adaptor complexes. FEBS Lett 581:629–636

11. Braun P, Tasan M, Dreze M et al (2009) An experimentally derived confidence score for binary protein-protein interactions. Nat Methods 6:91–97

12. Boxem M, Maliga Z, Klitgord N et al (2008) A protein domain-based interactome network for C. elegans early embryogenesis. Cell 134:534–545

13. Simonis N, Rual JF, Carvunis AR et al (2009) Empirically controlled mapping of the Caenorhabditis elegans protein-protein interactome network. Nat Methods 6:47–54

14. Simonis N, Rual JF, Lemmens I et al (2012) Host-pathogen interactome mapping for HTLV-1 and -2 retroviruses. Retrovirology 9:26

15. Venkatesan K, Rual JF, Vazquez A et al (2009) An empirical framework for binary interactome mapping. Nat Methods 6:83–90

16. Yu H, Braun P, Yildirim MA et al (2008) High-quality binary protein interaction map of the yeast interactome network. Science 322:104–110

17. Bovijn C, Ulrichts P, De Smet AS et al (2012) Identification of interaction sites for

dimerization and adapter recruitment in Toll/ interleukin-1 receptor (TIR) domain of Toll-like receptor 4. J Biol Chem 287:4088–4098

18. Uyttendaele I, Lavens D, Catteeuw D et al (2012) Random mutagenesis MAPPIT analysis identifies binding sites for Vif and Gag in both cytidine deaminase domains of Apobec3G. PLoS One 7:e44143

19. Bovijn C, De Smet AS, Uyttendaele I et al (2013) Identification of Binding Sites for Myeloid Differentiation Primary Response Gene 88 (MyD88) and Toll-like Receptor 4 in MyD88 Adapter-like (Mal). J Biol Chem 288:12054–12066

20. Lievens S, Vanderroost N, Defever D et al (2012) ArrayMAPPIT: a screening platform for human protein interactome analysis. Methods Mol Biol 812:283–294

21. Lievens S, Vanderroost N, Van der Heyden J et al (2009) Array MAPPIT: high-throughput interactome analysis in mammalian cells. J Proteome Res 8:877–886

22. Lievens S, Van der Heyden J, Vertenten E et al (2004) Design of a fluorescence-activated cell sorting-based Mammalian protein-protein interaction trap. Methods Mol Biol 263:293–310

23. Lievens S, Caligiuri M, Kley N et al (2012) The use of mammalian two-hybrid technologies for high-throughput drug screening. Methods 58:335–342

24. Caligiuri M, Molz L, Liu Q et al (2006) MASPIT: three-hybrid trap for quantitative proteome fingerprinting of small molecule-protein interactions in mammalian cells. Chem Biol 13:711–722

25. Lievens S, Peelman F, De Bosscher K et al (2011) MAPPIT: a protein interaction toolbox built on insights in cytokine receptor signaling. Cytokine Growth Factor Rev 22:321–329

26. Lemmens I, Lievens S, Tavernier J (2008) MAPPIT: a versatile tool to study cytokine receptor signalling. Biochem Soc Trans 36:1448–1451

27. Lemmens I, Eyckerman S, Zabeau L et al (2003) Heteromeric MAPPIT: a novel strategy to study modification-dependent protein-protein interactions in mammalian cells. Nucleic Acids Res 31:e75

28. Wauman J, De Smet AS, Catteeuw D et al (2008) Insulin receptor substrate 4 couples the leptin receptor to multiple signaling pathways. Mol Endocrinol 22:965–977

29. Montoye T, Piessevaux J, Lavens D et al (2006) Analysis of leptin signalling in hematopoietic cells using an adapted MAPPIT strategy. FEBS Lett 580:3301–3307

Chapter 30

Bioluminescence Resonance Energy Transfer to Detect Protein-Protein Interactions in Live Cells

Nicole E. Brown, Joe B. Blumer, and John R. Hepler

Abstract

Bioluminescence resonance energy transfer (BRET) is a valuable tool to detect protein-protein interactions. BRET utilizes bioluminescent and fluorescent protein tags with compatible emission and excitation properties, making it possible to examine resonance energy transfer when the tags are in close proximity (<10 nm) as a typical result of protein-protein interactions. Here we describe a protocol for detecting BRET from two known protein binding partners (Gαi1 and RGS14) in HEK 293 cells using *Renilla* luciferase and yellow fluorescent protein tags. We discuss the calculation of the acceptor/donor ratio as well as net BRET and demonstrate that BRET can be used as a platform to investigate the regulation of protein-protein interactions in live cells in real time.

Key words Bioluminescence resonance energy transfer (BRET), *Renilla* luciferase (RLuc), Yellow fluorescent protein (YFP), RGS14, G protein regulatory (GPR) motif, G protein

1 Introduction

Bioluminescence Resonance Energy Transfer (BRET) is a method of studying protein-protein interactions in live cells [1]. BRET utilizes non-radiative energy transfer between energy donor and energy acceptor protein tags. The energy transfer occurs when the protein tags are in close proximity, as described by the Förster distance [2]. As shown in Fig. 1, BRET serves as a molecular ruler, detecting protein-protein interactions under 10 nm (*see* **Note 1**). For a comprehensive review, *see* refs. [3, 4].

BRET makes use of a bioluminescent energy donor while the energy acceptor is a fluorophore. The choice of BRET pair is based on the overlap of the bioluminescent protein (donor) emission spectrum with the excitation spectrum of the fluorescent protein (acceptor). For BRET experiments, the most commonly chosen bioluminescent donor is luciferase from the sea pansy *Renilla reniformis*. *Renilla* luciferase (RLuc) catalyzes the oxidation of its substrate, coelenterazine, to produce blue light at 482 nm.

Cheryl L. Meyerkord and Haian Fu (eds.), *Protein-Protein Interactions: Methods and Applications*, Methods in Molecular Biology, vol. 1278, DOI 10.1007/978-1-4939-2425-7_30, © Springer Science+Business Media New York 2015

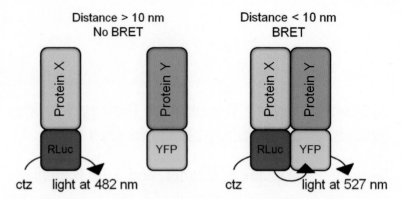

Fig. 1 BRET is dependent on the distance between the donor luciferase and the acceptor fluorophore. Addition of the cell permeant *Renilla* luciferase substrate coelenterazine (ctz) results in oxidation of the substrate to coelenteramide, which produces *blue light* at 482 nm. When protein-protein interactions between Protein X and Protein Y bring the donor luciferase (RLuc) and acceptor fluorophore (YFP) in close proximity (<10 nm), the energy from the donor can be transferred to the acceptor and light is produced at 527 nm. When the BRET tags are not in close enough proximity, light is only emitted at 482 nm

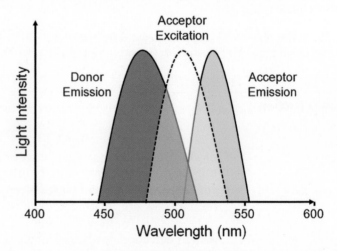

Fig. 2 The energy transfer between BRET pairs depends on the overlap of the donor emission spectrum with the excitation spectrum of the acceptor. For *Renilla* luciferase, oxidation of coelenterazine results in an emission peak at 482 nm. This emission overlaps well with the excitation spectrum of *yellow* fluorescent protein (excitation peak: 514 nm). The resulting energy transfer yields *yellow light* with an emission peak of 527 nm

The emission spectrum of RLuc overlaps well with the excitation spectra of the yellow fluorescent protein (YFP) family of proteins including the mutant YFP variants enhanced YFP (EYFP) and Venus [5] which emit light at ~527 nm (Fig. 2). For more information on BRET pairs, *see* **Note 2**.

BRET has distinct advantages over other techniques to detect protein-protein interactions. First, BRET is amenable to detecting interactions in live cells, thus proteins retain posttranslational modifications and cellular trafficking regulations that may be important for protein-protein interactions. BRET is readily adaptable to almost any cell type that allows expression of the donor and acceptor proteins. In live cells, protein-protein interactions can be monitored in real time over a time-course or for a fixed time interval in response to cellular treatments such as exposure to GPCR agonists, growth factors, or other drugs as an approach to define the regulation of protein complexes [6–12]. Additionally, fusion of BRET pairs to the same recombinant protein can be used to develop small molecule biosensors [13]. Methods have also been developed using BRET as a reporter for movement and subcellular location of target proteins [14]. Moreover, unlike the similar technique fluorescence resonance energy transfer (FRET), BRET does not require external excitation but instead relies on the addition of the cell permeant substrate coelenterazine to initiate the assay, thereby endowing the experimenter with temporal control over the assay and preventing unintentional activation of the acceptor fluorophore. Given these many advantages, BRET can be readily adapted for high throughput screening for small molecule modulators of protein-protein interactions. For review, *see* refs. [9, 15].

Below we describe a BRET experiment to explore the interactions between Regulator of G protein Signaling 14 (RGS14) and its binding partner Gαi1. RGS14 has previously been shown to interact with Gαi1 through its G protein regulatory (GPR) motif by traditional biochemical methods [16, 17]. We detail transfection of a C-terminal luciferase tagged RGS14 (RGS14-Luc) donor and internal YFP tagged Gαi1 (Gαi1-YFP) acceptor. We demonstrate a robust BRET signal between wild type RGS14 and Gαi1 that is disrupted with a mutant RGS14 (Q515A/R516A) that can no longer bind Gαi1. In our example, we show how to vary the acceptor protein expression level to achieve optimal net BRET signal. We describe how to calculate net BRET, acceptor/donor ratio, and fit the data using graphing software.

2   Materials

2.1   Cell Lines

1. Maintain HEK 293 cells in 1× Dulbecco's Modified Eagle Medium (DMEM) without phenol red indicator, supplemented with 2 mM L-glutamine, 100 U/mL penicillin, 100 mg/mL streptomycin, and 10 % fetal bovine serum (5 % for transfection). Grow cells in a humidified incubator with 5 % CO_2 at 37 °C.

2.2 Buffer Compositions/Stock Solutions

1. BRET buffer (Tyrode's solution): 140 mM NaCl, 5 mM KCl, 1 mM $MgCl_2$, 1 mM $CaCl_2$, 0.37 mM NaH_2PO_4, 24 mM $NaHCO_3$, 10 mM HEPES, pH 7.4, 0.1 % glucose.

2. Polyethylenimine (PEI) transfection reagent stock solution: Dissolve PEI (1 mg/mL) in dH_2O at 80 °C while stirring, cool and adjust to pH 7.2 using 0.1 N HCl and filter-sterilize. Aliquot and store at −80 °C. Use each aliquot only once.

3. Luciferase substrate stock solution: 2 mM benzyl coelenterazine H in 100 % ethanol containing 60 mM HCl, aliquot and store at −80 °C.

2.3 Instrumentation

1. Microplate reader with 485 nm emission and excitation filters, 530 nm emission filters, and compatible white-bottomed 96-well plates (*see* **Note 3**).

2. Compatible microplate reader, spreadsheet, and graphing software (*see* **Note 3**).

2.4 Plasmids

1. Donor plasmids can be constructed by inserting your gene of interest into a vector containing the humanized RLuc gene. For construction of RGS14-Luc constructs presented below, rat RGS14 cDNA was inserted into the phRLuc-N2 vector, which places the RLuc tag at the C-terminus of RGS14. Determining the optimal location of the Rluc tag relative to the protein of interest is an important parameter and must be determined empirically (*see* **Note 1**).

2. For many BRET experiments, acceptor plasmids can be constructed by inserting your gene of interest into a commercially available vector encoding YFP, EYFP or Venus. For construction of Gαi1-YFP used below, insertion of the YFP tag at either terminus compromised the function of Gαi1. Thus, the Gαi1-YFP construct was engineered by inserting the YFP coding sequence between the b and c helices of the helical domain of Gαi1 which was then expressed in a pcDNA3.1 vector [18].

3 Methods

3.1 Experimental Setup

1. Below we describe an experiment where the donor expression level is set and the acceptor expression level is varied. This experimental setup will allow the acceptor to saturate the donor and provide a maximal BRET signal. The expression level should be empirically determined; however, we typically use donor plasmid amounts that yield relative luminescence units (RLU) of 100,000–350,000 in our microplate reader (TriStar LB 941). In our example, 5 ng of phRLuc-N2:: RGS14 yields an RLU of ~100,000–300,000 but this will vary for other donor constructs depending on transfection

efficiency, the ability of the transfected cells to express the donor protein and the type of instrument used for detection. Additionally, on our microplate platform, we typically use acceptor plasmid amounts that yield relative fluorescence units (RFUs) of 30,000–200,000. In our example experiment, pcDNA3.1::Gαi1-YFP typically yields ~30,000–60,000 RFUs.

2. For the present experiment, 5 ng of the donor plasmid (RGS14-WT-Luc or RGS14-Q515A/R516A-Luc) was transfected with increasing amounts of acceptor (0, 10, 50, 100, 250, and 500 ng pcDNA3.1::Gαi1-YFP) (see **Note 4**).

3.2 Transient Transfection with Polyethylenimine (PEI) (See Note 5)

1. Seed 8×10^5 cells per well in six-well plates in 2 mL medium per well, grow in a humidified incubator at 37 °C overnight with 5 % CO_2.

2. Prior to transfection, change medium to $1\times$ DMEM containing 2 mM L-glutamine, 100 U/mL penicillin, 100 mg/mL streptomycin, and 5 % fetal bovine serum, 2 mL per well.

3. Generate *solution A* by adding 8 μL of 1 mg/mL PEI from stock to 92 μL of serum-free medium for each well, allow this solution to incubate for 3 min.

4. Generate *solution B* by adding up to 1.5 μg of DNA to 100 μL of serum-free medium for each condition in 1.5 mL microcentrifuge tubes. DNA amount is adjusted to a final concentration of 1.5 μg by adding empty pcDNA3.1 plasmid.

5. Add 100 μL of solution A to microcentrifuge tubes containing solution B to create *solution C.*

6. Cap the 1.5 mL microcentrifuge tube and immediately vortex for 3 s.

7. Incubate solution C at room temperature for 15 min.

8. Add solution C (~200 μL) dropwise to the appropriate well of cells in the six-well plates.

9. Allow cells to grow for 1–2 days (the medium does not need to be changed for PEI transfection).

3.3 BRET

1. Immediately prior to beginning the BRET experiment, prepare coelenterazine by diluting the stock solution to 50 μM in room temperature BRET buffer (see **Note 6**).

2. Aspirate the transfection medium from the six-well plates.

3. To each well, add 750 μL of room temperature BRET buffer, using a pipette to gently remove the cells from the plate.

4. Plate 90 μL of the cells in triplicate into white-bottomed 96-well plates.

5. Load plate into plate reader and detect fluorescence levels (excitation: 485 nm, emission: 530 nm) using microplate reader software (*see* **Note 7**).

6. Add 10 μL of coelenterazine solution to each well (5 μM final concentration of coelenterazine per well).

7. Incubate the cells with coelenterazine for 2 min at room temperature.

8. Take BRET readings by measuring luminescence at 485 ± 20 nm and fluorescence at 530 ± 20 nm (*see* **Note 8**).

3.4 Analysis

1. Export fluorescence and BRET data into spreadsheet software.

2. The BRET ratio can be determined by dividing fluorescence by luminescence (BRET readings at 530/485 nm).

3. Calculate net BRET by subtracting out background luminescence (BRET readings in cells expressing the donor without any acceptor).

4. Calculate the acceptor/donor ratio by dividing the initial fluorescence measurements (530 nm) by the luminescence measurements (485 nm).

5. Using graphing software, plot the acceptor/donor ratio against the net BRET as in Fig. 3. The data can then be fit using a nonlinear regression, (typically a one-site binding [hyperbola] is the most appropriate) to observe BRET saturation as a key indicator of signal specificity.

4 Notes

1. BRET efficiency (donor energy transfer to acceptor) is sensitive to the donor-acceptor proximity and is inversely proportional to the sixth power of distance between them. Thus, BRET signals generally reflect direct protein association; however, non-robust BRET signals can be detected due to close proximity without the occurrence of direct binding. For example, when a third intermediate protein brings the donor and acceptor into close proximity [10–12]. In addition to proximity, BRET also depends on the orientation of the protein tag dipoles. Inefficient dipole coupling can prevent energy transfer, despite protein-protein interactions. Thus, it is advantageous to use linkers (typically four or more Gly residues) inserted between the proteins of interest and the BRET tags to allow sufficient movement of the tags. Moreover, placement of tags must be considered when engineering recombinant proteins with BRET tags. Placement of the BRET tag at the N-terminus, internally, or at the C-terminus of the protein can have a profound impact on the observed BRET signal and

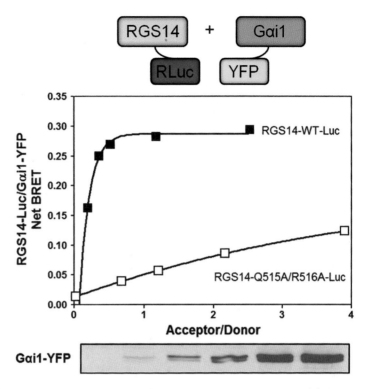

Fig. 3 HEK 293 cells were transfected with increasing amounts of Gαi1-YFP (0, 10, 50, 100, 250, and 500 ng) and either 5 ng RGS14-WT-Luc or RGS14-Q515A/R516A-Luc. Wild type RGS14 shows a robust BRET signal with Gαi1. Conversely, the RGS14 mutant (Q515A/R516A) that can no longer bind Gαi1 shows a drastically reduced maximal BRET signal indicating a disruption in the protein-protein interaction. The above data is representative of three independent experiments. *Curves* were generated with GraphPad Prism 5 using the one-site binding curve fitting function. Additionally, Gαi1-YFP expression levels were verified by immunoblot analysis

whether protein function is compromised. For example, placement of the luciferase tag at the N-terminus of RGS14 rather than the C-terminus results in a dramatic reduction in BRET signal with Gαi1-YFP. As another example, the acceptor YFP tag in Gαi1 cannot be placed at either termini without affecting protein function, and was placed internally between the b and c helices of the alpha helical domain with minimal consequences to Gαi1 function [18]. Due to these considerations, two interacting proteins may not always be detected by BRET due to the distance between the tags or abnormal protein function or localization through improper tag placement.

2. Additional BRET pairs and BRET substrates have been developed. Many of these BRET pairs can be used with the method

described above. For a comprehensive review of other BRET pairs and substrates, *see* ref. [19].

3. We use the TriStar LB 941 Multimode Microplate Reader (Berthold Technologies) for our BRET experiments; however, other plate readers can be used with similar results. The plate reader must detect light signals at two distinct wavelengths either simultaneously or sequentially. In our assay, we use Berthold Technologies filters at 485 nm to measure luminescence and at 530 nm to measure fluorescence though similar filters can be purchased from other vendors. To collect BRET data, we use the MikroWin 2000 software (Mikrotek). MikroWin is specialized for microplate experiments and optimized to run with a variety of instruments from various manufacturers. Additionally, data collected in MikroWin can be exported and further analyzed in spreadsheet software such as Microsoft Excel. Data can then be graphed in graphing software such as GraphPad Prism.

4. In order to calculate net BRET, it is necessary to include a donor-only control (donor transfected without any acceptor). The donor-only control is used to assess any background BRET observed in the absence of acceptor. BRET from donor-only controls are subtracted from observed BRET values to calculate the net BRET.

5. Other transfection reagents can be used with similar results. We choose to use PEI as it yields high transfection efficiency and reproducibility at very affordable cost.

6. Coelenterazine is light sensitive and should be kept away from light exposure until ready to use. Dilute coelenterazine stock immediately before performing BRET to prevent breakdown of the substrate.

7. Initial fluorescence levels are taken to determine the acceptor/donor ratio as well as an internal control to verify expression of the fluorescently tagged protein. For experiments where the amount of donor is held constant and the amount of acceptor is increased, corresponding increases of the acceptor should be observed in the fluorescence measurement.

8. As stated above, for detection on the TriStar platform, ideal luminescence should be about 100,000–350,000 relative luminescence units (RLUs). Ideal fluorescence should range between 30,000 and 200,000 relative fluorescence units (RFUs). As a corollary, typical acceptor/donor ratios range between 0 and 15; however, this number will vary depending on transfection efficiency and the expression level of the acceptor. In addition, lowering the level of donor expression will also inflate the acceptor/donor ratio.

References

1. Xu Y, Piston DW, Johnson CH (1999) A bioluminescence resonance energy transfer (BRET) system: application to interacting circadian clock proteins. Proc Natl Acad Sci U S A 96:151–156

2. Wu P, Brand L (1994) Resonance energy transfer: methods and applications. Anal Biochem 218:1–13

3. Xu Y, Kanauchi A, von Arnim AG et al (2003) Bioluminescence resonance energy transfer: monitoring protein-protein interactions in living cells. Methods Enzymol 360:289–301

4. Lohse MJ, Nuber S, Hoffmann C (2012) Fluorescence/bioluminescence resonance energy transfer techniques to study G-protein-coupled receptor activation and signaling. Pharmacol Rev 64:299–336

5. Nagai T, Ibata K, Park ES et al (2002) A variant of yellow fluorescent protein with fast and efficient maturation for cell-biological applications. Nat Biotechnol 20:87–90

6. Romero-Fernandez W, Borroto-Escuela DO, Tarakanov AO et al (2011) Agonist-induced formation of FGFR1 homodimers and signaling differ among members of the FGF family. Biochem Biophys Res Commun 409:764–768

7. Gales C, Rebois RV, Hogue M et al (2005) Real-time monitoring of receptor and G-protein interactions in living cells. Nat Methods 2:177–184

8. Angers S, Salahpour A, Joly E et al (2000) Detection of beta 2-adrenergic receptor dimerization in living cells using bioluminescence resonance energy transfer (BRET). Proc Natl Acad Sci U S A 97:3684–3689

9. Hamdan FF, Audet M, Garneau P et al (2005) High-throughput screening of G protein-coupled receptor antagonists using a bioluminescence resonance energy transfer 1-based beta-arrestin2 recruitment assay. J Biomol Screen 10:463–475

10. Vellano CP, Maher EM, Hepler JR et al (2011) G protein-coupled receptors and resistance to inhibitors of cholinesterase-8A (Ric-8A) both regulate the regulator of g protein signaling 14 RGS14.Galphai1 complex in live cells. J Biol Chem 286:38659–38669

11. Oner SS, Maher EM, Breton B et al (2010) Receptor-regulated interaction of activator of G-protein signaling-4 and Galphai. J Biol Chem 285:20588–20594

12. Oner SS, An N, Vural A et al (2010) Regulation of the AGS3.G{alpha}i signaling complex by a seven-transmembrane span receptor. J Biol Chem 285:33949–33958

13. Jiang LI, Collins J, Davis R et al (2007) Use of a cAMP BRET sensor to characterize a novel regulation of cAMP by the sphingosine 1-phosphate/G13 pathway. J Biol Chem 282:10576–10584

14. Lan TH, Liu Q, Li C et al (2012) Sensitive and high resolution localization and tracking of membrane proteins in live cells with BRET. Traffic 13:1450–1456

15. Couturier C, Deprez B (2012) Setting Up a Bioluminescence Resonance Energy Transfer High throughput Screening Assay to Search for Protein/Protein Interaction Inhibitors in Mammalian Cells. Front Endocrinol 3:100

16. Hollinger S, Taylor JB, Goldman EH et al (2001) RGS14 is a bifunctional regulator of Galphai/o activity that exists in multiple populations in brain. J Neurochem 79:941–949

17. Kimple RJ, De Vries L, Tronchere H et al (2001) RGS12 and RGS14 GoLoco motifs are G alpha(i) interaction sites with guanine nucleotide dissociation inhibitor Activity. J Biol Chem 276:29275–29281

18. Gibson SK, Gilman AG (2006) Gialpha and Gbeta subunits both define selectivity of G protein activation by alpha2-adrenergic receptors. Proc Natl Acad Sci U S A 103:212–217

19. Bacart J, Corbel C, Jockers R et al (2008) The BRET technology and its application to screening assays. Biotechnol J 3:311–324

Chapter 31

Mapping Biochemical Networks with Protein Fragment Complementation Assays

Ingrid Remy and Stephen W. Michnick

Abstract

Cellular biochemical machineries, what we call pathways, consist of dynamically assembling and disassembling macromolecular complexes. Although our models for the organization of biochemical machines are derived largely from in vitro experiments, do they reflect their organization in intact, living cells? We have developed a general experimental strategy that addresses this question by allowing the quantitative probing of molecular interactions in intact, living cells. The experimental strategy is based on Protein fragment Complementation Assays (PCA), a method whereby protein interactions are coupled to refolding of enzymes from cognate fragments where reconstitution of enzyme activity acts as the detector of a protein interaction. A biochemical machine or pathway is defined by grouping interacting proteins into those that are perturbed in the same way by common factors (hormones, metabolites, enzyme inhibitors, etc.). In this chapter we review some of the essential principles of PCA and provide details and protocols for applications of PCA, particularly in mammalian cells, based on three PCA reporters, dihydrofolate reductase, green fluorescent protein, and β-lactamase.

Key words Protein fragment Complementation Assays, Dihydrofolate reductase, Green fluorescent protein, TEM β-lactamase, Two-hybrid, Protein-protein interactions, Methotrexate, CCF2/AM, Nitrocefin, Fluorescein, Flow cytometry, CHO, COS, HEK 293 cells

1 Introduction

A first step in defining the function of a novel gene is to determine its interactions with other gene products in an appropriate context; that is, since proteins make specific interactions with other proteins as part of functional assemblies, an appropriate way to examine the function of the product of a novel gene is to determine its physical relationships with the products of other genes. This is the basis of the highly successful Yeast Two-Hybrid system [1–6]. The central problem with two-hybrid screening is that detection of protein-protein interactions occurs in a fixed context, the nucleus of *S. cerevisiae*, and the results of a screening must be validated as biologically relevant using other assays in appropriate cell, tissue, or organism models. Although this would be true for any screening

Cheryl L. Meyerkord and Haian Fu (eds.), *Protein-Protein Interactions: Methods and Applications*, Methods in Molecular Biology, vol. 1278, DOI 10.1007/978-1-4939-2425-7_31, © Springer Science+Business Media New York 2015

strategy, it would be advantageous if one could combine library screening with tests for biological relevance into a single strategy, thus tentatively validating a detected protein as biologically relevant and eliminating false-positive interactions immediately. It was with these challenges in mind that our laboratory developed the Protein-fragment Complementation Assays (PCA). In this strategy, the gene for an enzyme is rationally dissected into two pieces. Fusion proteins are constructed with two proteins that are thought to bind to each other, fused to either of the two probe fragments. Folding of the probe protein from its fragments is catalyzed by the binding of the test proteins to each other, and is detected as reconstitution of enzyme activity. We have already demonstrated that the PCA strategy has the following capabilities: (1) allows for the detection of protein-protein interactions in vivo and in vitro in any cell type; (2) allows for the detection of protein-protein interactions in appropriate subcellular compartments or organelles; (3) allows for the detection of induced versus constitutive protein-protein interactions that occur in developmental, nutritional, environmental, or hormone-induced signals; (4) allows for the detection of the kinetic and equilibrium aspects of protein assembly in these cells; (5) allows for screening of cDNA libraries for protein-protein interactions in any cell type.

In addition to the specific capabilities of PCA described above, there are special features of this approach that make it appropriate for screening of molecular interactions, including the following: (1) PCAs are not a single assay but a series of assays; an assay can be chosen because it works in a specific cell type appropriate for studying interactions of some class of proteins; (2) PCAs are inexpensive, requiring no specialized reagents beyond that necessary for a particular assay and off the shelf materials and technology; (3) PCAs can be automated and high-throughput screening could be done; (4) PCAs are designed at the level of the atomic structure of the enzymes used; because of this, there is additional flexibility in designing the probe fragments to control the sensitivity and stringencies of the assays; (5) PCAs can be based on enzymes for which the detection of protein-protein interactions can be determined differently including by dominant selection or production of a fluorescent or colored product.

The selection of enzymes and design of PCAs have been discussed in detail [7–19] and here we review only the most basic ideas. Polypeptides have evolved to code for all of the chemical information necessary to spontaneously fold into a stable, unique three-dimensional structure [20–22]. It logically follows that the folding reaction can be driven by the interaction of two peptides that together contain the entire sequence, and in the correct order of a single peptide that will fold. This was demonstrated in the classic experiments of Richards [23] and Taniuchi and Anfinsen [24]. In practice this does not easily work since the major driving

a Interaction-directed folding
from protein fragments (PCA)

b Weakly associating
subunits

Fig. 1 Two alternative strategies to achieve complementation. (**a**) The PCA strategy requires that unnatural peptide fragments be chosen that are unfolded prior to association of fused interacting proteins. This prevents spontaneous association of the fragments (pathway X) that can lead to a false signal. (**b**) Naturally occurring subunits that are already capable of folding can be mutated to interact with lower affinity. However, to some extent, this will always occur, requiring the selection of cells that express protein partner fusions at low enough levels that background is not detected

force for protein folding is the hydrophobic effect, so also is nonspecific aggregation of unfolded peptides. However, if one adds soluble interacting proteins to the fragments that by interacting increase the effective concentration of the fragments, correct folding could be favored over any other non-productive process [25–27]. If the protein that folds from its constitutive fragments is an enzyme, whose activity could be detected in vivo, then the reconstitution of its activity can be used as a measure of interaction of the interacting proteins (Fig. 1a). Furthermore, this binary, all or none folding event, provides for a very specific measure of protein interactions dependent on not mere proximity, but the absolute requirement that the peptides must be organized precisely in space to allow for folding of the enzyme from the cognate fragments. We select proteins to dissect into fragments that are not capable of spontaneously folding from their complementary fragment into a functional and complete protein. These facts distinguish the PCA strategy from complementation of naturally occurring and weakly associating subunits of enzymes [28], in which some spontaneous assembly occurs, as illustrated in Fig. 1b.

A number of PCAs have been developed based on dominant-selection, colorimetric, fluorescent, or luminescent outputs. In this chapter, we discuss three different PCAs, based on the enzymes murine dihydrofolate reductase (mDHFR) and TEM ß-lactamase, and also provide protocols for a PCA based on green fluorescent protein (GFP).

The DHFR PCA can be used in a variety of applications to perform both simple survival-selection as readout and simultaneously, a fluorescent assay allowing quantitative detection and the cellular localization of protein interactions can be performed [29–33]. The ß-lactamase assay can be used as a very sensitive in vivo or in vitro quantitative detector of protein interactions as, unlike DHFR, one measures the continuous conversion of substrate to colored or fluorescent product [34, 35]. However it should be noted that generation of a product by an enzyme does not guarantee that signal to background would be superior to that of fixed fluorophore reporters like GFP and fluorescein-conjugated methotrexate (fMTX) bound to DHFR. Observable signal to background depends, for example, on the quantum yield of the fluorophore, retention of fluorophore by a cell, the optical properties of the cells used, and the extent to which fluorophores are retained in individual cellular compartments. For instance, in spite of no enzymatic amplification, the DHFR fluorescence assay requires between only 1,000 and 3,000 molecules of reconstituted DHFR to clearly distinguish a positive response from background.

Reconstitution of DHFR activity can be monitored in vivo by cell survival in DHFR-negative cells (CHO-DUKX-B11, for example) grown in the absence of nucleotides. The principle of the DHFR PCA survival assay is that cells simultaneously expressing complementary fragments of DHFR fused to interacting proteins or peptides will survive in media depleted of nucleotides. This is an extraordinarily sensitive assay. In mammalian cells, survival is dependent only on the number of molecules of DHFR reassembled, and we have determined that this number is approximately 25 molecules of DHFR per cell [29]. The second approach is a fluorescence assay based on the detection of fMTX binding to reconstituted DHFR. The basis of the DHFR PCA fluorescence assay is that complementary fragments of DHFR, when expressed and reassembled in cells, will bind with high affinity ($K_d = 540$ pM) to fMTX in a 1:1 complex. fMTX is retained in cells by this complex, while the unbound fMTX is actively and rapidly transported out of the cells [29, 36, 37]. In addition, binding of fMTX to DHFR results in a 4.5-fold increase in quantum yield. Bound fMTX, and by inference reconstituted DHFR, can then be monitored by fluorescence microscopy, fluorescence-activated cell sorting (FACS), or spectroscopy [29–31]. It is important to note that, although fMTX binds to DHFR with high affinity, it does not induce DHFR folding from the fragments in the PCA. This is because the folding of DHFR from its fragments is obligatory; if binding of the oligomerization domains does not induce folding, no binding sites for fMTX are created. Therefore, the number of complexes observed as measured by number of fMTX molecules retained in the cell is a direct measure of the equilibrium number of oligomerization domain complexes formed,

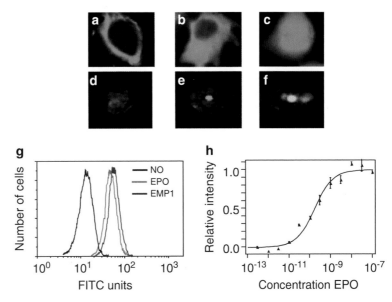

Fig. 2 Applications of the DHFR PCA to detecting the localization of protein complexes and quantitating protein interactions. (**a–c**) different protein pairs showing (**a**), plasma membrane, (**b**), cytosol, and (**c**), whole cell localization in transiently transfected COS cells. (**d–f**), localization of a protein complex in plant protoplasts. Potato protoplasts expressing two proteins implicated in response to salicylic acid (SA) fused to DHFR fragments. These are *NPR1/NIM1-DHFR (F[1,2])* and *TGA2-DHFR (F[3])* examined by fluorescent microscopy in the presence of fMTX and DAPI (4,6-diamidino-2-phenylindole). (**d**), A protoplast that has not been treated with SA or (**e**), treated with SA shows that that complex is induced to relocalize from cytosol to nucleus by SA. (**f**) Nuclear counterstaining with DAPI in the same protoplast. (**g**), FACS results of DHFR PCA. CHO cells expressing the erythropoietin receptor fused to complementary DHFR fragments. Receptor activation (conformation change) induced by erythropoietin (EPO) or a peptide agonist (EMP1) lead to an increase in fluorescence. (**h**) Dose–response curve for Epo-induced fluorescence as detected by FACS results in **g**

independent of binding of fMTX [29]. The other obvious application of the DHFR PCA fluorescence assay is in determining the location in the cell of interactions as illustrated in a number of cell types (Fig. 2). The GFP assay can be used for this purpose as well, but has the distinct advantage that no additional fluorophore is necessary to do this assay. However, readers should be cautioned that this assay is only appropriate for high-affinity, very stable complexes and applications to studying transient assembly and disassembly of protein complexes is very limited, due to the slow folding of the protein and maturation of the fluorophore.

ß-Lactamase is strictly a bacterial enzyme and has been genetically deleted from many standard *E. coli* strains. It is not present at all in eukaryotes. Thus, the ß-lactamase PCA can be used

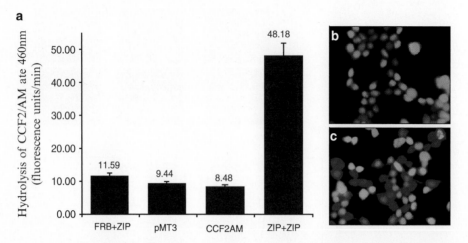

Fig. 3 ß-Lactamase PCA using the fluorescent substrate CCF2/AM. (**a**) ZIP (GCN4 leucine zipper-forming sequences) are tested in HEK 293 cells as described in the text. FRB (rapamycin-FKBP binding domain of FRAP) is used as a negative control. pMT3 is the expression vector alone and ZIP + ZIP is the positive control. Data recorded in white microtiter plates on a Perkin Elmer HTS 7000 plate reader. (**b**, **c**) Fluorescent micrographs of cells expressing ß-lactamase PCA showing negative (**b**, FRB + ZIP) or positive (**c**, ZIP + ZIP) response

universally in eukaryotic cells and many prokaryotes, without any intrinsic background. Also, assays are based on catalytic turnover of substrates with rapid accumulation of product. This enzymatic amplification should allow for relatively weak molecular interactions to be observed. The assay can be performed simultaneously or serially in a number of modes, such as the in vitro colorimetric assay or the in vivo fluorescence assay (Fig. 3) or the survival assay in bacteria. Assays can be performed independent of the measurement platform and can easily be adapted to high-throughput formats requiring only one pipetting step.

2   Materials

2.1   DHFR PCA Survival Assay

1. 12-well plates, tissue culture treated; six-well plates, tissue culture treated.
2. Minimum essential medium: alpha medium without ribonucleosides and deoxyribonucleosides (α-MEM).
3. Dialyzed fetal bovine serum (Hyclone, Cat. no: SH30079-03).
4. Adenosine; deoxyadenosine; thymidine.
5. Transfection reagent (e.g., Lipofectamine Plus reagent, Life Technologies).
6. Trypsin–EDTA.
7. Cloning cylinders.

2.2 DHFR PCA Fluorescence Assay

1. 12-well plates, tissue culture treated.
2. Dulbecco's modified Eagle medium (DMEM); minimum essential medium: alpha medium without ribonucleosides and deoxyribonucleosides (α-MEM).
3. Cosmic calf serum; dialyzed fetal bovine serum.
4. Transfection reagent (e.g., Lipofectamine Plus reagent, Life Technologies).
5. Fluorescein-conjugated methotrexate (fMTX) (Molecular Probes).
6. Dulbecco's Phosphate-Buffered Saline (PBS).
7. Aqueous mounting medium.
8. Trypsin–EDTA.
9. Micro cover glasses, 18 mm circles, No. 2.
10. Microscope slides, glass, $25 \times 75 \times 1.0$ mm.
11. 96-well white microtiter plates (Dynex no 7905).
12. Protein quantification reagents (e.g., Bio-Rad protein assay).

2.3 GFP PCA Fluorescence Assay

1. 12-well plates, tissue culture treated.
2. Dulbecco's modified Eagle medium (DMEM).
3. Cosmic calf serum.
4. Transfection reagent (e.g., Lipofectamine Plus reagent, Life Technologies).
5. Dulbecco's Phosphate-Buffered Saline (PBS).
6. Aqueous mounting medium.
7. Trypsin–EDTA.
8. Micro cover glasses, 18 mm circles, No 2.
9. Microscope slides, glass, $25 \times 75 \times 1.0$ mm.
10. 96-well black microtiter plates (Dynex no 7805).
11. Protein quantification reagents (e.g., Bio-Rad protein assay).

2.4 β-Lactamase PCA Colorimetric Assay

1. 12-well plates, tissue culture treated.
2. Dulbecco's modified Eagle medium (DMEM).
3. Cosmic calf serum.
4. Transfection reagent (e.g., FuGENE 6).
5. Trypsin–EDTA.
6. 100 mM Phosphate Buffer.
7. Nitrocefin.
8. 96-well plates.

**2.5 β-Lactamase
PCA Fluorometric
Assay**

1. 12-well plates, tissue culture treated.

2. Dulbecco's modified Eagle medium (DMEM).

3. Cosmic calf serum.

4. Transfection reagent (e.g., FuGENE).

5. Trypsin–EDTA.

6. Dulbecco's Phosphate-Buffered Saline (PBS).

7. CCF2-AM (kindly provided by Roger Tsien).

8. 96-well white microtiter plates (Dynex no 7905).

9. Normal saline: 140 mM, NaCl, 5 mM KCl, 2 mM $CaCl_2$, 10 mM Hepes, 6 mM sucrose, 10 mM glucose, pH 7.35.

10. Physiological saline solution: 10 mM HEPES, 6 mM sucrose, 10 mM glucose, 140 mM NaCl, 5 mM KCl, 2 mM $MgCl_2$, 2 mM $CaCl_2$, pH 7.35.

11. 15 mm glass coverslip.

3 Methods

**3.1 DHFR PCA
Survival Assay**

1. Twenty-four hours before transfection, plate 1×10^5 CHO DUKX-B11 cells (DHFR-negative; could also be done in other cells lines, *see* **Note 1**) in 12-well plates in α-MEM medium enriched with 10 % dialyzed fetal bovine serum and supplemented with 10 μg/ml of adenosine, deoxyadenosine, and thymidine.

2. Co-transfect cells with the PCA fusion partners (*see* **Note 2**) using a transfection reagent (e.g., Lipofectamine Plus reagent) according to the manufacturer's instructions.

3. Forty-eight hours after the beginning of the transfection, split cells at approximately 5×10^4 in 6-well plates in selective medium consisting of α-MEM enriched with dialyzed FBS but without addition of nucleotides (*see* **Notes 3** and **4**).

4. Change medium every 3 days. The appearance of distinct colonies usually occurs after 4–10 days of incubation in selective medium. Colonies are observed only for clones that simultaneously express both interacting proteins fused to one or the other complementary DHFR fragments. Only interacting proteins will be able to achieve normal cell division and colony formation.

   For further analysis of the interacting protein pair:

5. Isolate three to five colonies per interacting partners by trypsinization (trypsin–EDTA) using cloning cylinders and grow them separately.

6. Select the best expressing clone by immunoblot (Western blot) or using the DHFR PCA fluorescence assay (*see* Subheading 3.2). Amplification of the expressed gene using methotrexate resistance can be done afterwards if desired, to obtain clones with increased expression [38].

7. Carry out functional analysis of the clone stably expressing the interacting proteins pair fused to the complementary DHFR fragments by using the DHFR PCA fluorescence assay.

3.2   DHFR PCA Fluorescence Assay

3.2.1   Fluorescence Microscopy

1. Twenty-four hours before transfection, plate 1×10^5 COS cells (this assay can be performed using any other cell line, *see* **Note 5**) on 18 mm circular glass coverslips in 12-well plates in DMEM medium enriched with 10 % Cosmic calf serum.

2. Transiently co-transfect cells with the PCA fusion partners (*see* **Note 2**) using a transfection reagent (e.g., Lipofectamine Plus reagent) according to the manufacturer's instructions.

3. The next day, change medium and add fluorescein-conjugated methotrexate (fMTX) to the cells at a final concentration of 10 μM (*see* **Note 6**).

   For stable cell lines:

   For CHO DUKX-B11 cells (or other cell line) stably expressing PCA fusion partners, seed cells to approximately 2×10^5 on 18 mm glass coverslips in 12-well plates in α-MEM medium enriched with 10 % dialyzed FBS. The next day, fMTX is added to the cells at a final concentration of 10 μM.

4. After incubation with fMTX for 22 h at 37 °C, remove the medium and wash the cells with PBS and re-incubate for 15–20 min at 37 °C in the culture medium to allow for efflux of unbound fMTX (*see* **Note 7**). Remove medium and wash the cells four times with cold PBS on ice and finally mount the coverslips on microscope glass slides with an aqueous mounting medium.

5. Perform fluorescence microscopy on live cells (*see* **Note 8**). These experiments must be performed within 30 min of the wash procedure. If the negative control (untransfected cells treated with fMTX) is too fluorescent, the wash procedure must be modified (*see* **Note 9**).

3.2.2   Flow Cytometry Analysis

Preparation of cells for fluorescence-activated cell sorting (FACS) analysis is the same as described for fluorescence microscopy, except that following the PBS wash (twice in this case), cells are gently trypsinized (trypsin–EDTA), suspended in 500 μl of cold PBS and kept on ice prior to flow cytometric analysis within 30 min. Data are collected on a FACS analyzer with stimulation with an argon laser tuned to 488 nm with emission recorded through a 525 nm bandwidth filter.

3.2.3 Fluorometric Analysis

Preparation of cells for fluorometric analysis is the same as described for fluorescence microscopy, except that following the PBS wash (twice in this case), cells are gently trypsinized (trypsin–EDTA). Plates are put on ice and 100 μl of cold PBS is added to the cells. The total cell suspensions are transferred to 96-well white microtiter plates (Dynex) and keep on ice prior to fluorometric analysis. The assay can be performed on any microtiter plate reader; we use a Perkin-Elmer HTS 7000 Series Bio Assay Reader in the fluorescence mode. The excitation and emission wavelengths for the fMTX are 497 nm and 516 nm, respectively. Afterward, the data are normalized to total protein concentration in cell lysates.

3.3 GFP PCA Fluorescence Assay

All procedures describe for the DHFR PCA fluorescence assays are the same for GFP PCA, except that there are no use of fMTX and no washing steps. The wash procedure is obviously irrelevant in the case of the GFP PCA where the folded/reassembled protein is a fluorophore itself.

3.4 In Vitro ß-Lactamase PCA Colorimetric Assay

1. Twenty-four hours before transfection, plate 1×10^5 COS or HEK 293 T cells (this assay can be performed using any other cell line) in 12-well plates in DMEM medium enriched with 10 % Cosmic calf serum.

2. Transiently co-transfect cells with the PCA fusion partners (*see* **Note 10**) using a transfection reagent (e.g., FuGENE 6) according to the manufacturer's instructions.

3. Forty-eight hours after transfection, wash cells three times with cold PBS and resuspended in 300 μl of cold PBS and keep on ice. Centrifuge cells at 4 °C for 30 s, discard the supernatant, and resuspend cells in 100 μl of cold 100 mM, phosphate buffer pH 7.4 (ß-lactamase reaction buffer).

4. Freeze in dry ice/ethanol for 10 min and thaw in a water bath at 37 °C for 10 min, then lyse cells with 3 cycles of freeze and thaw. Remove cell membrane and debris are by centrifugation at 4 °C for 5 min ($10,000 \times g$). Collect the supernatant whole cell lysate and store at −20 °C until assays are performed.

5. Perform assays in 96-well microtiter plates. For testing ß-lactamase activity, 100 μl of 100 mM phosphate buffer is allocated into each well. Add 78 μl of H_2O and 2 μl of 10 mM Nitrocefin (final concentration of 100 μM). Finally, add 20 μl of unfrozen cell lysate (final buffer concentration of 60 μM).

6. The assays can be performed on any microtiter plate reader; we use a Perkin-Elmer HTS 7000 Series Bio Assay Reader in the absorption mode with a 492 nm measurement filter.

3.5 In Vivo Enzymatic Assay and Fluorescent Microscopy with CCF2/AM

1. Twenty-four hours before transfection, plate 1×10^5 COS or HEK293 T cells in 12-well plates in DMEM medium enriched with 10 % Cosmic calf serum.

2. Transiently co-transfect cells with the PCA fusion partners (*see* **Note 10**) using a transfection reagent (e.g., FuGENE 6) according to the manufacturer's instructions.

3. Twenty-four hours after transfection, split cells again to ensure 50 % confluency the following day (1.5×10^5) (*see* **Note 11**). Split cells either onto 12-well plates for suspension enzymatic assay or onto 15 mm glass coverslips for fluorescent microscopy.

4. Forty-eight hours after transfection, wash cells three times with PBS to remove all traces of serum (*see* **Note 12**).

5. Load cells with the following: 1 μM of CCF2/AM diluted into a physiologic saline solution for 1 h.

For in vivo enzymatic assay:

6. Wash cells washed twice with the physiologic saline and resuspend into the same solution. Aliquot 1×10^6 cells into a 96-well fluorescence white plate and measure blue fluorescence with a Perkin Elmer HTS 7000 Series Bio Assay Reader in the fluorescence Top reading mode with a 409 nm excitation filter and a 465 emission filter.

For fluorescence microscopy:

7. Wash cells twice with physiologic saline as in **step 5**, prior to examination under the microscope (*see* **Note 13**).

We have used two substrates to study the ß-lactamase PCA. The first one is the cephalosporin called Nitrocefin. This substrate is used in the in vitro colorimetric assay. ß-lactamase has a kcat/km of 1.7×10^4 mM$^{-1} \times$ s^{-1}. Substrate conversion can be easily observed by eye; the substrate is yellow in solution while the product is a distinct ruby red color. The rate of hydrolysis can be monitored quantitatively with any spectrophotometer by measuring the appearance of red at 492 nm. Signal to background, depending on the mode of measurement can be greater than 30 to 1.

We have also developed an in vivo fluorometric assay using the substrate CCF2/AM [39, 40]. Although not as good a substrate as nitrocefin (kcat/km of 1,260 mM$^{-1} \times$ s^{-1}), CCF2/AM has unique features that make it a useful reagent for in vivo PCA. First, CCF2/AM contains butyryl, acetyl and acetoxymethyl esters, allowing diffusion across the plasma membrane where cytoplasmic esterases catalyze the hydrolysis of its ester functionality releasing the polyanionic (4 anions) ß-lactamase substrate CCF2. Because of the negative charge of CCF2, the substrate becomes trapped in the cell. In the intact substrate, fluorescence resonance energy transfer (FRET) can occur between a coumarin donor and fluorescein acceptor pair covalently linked to the cephalosporin core. The coumarin donor can be excited at 409 nm with emission at 447 nm, which is within the excitation envelope of the fluorescence acceptor (maximum around 485 nm), leading to remission of green fluorescence at 535 nm. When ß-lactamase catalyzes hydrolysis of the substrate the fluorescein moiety is eliminated as a free thiol. Excitation of the coumarin donor at 409 nm then emits blue fluorescence at 447 nm whereas the acceptor (fluorescein) is quenched by the free thiol.

4 Notes

1. Alternatively, recessive selection can be achieved in eukaryotic cells by using DHFR fragments containing one or more of several mutations (for example F31S mutation, see below) that reduce the affinity of refolded DHFR to the anti-folate drug methotrexate and growing cells in the absence of nucleotides with selection for methotrexate resistance. This would obviously be necessary in working with mouse ES cells as, with all eukaryotes, DHFR activity is present.

2. The best orientations of the fusions for the DHFR PCA are: protein A-DHFR[1,2] + protein B-DHFR[3] or DHFR[1,2]-protein A + protein B-DHFR[3], where proteins A and B are the proteins to test for interaction. We typically insert a 10-amino-acid flexible polypeptide linker consisting of (Gly.Gly.Gly.Gly.Ser)$_2$ between the protein of interest and the DHFR fragment (for both fusions). DHFR[1,2] corresponds to amino acids 1–105, and DHFR[3] corresponds to amino acids 106–186 of murine DHFR. The DHFR[1,2] fragment that we use also contains a phenylalanine to serine mutation at position 31 (F31S), rendering the reconstituted DHFR resistant to methotrexate (MTX) treatment.

3. It is crucial that cell density is kept to a minimum and cells are well separated when split to avoid cells "harvesting" nutrients from adjacent cells on dense plates, or colonies might appear to be forming from clumps of cells that were not sufficiently separated during the splitting procedure.

4. The choice of dialyzed FBS manufacturer is crucial. Cells need very little nucleotide in the medium to propagate and this will result in false positives. The Hyclone dialyzed FBS has proven a particularly reliable source.

5. The fluorescence DHFR PCA assay is universal and in theory can be used in any cell type or organism. This assay has already been shown to work in several mammalian cell lines as well as in plant cells and insect cells.

6. A stock solution of 1 mM fMTX should be prepared as follows: Dissolve 1 mg of fMTX in 1 ml of dimethyl formamide (DMF). To facilitate the dissolution, incubate 15 min at 37 °C and mix by vortexing every 5 min. Protect the tube from light. Store at −20 °C.

7. Complementary fragments of DHFR fused to interacting protein partners, when expressed and reassembled in cells, will bind with high affinity ($K_d = 540$ pM) to fMTX in a 1:1 complex. fMTX is retained in cells by this complex, while the unbound fMTX is actively and rapidly transported out of the cells.

8. All of the work reported to date has been performed in live cells. While cells can be fixed, there is a significant reduction in observable fluorescence.

9. Particular attention must be given to optimizing the fMTX load and "wash" procedures. Important variables include the time of loading, temperatures at which each wash step is performed, the number and length of wash steps and the time between washing and visualization. Too little washing will mean that background cannot be distinguished from a positive result. One should scrutinize the relevant parameters in the same sense as one would for say, a Western blot. Results may also vary with the way the cells are plated and the types of cells used. Generally, as in other fluorescent microscopy procedures, the shape of cells and the localization of the fluorophore will result in better or worse results. For stable cell lines, the intensity of fluorescence will also depend on the levels of expression of the fusion proteins. The loading times and concentrations of fMTX (22 h, 10 µM) used may result in a nonspecific and punctate fluorescence that is observed with any filter set. We do not know the source of this background, but it should not be mistaken for the real fluorescence signal produced by the PCA, which should be observed strictly with a filter that is optimal for observation of fluorescein. We have observed that loading fMTX for between 2 and 5 h at lower (5 µM) concentrations prevents this nonspecific signal, although fewer cells are labeled. Loading times and concentrations must be optimized for specific cell types.

10. The best orientations of the fusions for the ß-lactamase PCA are: protein A-BLF[1] + protein B-BLF[2] or BLF[1]-protein A + protein B-BLF[2], where proteins A and B are the proteins to test for interaction. We typically insert a 15-amino-acid flexible polypeptide linker consisting of (Gly.Gly.Gly.Gly.Ser)$_3$ between the protein of interest and the ß-lactamase fragment (for both fusions). BLF[1] corresponds to amino acids 26–196 (Ambler numbering), and BLF[2] corresponds to amino acids 198–290 of TEM-1 ß-lactamase.

11. The maximum loading efficiency of CCF2-AM is observed at 50 % confluence.

12. Serum may contain esterases that can destroy the substrate.

13. We perform fluorescence microscopy on live HEK 293 or COS cells with an inverse Nikon Eclipse TE-200 (objective plan fluor 40× dry, numerically open at 0.75). Images were taken with a digital CCD cooled (−50 °C) camera, model Orca-II (Hamamatsu Photonics (exposure for 1 s, binning of 2 × 2 and digitalization 14 bits at 1.25 MHz). Source of light is a Xenon lamp Model DG4 (Sutter Instruments). Emission filters are changed by an emission filter switcher (model Quantoscope) (Stanford Photonics). Images are visualized with ISee software (Inovision Corporation) on an O2 Silicon Graphics computer. The following selected filters are used: Filter set #31016 (Chroma Technologies); Excitation filter: 405 nm (passing band of 20 nm); Dichroic Mirror: 425 nm DCLP; Emission filter #1: 460 nm (passing band of 50 nm); Emission filter #2: 515 nm (passing band of 20 nm).

References

1. Drees BL (1999) Progress and variations in two-hybrid and three-hybrid technologies. Curr Opin Chem Biol 3:64–70

2. Evangelista C, Lockshon D, Fields S (1996) The yeast two-hybrid system: prospects for protein linkage maps. Trends Cell Biol 6:196–199

3. Fields S, Song O (1989) A novel genetic system to detect protein-protein interactions. Nature 340:245–246

4. Vidal M, Legrain P (1999) Yeast forward and reverse 'n'-hybrid systems. Nucleic Acids Res 27:919–929

5. Walhout AJ, Sordella R, Lu X et al (2000) Protein interaction mapping in C. elegans using proteins involved in vulval development. Science 287:116–122

6. Uetz P, Giot L, Cagney G et al (2000) A comprehensive analysis of protein-protein interactions in Saccharomyces cerevisiae. Nature 403:623–627

7. Michnick SW, Remy I, Campbell-Valois FX et al (2000) Detection of protein-protein interactions by protein fragment complementation strategies. Methods Enzymol 328:208–230

8. Paulmurugan R, Umezawa Y, Gambhir SS (2002) Noninvasive imaging of protein-protein interactions in living subjects by using reporter protein complementation and reconstitution strategies. Proc Natl Acad Sci U S A 99:15608–15613

9. Hu CD, Kerppola TK (2003) Simultaneous visualization of multiple protein interactions in living cells using multicolor fluorescence complementation analysis. Nat Biotechnol 21:539–545

10. Paulmurugan R, Gambhir SS (2003) Monitoring protein-protein interactions using split synthetic renilla luciferase protein-fragment-assisted complementation. Anal Chem 75:1584–1589

11. Luker KE, Smith MC, Luker GD et al (2004) Kinetics of regulated protein-protein interactions revealed with firefly luciferase complementation imaging in cells and living animals. Proc Natl Acad Sci U S A 101:12288–12293

12. Magliery TJ, Wilson CG, Pan W et al (2005) Detecting protein-protein interactions with a green fluorescent protein fragment reassembly trap: scope and mechanism. J Am Chem Soc 127:146–157

13. Remy I, Michnick SW (2006) A highly sensitive protein-protein interaction assay based on Gaussia luciferase. Nat Methods 3:977–979

14. Wehr MC, Laage R, Bolz U et al (2006) Monitoring regulated protein-protein interactions using split TEV. Nat Methods 3:985–993

15. Michnick SW, Ear PH, Manderson EN et al (2007) Universal strategies in research and drug discovery based on protein-fragment complementation assays. Nat Rev Drug Discov 6:569–582

16. Remy I, Michnick SW (2007) Application of protein-fragment complementation assays in cell biology. Biotechniques 42:137, 139, 141 passim

17. Stefan E, Aquin S, Berger N et al (2007) Quantification of dynamic protein complexes using Renilla luciferase fragment complementation applied to protein kinase A activities in vivo. Proc Natl Acad Sci U S A 104:16916–16921

18. Michnick SW, Ear PH, Landry C et al (2010) A toolkit of protein-fragment complementation assays for studying and dissecting large-scale and dynamic protein-protein interactions in living cells. Methods Enzymol 470:335–368

19. Michnick SW, Ear PH, Landry C et al (2011) Protein-fragment complementation assays for large-scale analysis, functional dissection and dynamic studies of protein-protein interactions in living cells. Methods Mol Biol 756:395–425

20. Anfinsen CB, Haber E, Sela M et al (1961) The kinetics of formation of native ribonuclease during oxidation of the reduced polypeptide chain. Proc Natl Acad Sci U S A 47:1309–1314

21. Anfinsen CB (1973) Principles that govern the folding of protein chains. Science 181:223–230

22. Gutte B, Merrifield RB (1971) The synthesis of ribonuclease A. J Biol Chem 246:1922–1941

23. Richards FM (1958) On the Enzymic Activity of Subtilisin-Modified Ribonuclease. Proc Natl Acad Sci U S A 44:162–166

24. Taniuchi H, Anfinsen CB (1971) Simultaneous formation of two alternative enzymology active structures by complementation of two overlapping fragments of staphylococcal nuclease. J Biol Chem 246:2291–2301

25. Pelletier JN, Campbell-Valois FX, Michnick SW (1998) Oligomerization domain-directed reassembly of active dihydrofolate reductase from rationally designed fragments. Proc Natl Acad Sci U S A 95:12141–12146

26. Pelletier JN, Michnick SW (1997) A protein complementation assay for detection of protein-protein interactions in vivo. Protein Eng 10:89

27. Johnsson N, Varshavsky A (1994) Split ubiquitin as a sensor of protein interactions in vivo. Proc Natl Acad Sci U S A 91:10340–10344

28. Rossi F, Charlton CA, Blau HM (1997) Monitoring protein-protein interactions in intact eukaryotic cells by beta-galactosidase complementation. Proc Natl Acad Sci U S A 94:8405–8410

29. Remy I, Michnick SW (1999) Clonal selection and in vivo quantitation of protein interactions with protein-fragment complementation assays. Proc Natl Acad Sci U S A 96:5394–5399

30. Remy I, Wilson IA, Michnick SW (1999) Erythropoietin receptor activation by a ligand-induced conformation change. Science 283:990–993

31. Remy I, Michnick SW (2001) Visualization of biochemical networks in living cells. Proc Natl Acad Sci U S A 98:7678–7683

32. Remy I, Campbell-Valois FX, Michnick SW (2007) Detection of protein-protein interactions using a simple survival protein-fragment complementation assay based on the enzyme dihydrofolate reductase. Nat Protoc 2:2120–2125

33. Tarassov K, Messier V, Landry CR et al (2008) An in vivo map of the yeast protein interactome. Science 320:1465–1470

34. Galarneau A, Primeau M, Trudeau LE et al (2002) Beta-lactamase protein fragment complementation assays as in vivo and in vitro sensors of protein protein interactions. Nat Biotechnol 20:619–622

35. Remy I, Ghaddar G, Michnick SW (2007) Using the beta-lactamase protein-fragment complementation assay to probe dynamic protein-protein interactions. Nat Protoc 2:2302–2306

36. Israel DI, Kaufman RJ (1993) Dexamethasone negatively regulates the activity of a chimeric dihydrofolate reductase/glucocorticoid receptor protein. Proc Natl Acad Sci U S A 90:4290–4294

37. Kaufman RJ, Bertino JR, Schimke RT (1978) Quantitation of dihydrofolate reductase in individual parental and methotrexate-resistant murine cells. Use of a fluorescence activated cell sorter. J Biol Chem 253:5852–5860

38. Kaufman RJ (1990) Selection and coamplification of heterologous genes in mammalian cells. Methods Enzymol 185:537–566

39. Zlokarnik G, Negulescu PA, Knapp TE et al (1998) Quantitation of transcription and clonal selection of single living cells with beta-lactamase as reporter. Science 279:84–88

40. Zlokarnik G (2000) Fusions to beta-lactamase as a reporter for gene expression in live mammalian cells. Methods Enzymol 326:221–244

Chapter 32

Detection of Protein-Protein Interaction Using Bimolecular Fluorescence Complementation Assay

Cau D. Pham

Abstract

The bimolecular fluorescence complementation (BiFC) assay is a versatile technique for investigating protein-protein interaction (PPI) in living systems. The BiFC assay exploits the color-emitting moiety and the modular structure of fluorescent proteins to provide both temporal and spatial information of the PPI. The modular property of fluorescent proteins enables researchers to strategically partition a fluorescent protein into two nonfluorescent units, which can be independently fused to other proteins. When the fusion proteins interact with each other, the nonfluorescent fragments reconstitute to generate a fluorescence signal. PPI can then be detected by capturing the fluorescence signal with a fluorescence microscope. In this chapter, the Venus fluorescent protein is employed to demonstrate the application of the BiFC assay.

Key words Fluorescent proteins, BiFC technology, Protein-protein interaction, Protocol

1   Introduction

In 1962, the green fluorescent protein (GFP) of the *Aequorea victoria* jellyfish was the first protein identified to possess an intrinsic bioluminescence property [1]. Five decades later, more than 40 different species of fluorescent proteins are available for various applications with color spanning beyond the visible spectrum [2, 3]. Despite growing in number, fluorescent proteins share similar molecular weight and backbone features such as a barrel-like structure comprising of 11 β-sheets and an α-helix filament that traverses through the entire barrel at the center. The β-sheet barrel and the α-helix structure are essential for fluorescent protein maturation or chromophore formation, a necessary step for color-emission. Chromophore formation also requires oxygen, which facilitates the modification of at least four highly conserved amino acids on the fluorescent proteins: the tyrosine at position 66, the glycine at position 67, the arginine at position 96, and the glutamate at position 222 [3, 4].

Cheryl L. Meyerkord and Haian Fu (eds.), *Protein-Protein Interactions: Methods and Applications*, Methods in Molecular Biology, vol. 1278, DOI 10.1007/978-1-4939-2425-7_32, © Springer Science+Business Media New York 2015

Although GFP was discovered in 1962, it took three more decades for scientists to figure out how to harness the unique properties of GFP. In addition to being able to generate fluorescence signal, fluorescent proteins are nontoxic and their maturation is not dependent on any particular host [2, 3]. Fluorescent proteins can also be modified and/or linked to another protein without sacrificing their fluorescence signal. These traits enable scientists to utilize them as reporter proteins in different biological systems. In 1992, full length GFP was first cloned and used as a reporter protein to study gene expression in bacteria and nematodes [2]. From that time on, full length fluorescent proteins have been widely used to visualize protein expression, protein localization, and protein complexes in live cells [3, 4]. Furthermore, the bimolecular fluorescence complementary (BiFC) assay, a newly developed fluorescent protein-based method, has enabled scientists to visualize protein-protein interaction (PPI) in living systems [5–7].

The BiFC assay involves splitting a fluorescent protein into two fragments that are incapable of generating fluorescence signal on their own [5]. However, these fragments can reconstitute and produce fluorescence signal if they are brought together such as when they are fused to two proteins that interact with one another (Fig. 1). BiFC technology was successfully demonstrated by Hu et al. in 2002 by tethering Fos to the C-terminal fragment and Jun to the N-terminal fragment of the enhanced yellow fluorescent protein (EYFP) [5]. In addition to YFP, many other fluorescent proteins and multiple splitting positions have also been successfully developed for BiFC (Table 1).

The ability to examine PPI in living systems is essential to our understanding of biological processes [9, 10]. Toward this end, the BiFC assay holds several key advantages over other PPI detection methods such as yeast two-hybrid and co-immunoprecipitation [11, 12]. Firstly, the BiFC assay allows real-time visualization of PPI with minimal perturbation to their natural environment, such

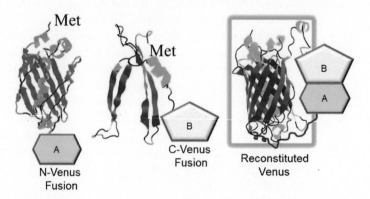

Fig. 1 The principle of Venus BiFC technology. The Venus fluorescent protein is strategically split into two nonfluorescent fragments. The reconstitution of Venus protein and signal from the nonfluorescent constituents can be mediated by two interacting proteins. Venus protein structure was drawn using the I-TASSER program [8]

Table 1
Fluorescent proteins commonly utilized for BiFC technology

Protein	Organism	Ex/Em	Color class	Percentage of brightness[a]	Brightness half-life[b] (in seconds)	Oligomerization of fully mature protein	BiFC split position	Reference
CFP	Aequorea victoria	433/475	Cyan	19	64	Monomer	154–155;172–173	[2, 6]
Cerulean	Aequorea victoria	433/475	Cyan	39	36	Weak dimer	154–155;172–173	[2, 6]
GFP	Aequorea victoria	485/500	Green	49	174	Weak dimer	157–158	[2, 6]
YFP	Aequorea victoria	514/527	Yellow	74	60	Weak dimer	154–155;172–173	[2, 6]
Venus	Aequorea victoria	515/528	Yellow	76	15	Weak dimer	154–155;172–173	[2, 6]
Citrine	Aequorea victoria	516/529	Yellow	85	49	Monomer	154–155;172–173	[2, 6]
mRFP1-Q66T	Discosoma sp.	549/570	Red	ND	ND	ND	154–155;168–169	[2, 6]
mCherry	Discosoma sp.	587/610	Red	23	96	Monomer	159–160	[2, 6]

A summary of the BiFC systems that have been developed in the last 10 years

[a]Percentage of brightness of intact fluorescent protein relative to fluorescein (100 %)

[b]The Brightness half-life was calculated by measuring the time it took to bleach the emission rate of 1,000 photon/s down to 500 photon/s [2]

as in live cells or animals [11, 12]. Secondly, the BiFC assay can provide valuable information regarding the sub-cellular location of the PPI [12]. Such information has good predictive value about the functional roles of the interaction. These and other aforementioned attributes are the reason why BiFC is a valuable tool to have in one's molecular cabinet. In this chapter, one will learn how to detect PPI in mammalian cells using Venus BiFC.

2   Materials

1. Thermocycler.

2. DNA polymerase and buffer (e.g., Phusion, New England BioLabs).

3. *P38* and *MKK3* cDNAs (Open Biosystems).

4. Gateway® entry vector (Life Technologies).

5. TE (10 mM Tris–HCl, 1 mM EDTA) pH 8.0.

6. BP Clonase™ II enzyme mix (Life Technologies).

7. Shaker-incubator and water bath.

8. Competent bacteria (e.g., NEB10β).

9. Bacterial growth media including S.O.C and LB. LB medium is supplemented with the appropriate antibiotic (100 μg/mL Ampicillin, 50 μg/mL Kanamycin).

10. Mini-prep/midi-prep kits.

11. BsrGI restriction endonuclease and buffer.

12. Agarose gel electrophoresis reagents (molecular grade agarose, Tris-Borate-EDTA buffer, ethidium bromide).

13. Gene-specific oligonucleotide primers for cloning and sequencing.

14. pSCM167-NV and pDEST26-CV mammalian expression vectors (developed in the lab of Dr. Haian Fu, Emory University).

15. LR Clonase™ II enzyme mix (Life Technologies).

16. A nucleic acid spectrophotometer (e.g., NanoDrop).

17. Cell growth medium for the specific cell line (e.g., DMEM supplemented with 5 % FBS and 1 % Penicillin–Streptomycin).

18. Cell line that can be transfected (e.g., the LN229 cell line).

19. Cell culture incubator with CO_2 supply.

20. 0.05 % Trypsin/0.53 mM EDTA.

21. Opti-MEM.

22. Transfection reagent (e.g., FuGENE® HD transfection reagent, Promega).

23. Fluorescence microscope and filters.

3 Methods

Described below are (1) the steps to create BiFC vectors and (2) how to examine PPI in mammalian cells (*see* **Note 1**). A previously known interacting protein pair MKK3 and P38 [13] will be used to demonstrate the use of a BiFC assay to monitor PPI. The BiFC reporter used in this example is the Venus fluorescent protein, a variant of EYFP (*see* **Note 2**). Venus fluoresces brighter, matures rapidly at 37 °C, and is less sensitive to environmental conditions [3, 14, 15]. Both Venus fragments are in the Gateway® vector system (Life Technologies) and will generate N-terminal fusion proteins (the Venus fragments fuse to the N-terminus of the protein of interest) (Fig. 2). The N-terminal fragment of Venus (NV),

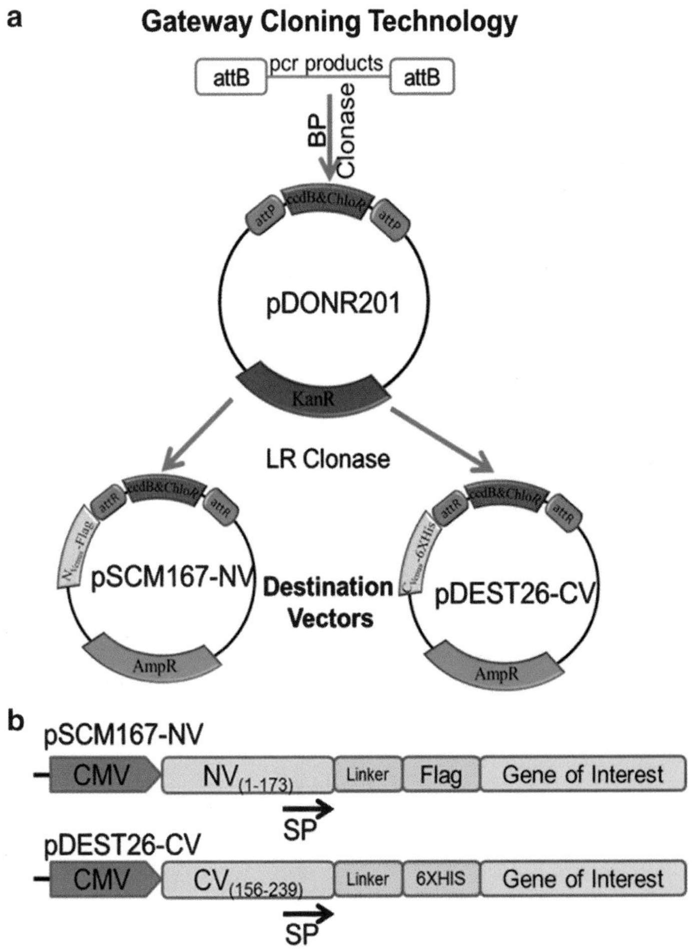

Fig. 2 Gateway® cloning technology and a schematic of the Venus BiFC constructs. (**a**) Gateway® vectors allow the shuttling of DNA fragments from one vector to another compatible vector via site-specific homologous recombination catalyzed by recombinase enzymes. (**b**) N-terminal fusion constructs regulated by the CMV promoter. For these vectors, the Venus fragments are attached to the protein of interest through a (2xGGGS) linker sequence

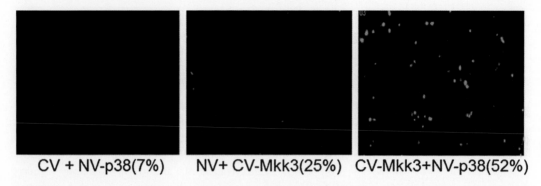

CV + NV-p38(7%)   NV+ CV-Mkk3(25%)   CV-Mkk3+NV-p38(52%)

Fig. 3 The interaction between P38 and MKK3 was used to demonstrate the Venus BiFC assay. Cells that expressed both P38 and MKK3 proteins fused to compatible Venus fragments exhibit strong Venus signals. The Venus signal in cells that expressed only one fusion protein along with its complementary Venus fragment was more than two times lower

residues 1–173, along with a linker (*see* **Note 3**), was inserted into the pSCM167 vector (Haian Fu lab) to generate the pSCM167-NV expression vector (*see* **Note 4**). The C-terminal fragment of Venus (CV), residues 156–239, along with a linker, was inserted into the pDEST26 vector (Life Technologies) to make the pDEST26-CV expression vector (*see* **Note 4**).

3.1 Creating BiFC Constructs for PPI Detection

Gateway® cloning technology can be used to clone DNA fragments into the Venus BiFC expression vectors (as an example, P38 and MKK3 were cloned into pSCM167-NV and pDEST26-CV, respectively) (Figs. 2 and 3). Gateway® technology employs a two-vector system, an entry and destination vector, to shuttle DNA fragments between vectors. To utilize this technology, a DNA fragment is first inserted into an entry vector. From the entry vector, the DNA fragment can be transferred to Gateway® destination/expression vectors via site-specific homologous recombination.

3.1.1 Cloning of DNA Fragments into a Gateway® Entry Vector

1. Use a polymerase chain reaction (PCR) to add attB sites to the 5′ and 3′ ends of the DNA fragment as described by the manufacturer (Life Technologies) (*see* **Note 5**). For example, *attB* sequences {5′-GGG GAC AAG TTT GTA CAA AAA AGC AGG CTT C-3′ and 5′-GGG GAC CAC TTT GTA CAA GAA AGC TGG GTC TTA CTA-3′} were added to the 5′ and 3′, respectively, of both P38 and MKK3. The *attB* sequences facilitate the insertion of DNA fragments into the Gateway® entry vectors via site-specific homologous recombination.

2. Perform the cloning of the DNA fragment, flanked by the *attB* sites, into the Gateway® entry vector (in this case pDONR201) using BP clonase enzyme and bacterial transformation as suggested by the manufacturer's instruction (with minor

modifications as described below). First, combine purified *attB*-PCR product (15–150 ng) and pDONR201 vector (150 ng) in a 1.5 mL microcentrifuge tube. Next, if necessary, adjust the reaction volume to 8 µL with TE buffer. Then, add 2 µL of BP Clonase™ II enzyme mix to the reaction, vortex to mix, brief centrifugation, and incubate the reaction at room temperature (25 °C) for 1 h. Finally, stop the reaction by adding 1 µL of proteinase K, vortex to mix, and incubate at 37 °C for 10 min. The BP reaction is now ready for bacterial transformation.

3. Bacterial transformation is achieved by adding 2 µL of the BP reaction to 20 µL of freshly thawed competent bacteria. Gently tap the tube to mix the reaction and follow with a 30 min incubation on ice. Next, perform a 30 s heat-shock by incubating the reaction in a 42 °C water bath. Then, place the reaction back on ice and add 250 µL of S.O.C medium. Transfer the reaction to a 37 °C shaker-incubator and incubate (with shaking) for 1 h. Finally, spread about 100–200 µL of the transformation mixture onto a LB plate that contains 50 µg/mL of Kanamycin and incubate the plate overnight at 37 °C (*see* **Note 6**).

4. To determine the correct clone, pick a few colonies from the transformation plate and inoculate each of them in 4 mL of LB broth containing 50 µg/mL of Kanamycin. Allow the bacteria to populate overnight at 37 °C in a shaker-incubator. Next, use a miniprep kit (follow the manufacturer's protocol) to extract the plasmid DNA from the bacteria. Then, perform BsrGI restriction digestion (one BsrGI restriction site is present at each *att* site) of the plasmid followed by agarose gel electrophoresis analysis to verify the insertion of the DNA fragment into the entry vector. Finally, perform sequence analysis to check the fidelity and the orientation of the inserted DNA fragment.

3.1.2 Cloning of the DNA Fragment into the pSCM167-NV and pDEST26-CV Vectors

Both the pSCM167-NV and the pDEST26-CV are Gateway®-based expression vectors. This enables the transfer of DNA fragments from entry vectors (such as pDONR201) directly into the pSCM167-NV and pDEST26-CV vectors via site-specific homologous recombination. The transferring of a DNA fragment from Gateway® entry vectors into Gateway® destination vectors is catalyzed by the LR Clonase™ enzyme and is performed as instructed by the manufacturer (with minor modifications).

1. First, combine the purified entry vector (15–150 ng) and 150 ng of the destination vector (pSCM167-NV or pDEST26-CV) in a 1.5 mL microcentrifuge tube. Next, if necessary, adjust the reaction volume to 8 µL with TE buffer. Then, add 2 µL of LR Clonase™ II enzyme mix to the reaction,

vortex to mix, briefly centrifuge, and incubate the reaction at room temperature (25 °C) for 1 h. Finally, stop the reaction by adding 1 μL of proteinase K, vortex to mix, and incubate at 37 °C for 10 min. The LR reaction is now ready for bacterial transformation.

2. Transform competent bacteria, as described above, with the product of the LR reaction. Positive clones can be selected by plating the bacteria on LB plates containing 100 μg/mL of Ampicillin. Finally, perform BsrGI restriction digestion and DNA sequence analysis to verify the insertion of the DNA fragment in the destination vector (*see* **Note 7**).

3. Vectors with the correct fusion of the DNA fragment to the respective Venus fragments can then be used in subsequent BiFC assays.

3.2 Performing the BiFC Assay

In order to perform the BiFC assay, cells must uptake the expression vectors. Prior knowledge about the model system is essential (*see* **Note 1**). Prior to perform BiFC assay, one needs to confirm the expression level of the fusion proteins. This can be determined by Western blot (*see* **Note 8**). One also needs to consider checking the folding of the proteins of interest. Incorrect folding of the proteins of interest caused by the fused fluorescent protein fragments could affect the stability and localization of the protein [11] (*see* **Note 9**). Localization of fusion proteins and untagged wild-type proteins can be assessed by indirect immunofluorescence assay [11].

3.2.1 Transfecting the Expression Vectors into a Cell Line

As an example, pSCM167-NV-*P38* and pDEST26-CV-*MKK3* vectors were transiently co-transfected into LN229 cells using FuGENE® HD transfection reagent. Likewise, pSCM167-NV-*P38* and pDEST26-CV as well as pDEST26-CV-*MKK3* and pSCM167-NV vector combinations were also co-transfected into the LN229 cell line to serve as negative controls (*see* **Note 10**). In addition to negative controls, one should also consider a positive control and a transfection control to monitor PPI and transfection efficiency, respectively (*see* **Note 10**). The transfection protocol below was adapted from the FuGENE® HD manufacturer's instructions and was optimized for the LN229 cell line. All plasmid DNAs were prepared using Qiagen midi-prep kit as described in the manufacturer's protocol. DNAs were quantified using a NanoDrop (*see* **Note 11**).

3.2.2 Cell Growth Conditions, Cell Seeding, and Cell Transfection

In this example, low passage LN229 cells were cultured in DMEM supplemented with 5 % FBS and 1 % Penicillin/Streptomycin (please note that the cell culture and transfection conditions will need to be adjusted for the cell lines being used). Cells were kept in a water-jacketed incubator at 37 °C and supplemented with 5 % CO_2 humidity. For transfection, the following steps were taken.

1. Pre-culture 25 mL of LN229 cells in a 75 mL flask until the growing surface of the flask is about 90 % covered with cells.

2. Harvest the cells by trypsinization.

3. Completely remove all growth medium.

4. Add 1.5 mL of 0.05 % Trypsin/0.53 mM EDTA.

5. Incubate at room temperature for 3–5 min or until most cells have detached from the surface of the plate.

6. Add 10 mL of cell growth medium to the flask to deactivate the trypsin (see **Note 12**).

7. Pipet the cells eight to ten times with a 10 mL transfer pipet to dissociate the cells, and determine the cell concentration using a hemocytometer.

8. Prepare a six-well plate of cells for transfection by adding 2 mL (300,000 cell/mL) of trypsinized cells, in growth medium, to each well. Then, gently swirl the plate to evenly disperse the cells around the wells.

9. Allow the cells to cover about 60 % of the well's surface by keeping the plate at 37 °C and 5 % CO_2 humidity in a water-jacketed incubator (usually takes about 18 h). When the cells are about 60 % confluent, they are ready for transfection.

10. Prepare the transfection reaction by performing the following steps. First, transfer 200 µL of Opti-MEM to a 1.5 mL sterile microcentrifuge tube. Next, add 320 ng of the pSCM167-NV and 320 ng of the pDEST26-CV expression vectors (see **Note 13**). Gently tap the tube to mix the DNAs. Finally, add 1.92 µL of transfection reagent directly to the solution (see **Note 14**) and gently tap the tube to mix the reaction.

11. Incubate the reaction at room temperature (25 °C) for 15 min.

12. Use a micropipettor and slowly dispense 100 µL of the transfection reaction drop-by-drop over the whole well.

13. Gently swirl the plate to facilitate diffusion of the transfection reagent.

14. Place the plate back into the cell culture incubator.

3.2.3 Detecting PPI with a Fluorescence Microscope

Proteins that show strong interaction can be detected 24 h post-transfection (see **Note 15**). Once the cells have been transfected with the appropriate plasmids, PPI can be observed using a fluorescence microscope with the correct filter. The YFP filter set can be used to detect Venus signal. As demonstrated in Fig. 3, strong Venus signal was observed in cells that were co-transfected with the pSCM167-NV-*P38* and the pDEST26-CV-*MKK3* vectors. Noticeably, a low Venus signal was also observed in cells that were co-transfected with control vectors; the pSCM167-NV-*P38* and pDEST26-CV pair as well as pDEST26-CV-*MKK3* and pSCM167-NV pair (*see* **Notes 10** and **16**). However, the signals from these cells were much lower than the cells that harbored both the P38 and MKK3 fusion proteins (Fig. 3). For a novel protein interaction pair, the interaction can be further validated using a co-immunoprecipitation assay or tag pull-down assay with GST or FLAG tags.

4   Notes

1. The model system must able to uptake foreign DNA such as plasmids. Also keep in mind that fluorescent protein maturation requires oxygen. Therefore, this assay will not work in organisms that are sensitive to oxygen or in an environment that lacks oxygen. Temperature can also affect the maturation process of fluorescent proteins. For example, YFP-BiFC optimal maturation temperature is 30 °C while other fluorescent proteins typically mature at 37 °C [11].

2. Keep in mind that the signal from the reconstituted fluorescent proteins is not as bright as intact fluorescent proteins. For weak interacting proteins, it is best to use a strong fluorescence signal BiFC system such as Venus.

3. A linker is a short peptide that connects the fluorescent protein fragment to the protein of interest. The linker provides limited independent movement between the fluorescent protein fragment and its fusion protein. The length and the amino acid sequence of the linkers have been known to affect PPI detection. Linker sequences such as RSIAT, RPACK-IPNDLKQKVMNH, AAANSSIDLISVPVDSR, and (GGGS)n have been successfully employed by many groups [11]. The GGGSGGGS linker sequence was used for the example experiment.

4. When choosing expression vectors, make sure that the vector system is compatible with model system (i.e., replication, promoter, etc.). Both pSCM167-NV and pDEST26-CV are expression vectors for mammalian cells and protein expression is regulated by the CMV promoter. Using these vectors, the proteins of interest will be fused to the Venus fragments at their N-terminus. Both Venus fragments have a start codon and no stop codon. Both vectors have an additional tag. The pSCM167-NV vector has a FLAG tag while the pDEST26-CV has a 6XHis tag. These tags don't interfere with the BiFC assay, but do provide another platform for detecting PPI (i.e., FLAG or HIS pull-down).

5. The *attB* sequences are added to the PCR-primers during primer synthesis. Make sure that the *attB* sequences as well as

their orientation are correct when designing primers for Gateway® cloning. If the DNA fragment is intended for expression, it is essential to check the reading frame and the presence of a start as well as a stop codon.

6. According to the manufacturer, Gateway® cloning is highly efficient. Gateway® vectors (i.e., PDONR201, pSCM167-NV, and pDEST26-CV) contain the *ccdB* gene that encodes for a bacterial toxin (use *ccdB* resistant bacteria to propagate such plasmids). The *ccdB* gene is replaced by the target DNA fragment during the BP and LR clonase reaction. Therefore, one only needs to check a few colonies to find a clone with the correct insert. If needed, increasing the incubation time (i.e., overnight) of BP and LR reactions can increase the number of colonies with the desired insert.

7. The sequencing primers should be located within the Venus fragments and about 100 base-pairs upstream from the protein of interest. Orientation and reading frame alignment can both be determined using these primers.

8. The assay will not work if either fusion protein fails to express. On the other hand, high protein expression of the fusion proteins could lead to nonspecific interaction [15]. Depending on the location of the epitope used to generate the antibody, a GFP antibody can be used to detect both the N- and C-Venus fragments. Protein specific antibodies can also be used to check for fusion protein expression.

9. If protein folding or PPI is affected by the fused reporter fragment(s), the problem can be mitigated by changing the fusion site and/or reporter fragment. There are 8 possible fusion combinations for each pair of proteins (e.g., NV-proteinA and CV-proteinB, CV-proteinA and NV-proteinB, NV-proteinA and proteinB-CV). For a novel pair of proteins, it is imperative to try all eight pairing combinations before deciding that the proteins do not interact.

10. Controls are essential to assess the success of the BiFC assay. Negative controls are necessary to rule out false-positive, strong fluorescence signals from a pair of non-interacting proteins. Positive controls allow the researcher to determine whether the assay worked or did not work. If possible, choose a protein pair that can serve as positive PPI (wild type proteins) as well as negative PPI controls (non-interacting mutants). One can also use the BiFC fragments (such as pSCM167-NV-*P38* and pDEST26-CV, pDEST26-CV-*MKK3* and pSCM167-NV, and pDEST26-CV and pSCM167-NV) as negative controls to monitor self-assembly of the BiFC fragments. To monitor transfection efficiency, one also needs to consider using an internal control reporter-plasmid (i.e., full length fluorescent protein of a different color, luciferase reporter proteins, etc.).

11. Good quality DNA (260/280 ratio close to 2.0) is essential for a successful assay. Poor quality DNA can reduce transfection efficiency or can be detrimental to the cell.

12. Make sure that FBS is present in the growth medium. FBS inhibits trypsin activity [16].

13. Too much DNA can be toxic to some cell lines. Moreover, too much DNA could lead to high background noise [15].

14. Various transfection reagents may be toxic to some cell lines. If using FuGENE® HD reagent, minimize contact of the reagent with plastic surface (i.e., microcentrifuge tube's wall) because this could lead to a decrease in transfection efficiency.

15. Strong interacting pairs of proteins can be detected as early as 24 h post-transfection. Long incubation times can lead to high background signal.

16. Self-assembly of the Venus BiFC fragments as well as of other BiFC fragments is often a drawback for this assay. Self-assembly of the BiFC fragments is a common contributor of false-positive results. Self-assembly signals are typically lower than protein-interaction assisted reconstitution signals for many PPIs. However, self-assembly signals may pose a problem for detecting weak PPIs. Replacing certain amino acids in BiFC systems (i.e., Venus) can reduce self-assembly [14].

Acknowledgements

I would like to thank Drs. Haian Fu and Jonathan Havel for their generous gifts of the pSCM167-NV and pDEST26-CV Gateway® destination vectors as well as constructive input to make this assay work.

References

1. Shimomura O, Johnson FH, Saiga Y (1962) Extraction, purification, and properties of aequorin, a bioluminescent protein from the luminous hydromedusan, Aequorea. J Cell Comp Physiol 59:223–239

2. Shaner NC, Steinbach PA, Tsien RY (2005) A guide to choosing fluorescent proteins. Nat Methods 2:905–909

3. Kremers G-J, Gilbert SG, Cranfill PJ, Davidson MW, Piston DW (2011) Fluorescent proteins at a glance. J Cell Sci 124:157–160

4. Remington SJ (2006) Fluorescent proteins: maturation, photochemistry, and photophysics. Curr Opin Struct Biol 16:714–721

5. Hu CD, Chinenov Y, Kerppola TK (2002) Visualization of interactions among bZIP and Rel family proteins in living cells using bimolecular fluorescence complementation. Mol Cell 9:789–798

6. Vidi PA, Watts VJ (2009) Fluorescent and bioluminescent protein-fragment complementation assays in the study of G protein-coupled receptor oligomerization and signaling. Mol Pharmacol 75:733–739

7. Ozawa T (2006) Designing split reporter proteins for analytical tools. Anal Chim Acta 556:58–68

8. Zhang Y (2008) I-TASSER server for protein 3D structure prediction. BMC Bioinformatics 9:40

9. Remy I, Michnick SW (2007) Application of protein-fragment complementation assays in cell biology. Biotechniques 42:137–145

10. Gandhi TK, Zhong J, Mathivanan S, Karthick L, Chandrika KN, Mohan SS, Sharma S, Pinkert S, Nagaraju S, Periaswamy B, Mishra G, Nandakumar K, Shen B, Deshpande N, Nayak R, Sarker M, Boeke JD, Parmigiani G, Schultz J, Bader JS, Pandey A (2006) Analysis of the human protein interactome and comparison with yeast, worm and fly interaction datasets. Nat Genet 38:285–293

11. Kerppola TK (2006) Design and implementation of bimolecular fluorescence complementation (BiFC) assays for the visualization of protein interactions in living cells. Nat Protoc 1:1278–1286

12. Michnick SW (2003) Protein fragment complementation strategies for biochemical network mapping. Curr Opin Biotechnol 14:610–6177

13. Raingeau J, Whitmarsh AJ, Barrett T, Derijard B, Davis RJ (1996) MKK3- and MKK6-regulated gene expression is mediated by the p38 mitogen-activated protein kinase signal transduction pathway. Mol Cell Biol 16:1247–1255

14. Kodama Y, Hu CD (2010) An improved bimolecular fluorescence complementation assay with a high signal-to-noise ratio. Biotechniques 49:793–805

15. Shyu YJ, Liu H, Deng X, Hu CD (2006) Identification of new fluorescent protein fragments for bimolecular fluorescence complementation analysis under physiological condition. Biotechniques 40:61–66

16. Rogers M, Dani JA (1995) Comparison of quantitative calcium flux through NMDA, ATP, and ACh receptor channels. Biophys J 68:501–506

Chapter 33

Split-Luciferase Complementation Assay to Detect Channel–Protein Interactions in Live Cells

Alexander S. Shavkunov, Syed R. Ali, Neli I. Panova-Elektronova, and Fernanda Laezza

Abstract

The understanding of ion channel function continues to be a significant driver in molecular pharmacology. In this field of study, protein-protein interactions are emerging as fundamental molecular determinants of ion channel function and as such are becoming an attractive source of highly specific targets for drug development. The investigation of ion channel macromolecular complexes, however, still relies on conventional methods that are usually technically challenging and time-consuming, significantly hampering our ability to identify, characterize and modify ion channel function through targeted molecular approaches. As a response to the urgent need of developing rapid and albeit accurate technologies to survey ion channel molecular complexes, we describe a new application of the split-luciferase complementation assay to study the interaction of the voltage-gated Na + channel with the intracellular fibroblast growth factor 14 and its dynamic regulation in live cells. We envision that the flexibility and accessibility of this assay will have a broad impact in the ion channel field complementing structural and functional studies, enabling the interrogation of protein–channel dynamic interactions in complex cellular contexts and laying the basis for new frameworks in drug discovery campaigns.

Key words Protein-protein interactions, Protein fragment complementation, Split luciferase complementation, Firefly luciferase, Bioluminescence, Two-way protein fragment complementation assay, Application of LCA, Sub-cloning of ion channel

1 Introduction

The split-luciferase complementation assay is a flexible and convenient method that allows the interrogation of a variety of protein-protein interactions in live cells [1–11]. Construction of a split-luciferase complementation assay requires cloning of the interacting proteins of interest into suitable mammalian expression vectors. The vectors are designed to express the target proteins in frame with complementary luciferase fragments spaced by a flexible linker region. The resulting DNA constructs are then cotransfected into mammalian cells, which are allowed to grow and

Cheryl L. Meyerkord and Haian Fu (eds.), *Protein-Protein Interactions: Methods and Applications*, Methods in Molecular Biology, vol. 1278, DOI 10.1007/978-1-4939-2425-7_33, © Springer Science+Business Media New York 2015

express the fusion proteins. The cells are then re-plated onto multi-well plates, substrate is added, and luminescence resulting from complementation of luciferase fragments is measured as a function of the protein-protein interaction of interest. At this step, additional experimental manipulations (e.g., pharmacological) can be introduced to test the effect on the protein-protein interaction of interest in live cells. Additional control experiments are utilized to rule out possible artifacts that may arise from changes in the reporter construct abundance (expression and/or degradation levels), viability and metabolic state of transfected cells, or direct impact on the reporter enzymatic activity non-related to the efficiency of the complex assembly. In this chapter, we describe a new application of LCA for the study of transmembrane ion channels.

Recent advancements in biology and pharmacology highlight the crucial role of protein-protein interactions for ion channel proper functioning and regulation [12–14]. The protein interfaces through which these interactions are mediated present a novel class of promising drug targets with high potential for improved specificity of therapeutic effects [15–19]. However, characterization of the interactions of ion channels with their regulatory proteins, as well as high-throughput screening for potential modulators of these interactions, proves to be quite challenging with conventional techniques [20, 21]. Direct analysis of protein-protein interactions by traditional biochemical methods, like enzyme-linked immunosorbent assays (ELISA), surface plasmon resonance (SPR), and fluorescence polarization (FP), is complicated by the nature of ion channels, which are difficult to express, purify, and reconstitute in vitro. The functional effect of protein binding to ion channels has been also studied using manual and/or automated patch-clamp electrophysiology [22], fluorescence-based methods [23], or ion flux assays [24]. Yet, despite certain advantages of these methods over biochemical assays (higher cost efficiency, no need for protein purification, measurements done in the cellular context), they possess inherent shortages and limitations of their own, such as lower processivity, complicated manual operation, etc.

These considerations stimulated us to seek alternative strategies to characterize the interaction between the voltage-gated sodium channel Nav1.6 and the intracellular fibroblast growth factor 14 (FGF14), a biologically relevant regulatory protein which controls gating, stability, and targeting of Nav channel α subunits (Nav1.1–Nav1.9) through a high affinity interaction with the intracellular C-terminal tail [25–30]. Our approach relied on the bioluminescence-based split luciferase complementation assay (LCA) introduced by Luker et al. [31]. This method is based on functional complementation between two separated fragments of the *Photinus pyralis* firefly luciferase which are fused to the pair of interacting proteins of interest. The interaction of the respective binding partners drives the complementation of the luciferase

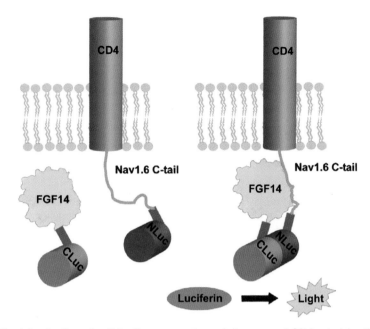

Fig. 1 Application of split luciferase complementation assay (LCA) to studying the FGF14-Nav interaction. The principle of the split luciferase complementation assay (LCA) for detection of the FGF14-Nav1.6Ctail channel complex formation

fragments leading to recovery of enzymatic activity and generation of luminescent signal in the presence of the substrate. The LCA provides a quantitative and reversible real-time readout of protein-protein interactions in vitro and in vivo and is the emerging alternative to fluorescence-based assays (for example, FRET [23]) that are limited by cellular autofluorescence, low signal-to-noise ratio, and narrow dynamic range [1, 32, 33]. It combines the processivity and scalability of conventional biochemical assays with the dynamic readout, more adequate biological context and higher cost efficiency of electrophysiological and other *in cellulo* approaches. The format of the assay and its reliability allow its application for high-throughput screening experiments.

We have successfully adapted the LCA to detect the assembly of the C-terminal tail of the Nav1.6 and FGF14 in live cells (Fig. 1) and demonstrated its applicability for identification of intracellular signaling pathways which modulate this interaction [10, 11].

2 Materials

2.1 Cell Lines and Bacterial Strains

1. HEK-293 cells (or another cell line that is easily transfected).

2. Chemically competent *E. coli* (such as One Shot TOP10 Chemically Competent *E. coli*, Invitrogen).

2.2 Tissue Culture

1. Dulbecco's Modified Eagle's Medium (DMEM).

2. DMEM without L-glutamate and Phenol Red.

3. F12 nutrient mixture.

4. Fetal bovine serum.

5. Penicillin–Streptomycin, liquid.

6. 10× Stock solution phosphate-buffered saline (PBS): 0.2 M phosphate (0.038 M NaH_2PO_4, 0.162 M Na_2HPO_4), 1.5 M NaCl, pH 7.4, titrated with HCl.

7. 0.25 % Trypsin–EDTA.

8. 25 cm^2 or 75 cm^2 cell culture flask.

9. 24-well tissue culture plates.

10. 96-well tissue culture white/μClear® plates (Greiner Bio-One).

11. Transfection reagent (such as Lipofectamine 2000, Invitrogen).

2.3 Cloning and DNA Preparation

1. FRB-NLuc (Gift from Dr. Piwnica-Worms, Washington University, St. Louis, MO [1]).

2. CLuc-FKBP (Gift from Dr. Piwnica-Worms, Washington University, St. Louis, MO [1]).

3. CD4-Nav1.6Ctail (Gift of Dr. Benedict Dargent, INSERM, France [34]).

4. FGF14-1b-GFP [28].

5. pGL3 firefly luciferase plasmid (Promega, Madison, WI).

6. Phusion High-Fidelity DNA Polymerase.

7. DNA restriction enzymes (*see* **Note 1**).

8. Primers.

9. T4 DNA ligase.

10. Gel Extraction Kit.

11. LB broth: 10 g of Bacto-Tryptone, 5 g of yeast extract, 10 g of NaCl dissolved in 1 l of distilled or deionized H_2O and sterilized by autoclaving.

12. LB agar: 10 g of Bacto-Tryptone, 5 g of yeast extract, 10 g of NaCl, 15 g of agar dissolved in 1 l of distilled or deionized H_2O with heating and sterilized by autoclaving.

13. Ampicillin.

14. Selection media: LB agar plates supplemented with 100 μg/ml ampicillin.

15. Miniprep Kit.

16. EndoFree Plasmid Maxi Kit.

2.4 Luciferase Assay

1. D-luciferin.

2. SP600125 (1,9-pyrazoloanthrone).

3. BAY 11-7082 ((E)3-[(4-methylphenyl) sulfonyl]-2-propenenitrile).

4. Dimethyl sulfoxide (DMSO).

2.5 Cell Viability Assay

1. Cell Proliferation Assay Kit (such as CyQUANT, Invitrogen).

2.6 Western Blot

1. Cell scraper.

2. 7.5 % SDS-PAGE gel.

3. PVDF membranes.

4. Lysis buffer: 20 mM Tris base, 150 mM NaCl, 1 % NP-40.

5. Protease inhibitor cocktail for mammalian cells.

6. 10× Tris/Glycine/SDS running buffer: 250 mM Tris–HCl, 1.92 M glycine, 1 % SDS, pH 8.3.

7. Transfer buffer: 20 mM Tris–HCl, 150 mM glycine, pH 8.0.

8. 4× SDS-PAGE sample buffer: 250 mM Tris–HCl, 40 % glycerol, 4 % SDS, 0.002 % bromophenol blue, pH 6.8.

9. Bond-Breaker TCEP solution: 50 mM tris(2-carboxyethyl) phosphine.

10. TBS-T: 20 mM Tris–HCl, 137 mM NaCl, pH 7.6, 0.1 % Tween 20.

11. Blocking buffer: 2.5 % nonfat dry milk in TBS-T buffer.

12. Anti-luciferase goat polyclonal antibody.

13. Anti-calnexin rabbit polyclonal antibody.

14. Peroxidase horse anti-goat IgG antibody.

15. Peroxidase goat anti-rabbit IgG antibody.

16. ECL Advance Western Blotting Detection kit.

2.7 Instrumentation

1. Thermal cycler.

2. UV/Vis spectrophotometer.

3. Sonicator.

4. Synergy H4 Multi-Mode Microplate Reader.

5. Power source.

6. Electrophoresis unit.

2.8 Analysis Software

1. Microsoft Excel.

2. Origin 8.6 (Origin Lab Corporation).

3. FluorChem HD2 System.

4. AlphaView 3.1 software (ProteinSimple).

3 Methods

3.1 DNA Manipulation

One of the major advantages of LCA is the possibility to monitor protein-protein interactions in live cells. This usually requires transfection of mammalian cells with DNA constructs expressing hybrid proteins bearing complementary fragments of the luciferase reporter. As an example, we describe the sub-cloning of the DNA coding sequences of our proteins of interest into the expression vectors encoding complementary *Photinus pyralis* luciferase fragments which were designed and constructed by Luker et al. [31]. The resulting plasmids were used to express hybrid protein constructs CLuc-FGF14, containing the C-terminus fragment (CLuc, aa. 398–550) of the luciferase and the full-length human FGF14 isoform 1b, and CD4-Nav1.6-NLuc, containing a chimera of the CD4ΔCtail (aa. 1–395) and the Nav1.6 C-tail (aa. 1763–1976) fused to the N-terminus fragment (NLuc, aa. 2–416) of the luciferase (Figs. 1 and 2). Using the CD4-Nav1.6 chimera has a number advantages, as it is expressed more efficiently than the full-length Nav1.6 in a heterologous system, ensures proper membrane targeting [34] and exposes the Nav channel C-tail on the plasma membrane in the correct orientation. This also allows for the isolation of the C-tail from the rest of the Nav channel, limiting any potential indirect modulatory effects on the protein–channel complex induced by other intracellular domains of the Nav channel. The design of fusions between the reporter fragments and the proteins of interest may vary depending on the particular pair of interaction partners. The N-Luc or the C-Luc fragment of the reporter may be placed at either terminus of the polypeptide chain of each of the proteins of interest. Important factors to consider are the spatial orientation of the interaction partners, in case structural

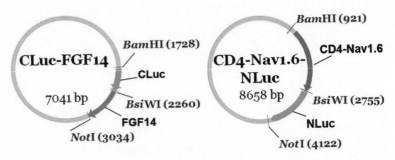

Fig. 2 Map of the expression vectors used for of the LCA. The detailed description of the cloning procedure can be found in Subheading 3.1

information is available, as well as their cellular localization and membrane topology (in the case of peripheral or integral membrane proteins).

1. Amplify the coding sequence of FGF14-1b by PCR using the following primers containing 5'-*BsiW*I and 3'-*Not*I restriction sites:

   Sense: 5'-CTCGTACGCGTCCCGGGGCGTAAAACCGGT GCCCCTCTTC-3'.
   Antisense: 5'-GTTTAGCGGCCGCCTATGTTGTCTTACTC TTGTTGACTGG-3'.
   Use the following conditions:

Segment	# of cycles	Temperature (°C)	Duration
1	1	95	2 min
2	30	95	30 s
		60	30 s
		72	1 min
3	1	72	10 min
4	Hold	4	∞

2. Amplify the coding sequence of CD4-Nav1.6 by PCR using the following primers containing 5'-*BamH*I and 3'-*BsiW*I restriction sites:

   Sense: 5'-CGGGGTACCCAAGCCCAGAGCCCTGCCATTT-CTGTGGGCTCAGGT-3.
   Antisense: 5'-CGCGTACGAGATCTGGCACTTGGACTCC-CTGACCT CTTTTTGCCT-3'.
   Use the following conditions:

Segment	# of cycles	Temperature (°C)	Duration
1	1	95	2 min
2	30	95	30 s
		65	30 s
		72	3.5 min
3	1	72	10 min
4	Hold	4	∞

3. Resolve the PCR products by electrophoresis in 1 % agarose gel and purify them using a gel extraction kit according to the manufacturer's instructions.

4. Measure the concentrations of the purified PCR products by UV spectrophotometry (optical density at 260 nm).

5. Digest 50 ng of the FGF14-1b PCR product with 5 units of *Not*I for 8 h at 37 °C; add 5 units of *BsiW*I and continue digestion overnight at 55 °C (*see* **Note 1**).

6. Digest 1 μg of CLuc-FKBP with 5 units of *Not*I for 8 h at 37 °C; add 5 units of *Bsi*WI and continue digestion overnight at 55 °C.

7. Digest 50 ng of the CD4-Nav1.6 PCR product with 5 units of *Bam*HI for 8 h at 37 °C; add 5 units of *Bsi*WI and continue digestion overnight at 55 °C.

8. Digest 1 μg of FRB-NLuc with 5 units of *Bam*HI for 8 h at 37 °C; add 5 units of *Bsi*WI and continue digestion overnight at 55 °C.

9. Resolve the digested fragments by electrophoresis in 1 % agarose gel and purify the required fragments using a gel extraction kit according to the manufacturer's instructions.

10. Measure the concentrations of the purified DNA fragments by UV spectrophotometry (optical density at 260 nm).

11. Set up ligation reactions for FGF14 coding sequence with CLuc and CD4-Nav1.6 with NLuc in 10 μl final volume using 400 units of T4 DNA ligase per reaction according to the manufacturer's instructions. A typical reaction contains 60 ng (CD4-Nav1.6) or 33 ng (FGF14-1b) of the insert and 100 ng of the corresponding plasmid vector, with the vector and the insert at 1:3 molar ratio (*see* **Note 2**). Incubate overnight at 16 °C.

12. Use 0.5–5 μl of the ligation reaction mixture to transform 20 μl of competent *E. coli* cells, add 200 μl of LB broth, and incubate at 37 °C for 1 h prior to plating on selective LB agar plates supplemented with 100 μg/ml ampicillin.

13. Select individual colonies and grow 5-ml overnight cultures in LB broth containing 100 μg/ml ampicillin.

14. Purify plasmid DNA using a Miniprep Kit and confirm the identity of recombinant plasmids by restriction digestion and DNA sequencing.

15. Select the colonies which yielded correct recombinant plasmids and grow 500-ml cultures in LB broth containing 100 μg/ml ampicillin.

16. Purify plasmid DNA using EndoFree Plasmid Maxi Kit.

3.2 Transient Transfection of HEK293 Cells

1. Grow and maintain HEK293 cells in 25 cm^2- or 75 cm^2-tissue culture flasks at 37 °C with 5 % CO_2 in medium composed of equal volumes of DMEM and F12 supplemented with 10 % fetal bovine serum, 0.05 % glucose, 0.5 mM pyruvate, 100 U/ml penicillin, and 100 μg/ml streptomycin (supplemented DMEM/F12).

2. Plate 4.5×10^5 HEK293 cells per each well of a 24-well tissue culture plate and incubate overnight to give monolayers at 90–100 % confluency (*see* **Note 3**).

3. The next day, pre-warm sterile PBS and DMEM/F12 without serum and antibiotics at 37 °C.

4. Prepare two Eppendorf tubes with equal volumes of DMEM/F12 without serum or antibiotics (50 μl of medium per each well of cells in the 24-well plate).

5. Add 1 μg of CD4-Nav1.6-NLuc and 1 μg of CLuc-FGF14 plasmid DNA per each well to be transfected into tube 1.

6. Add 2 μl of transfection reagent (e.g., Lipofectamine 2000) per each well to be transfected (1 μl of the transfection reagent per 1 μg of plasmid DNA) into tube 2. Incubate for 5 min at room temperature (*see* **Note 4**).

7. Transfer the medium containing DNA from tube 1 into tube 2 with the medium containing the transfection reagent. Mix by gentle tapping (do not vortex). Incubate the transfection mixture for 20 min at room temperature.

8. During incubation, aspirate the medium from the wells of the 24-well plate with HEK293 cells, wash briefly with PBS and replace with 100 μl per well of DMEM/F12 without serum or antibiotics.

9. Carefully dispense 100 μl of the transfection mixture to the respective well of the 24-well plate containing cells (*see* **Note 5**).

10. Incubate the plate with the cells for 6 h at 37 °C with 5 % CO_2.

11. Add 800 μl of supplemented DMEM/F12 per well and let the cells grow for 24 h (*see* **Note 6**).

3.3 Preparation of Transiently Transfected HEK293 Cells for Luminescence Measurement

1. 24 h post transfection aspirate the culture medium from the 24-well plate with transfected HEK293 cells (*see* **Note 7**).

2. Wash each well briefly with 400 μl of PBS and dispense 100 μl of 0.04 % trypsin solution (0.25 % trypsin diluted with PBS) per well (*see* **Note 8**).

3. Place the plate in the CO_2 incubator and monitor the cells periodically under a microscope (*see* **Note 9**).

4. As soon as the cells detach from the plastic (typically 3–5 min) add 800 μl of supplemented DMEM/F12 per each well to stop trypsinization.

5. Wash the cells off the plastic by careful pipetting and transfer the cell suspension from individual wells into a 15-ml tube with a conical bottom.

6. Centrifuge for 5 min at $800 \times g$.

7. Aspirate the medium and resuspend the cells in fresh supplemented DMEM/F12 (800 μl of medium per each well of transfected cells from the 24-well plate).

8. Dispense 200 μl of the cell suspension into each well of a 96-well plate with white walls and clear flat bottom (*see* **Note 10**).

9. Place the plate on a rotary shaker at low speed for 2 min. Check the wells under microscope to make sure that the cells are evenly distributed across the bottom.

10. Let the cells grow for 48 h.

3.4 Luminescence Reader Method Setup and Data Collection

1. Set up the experimental protocol for the Synergy H4 Multi-Mode Microplate Reader with the following parameters

   • Maintain temperature at 37 °C.

   • Dispense 100 μl of substrate solution per well prior to measurement.

   • Shake the plate for 3 s.

   • Perform luminescence readings for 30 min at 2 min intervals (open hole; integration time 0.5 s, sensitivity 245).

2. Replace the culture medium in the wells of the 96-well plate with HEK293 cells (prepared as described in Subheadings 3.2 and 3.3) with 100 μl of DMEM/F12 without Phenol Red (*see* **Note 11**).

3. Prepare working solution of 1.5 mg/ml of D-luciferin in PBS (*see* **Note 12**).

4. Prime the substrate dispenser of the Synergy H4 Multi-Mode Microplate Reader with 1,200 μl of D-luciferin working solution.

5. Load the 96-well plate with cells into the Synergy H4 and initiate the experimental protocol sequence.

3.5 Identification of Compounds Modulating the Protein-Protein Interaction Using the LCA

The LCA provides opportunities to test the effects of various pharmacological agents on the stability of a protein complex [1]. We have successfully applied this methodology to identify protein kinase pathways which modulate the interaction between FGF14 and the Nav channel (Fig. 3a, b) [10, 11]. It is important to exclude false positive hits due to direct influence of the compound on the luciferase enzymatic activity; this type of artifact has been reported previously with conventional luciferase reporter systems [5, 35]. We address this issue by parallel testing of candidate inhibitors on transfected cells expressing full length *Photinus pyralis* luciferase (from which the NLuc and CLuc fragments of the reporter have been engineered) treated under the same conditions as in the LCA experiment (Fig. 3c).

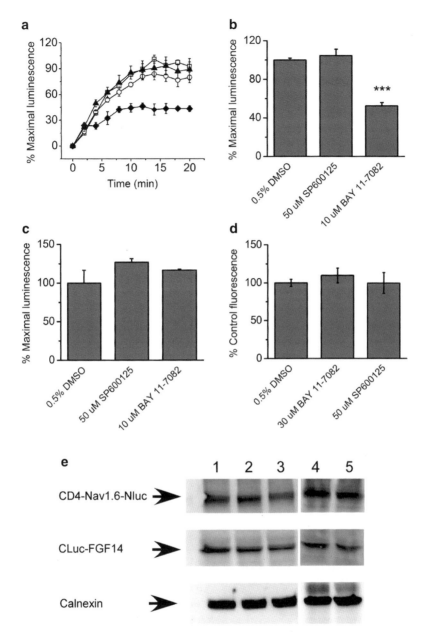

Fig. 3 The FGF14-Nav1.6Ctail complex formation is regulated by specific kinase inhibitors. (**a**) HEK293 cells were transiently transfected with CLuc-FGF14 and CD4-Nav1.6-NLuc and treated with the c-JNK inhibitor SP600125 (50 µM; *closed triangle*), the IKK inhibitor, BAY 11-7082 (10 µM; *closed diamond*) or dimethylsulf-oxide (DMSO; 0.5 % control; *open circle*); control without treatment is also shown (*open square*). Assembly of the LCA pair is detected as luminescence upon the addition of the D-luciferin substrate at time zero and normalized to % maximal luminescence signal in the untreated control; data are mean ± SEM, representing quadruplicates from one representative experiment. (**b**) Bar graph represents mean ± SEM expressed as % maximal luminescence of control (0.5 % DMSO) from at least three independent experiments. The graphs illustrates the effect of 50 µM SP600125 and 10 µM BAY 11-7082 on the FGF14-Nav1.6 channel C-tail assembly. BAY 11-7082 causes a twofold reduction of the LCA reporter assembly and luminescence output, while SP600125 has no significant impact. (**c**) HEK293 cells were transiently transfected pGL3 expressing

1. Transfect HEK293 cells with the vectors expressing CLuc-FGF14 and CD4-Nav1.6-NLuc and prepare them for luciferase activity measurement as described above in Subheadings 3.2 and 3.3.

2. Transfect HEK293 cells with the pGL3 plasmid expressing full length firefly luciferase and prepare them following the same procedure; use 1 μg of plasmid DNA and 1 μl of transfection reagent per each well of a 24-well plate.

3. Wash the cells briefly with pre-warmed sterile PBS and change the medium to DMEM/F12 without Phenol Red, serum and antibiotics, 100 μl per well of the 96-well plate.

4. Prepare dilutions of protein kinase inhibitors in DMSO. The inhibitor concentration in the intermediate stocks should be 200× the intended final concentration in the culture medium (*see* **Note 13**).

5. Dispense 0.5 μl of protein kinase inhibitor solution into each experimental well and 0.5 μl of DMSO into each control well of the 96-well plate with transfected cells (*see* **Note 14**).

6. Gently pipette the culture medium in each well one to two times to ensure uniform concentration of the inhibitor throughout the volume and to prevent it from precipitating.

7. Incubate the plate for 1 h at 37 °C with 5 % CO_2.

8. Proceed to luminescence measurement (as described in Subheading 3.4).

3.6 Data Analysis

1. Export the relative luminescence values measured by Synergy H4 and organized by well position and time point into Microsoft Excel.

2. Plot the luminescence values of the corresponding experimental wells as a function of time.

3. To compare the luminescence levels across experimental conditions, calculate maximal signal intensity for each well as the

Fig. 3 (Continued) full-length *Photinus pyralis* firefly luciferase holoenzyme and treated for 1 h prior to the assay with the indicated compounds. Bar graph expressed as % maximal luminescence illustrates the effect of the indicated compounds on the intrinsic enzymatic activity of luciferase; no statistically significant change in full-length lucidferase activity was observed. (**d**) HEK293 cells were transiently transfected with CLuc-FGF14 and CD4-Nav1.6-NLuc and treated with the indicated compounds. The effect of compounds on cell viability was determined using a CyQUANT Cell Proliferation Assay Kit (Promega). Bar graph expressed as % control fluorescence illustrates the effect of the indicated compounds on cell viability. Data are mean ± SEM; ***$p < 0.001$, Student's t-test. (**e**) Representative example of Western blots of lysates (equal amount of protein per lane) from cells transfected with CLuc-FGF14 + CD4-Nav1.6-NLuc and treated for 1 h with 0.5 % DMSO (2, 4), 50 μM SP600125 (3), and 10 μM BAY 11-7082 (5). Untreated control is also shown (1). Western blots were probed with a polyclonal anti-luciferase antibody; immunodetection of calnexin was used as loading control

average of maximum luminescence values at three consecutive time points (*see* **Note 15**).

4. Plot the results using appropriate graphical software (e.g., Origin 8.6).

3.7 Cell Viability Testing

Cell toxicity of test compounds is one of the factors which may affect the LCA output and produce false-positive results. To account for this type of artifact, positive identification of compounds by the LCA should be followed by tests on cell viability under conditions identical to the LCA (Fig. 3d).

1. Transfect the cells with the LCA reporter constructs as described in Subheading 3.2 and re-plate them onto a 96-well plate as in Subheading 3.3.

2. Treat the cells with the test compounds as described in Subheading 3.4, under the same conditions as in the LCA experiment.

3. After incubation with the test compounds carefully aspirate the culture medium and wash the cells with PBS.

4. Aspirate the PBS, freeze the cells, and store the plate at $-80\,^\circ\mathrm{C}$ until the samples are to be assayed (*see* **Note 16**).

5. Prepare the working solution for the cell proliferation assay (the procedure below is for the CyQUANT assay).
   - Dilute Component B (Cell lysis buffer) 1:20 in nuclease-free distilled H_2O (1 ml +19 ml).
   - Thaw Component A (CyQUANT GR Dye) at room temperature; add 250 µl (1:80) to the solution of Component B prepared as above (*see* **Note 17**).

6. Thaw the cells at room temperature and add 200 µl of the working solution to each well, including empty wells as blanks.

7. Incubate for 2–5 min at room temperature protected from light.

8. Measure fluorescence using a Synergy™ H4 Multi-Mode Microplate Reader (excitation $\lambda = 485$ nm, emission $\lambda = 528$ nm) or any suitable microplate reader with filters appropriate for ~480 nm excitation and ~520 nm emission maxima.

9. Subtract the mean fluorescence value of the blank from fluorescence values of the wells containing cells; calculate relative cell viability as percent mean fluorescent signal intensity in the mock-treated control samples from the same experimental plate.

4 SDS-Polyacrylamide Gel Electrophoresis and Western Blotting

One of the specific concerns when applying LCA for compound screening is a possibility that the compounds may affect reporter activity by changing the expression levels or degradation rates of the reporter proteins. Thus, it is important to verify the reporter expression level in HEK293 cells after treatment with inhibitors positively identified by LCA under identical conditions (Fig. 3e). Polyclonal anti-luciferase antibody allows for the detection of both reporter constructs through their corresponding luciferase fragments; thus, no specific antibodies are required for each new pair of constructs, and the relative expression level of different constructs can be compared more easily (*see* **Note 18**).

1. Wash transfected HEK293 cells in the wells of the 24-well plate with warm PBS (*see* **Note 19**).

2. Add 50 μl of lysis buffer freshly supplemented with protease inhibitor cocktail for mammalian cells.

3. Incubate the plate with cells on ice for 5 min with periodic stirring.

4. Scrape the cells off the plastic with a cell scraper, triturate by careful pipetting and transfer lysate from each well into a microcentrifuge tube.

5. Sonicate the cell suspensions on ice for 20 s with 1 s pulses.

6. Clear the lysates by centrifugation ($15,000 \times g$, 15 min at 4 °C) and transfer the supernatant into new tubes.

7. Add 33 μl of 4× SDS-PAGE sample buffer and 11 μl of Bond-Breaker TCEP solution to each sample.

8. Incubate the samples at 65 °C for 10 min; let them cool down to room temperature and spin briefly.

9. Load 50 μl of each sample and 5 μl of a protein molecular mass standard onto a 7.5 % SDS-PAGE gel and run electrophoresis at 120 V constant.

10. Open the cassette with the gel, assemble it with the PVDF membrane and load into an electrophoretic transfer cell with prechilled transfer buffer. Place the transfer cell in an ice bucket.

11. Run electrophoretic transfer for 2 h at 75 V constant.

12. Incubate the membrane with the transferred protein in blocking buffer for 30 min at room temperature.

13. Incubate the membrane overnight with anti-luciferase goat polyclonal antibody or anti-calnexin rabbit polyclonal antibody (1:1,000 final dilution) in blocking buffer at 4 °C.

14. Wash the membrane with TBS-T three times for 10 min.

15. Incubate the membrane with HRP-labeled goat anti-rabbit or horse anti-goat secondary antibody (1:5,000).

16. Wash the membrane with TBS-T three times for 10 min.

17. Incubate the membrane with luminescent substrate solution for 5 min at room temperature.

18. Visualize luminescent signal and record the blot image using FluorChem HD2 System and AlphaView 3.1 software.

19. Measure the pixel intensity of protein bands using AlphaView 3.1. Normalize the raw values of the bands of interest in each sample to the corresponding values of calnexin (loading control).

5   Future Directions: Dual-Color Split Luciferase Complementation Assay (DC-LCA)

A complex problem in protein-protein interaction studies is to characterize two or more interactions occurring simultaneously. Advancement in this direction is the introduction of a new type of LCA, dual-color split luciferase complementation assay (DC-LCA), which allows the interaction of a particular protein with two binding partners to be simultaneously monitored [36]. This approach utilizes two N-terminal fragments derived from structurally related luciferases of different click beetle species, one with green light and the other one with red light emission, and a common C-terminal fragment originating from the green-emitting luciferase. The C-terminal fragment fused to the protein of interest produces luminescence in the presence of D-luciferin upon interaction with either of the two N-terminal fragments fused to the protein binding partners, resulting in red or green luminescence production depending on the type of protein pair driving the reporter assembly. The DC-LCA method could resolve the important question of whether FGF14 binds to the Nav channel as a monomer, as suggested by structural studies [26], or in a dimeric form, which would greatly advance our knowledge of the mechanisms underlying ion channel regulation.

6   Notes

1. We recommend High Fidelity BamHI and NotI enzymes to avoid star activity.

2. Calculations for the optimal vector-insert ratio can be found at http://www.insilico.uni-duesseldorf.de/Lig_Input.html.

3. Typically, three 25-cm^2 tissue culture flasks or one 75-cm^2 tissue culture flask yields enough cells for two 24-well tissue culture plates.

4. You must proceed to the next step within 25 min.

5. HEK293 cells easily come off the plastic, and therefore it is important to dispense the media very carefully. Cell loss

can be minimized if the media is slowly poured along the wall of the wells.

6. It is normal to observe some cell loss after transfection. However, it is recommended to check the plate with transfected cells under a microscope and discard wells with abnormally high cell loss.

7. Do not leave the cells without media. This becomes critical when cells are split (passaged) into the 96-well plate. We recommend aspirating one row and then dispensing the media before proceeding to the second row of the 96-well plates.

8. To minimize cell loss from aspiration, we recommend using a 10× pipette tip fitted on a glass Pasteur pipette to reduce the suction of liquid flow.

9. Avoid long exposure of cells to trypsin.

10. It is recommended to use 96-well plates with non-transparent (black or white) walls to avoid the bleed-through of the luminescent signal between the adjacent wells. A clear (transparent) bottom well enables routine microscopic examination of the cell monolayer quality.

11. Using medium without Phenol Red helps to maximize the luminescence output.

12. Prepare stock solution of D-luciferin in PBS (30 mg/ml) and store aliquots of appropriate size at −20 °C. Keep solutions containing luciferin protected from light.

13. Protein kinase inhibitors are generally better soluble and more stable in DMSO than in aqueous solutions. It is recommended to prepare a stock solution of the inhibitor in DMSO and store it at −80 °C in aliquots of appropriate size (e.g., 10 μl) to avoid multiple freezing and thawing. Using a 20 mM stock of a kinase inhibitor is convenient for most applications.

   Addition of DMSO to the culture medium helps to prevent kinase inhibitors from precipitating and facilitates intracellular penetration. However, high concentrations of DMSO may be detrimental, largely due to compromised integrity of the cell membrane. In our experiments, concentrations of DMSO in the culture medium up to 1 % had no significant effect on the split luciferase reporter activity after 1 h of incubation (unpublished data); however, in cell-based assays it is generally not recommended to exceed 0.5 % of DMSO in the culture medium [37]. Using 200× intermediate dilutions of the inhibitors provides a 0.5 % final concentration of DMSO in the medium with different inhibitor final concentrations. It is recommended to keep the DMSO and protein kinase inhibitor solutions protected from direct light.

14. It is recommended to have several identically treated wells of a 96-well plate per each experimental condition. We typically use 4 wells per condition (quadruplicate).

15. Normalization of signal intensities across each experimental plate to the average luminescence value of the untreated control wells in a corresponding plate helps to improve reproducibility of results between individual experiments and minimize variability arising from non-related factors (e.g., overall transfection efficiency).

16. Samples for the CyQUANT Cell Proliferation Assay can be stored at -80 °C for up to 4 weeks.

17. The working solution can be prepared in advance and stored for up to 2–3 h at room temperature protected from light.

18. Because HEK293 cells easily detach from plastic during consecutive steps of washing and changing solutions (*see* **Note 5**) in-cell Western assay and similar techniques are not optimal for this purpose. Besides, this type of assay does not allow discrimination between the two reporter constructs when immunodetection is done using anti-luciferase polyclonal antibody.

19. It is preferable to pre-warm PBS to 37 °C to prevent cells coming off the plastic.

References

1. Luker KE, Smith MC, Luker GD et al (2004) Kinetics of regulated protein–protein interactions revealed with firefly luciferase complementation imaging in cells and living animals. Proc Natl Acad Sci U S A 101:12288–12293

2. Misawa N, Kafi AK, Hattori M et al (2010) Rapid and high-sensitivity cell-based assays of protein–protein interactions using split click beetle luciferase complementation: an approach to the study of G-protein-coupled receptors. Anal Chem 82:2552–2560

3. Paulmurugan R, Gambhir SS (2003) Monitoring protein–protein interactions using split synthetic renilla luciferase protein-fragment-assisted complementation. Anal Chem 75:1584–1589

4. Paulmurugan R, Umezawa Y, Gambhir SS (2002) Noninvasive imaging of protein–protein interactions in living subjects by using reporter protein complementation and reconstitution strategies. Proc Natl Acad Sci 99:15608–15613

5. Herbst KJ, Allen MD, Zhang J (2009) The cAMP-dependent protein kinase inhibitor H-89 attenuates the bioluminescence signal produced by *Renilla* luciferase. PLoS One 4:e5642

6. Herbst KJ, Allen MD, Zhang J (2011) Luminescent kinase activity biosensors based on a versatile bimolecular switch. J Am Chem Soc 133:5676–5679

7. Stefan E, Aquin S, Berger N et al (2007) Quantification of dynamic protein complexes using Renilla luciferase fragment complementation applied to protein kinase A activities in vivo. Proc Natl Acad Sci U S A 104:16916–16921

8. Thorne N, Inglese J, Auld DS (2010) Illuminating insights into firefly luciferase and other bioluminescent reporters used in chemical biology. Chem Biol 17:646–657

9. Auld DS, Thorne N, Nguyen DT et al (2008) A specific mechanism for nonspecific activation in reporter-gene assays. ACS Chem Biol 3:463–470

10. Shavkunov A, Panova N, Prasai A et al (2012) Bioluminescence methodology for the detection of protein–protein interactions within the voltage-gated sodium channel macromolecular complex. Assay Drug Dev Technol 10:148–160

11. Shavkunov AS, Wildburger NC, Nenov MN et al (2013) The fibroblast growth factor 14 (FGF14)/voltage-gated sodium channel complex is a new target of glycogen synthase kinase

3 (GSK3). J Biol Chem 288 (27):19370–19385

12. Thayer DA, Jan LY (2010) Mechanisms of distribution and targeting of neuronal ion channels. Curr Opin Drug Discov Devel 13:559–567

13. Leterrier C, Brachet A, Fache MP et al (2010) Voltage-gated sodium channel organization in neurons:protein interactions and trafficking pathways. Neurosci Lett 486:92–100

14. Catterall WA (2010) Signaling complexes of voltage-gated sodium and calcium channels. Neurosci Lett 486:107–116

15. Jubb H, Higueruelo AP, Winter A et al (2012) Structural biology and drug discovery for protein–protein interactions. Trends Pharmacol Sci 33:241–248

16. Smith MC, Gestwicki JE (2012) Features of protein–protein interactions that translate into potent inhibitors: topology, surface area and affinity. Expert Rev Mol Med 14:e16

17. Wells JA, McClendon CL (2007) Reaching for high-hanging fruit in drug discovery at protein–protein interfaces. Nature 450:1001–1009

18. Buchwald P (2010) Small-molecule protein–protein interaction inhibitors: therapeutic potential in light of molecular size, chemical space, and ligand binding efficiency considerations. IUBMB Life 62:724–731

19. Zinzalla G, Thurston DE (2009) Targeting protein–protein interactions for therapeutic intervention: a challenge for the future. Future Med Chem 1:65–93

20. Clare JJ (2010) Targeting ion channels for drug discovery. Discov Med 9:253–260

21. Wickenden A, Priest B, Erdemli G (2012) Ion channel drug discovery: challenges and future directions. Future Med Chem 4:661–679

22. Kiss L, Bennett PB, Uebele VN et al (2003) High throughput ion-channel pharmacology: planar-array-based voltage clamp. Assay Drug Dev Technol 1:127–135

23. Masi A, Cicchi R, Carloni A et al (2010) Optical methods in the study of protein–protein interactions. Adv Exp Med Biol 674:33–42

24. Terstappen GC (2004) Nonradioactive rubidium ion efflux assay and its applications in drug discovery and development. Assay Drug Dev Technol 2:553–559

25. Laezza F, Lampert A, Kozel MA et al (2009) FGF14 N-terminal splice variants differentially modulate Nav1.2 and Nav1.6-encoded sodium channels. Mol Cell Neurosci 42:90–101

26. Goetz R, Dover K, Laezza F et al (2009) Crystal structure of a fibroblast growth factor homologous factor (FHF) defines a conserved surface on FHFs for binding and modulation of voltage-gated sodium channels. J Biol Chem 284:17883–17896

27. Laezza F, Gerber BR, Lou JY et al (2007) The FGF14(F145S) mutation disrupts the interaction of FGF14 with voltage-gated Na + channels and impairs neuronal excitability. J Neurosci 27:12033–12044

28. Lou JY, Laezza F, Gerber BR et al (2005) Fibroblast growth factor 14 is an intracellular modulator of voltage-gated sodium channels. J Physiol 569:179–193

29. Goldfarb M, Schoorlemmer J, Williams A et al (2007) Fibroblast growth factor homologous factors control neuronal excitability through modulation of voltage-gated sodium channels. Neuron 55:449–463

30. Dover K, Solinas S, D'Angelo E et al (2010) Long-term inactivation particle for voltage-gated sodium channels. J Physiol 589(Pt 6):1505

31. Luker KE, Piwnica-Worms D (2004) Optimizing luciferase protein fragment complementation for bioluminescent imaging of protein–protein interactions in live cells and animals. Methods Enzymol 385:349–360

32. Yang KS, Ilagan MX, Piwnica-Worms D et al (2009) Luciferase fragment complementation imaging of conformational changes in the epidermal growth factor receptor. J Biol Chem 284:7474–7482

33. Ilagan MX, Lim S, Fulbright M et al (2011) Real-time imaging of notch activation with a luciferase complementation-based reporter. Sci Signal 4:rs7

34. Garrido JJ, Fernandes F, Giraud P et al (2001) Identification of an axonal determinant in the C-terminus of the sodium channel Na(v)1.2. EMBO J 20:5950–5961

35. Thorne N, Auld DS, Inglese J (2010) Apparent activity in high-throughput screening: origins of compound-dependent assay interference. Curr Opin Chem Biol 14:315–324

36. Villalobos V, Naik S, Bruinsma M et al (2010) Dual-color click beetle luciferase heteroprotein fragment complementation assays. Chem Biol 17:1018–1029

37. Lazo JS, Brady LS, Dingledine R (2007) Building a pharmacological lexicon: small molecule discovery in academia. Mol Pharmacol 72:1–7

Chapter 34

Confocal Microscopy for Intracellular Co-localization of Proteins

Toshiyuki Miyashita

Abstract

Confocal laser scanning microscopy is the best method to visualize intracellular co-localization of proteins in intact cells. Because of the point scan/pinhole detection system, light contribution from the neighborhood of the scanning spot in the specimen can be eliminated, allowing high Z-axis resolution. Fluorescence detection by sensitive photomultiplier tubes allows the usage of filters with a narrow bandpath, resulting in minimal cross-talk (overlap) between two spectra. This is particularly important in demonstrating co-localization of proteins with multicolor labeling. Here, the methods outlining the detection of transiently expressed tagged proteins and the detection of endogenous proteins are described. Ideally, the intracellular co-localization of two endogenous proteins should be demonstrated. However, when antibodies raised against the protein of interest are unavailable for immunofluorescence or the available cell lines do not express the protein of interest sufficiently enough for immunofluorescence, an alternative method is to transfect cells with expression plasmids that encode tagged proteins and stain the cells with anti-tag antibodies. However, it should be noted that the tagging of proteins of interest or their overexpression could potentially alter the intracellular localization or the function of the target protein.

Key words Confocal microscopy, Immunofluorescence, Multicolor labeling, Laser, Fluorophore

1 Introduction

Confocal laser scanning microscopy is the best method to visualize intracellular co-localization of proteins in intact cells. Confocal laser scanning microscopy offers significant advantages for viewing sub-cellular localization of proteins compared with conventional fluorescence microscopy.

First of all, the point scan/pinhole detection system eliminates light contribution from the neighborhood of the scanning spot in the specimen (Fig. 1). Therefore, high Z-axis resolution as well as an X-Y image can be obtained. Fluorescence detection by sensitive photomultiplier tubes (PMT) allows the usage of filters with a narrow bandpath, resulting in minimal cross-talk (overlap) between two spectra. This is particularly important in demonstrating the

Cheryl L. Meyerkord and Haian Fu (eds.), *Protein-Protein Interactions: Methods and Applications*, Methods in Molecular Biology, vol. 1278, DOI 10.1007/978-1-4939-2425-7_34, © Springer Science+Business Media New York 2015

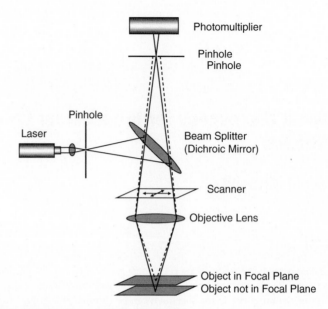

Fig. 1 Schematic diagram of the light path

co-localization of proteins with multicolor labeling. In addition, digital images obtained with PMT are easy to record, modify, and transfer. Transmitted light images can be obtained by use of a transmitted light detector, but the images will not be confocal images. Such transmitted light image can be used in combination with a fluorescence image to demonstrate fluorescent localization. Notably, co-localization of two proteins does not necessarily mean that there is a physical association between the proteins. Other methodologies, such as co-immunoprecipitation, should be used to confirm the interaction.

The procedure described in this chapter includes transient transfection of constructs which encode GFP- and HA-tagged proteins, followed by immunofluorescence using anti-HA antibody. Finally, the interaction of the two proteins of interest is observed under a confocal laser scanning microscope by demonstrating the intracellular co-localization after a merged image is generated. The procedure to detect the intracellular co-localization of two endogenous proteins by multicolor labeling is also described.

2 Materials

1. Confocal laser scanning microscope (CLSM) (e.g., Olympus FLUOVIEW FV300 or equivalent).

2. Tissue culture equipment.

3. HeLa cells.

4. DMEM medium with 10 % fetal bovine serum.

5. Constructs encoding GFP-tagged or HA-tagged proteins of interest.

6. Chamber slide with cover.

7. Transfection reagent (Effectene transfection reagent (QIAGEN) is used as an example).

8. PBS (phosphate-buffered saline).

9. 4 % Paraformaldehyde in PBS.

10. Permeabilization solution: 0.1 % TritonX-100 in PBS.

11. Preblock solution: 10 mM Tris–HCl, pH 8.0, 150 mM NaCl, 0.1 % Tween 20, 5 % skim milk, 2 % bovine serum albumin (should be filtered through filter paper just before use).

12. Anti-HA mouse monoclonal antibody (HA.11, MMS-101R, CRP Inc.) (*see* **Note 1**).

13. Secondary antibodies: TRITC (tetramethylrhodamine-isothiocyanate)-labeled rabbit anti-mouse immunoglobulin (R0270, DAKO), Alexa Fluor 488-labeled goat anti-mouse immunoglobulin (A-11029, Molecular Probes), Alexa Fluor 546-labeled goat anti-rabbit immunoglobulin (A-11035, Molecular Probes).

14. Vectashield® mounting medium (Vector laboratories, Inc.).

15. Micro cover glass.

16. Rubber cement or nail polish.

3 Methods

The methods described below outline (1) the detection of transiently expressed tagged proteins and (2) the detection of endogenous proteins. To obtain multicolor images with optimum contrast and minimal cross-talk, appropriate combinations of fluorophores should be selected. The spectral properties of frequently used fluorophores and recommended lasers are listed in Table 1.

3.1 Detection of Transiently Expressed GFP- and HA-Tagged Proteins

When antibodies raised against a protein of interest are unavailable for immunofluorescence or the available cell lines do not express the protein of interest sufficiently enough for immunofluorescence, an alternative method is to transfect cells with expression plasmids that encode tagged proteins and stain the cells with anti-tag antibodies. In addition, the protein of interest can be fused with a fluorophore for direct detection. Currently, a wide variety of fluorescent proteins have been developed for generating fusion proteins. By selecting a proper combination (e.g., EGFP and DsRed2-Monomer), it is possible to investigate the subcellular

Table 1
Approximate spectral properties of representative fluorophores used for multicolor labeling

Fluorophore	Absorption (nm)	Emission (nm)	Laser	Notes
Group 1 (green)				
FITC	490	512	Ar.488	Most widely used green fluorescent dye, easy to photobleach
EGFP	488	507	Ar.488	Brighter than wild-type GFP, for generating fusion proteins
Alexa Fluor 488	495	519	Ar.488	Brighter and more photostable than FITC
Group 2 (red)				
TRITC	541	572	HeNe543 or Ar.Kr.568	Commonly used red fluorescent dye in combination with FITC
Texas red	596	620	HeNe543 or Ar.Kr.568	Good spectral separation from FITC, slightly higher background staining [4]
DsRed2-monomer	556	586	HeNe543 or Ar.Kr.568	For generating fusion proteins
Cy3	552	565	HeNe543 or Ar.Kr.568	Brighter than TRITC
Alexa Fluor 546	556	573	HeNe543 or Ar.Kr.568	Brighter than TRITC or Cy3
Propidium iodide	530	615	HeNe543 or Ar.Kr.568	For DNA/RNA staining
Group 3 (blue)				
DAPI	345	425	UV.Ar.364	For nuclear staining
Hoechst33342	355	465	UV.Ar.364	For nuclear staining

For multicolor labeling, choose one fluorophore from each group depending on the laser sources equipped in the CLSM, thus allowing up to three-color labeling

localization of two proteins of interest in living cells [1]. Furthermore, when the CLSM is equipped with a proper stage incubator in which temperature and CO_2 are precisely controlled, long-term time lapse imaging can be obtained [2].

When overexpressing a tagged protein of interest, it is mandatory to verify the plasmid by DNA sequencing and confirm the proper molecular weight of the expressed protein by Western blotting [3]. It should also be noted that the tagging of proteins of interest or their overexpression could potentially alter the intracellular localization or the function of the target protein. To demonstrate that the protein of interest is localized in a particular organelle, a number of organelle-specific probes or vectors as well as organelle-specific monoclonal antibodies are available.

To visualize mitochondria, for example, MitoTracker probes (Molecular Probes) and mitochondria localization vectors, such as pECFP-Mito (Clontech; Palo Alto, CA), can be used.

3.1.1 Transfection

1. The day before transfection, seed HeLa cells at a density of 7×10^4 cells/well (when two well chamber slides are used).

2. Incubate the cells at 37 °C and 5 % CO_2 in DMEM medium.

3. Prepare 0.5 μg of each plasmid that encodes GFP- or HA-tagged protein. The minimum DNA concentration should be 0.1 μg/μl.

4. Dilute the DNA in 60 μl of Buffer EC (Effectene transfection reagent kit). Add 6 μl of Enhancer and mix by vortexing for 1 s (*see* **Note 2**).

5. Incubate at room temperature for 2–5 min and spin down the mixture for a few seconds.

6. Add 10 μl of Effectene Transfection Reagent to the mixture. Mix by vortexing for 10 s.

7. Incubate the samples for 5–10 min at room temperature to allow complex formation.

8. During **step 7**, aspirate the medium from the slides and wash the cells once with PBS. Add 1 ml of fresh growth medium to each well.

9. Add the transfection complexes drop-wise onto the cells. Gently swirl the plates to ensure uniform distribution of the complexes.

10. Incubate the cells with the complexes at 37 °C and 5 % CO_2 for 24 h to allow for gene expression.

3.1.2 Immunofluorescence

1. Carefully aspirate the medium and add 0.5 ml/well of 4 % paraformaldehyde in PBS drop-wise onto the cells (*see* **Note 3**).

2. To fix the cells, incubate the samples for 1 h at 4 °C.

3. Carefully replace the solution with 0.5 ml/well of permeabilization solution.

4. Incubate the samples for 5 min at room temperature.

5. Carefully replace the solution with 0.5 ml/well of preblock solution and incubate the samples for 1 h at room temperature.

6. Carefully aspirate the preblock solution and add 0.5 ml/well of anti-HA mouse monoclonal antibody diluted with preblock solution 1:200.

7. Incubate the samples for 1 h at room temperature (*see* **Note 4**).

8. Carefully aspirate the antibody solution and add 1 ml/well of PBS.

9. Incubate the samples for 10 min at room temperature. Gentle swirling facilitates washing efficiency.

10. Replace the solution with another 1 ml/well of fresh PBS and repeat **step 9**.

11. Carefully aspirate PBS and add 0.5 ml/well of TRITC-labeled rabbit anti-mouse immunoglobulin diluted with preblock solution. A 1:20 dilution is recommended.

12. Incubate the samples for 1 h at room temperature.

13. Carefully aspirate the antibody solution and wash the samples three times as described in **steps 8** and **9** (*see* **Note 5**).

14. Blot off excess PBS with a tissue, taking care not to allow the cells to dry out.

15. Place one drop of Vectashield® over each well of cells and then place a micro cover glass over the entire well, taking care to minimize air bubbles.

16. Secure the cover glass with rubber cement or nail polish.

3.1.3 Confocal Laser Scanning Microscopy

Since the operation procedure of a CLSM depends on manufacturer and systems, it is not possible to generalize this procedure. This chapter describes techniques for the Olympus CLSM. The manufacturer's manual should be consulted for details.

1. Choose the correct combination of laser, barrier filters, and excitation dichroic mirrors for the dyes in use according to the manufacturer's recommendation (for GFP/TRITC, the following combination is appropriate. Laser combination; Ar488nm + HeNe543nm, Barrier filters; BA510IF + BA530RIF for channel 1, BA565IF + BA590 for channel 2, dichroic mirror; DM488/543).

2. Set the power switch of each unit to ON. To stabilize the laser beam output, allow the system to warm up for at least 10 min after turning the laser power ON.

3. Start the FLUOVIEW software by double-clicking the FLUOVIEW icon on the desktop.

4. Place the specimen in an inverted position (in combination with an inverted microscope) on the microscope stage and turn the light path selector to the "Binocular" section.

5. Engage the optimum cube for specimen dye by operating the cube turret and focus on the specimen initially using transmitted light and then quickly observe the cells with fluorescence.

6. When the area to be observed with confocal microscopy is determined, turn the light path selector to the "Side port" position and rotate the cube turret so that no cube is engaged.

7. Choose the highest scan speed and set the zoom ratio to "X1." Set the channel to be acquired (brightness should be adjusted one channel at a time).

8. Click the "Focus" button to acquire repeated images at a high speed.

9. Focus on the cells of interest and adjust the image brightness. Parameters to be considered at this step are "PMT voltage," "Offset," and "Gain," each of which can be adjusted independently. The ND filters can also be selected using the LASER INTENSITY turret. Select the optimum ND filters according to the brightness of the specimen. Please note that a strong laser intensity will make the fluorescence fade quickly.

10. When it is necessary to observe the detail of a specific area, the image of a limited area can be acquired by using the "Zoom" scale and the "Pan" buttons.

11. Once the appropriate parameters have been determined, stop repeated scanning, set a lower scan speed and acquire the image by selecting the "Once" button (*see* **Notes 6** and 7).

12. After the desired image has been acquired, save it to a disk. It is possible to save images acquired with more than one channel at a time. The user can select the file types used for saving an image in a file. Fluoview Multi Tiff format is used for image analysis and processing on FLUOVIEW. Single TIF is a format for maximum portability between Macintosh and IBM PCs and programs such as Adobe Photoshop. To demonstrate the co-localization of two proteins, the merged image of the two proteins in a single view should be shown as well as displaying images of the two channels side by side. For example, when the expression of a GFP-fused protein is displayed in green and that of an HA-tagged protein is shown in red, then co-localization of the two proteins is visualized in yellow in the merged image (Fig. 2a–c).

3.2 Detection of Endogenous Proteins Using Labeled Secondary Antibodies

Demonstrating the intracellular co-localization of two endogenous proteins is more convincing than showing subcellular localizations of epitope- or GFP-tagged proteins. However, results should be obtained with highly specific primary antibodies. For, example antibodies that detect major cross-reactive bands in Western blotting are not suitable for immunofluorescence. Antibodies should also be affinity-purified prior to use to reduce background. In addition, the cell lines used for this analysis should express the proteins of interest sufficiently enough for immunofluorescence. The two primary antibodies used in these experiments should be raised in two different species. Secondary antibodies used for simultaneous detection of more than one antigen should also be carefully selected according to the following criteria. The secondary

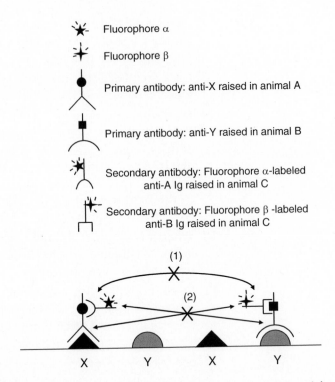

Fig. 2 Schematic diagram of multicolor detection of endogenous proteins

antibodies should (1) be derived from the same host species so that they do not recognize one another (Fig. 2 (1)), (2) not cross-react with other primary antibodies used in the assay system (Fig. 2 (2)), (3) not cross-react with endogenous proteins present in the cell lines under investigation. Secondary antibodies absorbed against the sera of a number of species to minimize cross-reactivity are commercially available and are usually marked as "highly cross-absorbed" or "for multiple labeling." When cells growing in suspension are to be stained, they should be cytocentrifuged prior to the fixation using a Cytospin or equivalent that allows low-speed centrifugal force to separate the cells on slides while maintaining cellular integrity.

1. Carefully aspirate the medium and add 0.5 ml/well of 4 % paraformaldehyde in PBS drop-wise onto the cells (*see* **Note 3**).

2. To fix the cells, incubate the samples for 1 h at 4 °C.

3. Carefully replace the solution with 0.5 ml/well of permeabilization solution.

4. Incubate the samples for 5 min at room temperature.

5. Carefully replace the solution with 0.5 ml/well of preblock solution and incubate the samples for 1 h at room temperature.

6. Carefully aspirate the preblock solution and add 0.5 ml/well of anti-protein X mouse monoclonal antibody diluted with

preblock solution. Because staining protocols vary with application, the appropriate dilution of antibody should be determined empirically.

7. Incubate the samples for 1 h at room temperature.

8. Carefully aspirate the antibody solution and add 1 ml/well of PBS.

9. Incubate the samples for 10 min at room temperature. Gentle swirling facilitates washing efficiency.

10. Replace the solution with another 1 ml/well of fresh PBS and repeat **step 9**.

11. Carefully aspirate PBS and add 0.5 ml/well of Alexa Fluor 488-labeled goat anti-mouse immunoglobulin diluted with preblock solution (1:200 or higher dilution).

12. Repeat **steps 7–10**.

13. Carefully replace the solution with 0.5 ml/well of preblock solution and incubate the samples for 1 h at room temperature. This step can be omitted depending on the antibody.

14. Carefully aspirate the preblock solution and add 0.5 ml/well of anti-protein Y rabbit polyclonal antibody diluted appropriately with preblock solution.

15. Repeat **steps 7–10**.

16. Carefully aspirate PBS and add 0.5 ml/well of Alexa Fluor 546-labeled goat anti-rabbit immunoglobulin diluted with preblock solution (1:200 or higher dilution).

17. Incubate the samples for 1 h at room temperature.

18. Carefully aspirate the antibody solution and wash the samples three times as described at **steps 8** and **9** (*see* **Note 5**).

19. Blot off excess PBS with a tissue, taking care not to allow the cells to dry out.

20. Place one drop of Vectashield® over each well of cells and then place a micro cover glass over the entire well, taking care to minimize air bubbles.

21. Secure the cover glass with rubber cement or nail polish.

Confocal laser scanning microscopy can be performed using the same combinations of lasers, barrier filters, and dichroic mirror as described in Subheading 3.1.3 (channel 1 for Alexa Fluor 488, channel 2 for Alexa Fluor 546). Examples of multicolor labeling of HeLa cells are demonstrated in Fig. 3.

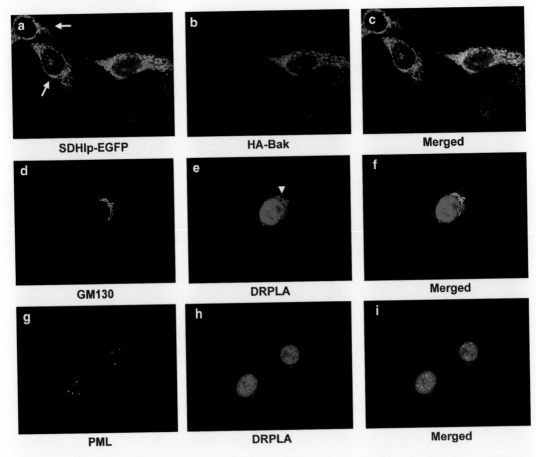

Fig. 3 Multicolor labeling of HeLa cells. (**a–c**) HeLa cells were transfected with pSDHIp-EGFP and pHA-Bak that encode the EGFP-tagged iron sulfur subunit of human succinate dehydrogenase [5] and HA-tagged proapoptotic Bcl-2 family member, Bak [6, 7], respectively. At 24 h after transfection, cells were stained with anti-HA antibody followed by TRITC-labeled secondary antibody. Panels (**a**) and (**b**) show the subcellular localizations of SDHIp and Bak, respectively. A merged picture is shown in panel (**c**). SDHIp is a mitochondrial protein and Bak is shown to co-localize with SDHIp. *Arrows* indicate cells transfected with only pSDHIp-EGFP. (**d–f**) HeLa cells were incubated with mouse anti-GM130 monoclonal antibody (BD transduction Laboratories) and rabbit anti-DRPLA antibody [8]. The primary antibodies were detected by secondary antibodies conjugated to Alexa Fluor 488 or 546. DRPLA protein, a product of the gene responsible for a genetic neurodegenerative disease, dentatorubral-pallidoluysian atrophy (DRPLA) [9], is localized in the nucleus [10]. However, a closer observation reveals that it is also localized in the juxtanuclear region (*arrowhead*) where it co-localizes with GM130 (**f**), which is a *cis*-Golgi network marker [11]. (**g–i**) HeLa cells were treated similarly as in (**d–f**) except that anti-PML mouse monoclonal antibody (Santa Cruz) was used as one of the primary antibodies. The promyelocytic leukemia (PML) protein is a major component of nuclear dot-like structures known as PML nuclear bodies (NBs) or PML oncogenic domains (POD) [12, 13] (**g**). These panels show that the nuclear DRPLA protein is not recruited to NBs or POD

4 Notes

1. Other representative anti-tag antibodies are described below. Recommended starting dilutions are indicated in parentheses.

   Anti-FLAG (M2) (mouse monoclonal) (1:100) (F3165, SIGMA; St. Louis, MO).

   Anti-His (His-probe, H-15) (rabbit polyclonal) (1:200) (sc-803, Santa Cruz; Santa Cruz, CA).

   Anti-c-Myc (9E10) (mouse monoclonal) (1:100) (sc-40, Santa Cruz).

2. To achieve optimal transfection efficiency for every new cell line/plasmid DNA combination, it is recommended to optimize the amounts of transfection reagent, DNA, and the cell number prior to transfection according to the manufacturer's protocol.

3. The alternative method to fix and permeabilize cells is to use methanol or acetone. The method should be selected empirically depending on the cell lines and antibodies.

4. Fluorophore-conjugated anti-tag antibodies are also available. In this case, go to **step 13**.

5. If the confocal laser scanning microscope is equipped with a UV laser, it is recommended to add 1.25 μM of Hoechst33342 (Table 1) to the second wash for nuclear counterstaining. This allows observation of the location of the nucleus.

6. To improve the image quality by reduction of noise, Kalman accumulation is recommended. It allows acquiring of images for a specified number of times while averaging the images.

7. For a double-labeled specimen, there are two alternatives to obtain the image, sequential acquisition and simultaneous acquisition. With the former image capturing method, an image slice from each laser excitation is obtained using one PMT at a time. This scan is advantageous in acquiring an image with a lower cross-talk between two wavelengths. In contrast, with a simultaneous scan, images are obtained without a time lag between the two wavelengths.

Acknowledgments

I thank Yuko Ohtsuka and Mami U for their technical assistance. I am also grateful to Drs. Yoshiaki Shikama and Yuko Okamura-Oho for their valuable discussions.

References

1. Abe Y, Oka A, Mizuguchi M, Igarashi T, Ishikawa S, Aburatani H, Yokoyama S, Asahara H, Nagao K, Yamada M, Miyashita T (2009) *EYA4*, deleted in a case with middle interhemispheric variant of holoprosencephaly, interacts with *SIX3* both physically and functionally. Hum Mutat 30:E946–E955

2. Yamagata K, Suetsugu R, Wakayama T (2009) Long-term, six-dimensional live-cell imaging for the mouse preimplantation embryo that does not affect full-term development. J Reprod Dev 55:343–350

3. Sambrook J, Russell DW (2001) Molecular cloning, a laboratory manual, 3rd edn. Cold Spring Harbor Laboratory Press, Cold Spring Harbor, NY

4. Wessendorf MW, Brelje TC (1992) Which fluorophore is brightest? A comparison of the staining obtained using fluorescein, tetramethylrhodamine, lissamine rhodamine, Texas red, and cyanine 3.18. Histochemistry 98:81–85

5. Kita K, Oya H, Gennis RB, Ackrell BC, Kasahara M (1990) Human complex II (succinate-ubiquinone oxidoreductase): cDNA cloning of iron sulfur (Ip) subunit of liver mitochondria. Biochem Biophys Res Commun 166:101–108

6. Chittenden T, Harrington EA, O'Connor R, Flemington C, Lutz RJ, Evan GI, Guild BC (1995) Induction of apoptosis by the Bcl-2-homologue Bak. Nature 374:733–736

7. Kiefer MC, Brauer MJ, Powers VC, Wu JJ, Umansky SR, Tomei LD, Barr PJ (1995) Modulation of apoptosis by the widely distributed Bcl-2 homologue Bak. Nature 374:736–739

8. Miyashita T, Okamura-Oho Y, Mito Y, Nagafuchi S, Yamada M (1997) Dentatorubral pallidoluysian atrophy (DRPLA) protein is cleaved by caspase-3 during apoptosis. J Biol Chem 272:29238–29242

9. Nagafuchi S, Yanagisawa H, Ohsaki E, Shirayama T, Tadokoro K, Inoue T, Yamada M (1994) Structure and expression of the gene responsible for the triplet repeat disorder, dentatorubral and pallidoluysian atrophy (DRPLA). Nat Genet 8:177–182

10. Miyashita T, Nagao K, Ohmi K, Yanagisawa H, Okamura-Oho Y, Yamada M (1998) Intracellular aggregate formation of dentatorubral-pallidoluysian atrophy (DRPLA) protein with the extended polyglutamine. Biochem Biophys Res Commun 249:96–102

11. Nakamura N, Rabouille C, Watson R, Nilsson T, Hui N, Slusarewicz P, Kreis TE, Warren G (1995) Characterization of a cis-Golgi matrix protein, GM130. J Cell Biol 131:1715–1726

12. Dyck JA, Maul GG, Miller WHJ, Chen JD, Kakizuka A, Evans RM (1994) A novel macromolecular structure is a target of the promyelocyte-retinoic acid receptor oncoprotein. Cell 76:333–343

13. Weis K, Rambaud S, Lavau C, Jansen J, Carvalho T, Carmo-Fonseca M, Lamond A, Dejean A (1994) Retinoic acid regulates aberrant nuclear localization of PML-RAR alpha in acute promyelocytic leukemia cells. Cell 76:345–356

Part V

High Throughput Screening Assays for Protein-Protein Interactions: Case Studies

Chapter 35

Fluorescence Polarization Assay to Quantify Protein-Protein Interactions in an HTS Format

Yuhong Du

Abstract

Fluorescence polarization (FP) technology is based on the measurement of molecule rotation, and has been widely used to study molecular interactions in solution. This method can be used to measure binding and dissociation between two molecules if one of the binding molecules is relatively small and fluorescent. The fluorescently labeled small molecule (such as a small peptide) rotates rapidly in the solution. Upon excitation by polarized light, the emitted light remains depolarized and gives rise to a low FP signal. When the fluorescent small molecules in solution are bound to bigger molecules (such as a protein), the movement of the complex becomes slower. When such a complex is excited with polarized light, much of the emitted light is polarized because of the slow movement of the complex. Thus, the binding of a fluorescently labeled small molecule to a bigger molecule can be monitored by the change in polarization and measured by the generation of an increased FP signal. This chapter aims to provide a step-by-step practical procedure for developing an FP assay in a multi-well plate format to monitor protein-protein interaction (PPI) in a homogenous format.

Key words Fluorescence polarization (FP), Protein-protein interaction (PPI), Small molecule, Probe, Multi-well plate, Plate reader, Dissociation constant (Kd), Competition assay

1 Introduction

Fluorescence polarization (FP) is a very powerful and sensitive nonradioactive technique for the study of molecular interactions in solution. First described by Perrin in 1926 [1], FP is based on the observation of the molecular movement of fluorescent molecules in solution. When a fluorescently labeled molecule is excited by polarized light, it emits light with a degree of polarization that is inversely proportional to the rate of molecular rotation. This property can be used to measure binding and dissociation between two molecules if one of the binding molecules is relatively small and fluorescent. Molecular rotation is largely dependent on molecular mass, with larger masses showing slower rotation. The basic principle of FP is depicted in Fig. 1. When a fluorescently labeled small

Cheryl L. Meyerkord and Haian Fu (eds.), *Protein-Protein Interactions: Methods and Applications*, Methods in Molecular Biology, vol. 1278, DOI 10.1007/978-1-4939-2425-7_35, © Springer Science+Business Media New York 2015

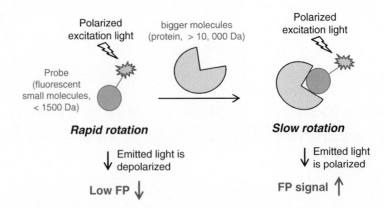

Fig. 1 Schematic of the theory behind FP assays. A rapidly rotating small molecule fluorophore gives a low FP signal (low mP). The association of a relatively large molecule, such as a protein, with the small molecule fluorophore slows down the rotation of the fluorophore, leading to an increased FP signal

molecule (typically <1,500 Da), such as a peptide or ligand, is free in solution, it rotates rapidly. Upon excitation by polarized light, the emitted light will be largely depolarized light. When this fluorescent small molecule is bound to a bigger molecule (typically >10 kDa), such as a protein, the rotational movement of the fluorophore becomes slower due to the significantly reduced rotational speed of the complex. Thus, the emitted light will remain polarized. Therefore, the binding of a fluorescently labeled small molecule to a protein can be monitored by the change in polarization from low to high.

Protein-protein interactions (PPI) can be monitored by FP if one of the components of the PPI is small and fluorescent. Therefore, the first step to develop an FP binding assay for PPI is to identify a small binding epitope or peptide for one of the partners, which can then be labeled with a fluorophore dye. Fluorescently labeled peptides, which can be synthesized by commercial services, are typically referred to as a "tracer" or "probe" in an FP assay. Commonly used fluorophores for FP tracers include fluorescein, rhodamine, BODIPY, and Cy5 dyes, or their derivatives. The molecular weight of fluorescently labeled small molecules for FP assay is typically less than 1,500 Da, although up to 5,000 Da can be acceptable if the binding partner is very large.

Once the probe has been designed and synthesized, the assay procedure for FP with modern instrumentation that is capable of measuring an FP signal in a multiwell plate (96/384/1,536) is quite simple and easy. An FP binding experiment is performed by titrating a protein of interest with a fixed concentration of tracer in a multiwell plate. The FP signal, which is expressed as millipolarization units (mP), is then measured with a multimode plate reader equipped with a proper set of filters for the specific fluorophore.

The tracer alone without protein should give rise to only minimal FP signal for the free tracer. Increasing FP signal with increasing protein concentrations in the presence of tracer should be observed if the protein binds to the tracer. At higher protein concentrations, the FP signal should rise to a plateau that corresponds to complete binding of all tracers. A binding curve can be generated by plotting FP signal against protein concentration, and used for dissociation constant (Kd) calculation.

The following protocol describes a general step-by-step procedure of developing an FP binding assay to monitor the interaction of two biomolecules. The results of FP assay development for monitoring the interaction of 14-3-3 protein with a fluorescently labeled phosphopeptide binding partner are presented as a case study [2]. The 14-3-3 proteins mediate phosphorylation-dependent protein-protein interactions. Through binding to numerous client proteins, 14-3-3 controls a wide range of physiological processes and has been implicated in a variety of diseases, including cancer and neurodegenerative disorders [3]. In order to better understand the structure and function of 14-3-3 proteins and to develop small molecule modulators of 14-3-3 proteins for physiological studies and potential therapeutic interventions, a highly sensitive FP-based 14-3-3 assay was designed and optimized. Using the interaction of 14-3-3 with a fluorescently labeled phosphopeptide derived from Raf-1 as a model system (*see* Fig. 3a), the detailed procedure to develop an FP assay for monitoring the interaction of 14-3-3 protein with its binding partners is described. This 14-3-3 FP binding assay is a simple one-step "mix-and-read" method for analyzing 14-3-3 proteins. This solution based, versatile method can be used to monitor the binding of 14-3-3 with a variety of client proteins. While this chapter uses 14-3-3 protein interactions as an example, the procedure described in this chapter can be adapted and applied to any FP-suited system for monitoring the interaction of two molecules.

2   Materials

1. Protein (14-3-3γ): recombinant GST-14-3-3 protein was expressed in *Escherichia coli* strain BL21 (*DE3*) as a GST-tagged product and purified as described [2].

2. LB medium for auto-induction.

3. Assay buffer: HEPES buffer containing 10 mM HEPES, 150 mM NaCl, 0.05 % Tween-20, and 0.5 mM DTT, pH 7.4 (*see* **Note 1**).

4. Resuspension buffer for protein purification: phosphate-buffered saline (PBS) containing 1.0 mM phenylmethylsulfonyl fluoride (PMSF), 1 μg/ml leupeptin, and 1 μg/ml aprotinin.

5. Glutathione-Sepharose beads.

6. Elution buffer: PBS, pH 9, containing 20 mM reduced gluta-thione, 0.25 mM PMSF, 5 mM dithiothreitol (DTT), and 0.1 % Triton X-100.

7. De-salt gel filtration column (such as a PD-10 gel filtration column from GE Healthcare).

8. Fluorescent Probe (TMR-pS259-Raf): a phosphopeptide derived from a well-studied 14-3-3 binding protein, Raf-1, was synthesized and labeled with 5/6 carboxytetramethylrho-damine (TMR). The TMR-pS259-Raf contains 15 residues: 5/6-TMR-LSQRQRST[pS]TPNVHM, and was dissolved in HEPES buffer as 200 µM stock and stored at −20 °C before use.

9. Unlabeled antagonist or control peptides: R18 (PHCVPKNLSWLNLEAN MCLP) [4] and mutated R18, R18Lys, where D12 and E14 were changed to K [5], were dissolved in HEPES buffer as 200 µM stock and stored at −20 °C before use.

10. Plate: 384-well solid bottom black plate (*see* **Note 2**).

11. Plate reader for FP signal measurements (*see* **Note 3**): For 14-3-3 FP assay development, FP measurements were performed on an Analyst HT plate reader (Molecular Devices) using an FP protocol. For the tetramethylrhodamine (TMR)-labeled probe (Ex: 545 nm; Em: 610 nm), a dichroic mirror of 565 nm was used.

3   Methods

3.1   Protein Purification

E. coli BL21 (DE3) strains were grown in LB medium with ampi-cillin (100 µg/ml) for protein auto-induction. After overnight incubation with shaking at 37 °C, the cells were harvested and the pellet was resuspended in ice-cold resuspension buffer. The GST-14-3-3 γ protein was then purified by using Glutathione-Sepharose beads, and the bound protein was eluted in elution buffer. Salts were removed from the pooled elution fractions containing the GST-14-3-3γ fusion protein using a PD-10 gel filtration column equilibrated in assay buffer without DTT. The protein concentra-tion was measured and GST-14-3-3γ protein was stored at −20 °C before use.

3.2   Determination of the Concentration of Fluorescent Probe (See Note 4)

To determine the concentration of TMR-pS259-Raf peptide for the 14-3-3 binding FP assay, the peptide was serially diluted in assay buffer in a 384-well plate. An example plate format for the serial dilution of peptide is shown in Fig. 2a. The final volume for each well is 30 µl.

a   Peptide titration plate format:

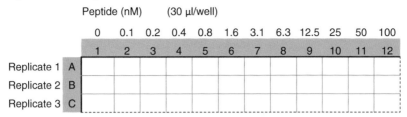

b   Protein titration plate format:

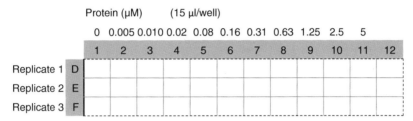

Fig. 2 An example plate format for probe (**a**) and protein (**b**) titrations for FP assay development in a 384-well plate format. The FP binding assay is then preformed by adding 15 μl of probe at a selected concentration to the serially diluted protein. The final volume for each well of the 384-well plate is 30 μl

1. Add 30 μl of assay buffer to columns 1–11 in rows A, B, and C, as triplicates.

2. In a 1.5 ml microcentrifuge tube, make 200 μl of the highest concentration of peptide tested (e.g., 100 nM of TMR-pS259-Raf peptide) in assay buffer.

3. Pipette 60 μl of the highest concentration of peptide (100 nM of TMR-pS259-Raf) to column 12 of rows A, B, and C.

4. Load 3 tips using a multichannel pipette for 384-well plates, transfer 30 μl of the highest concentration of peptide in triplicate (column 12, rows A, B, and C) to the next column (column 11, rows A, B, and C) containing 30 μl of assay buffer. The same three tips may be used for the entire serial dilution process.

5. Mix the solutions in column 11 (rows A, B, and C) by pipetting 30 μl of the solution up and down at least five times.

6. Transfer 30 μl of the mixture in column 11 (rows A, B, and C) to the next column (column 10, rows A, B, and C) containing 30 μl of assay buffer.

7. Repeat the above process until reaching column 2. After mixing the solution in column 2, aspirate and discard 30 μl of solution, so the final volume of the wells in column 2 is 30 μl. At this point, all wells contain an equal volume and the concentrations of peptide range from 0 (column 1) to 100 nM (column 12) with triplicates for each concentration.

8. Centrifuge the plate at 1,000 rpm for 2 min to get rid of any bubbles in the wells. This step is important especially for handling low volumes, as the bubbles will affect the reading.

9. Load the plate on the plate reader.

10. Set up the proper settings for the instrument software for the FP measurement.

   For 14-3-3 FP assay development, an Analyst HT plate reader was used. An integration time of 100 ms was used, and the Z height was set at 2.15 mm (middle). The excitation polarization was set at "static," and emission polarization was set at "dynamic." For the TMR-labeled probe, an excitation filter at 545 nm and an emission filter at 610–675 nm were used with a dichroic mirror of 565 nm. Three panels of the data were obtained from the FP measurement (*see* **Note 3**).

11. Data analysis for peptide titration:

   Caluculate the average fluorescence intensity (FI) in parallel channel for each sample. The concentrations of peptide that exhibit about ten times or more FI signal compared to buffer-only background controls should be selected for the following FP binding assay development (*see* **Note 4**).

   For the 14-3-3 FP binding assay, 1 nM TMR-pS259-Raf peptide was chosen based on the observation that this concentration of the TMR-labeled peptide exhibited about ten times more FI signal compared to buffer-only background controls.

3.3 Development of the FP Binding Assay (See Note 5)

After selecting the probe concentration, the binding of 14-3-3 protein to TMR-pS259-Raf peptide was performed by titrating 14-3-3 protein using a fixed concentration of TMR-pS259-Raf peptide (1 nM) (*see* **Note 6**). Serial dilution of protein was carried out using a similar procedure as the peptide serial dilution except the initial volume for each well was different. An example plate format for protein dilution is shown in Fig. 2b, the final volume for each well was 15 μl. The FP binding reaction was then performed by adding 15 μl of peptide at the selected final concentration to all wells containing increasing concentrations of protein.

1. Add 15 μl of assay buffer to columns 1–11, rows D, E, and F, as triplicates, in the same plate as the peptide titration.

2. In a 1.5 ml microcentrifuge tube, make 100 μl of 2× of the highest concentration of protein tested (e.g., 5 μM of GST-14-3-3γ protein) in assay buffer.

3. Pipette 30 μl of 5 μM GST-14-3-3γ protein to column 12, rows D, E, and F as triplicates.

4. Load three tips using a multichannel pipette for 384-well plates, transfer 15 μl of the highest concentration of protein

in triplicate (column 12, rows D,E, and F) to the next column (column 11, rows D,E, and F) containing 15 μl of assay buffer.

5. Mix the solutions in column 11 (rows D, E, and F) by pipetting 15 μl of the solution up and down at least five times.

6. Transfer 15 μl of the mixture in column 11 (rows D, E, and F) to the next column (column 10, rows D, E, and F) containing 15 μl of assay buffer.

7. Repeat the above process until reaching column 2. After mixing the solution in column 2, aspirate and discard 15 μl of solution, so the final volume of the wells in column 2 is 15 μl. At this point, all wells contain an equal volume (15 μl) and the concentrations of protein range from 0 (column 1) to 5 μM (column 12) with triplicates (rows D, E, and F) for each concentration.

8. In a 1.5 ml microcentrifuge tube, make 600 μl of 2× peptide (2 nM of TMR-pS259-Raf) in assay buffer and transfer to a reagent reservoir.

9. Using a 384-multichannel pipette, add 15 μl of 2 nM TMR-pS259-Raf peptide to the serially diluted protein in columns 1–12, rows D, E, and F. At this point, all wells contain an equal volume (30 μl) with the same concentration of peptide (final: 1 nM). The final concentrations of protein range from 0 (column 1) to 2.5 μM (column 12) with triplicates (rows D, E, and F) for each concentration.

10. Centrifuge the plate at 1,000 rpm for 2 min.

11. Load the plate on the plate reader.

12. After incubating the plate at room temperature (RT) for various time periods (*see* **Note** 7), read the FP signal using the proper settings.

3.4 Data Analysis for the FP Binding Assay

The dynamic range of the FP binding assay (i.e., the FP assay window) is defined by the difference of the measured FP signal between bound fluorescent probe in the presence of protein and unbound (free) probe in the absence of protein.

1. Calculate the FP assay window for each protein concentration using the following equation:

$$\text{FP assay window} = \text{mPb} - \text{mPf}$$

mPb is the recorded mP value for the wells containing TMR-pS259-Raf peptide and 14-3-3 protein, which defines the binding FP signal from bound peptide. mPf is the recorded mP value for the samples with TMR-pS259-Raf peptide only in the absence of 14-3-3 protein, which defines the minimal, or background, FP signal from unbound peptide.

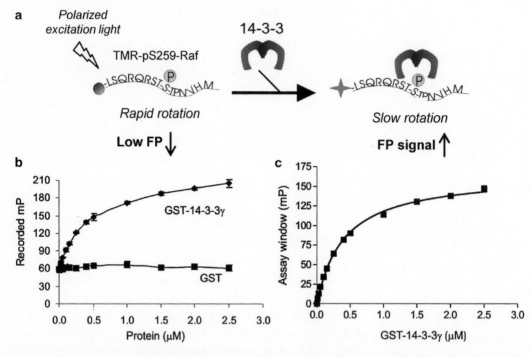

Fig. 3 Development of an FP binding assay for monitoring 14-3-3 protein interactions. (**a**) 14-3-3 binding FP assay design. The 14-3-3 binding peptide, pS259-Raf, is labeled with 5/6 carboxytetramethylrhodamine (TMR). The unbound peptide (fluorescent small molecule) rotates fast in solution and gives a low FP signal when excited with polarized light. The association of 14-3-3 protein (large molecule) slows down the rotation of the peptide, leading to an increased FP signal. (**b**) Interaction of 14-3-3 TMR-pS259-Raf peptide induces the generation of an FP signal. TMR-pS259-Raf peptide (1 nM, final) was incubated with increasing concentrations of GST-14-3-3γ or GST proteins in a 384-well black microplate. The FP signals were recorded after 1 h of incubation using an Analyst HT. (**c**) The assay window was calculated by subtracting free peptide polarization values from values of bound peptide recorded in the presence of specified protein concentrations of GST-14-3-3. The data were fit to the model FP = ([14-3-3] × Bmax)/([14-3-3] + Kd) using nonlinear regression analysis, where Bmax is the maximal binding (Prism 4.0; Graphpad). Data shown are the average values from triplicate samples with standard deviation (SD)

2. Generate a binding isotherm for the calculation of association parameters such as Kd and maximal binding for the interaction of 14-3-3 with TMR-pS259-Raf peptide. The obtained FP assay window from the protein titration should be plotted against the increasing concentrations of 14-3-3 protein. The Kd for the binding can be calculated using Graphpad Prism 4.0 software nonlinear regression analysis or another curve fitting software.

   As an example, results from the 14-3-3/TMR-pS259-Raf FP assay are shown in Fig. 3b. Interaction of 14-3-3 with TMR-pS259-Raf gave rise to a significant FP signal with minimal background polarization with the peptide probe alone. With increasing amounts of GST-14-3-3γ protein, polarization values progressively increased to reach saturation, suggesting that a greater fraction of fluorescent peptide was bound to the 14-3-3 protein. This FP signal is likely due to the specific interaction of TMR-pS259-Raf with 14-3-3γ because incubation of TMR-pS259-Raf with GST alone was incapable of generating an increase in FP signal. The maximum FP assay window reached approximately 150 mP with an estimated dissociation constant, Kd, of 0.41 µM for the Raf peptide (Fig. 3c). This procedure provides a simple, quantitative "mix-and-measure" FP assay for studying 14-3-3 protein interactions.

3.5 Development of an FP Competition Assay to Validate the Specificity of the FP Binding Signal

Once the FP binding assay has been developed with a reasonable maximum FP signal window (above ~60 mP), the specificity of the FP signal should be evaluated in a competition FP binding assay format. Such an assay works by measurement of the decrease in FP signal caused by a competitor, e.g., an un-labeled or antagonist ligand that displaces the binding of the labeled probe to the protein. To establish the competition assay, titration of a competitor to a reaction mixture containing fixed concentrations of protein and probe is performed as follows.

1. Select the protein concentration for the competition assay.

   A concentration of the protein should be chosen that generates a signal between the EC_{50} to EC_{80} of the maximum FP signal window based on the established FP binding curve. Higher concentrations of the protein will lie in the plateau region of the binding curve, which may lead to the competition assay being insensitive to competitors. Lower concentrations of the protein will not provide a good assay window for measurement of the decreased signal by displacement of the probe.

   Based on the 14-3-3 FP binding curve (Fig. 3b), 0.5 µM GST-14-3-3γ was selected for the competition assay. This concentration of the protein produced ~70 mP FP assay window and was close to its Kd or EC_{50} concentration (0.4 µM).

2. Prepare serially diluted competitors in a separate plate (*see* **Note 8**).

   In a 384-well round bottom plate, make 5× serially diluted antagonist or control peptide in assay buffer using the procedures described in Subheadings 3.2 and 3.3. An example plate format is given in Fig. 4a.

3. Prepare two mixtures containing 1.25× the selected concentration of peptide and protein in two 1.5 ml microcentrifuge tubes. Mixture 1 should contain 1.25 nM TMR-pS259-Raf peptide only and mixture 2 should contain both the peptide and protein at 1.25× the selected concentrations (1.25 nM of TMR-pS259-Raf and 0.63 µM of GST-14-3-3γ).

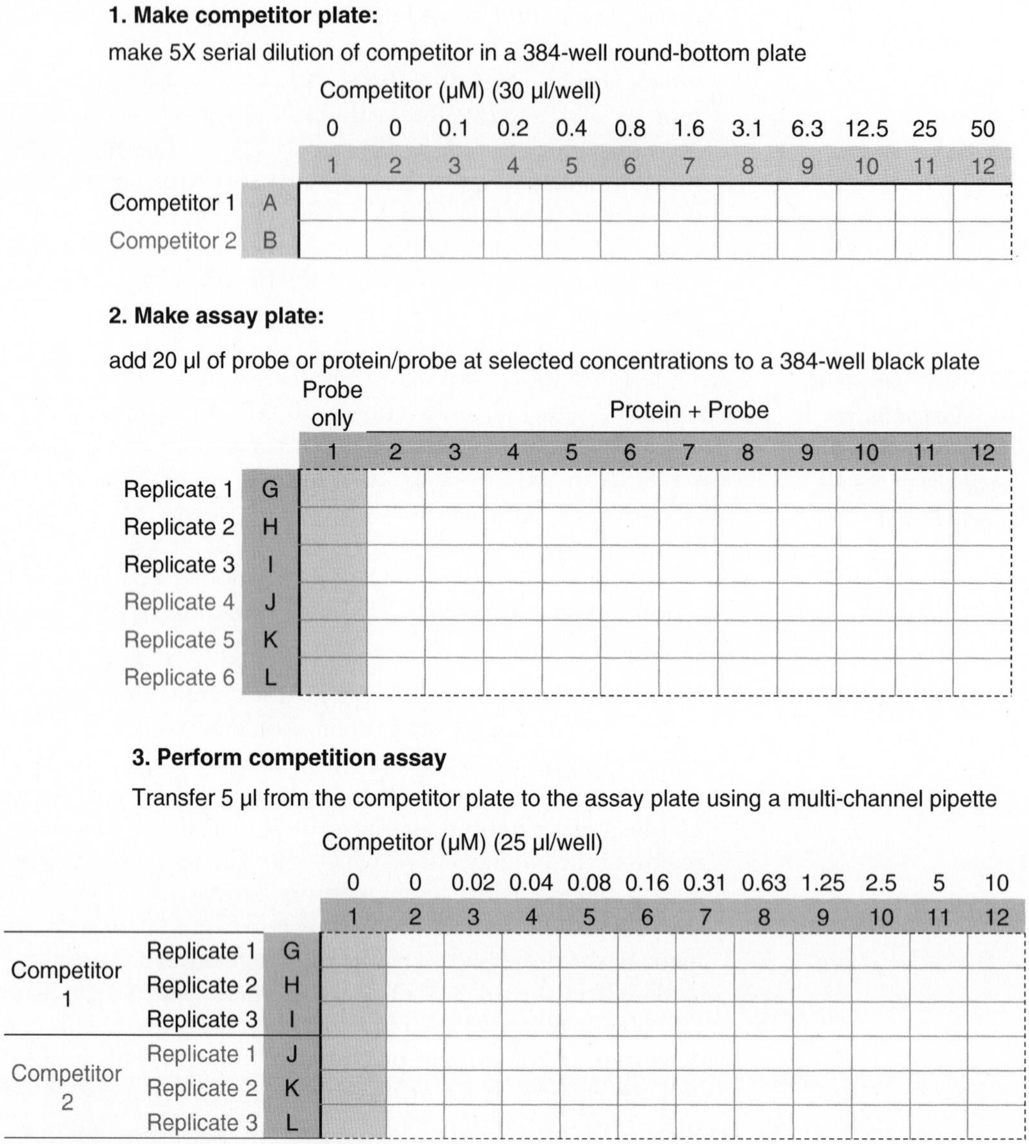

1. Make competitor plate:

make 5X serial dilution of competitor in a 384-well round-bottom plate

Competitor (μM) (30 μl/well)

		0	0	0.1	0.2	0.4	0.8	1.6	3.1	6.3	12.5	25	50
		1	2	3	4	5	6	7	8	9	10	11	12
Competitor 1	A												
Competitor 2	B												

2. Make assay plate:

add 20 μl of probe or protein/probe at selected concentrations to a 384-well black plate

		Probe only	Protein + Probe										
		1	2	3	4	5	6	7	8	9	10	11	12
Replicate 1	G												
Replicate 2	H												
Replicate 3	I												
Replicate 4	J												
Replicate 5	K												
Replicate 6	L												

3. Perform competition assay

Transfer 5 μl from the competitor plate to the assay plate using a multi-channel pipette

Competitor (μM) (25 μl/well)

			0	0	0.02	0.04	0.08	0.16	0.31	0.63	1.25	2.5	5	10
			1	2	3	4	5	6	7	8	9	10	11	12
Competitor 1	Replicate 1	G												
	Replicate 2	H												
	Replicate 3	I												
Competitor 2	Replicate 1	J												
	Replicate 2	K												
	Replicate 3	L												

Fig. 4 An example plate format and procedure for developing a competition FP assay. (1) Make serial dilutions of competitors in a 384-well round-bottom plate. (2) Prepare an assay plate in a 384-well black plate containing selected concentrations of probe and protein. Probe only without protein is included as a background control which gives rise to minimal FP signals. (3) Perform the competition assay by transferring the competitor to assay plate with triplicate samples

4. As shown in the competition assay plate format (Fig. 4), pipette 20 μl of mixture 1 (peptide only) to column 1, rows G, H, and I as triplicates. Add 20 μl of mixture 2 (peptide and protein) to the rest of the wells (columns 2-12, rows G, H, and I).

5. Load a multichannel pipette with 12 pipette tips. Transfer 5 μl/ well of serially diluted competitor in row A (column 1–12) in

the competitor plate to the corresponding well of the assay plate (row G). Repeat the procedure to transfer the diluted competitor to rows H and I of the assay plate to generate triplicate samples for each point (*see* **Note 9**).

6. Incubate the plate at RT for a period of time (*see* **Note 10**).

7. Load the plate on the plate reader.

8. Read the FP signal using the proper instrument settings.

9. Analyze the data for the competition assay

   (a) Calculate percentage of control (% of control) for each dose of competitor:

   The effect of the competitors on the disruption of the interaction is typically expressed as % of control and calculated as the following equation:

   $$\% \text{ of control} = (\text{recorded FP signal}_{competitor} \\ - \text{recorded FP signal}_{blank})/ \\ (\text{recorded FP signal}_{control} \\ - \text{recorded FP signal}_{blank}) \times 100$$

   Where recorded FP signal$_{blank}$ is the average measured FP signal from peptide only wells (triplicates), which defines the minimal FP signal. The recorded FP signal$_{control}$ is the average FP signal from wells containing protein and peptide, which defines the maximal FP signal, and the recorded FP signal$_{competitor}$ is the FP signal from wells containing protein and peptide in the presence of competitors.

   (b) Calculate the IC$_{50}$ of the competitor:

   Plot the % of control against the competitor concentration using Graphpad software, or another graphing software with curve fitting, to obtain the IC$_{50}$, the competitor concentration that leads to displacement of 50 % of the maximal binding.

10. To obtain the optimal conditions, the competition assay should be repeated with several different protein concentrations or more dose points for the competitor tested (*see* **Note 11**).

Figure 5a shows an example competition curve from the 14-3-3 FP competition assay. A well-characterized 14-3-3 peptide antagonist, R18, was used as a competitor in the assay. The addition of R18 dose-dependently decreased the FP signal for the binding of 14-3-3 to the TMR-pS259-Raf peptide. The negative control peptide, R18Lys, which is a mutated form of R18 peptide, did not have any effect on the FP signal. These results validate the specificity of the 14-3-3 FP binding assay for studying 14-3-3 protein-protein interactions.

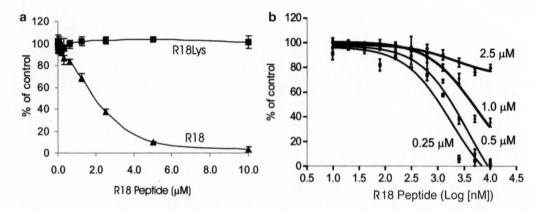

Fig. 5 Competition FP assay. (**a**) Competitive inhibition of 14-3-3 antagonist peptide, R18, on the 14-3-3/Raf peptide interaction FP assay. A mutant derivative of R18, R18Lys, which cannot bind 14-3-3, was used as a negative control. Increasing concentrations of R18 or R18Lys were added to the reaction containing 1 nM TMR-pS259-Raf peptide and 0.5 μM GST-14-3-3γ. After incubation for 1 h at room temperature, the FP signals were recorded. Percent of control was calculated as described in Subheading 3.5 and plotted against peptide concentration. (**b**) Competition of R18 peptide for the binding of TMR-pS259-Raf peptide to increasing concentrations of GST-14-3-3γ. Increasing concentrations of R18 peptide were added to four panels of reactions containing 0.25, 0.5, 1, and 2.5 μM of GST-14-3-3γ and 1 nM of TMR-pS259-Raf peptide. The FP signals were recorded after 1 h incubation at RT. Percent of control were calculated and plotted against R18 peptide concentrations. Data shown are the average signal from triplicates with SD

4   Notes

1. The buffer used for an FP assay must have low fluorescence background. Also, the chosen buffer must be one in which the protein and peptide are stable. Frequently used buffers have a neutral pH such as PBS, HEPES, or Tris.

2. Other types of multiple-well plates can also be used depending on the working volume. For example, a minimum of 50 μl solution per well is required for 96-well plates, 15–50 μl per well is required for a 384-well plate, and 2–5 μl per well is sufficient for a 1,536-well plate. To avoid fluorescence crosstalk, the use of solid-bottom black plates is recommended. Typically, one 384-well plate should be sufficient for developing an FP binding assay using the example plate format/layout as presented in Figs. 2 and 3.

3. Many commercially available instruments are capable of measuring an FP signal from solution in 96/384/1,536-well microtiter plate format. The fluorescence is measured using polarized excitation and emission filters. Two measurements are performed on every well and fluorescence polarization is defined and calculated as:

$$\text{Polarization} = P = (I_{\text{vertical}} - I_{\text{horizontal}})/(I_{\text{vertical}} + I_{\text{horizontal}})$$

Where $I_{vertical}$ is the intensity of the emission light parallel to the excitation light plane and $I_{horizontal}$ is the intensity of the emission light perpendicular to the excitation light plane [6]. All polarization values are expressed as millipolarization units (mP).

All commercial microplate readers have built-in software for mP calculation. Depending on the instrument used, three sets of data are generally reported: (1) calculated mP values; (2) raw fluorescence intensity counts of vertical (or parallel/S-channel); and (3) horizontal (or perpendicular/P-channel) measurements for each well.

mP calculation for different instruments requires the proper use of measured fluorescence intensity of parallel/S-channel and perpendicular/P-channel. As optical parts of fluorometers possess unequal transmission or varying sensitivities for vertically or horizontally polarized light, such instrument artifacts should be corrected for accurate calculation of the absolute polarization state of the molecule using fluorescent readers. This correction factor is known as the "G Factor" which is instrument-dependent. The G factor corrects for any bias towards the horizontal (or perpendicular/P-channel) measurement. Most commercially available instruments have an option for correcting a single-point polarization measurement with a G factor. For example, the mP values for an FP measurement with an Envision Multilabel plate reader are calculated as:

$$mP = 1000 \times (S - G \times P)/(S + G \times P)$$

In practice for HTS applications, however, it is unnecessary to measure absolute polarization states; the assay window is the important factor. The assay window is not significantly changed by G factor variation.

4. In order to select the proper concentration of fluorescent probe for an FP binding assay, increasing concentrations of fluorescent probe are prepared in assay buffer without the binding protein using serial dilution. The background control is assay buffer alone without peptide. In general, the concentration of peptide that generates a fluorescence intensity (FI) signal of more than about ten times the "buffer-only" control sample is selected for developing the FP binding assay. Notice that the FP signal is expressed as a ratio of fluorescence intensities. Thus, the FP signal is not influenced by changes in intensity brought about by changes in the probe concentration. Typically, low nM fluorescent probe concentrations should be sufficient for FP binding assay development.

5. For a protein titration experiment, it is recommended to start with the highest protein concentration possible to make sure the plateau region of the binding curve is well defined.

Depending on the solubility of the protein, the starting concentration of a protein can be in the µM range (e.g., 40–100 µM). More protein dilution points that cover low nM to high µM are necessary to obtain the initial portion of the binding curve, which is especially important if the affinity of the binding is very high.

6. To determine the specificity of the FP binding signal for a PPI, it is recommended to include a non-binding protein as a negative control that should generate only minimal FP signal. In the case of the 14-3-3 binding assay, GST protein was used as a negative control which did not give rise to any FP signal in the presence of the TMR-pS259-Raf peptide (Fig. 3b).

7. After mixing the peptide with the protein, it is recommended to read the FP binding signal over a period of time, for example, 5 min, 30 min, 1 h, 2 h, 4 h, 18 h. Depending on the interaction, the incubation time to reach the equilibration may vary.

8. The final volume for the diluted competitors in the plate will be determined based on the number of assay wells for each sample. Depending on the type of plate, 5–10 µl of dead volume per well should be prepared for transferring the diluted competitors to an assay plate containing the reaction mixture. Multiple competitors or controls for the competition assay may be tested at the same time by adding more rows of dilution for additional competitors/controls. Repeating the assay using a different concentration range of competitor may be necessary.

9. If testing multiple competitors, the assay wells should be increased accordingly. Figure 4 shows an example assay format for testing a second competitor (competitor 2) or control.

10. The FP signal should be measured at various time points after addition of antagonists, e.g., 30 min, 1 h, 2 h, and overnight.

11. For competition assay development for high throughput screening (HTS) of PPI inhibitors, it is recommended to test the effect of competitors in an FP assay using several different concentrations of protein with reasonable FP signal windows (above 60 mP). As an example, Fig. 5b shows the effect of the 14-3-3 antagonist peptide on the 14-3-3 competition assay when the concentration of the protein ranges from EC_{20} to EC_{100} (maximal) of the binding curve. As expected, the sensitivity to the antagonist peptide, R18, decreased as the concentration of protein increased as reflected by the decreased IC_{50} values with the use of increasing concentrations of GST-14-3-3γ protein in the FP assay. The IC_{50} values of 1.4, 2.4, 3.5, and greater than 10 µM were observed when 0.25, 0.5, 1.0, and 2.5 µM of GST-14-3-3γ were used, respectively. Although the assay window was increased with increasing concentrations of

14-3-3 protein (Fig. 3c), the sensitivity to the antagonist was decreased. This aspect is especially important for selecting FP assay conditions for HTS. The protein concentration selected in an FP competition assay should strike a balance between large dynamic range and sensitivity to inhibitors.

5 Summary

FP-based technology has a number of key advantages for monitoring bimolecular interactions. It is nonradioactive in a homogenous "mix-and-read" format without wash steps, multiple incubations, or separation required. FP measurement is directly carried out in solution; no perturbation of the sample is required, making the measurement faster and more quantitative than plate-based methods for bimolecular interactions. It is readily adaptable to low volume (20–30 µl per well for a 384-well plate or 2–5 µl per well for a 1,536-well plate). Due to its technical simplicity and low cost, the FP assay is well-suited for HTS applications to discover PPI modulators. In addition to monitoring PPI, the FP assay has been successfully used to study a wide variety of targets including kinases, phosphatases, proteases, G-protein-coupled receptors (GPCRs), and nuclear receptors [7, 8]. The procedure described here is generally applicable to develop an FP assay for various applications.

Acknowledgments

I thank Dr. Haian Fu for his insightful comments and members of the Fu laboratory for constructive suggestions. This work was supported in part by grants from the National Institutes of Health/National Institute of General Medical Sciences (R01 GM53165 to H.F.) and a Lung Cancer program seed grant from the Winship Cancer Institute. Yuhong Du is a recipient of the Emory Drug Development and Pharmacogenomics Academy Research Fellowship.

References

1. Perrin F (1926) Polarization of light of fluorescence, average life of molecules. J Phys Radium 7:390–401
2. Du Y, Masters SC, Khuri FR et al (2006) Monitoring 14-3-3 protein interactions with a homogeneous fluorescence polarization assay. J Biomol Screen 11:269–276
3. Fu H, Subramanian RR, Masters SC (2000) 14-3-3 proteins: structure, function, and regulation. Annu Rev Pharmacol Toxicol 40:617–647
4. Wang B, Yang H, Liu YC et al (1999) Isolation of high-affinity peptide antagonists of 14-3-3 proteins by phage display. Biochemistry 38:12499–12504
5. Masters SC, Fu H (2001) 14-3-3 proteins mediate an essential anti-apoptotic signal. J Biol Chem 276:45193–45200

6. Jameson DM, Croney JC (2003) Fluorescence polarization: past, present and future. Comb Chem High Throughput Screen 6:167–173

7. Owicki JC (2000) Fluorescence polarization and anisotropy in high throughput screening: perspectives and primer. J Biomol Screen 5:297–306

8. Burke TJ, Loniello KR, Beebe JA et al (2003) Development and application of fluorescence polarization assays in drug discovery. Comb Chem High Throughput Screen 6:183–194

Chapter 36

Estrogen Receptor Alpha/Co-activator Interaction Assay: TR-FRET

Terry W. Moore, Jillian R. Gunther, and John A. Katzenellenbogen

Abstract

Time-resolved fluorescence resonance energy transfer, TR-FRET, is a time-gated fluorescence intensity measurement which defines the relative proximity of two biomolecules (e.g., proteins, peptides, or DNA) based on the extent of non-radiative energy transfer between two fluorophores with overlapping emission/excitation spectra. In these assays, an excited lanthanide ion acts as a "donor" that transfers energy to an "acceptor" fluorophore through dipole–dipole interactions. A FRET signal is reported as the ratio of acceptor to donor emission following donor excitation. When a donor-conjugated protein interacts with an acceptor-conjugated protein, the donor and acceptor fluorophores are brought in close proximity allowing energy transfer from the donor to the acceptor resulting in a FRET signal. Because the lanthanide donors have a long emission half-life, the energy transfer measurement can be time-gated, which dramatically reduces assay interference (due to background autofluorescence and direct acceptor excitation) and thereby increases data quality. Here, we describe a TR-FRET assay that monitors the interaction of the estrogen receptor (ER) α ligand binding domain (labeled with a terbium chelate via a streptavidin–biotin interaction) with a sequence of coactivator protein SRC3 (labeled directly with fluorescein) and the disruption of this interaction with a peptide and a small molecule inhibitor.

Key words TR-FRET, Protein-protein interaction, Lanthanide, Long-lifetime donor, Fluorophore, Protein labeling

1 Introduction

The transcription-regulating function of the estrogen receptors (ERs), ERα and ERβ, relies on their interaction with coactivator proteins. The best studied coactivators are members of the p160 class of steroid receptor coactivators (SRCs) that functionally link ER with modification of chromatin structure and activation of the basal transcriptional machinery [1]. The interaction of the SRCs with ER is regulated by ligand-induced conformation of the ER ligand-binding domain (LBD) where coactivator proteins are bound to ER in the presence of agonist ligands, such as estradiol. Thus, it is intuitive that a reliable and robust assay able to probe the interaction state of the estrogen receptor with its coactivator

Cheryl L. Meyerkord and Haian Fu (eds.), *Protein-Protein Interactions: Methods and Applications*, Methods in Molecular Biology, vol. 1278, DOI 10.1007/978-1-4939-2425-7_36, © Springer Science+Business Media New York 2015

protein would allow further elucidation of the dynamics of this interaction, as well as provide a tool which could be used in high-throughput screening for the identification and development of inhibitors of these two biologically relevant proteins.

A number of assays have been developed to study receptor-coactivator interactions. For instance, glutathione S-transferase (GST)-pull down and related assays have been used to study receptor-coactivator interactions, but these assays are rather labor intensive [2, 3]. An easily employed assay that we [4] and others [5, 6] have described is based on the principle of fluorescence polarization (FP) and monitors the interaction of a fluorophore-labeled SRC LXXLL peptide with the ER LBD. Unfortunately, this assay has low dynamic range and also requires high ER concentrations (200 nM), which make accurate determination of K_i values difficult and costly. Some groups have reported FRET-based assays (*see* chapter 20 for more details regarding FRET assays) to examine nuclear receptor-coactivator interactions, but we have found certain features of these assays make them less than ideal: blocking and washing steps, expensive lanthanide-conjugated antibodies [7–9], or expensive biologic fluorophores [8].

As we found the state-of-the-art assays that were available to study ER/coactivator interactions less than optimal, we developed a TR-FRET assay that is amenable to a high-throughput screening format [10, 11]. The assay we developed uses TR-FRET to monitor the interaction between the ER LBD labeled (via a streptavidin-biotin interaction) with a terbium chelate and a fluorescein-labeled sequence of the SRC3 coactivator protein (*see* Fig. 1). Terbium functions as a long-lifetime (ca. millisecond) luminescent donor, and the fluorescein serves as the TR-FRET acceptor. This assay is superior to organic dye FRET because the emission half-life of fluorescein is short (nsec) relative to that of the terbium complex (msec half-life) [12]. Background emission stemming from direct excitation of fluorescein or endogenous cellular fluorophores can thus be eliminated by pulsing the terbium complex at the excitation wavelength and gating the emission with a 50-μs delay. When properly optimized, the TR-FRET method gives a good signal-to-noise ratio and can be run in a straightforward, mix-and-measure format with very low concentrations of terbium-labeled streptavidin and biotin-labeled ER-LBD.

We note that, in a previous publication [11], we detailed the use of a Cy5-europium pair that was developed in collaboration with our colleagues at Emory University. Using this pair is advantageous because it allows the monitoring of acceptor emissions at longer wavelengths than fluorescein. Autofluorescent compounds found in libraries typically emit at wavelengths shorter than 550 nm; thus, when used in a high-throughput format, the Eu-Cy5 system is a better choice for minimizing false positives arising from interfering emission patterns. The reason we have detailed the

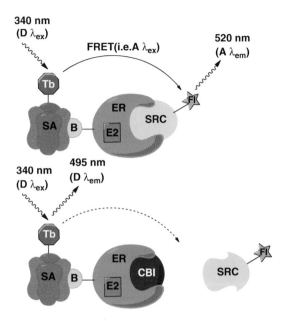

Fig. 1 Schematic of the time-resolved fluorescence resonance energy transfer (FRET) assay. (*Top*) In general, FRET occurs when an emission wavelength of a donor molecule (D λ_{em}; e.g., 495 nm) overlaps with the excitation wavelength of a nearby acceptor (A λ_{ex}), resulting in an emission signal from the acceptor (A λ_{em}; e.g., 520 nm). FRET occurs between the streptavidin-terbium (SA-Tb) donor and the fluorescein-steroid receptor coactivator (SRC-Fl) acceptor when SRC-Fl is recruited to the biotin-labeled estrogen receptor (B-ER) bound with the agonist ligand 17 β-estradiol (E2). (*Bottom*) In the presence of coactivator binding inhibitor (CBI), this assembly is disrupted, and the FRET signal decreases

terbium-fluorescein pair here is because we found it to give better signal-to-noise than the Eu-Cy5 pair when using our particular plate reader (VICTOR multi-label plate reader) for routine dose–response assays. In general, if an assay is needed for a high-throughput screen, we would recommend the Eu-Cy5 pair.

We have developed this assay and explained in detail below the steps necessary to replicate it using ER alpha. We [13–16] and others [17–21] have since generalized the assay to other nuclear hormone receptors and coactivator protein segments, and we encourage other users to do the same. This does sometimes require fine-tuning of assay component concentrations, but, generally, the results are very accurate and reliable.

2   Materials

Prepare all solutions using autoclaved, deionized water and analytical grade reagents. Prepare and store all reagents at room temperature (unless indicated otherwise).

1. SRC1-Box II peptide. Store wrapped in foil at −20 °C (*see* Note 1).

2. Pyrimidine coactivator binding inhibitor (CBI) 1. Store at −20 °C (*see* Note 2).

3. 20 mM solutions of test compounds in DMSO (or DMF). Store at −20 °C.

4. ERα-417-biotin and SRC-3-NRD-fluorescein. Store at −20 °C (*see* Note 3).

5. 200 nM fluorescein-SRC-3-NRD. Store at 4 °C (*see* Note 4).

6. 1 mM solution of 17β-estradiol in DMF. Store at −20 °C.

7. LanthaScreen™ Streptavidin-Terbium (Invitrogen). Store at −20 °C.

8. TR-FRET buffer: 20 mM Tris–HCl, pH 7.5, 0.01 % NP40, 50 mM NaCl; adjust pH with conc. HCl to pH 7.5 (*see* Notes 5–8).

9. 96-Well black HE high efficiency microplate (Molecular Devices) (*see* Note 9).

10. 96-Well polypropylene plate (*see* Note 10).

11. Eight-channel autopipettor with tips (0–20 μL capacity) (*see* Note 11).

12. Single-channel manual pipettors with tips (0–1,000 μL capacity) (*see* Note 12).

13. VICTOR™ Multilabel plate reader (Perkin-Elmer) (*see* Note 13).

3   Methods

1. Prepare a 4× stock solution (Solution A) containing the following components at the following concentrations: ERα-417 (8 nM), 17β-estradiol (4 μM), and LanthaScreen™ Streptavidin-Terbium (2 nM) in TR-FRET buffer. Keep this solution at 0–5 °C (*see* Notes 14 and 15).

2. In a 96-well colorless polypropylene plate, serially dilute (*see* Note 16) each 20 mM solution of test compound, including positive controls pyrimidine CBI 1 and SRC1-Box II peptide, using DMF (*see* Note 17).

3. In a second 96-well colorless polypropylene plate, dilute (*see* Note 18) the previously prepared DMF-diluted test compounds in a 1:10 fashion into TR-FRET buffer (*see* Note 19).

4. Add 10 μL of each of the buffer-diluted compound solutions to a black 96-well plate, starting with the highest concentration at the top left hand corner of the plate, with decreasing

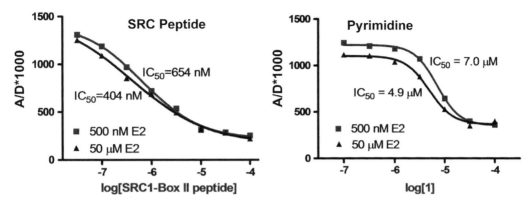

Fig. 2 Representative data from a fluorescence resonance energy transfer (FRET) assay. By plotting the ratio of the emission intensities of the acceptor to the donor (A/D × 1,000) against the log of ligand molar concentration, dose–response curves for displacement of SRC-3-NRD-fluorescein by the steroid receptor coactivator peptide (*left*) and pyrimidine coactivator binding inhibitor **1** (*right*) can be generated. Varying the concentration of the agonist 17β-estradiol (E2; 500 nM (*filled square*) and 50 μM (*filled triangle*)) has no substantial effect on the IC$_{50}$ values of the compounds, implying that these positive control compounds do not compete with E2 for the ligand-binding pocket but act by direct displacement of the SRC-3-NRD-fluorescein

concentrations down the plate. For every compound tested, the eight concentration points are tested in duplicate (two full columns for each test compound). Add 5 μL of Solution A to each well of the black 96-well plate.

5. Incubate the plate at 0–5 °C (on a foil-covered ice bucket or in the refrigerator) for 20 min.

6. Add 5 μL of 200 nM fluorescein-SRC-3-NRD to each well of the black plate (*see* **Notes 20** and **21**).

7. Incubate the plate at room temperature for 1 h in the dark (*see* **Notes 22** and **23**).

8. Measure the TR-FRET signal using an excitation filter at 340/10 nm (*see* **Note 24**), and emission filters for terbium and fluorescein at 495/20 and 520/25 nm, respectively, with an acquisition gated with a 50-μs delay (*see* **Notes 25–28**) (*see* Fig. 2 for an example of data obtained).

4   Notes

1. The sequence of the SRC1-Box II peptide is NH$_2$-CLTERH-KILHRLLQE-CO$_2$H. It was synthesized at a private protein sciences facility. Fluorescein labeling was through the N-terminal cysteine residue.

2. The pyrimidine CBI **1** positive control was synthesized by our laboratory as outlined in references [4, 16] below. When designing a TR-FRET assay to monitor a desired protein-protein interaction, it is best to include a known positive control.

pyrimidine CBI 1

3. The mutant ER protein labeled with biotin and the SRC-3-NRD protein fragment labeled with fluorescein were made in our laboratory, as outlined in the references below [16, 22], and stored as a 1:1 glycerol–buffer mixture at −20 °C. When working with these proteins, try to minimize the time at room temperature or even on the cooling block. It is also advisable to aliquot these proteins so that some can be stored untouched in a freezer without daily disturbance.

4. Diluted fluorescein-SRC-3-NRD can be made by diluting stock protein into TR-FRET buffer. It should be made prior to use, and any extra should be discarded.

5. Stock solutions of 1 M Tris at various pH values are used. For example, to make this buffer, 20 mL of 1 M Tris was added to a 1 L bottle with NP40 and NaCl, and filled to 1 L.

6. NP40 is very viscous. Take time to ensure correct measurement. Thoroughly rinse the graduated cylinder used for measuring and add rinse to the buffer bottle.

7. This concentration was achieved by diluting from a stock solution of 0.5 M NaCl.

8. TR-FRET buffer is very stable and can be stored at room temperature.

9. Other 96-well, 20-μL plates may work, but we have found these to give optimal results.

10. These plates are used for dilution. Any plates that can handle solvents such as DMF or DMSO without breakdown and can accommodate the necessary volumes could be used.

11. Autopipettors with low μL capability reduce the error associated with this assay and are highly recommended.

12. Use caution when adding low μL volumes to plates or even to stock solutions. It is very difficult to see if small volumes have been added to a black plate. This can be checked using a manual pipettor to aspirate one of the wells and determine the μL volume in the well, but it is generally not recommended as some solution will be lost on the pipet tip.

13. This assay has been performed on other plate readers with TR-FRET capabilities. Concentrations of reagents may need to be adjusted to generate optimal signal-to-noise results.

14. This stock solution can be divided into eight wells of a dilution plate so that it can be dispensed into the final black plate using an autopipettor.

15. This solution should be kept in a cooling block on ice. Any extra should be discarded at the end of the day.

16. We prefer eight concentrations for dose–response assays using 1:10 and 1:3 serial dilutions. For example, to make approximately 100 μL of each concentration point, put 100 μL of 20 mM solution in the first well. Make the second well by adding 30 μL of 20 mM stock to 70 μL of DMF. Then serially dilute each of these two wells 1:10 into DMF. In this case, the DMF concentrations would be 20, 6, 2, 0.6, 0.2, 0.06, 0.02, and 0.006 mM.

17. We often use DMF to prepare stock solutions because DMF does not freeze and is not as hygroscopic as DMSO. DMF is, however, less inert than DMSO, and, in other systems, it may denature proteins. We have found that, in this assay, either DMSO or DMF can be used with no noticeable effect on assay performance.

18. For example, add 10 μL of DMF solution to 90 μL TR-FRET buffer. At this point, the compound concentrations are 2,000, 600, 200, 60, 20, 6, 2, and 0.6 μM, with 10 % DMF.

19. It is not recommended to keep the diluted compounds either in solvent or in buffer. Even with the most careful covering with acetate covers and foil wrap, some evaporation does occur, changing the concentration of the compounds. If the dilution plates are to be used for several hours, keep covered using a solvent-resistant plate cover and wrap the plate in foil. Keep the plate on a chilled block.

20. The 200 nM fluorescein-SRC-3-NRD stock solution can be divided into eight wells of a dilution plate so that it can be dispensed into the final black plate using an autopipettor.

21. At this final point, the compound concentrations are 1,000, 300, 100, 30, 10, 3, 1, and 0.3 μM, with 5 % DMF. The concentrations of each assay component are, for ERα-417, 2 nM; for 17β-estradiol, 1 μM; and for LanthaScreen™ Streptavidin-Terbium, 0.5 nM.

22. Although we recommend waiting for an hour for incubation of the assay components before reading, the final dose–response curve and, therefore, K_i value are nearly identical after incubation for only 5 min. Taking a measurement at an earlier timepoint can be helpful if protein viability needs to be checked or preliminary results are needed extremely quickly.

23. Fluorescein is a light-sensitive fluorophore. Covering the plate with aluminum foil is typically sufficient to remove ambient light.

24. The nomenclature "340/10 nm" refers to the allowance of the filter used; thus, "340/10 nm" implies that light of wavelengths from 335 to 345 nm is allowed to pass through the filter.

25. The plate reader should output data giving emission intensities for the donor (D) and acceptor (A). We calculate the final ratio used with the following formula: $A/D \times 1,000$. This number can then be plotted against concentration to give a dose–response curve.

26. If a compound or compound series does show significant interference with the fluorescence filters used in this assay, consider switching to a different acceptor, or donor/acceptor pair (easily researched on many commercial websites).

27. If unexpected results occur (e.g., a compound gives increasing interaction with increasing dose) it is likely due to fluorescent interference of the compound instead of a true increased association of the proteins or stabilizing effect. Check the fluorescence spectrum of the compound for any interference.

28. If this assay does not produce reliable, robust results, it is most commonly due to degradation of protein over time. Realistically, the proteins should last months when stored and handled appropriately. Replace the proteins when the signal-to-noise ratio begins to decrease or the K_i values of the standard peptide start to drift.

References

1. Tamrazi A, Carlson KE, Rodriguez AL et al (2005) Coactivator proteins as determinants of estrogen receptor structure and function: spectroscopic evidence for a novel coactivator-stabilized receptor conformation. Mol Endocrinol 19:1516–1528

2. Bramlett KS, Wu YF, Burris TP (2001) Ligands specify coactivator nuclear receptor (NR) box affinity for estrogen receptor subtypes. Mol Endocrinol 15:909–922

3. Mueller SO (2004) Xenoestrogens: mechanisms of action and detection methods. Anal Bioanal Chem 378:582–587

4. Rodriguez AL, Tamrazi A, Collins ML et al (2004) Design, synthesis, and in vitro biological evaluation of small molecule inhibitors of estrogen receptor a coactivator binding. J Med Chem 47:600–611

5. Becerril J, Hamilton AD (2007) Helix mimetics as inhibitors of the interaction of the estrogen receptor with coactivator peptides. Angew Chem Int Ed 46:4471–4473

6. Ozers MS, Ervin KM, Steffen CL et al (2005) Analysis of ligand-dependent recruitment of coactivator peptides to estrogen receptor using fluorescence polarization. Mol Endocrinol 19:25–34

7. Gowda K, Marks BD, Zielinski TK et al (2006) Development of a coactivator displacement assay for the orphan receptor estrogen-related receptor-gamma using time-resolved fluorescence resonance energy transfer. Anal Biochem 357:105–115

8. Liu JW, Knappenberger KS, Kack H et al (2003) A homogeneous in vitro functional assay for estrogen receptors: coactivator recruitment. Mol Endocrinol 17:346–355

9. Zhou GC, Cummings R, Li Y et al (1998) Nuclear receptors have distinct affinities for coactivators: characterization by fluorescence

resonance energy transfer. Mol Endocrinol 12:1594–1604

10. Gunther JR, Moore TW, Collins ML et al (2008) Amphipathic benzenes are designed inhibitors of the estrogen receptor α/steroid receptor coactivator interaction. ACS Chem Biol 3:282–286

11. Gunther JR, Du Y, Rhoden E et al (2009) A set of time-resolved fluorescence resonance energy transfer assays for the discovery of inhibitors of estrogen receptor-coactivator binding. J Biomol Screen 14:181–193

12. Mathis G (1995) Probing molecular-interactions with homogeneous techniques based on rare-earth cryptates and fluorescence energy-transfer. Clin Chem 41:1391–1397

13. Kim SH, Gunther JR, Katzenellenbogen JA (2010) Monitoring a coordinated exchange process in a four-component biological interaction system: development of a time-resolved terbium-based one-donor/three-acceptor multicolor FRET system. J Am Chem Soc 132:4685–4692

14. Jeyakunnar M, Carlson KE, Gunther JR et al (2011) Exploration of dimensions of estrogen potency parsing ligand binding and coactivator binding affinities. J Biol Chem 286: 12971–12982

15. Gunther JR, Parent AA, Katzenellenbogen JA (2009) Alternative inhibition of androgen receptor signaling: peptidomimetic pyrimidines as direct androgen receptor/coactivator disruptors. ACS Chem Biol 4:435–440

16. Parent AA, Gunther JR, Katzenellenbogen JA (2008) Blocking estrogen signaling after the hormone: pyrimidine-core inhibitors of estrogen receptor-coactivator binding. J Med Chem 51:6512–6530

17. Shukla SJ, Nguyen DT, MacArthur R et al (2009) Identification of pregnane X receptor ligands using time-resolved fluorescence resonance energy transfer and quantitative high-throughput screening. Assay Drug Dev Technol 7:143–169

18. Wang Y, Yang DZ, Chang A et al (2012) Synthesis of a ligand-quencher conjugate for the ligand binding study of the aryl hydrocarbon receptor using a FRET assay. Med Chem Res 21:711–721

19. Schaufele F (2011) FRET analysis of androgen receptor structure and biochemistry in living cells. Methods Mol Biol 776:147–166

20. Vogel KW, Marks BD, Kupcho KR et al (2010) Improved nuclear receptor binding assays for HTS by conversion from FP to TR-FRET read-out. Endocr Rev 31

21. Hilal T, Puetter V, Otto C et al (2010) A dual estrogen receptor TR-FRET assay for simultaneous measurement of steroid site binding and coactivator recruitment. J Biomol Screen 15:268–278

22. Tamrazi A, Carlson KE, Daniels JR et al (2002) Estrogen receptor dimerization: ligand binding regulates dimer affinity and dimer dissociation rate. Mol Endocrinol 16:2706–2719

Chapter 37

High Content Screening Biosensor Assay to Identify Disruptors of p53–hDM2 Protein-Protein Interactions

Yun Hua, Christopher J. Strock, and Paul A. Johnston

Abstract

This chapter describes the implementation of the p53–hDM2 protein-protein interaction (PPI) biosensor (PPIB) HCS assay to identify disruptors of p53–hDM2 PPIs. Recombinant adenovirus expression constructs were generated bearing the individual p53–GFP and hDM2–RFP PPI partners. The N-terminal p53 transactivating domain that contains the binding site for hDM2 is expressed as a GFP fusion protein that is targeted and anchored in the nucleolus of infected cells by a nuclear localization (NLS) sequence. The p53–GFP biosensor is localized to the nucleolus to enhance and facilitate the image acquisition and analysis of the PPIs. The N-terminus of hDM2 encodes the domain for binding to the transactivating domain of p53, and is expressed as a RFP fusion protein that includes both an NLS and a nuclear export sequence (NES). In U-2 OS cells co-infected with both adenovirus constructs, the binding interactions between hDM2 and p53 result in both biosensors becoming co-localized within the nucleolus. Upon disruption of the p53–hDM2 PPIs, the p53–GFP biosensor remains in the nucleolus while the shuttling hDM2–RFP biosensor redistributes into the cytoplasm. p53–hDM2 PPIs are measured by acquiring fluorescent images of cells co-infected with both adenovirus biosensors on an automated HCS imaging platform and using an image analysis algorithm to quantify the relative distribution of the hDM2–RFP shuttling component of the biosensor between the cytoplasm and nuclear regions of compound treated cells.

Key words Protein-protein interaction biosensors, High content screening, Imaging, Image analysis

1 Introduction

The combination of high content screening (HCS) with fluorescent biosensors allows for high spatial and temporal resolution of protein-protein interaction (PPI) partners that have been fused to spectrally distinct fluorescent reporter proteins [1–9]. The distribution of macromolecules within compartments of the cell is a tightly regulated process, exemplified by the passage of proteins through the nuclear pore complexes (NPCs) in the nuclear envelope that separates the cytoplasm and nucleus [10]. While small molecules pass through NPCs by passive diffusion, the passage of cargos larger than ~40 kDa require a facilitated active import and

Cheryl L. Meyerkord and Haian Fu (eds.), *Protein-Protein Interactions: Methods and Applications*, Methods in Molecular Biology, vol. 1278, DOI 10.1007/978-1-4939-2425-7_37, © Springer Science+Business Media New York 2015

export process mediated by specific receptor proteins [10]. Passage through the NPC involves the assembly of a trimeric import-complex in the cytoplasm between an importin-α adaptor receptor that recognizes the nuclear localization sequence (NLS) of the cargo and the importin-β transport receptor that facilitates docking interactions with the nucleoporins [10, 11]. Since the nucleus and cytoplasm represent subcellular compartments that can readily be distinguished by HCS methods, investigators have taken advantage of the regulated nucleocytoplasmic transport of proteins to design biosensors to measure PPIs [1, 2, 4, 5]. To construct translocation or positional biosensors, the "bait" biosensor is typically anchored to a specific location within the cell while the "prey" biosensor is designed to shuttle between different compartments or regions of the cell, and productive PPIs are required for the co-localization of both biosensors [1, 2, 4, 5].

In an example involving p53 and hDM2, Stauber and colleagues generated a "bait" biosensor encoding the interacting domain of p53 fused to HIV-1 Rev and blue fluorescent protein (BFP) that when transfected into cells was expressed in the nucleolus region of the nucleus [5]. They constructed a shuttling "prey" biosensor encoding the interacting domain of mdm2 fused with a SV40 NLS sequence, an HIV-1 Rev NES sequence, GST, and green fluorescent protein (GFP) to generate an NLS-GFP-GST-mdm2-NES expression plasmid that when transfected into cells was predominantly localized in the cytoplasm [5]. Co-transfection of cells with both the p53 "bait" and mdm2 "prey" biosensors resulted in translocation of the mdm2 biosensor into the nucleus and co-localization with the p53 biosensor in the nucleolus [5]. Nutlin-3 treatment of cells co-transfected with both biosensors caused a redistribution of the mdm2 prey biosensor into the cytoplasm [5]. We have recently described the development and implementation of a similarly designed p53–hDM2 protein-protein interaction biosensor (PPIB) HCS assay in which we screened 220,017 of the NIH's compound library and identified three novel structurally related methylbenzo-naphthyridin-5-amine (MBNA) hits [1, 2]. Hit follow-up studies revealed that the MBNAs triggered the expected biological responses, including enhanced p53 protein and p53 target gene levels, p53-dependent cell cycle arrest in G1, apoptosis, and growth inhibition with IC_{50}s ~4 μM [2].

This chapter describes the employment of the p53–hDM2 PPIB HCS assay to identify p53–hDM2 PPI disruptors [1, 2]. Recombinant adenovirus expression constructs were created for the individual p53–GFP and hDM2–RFP PPI partners (Fig. 1). The N-terminal residues 1–131 of p53 include the p53 transactivating domain that contains the binding site for hDM2. This GFP fusion protein fragment is targeted to and anchored in the nucleolus by the inclusion of nuclear localization (NLS) and nucleolus

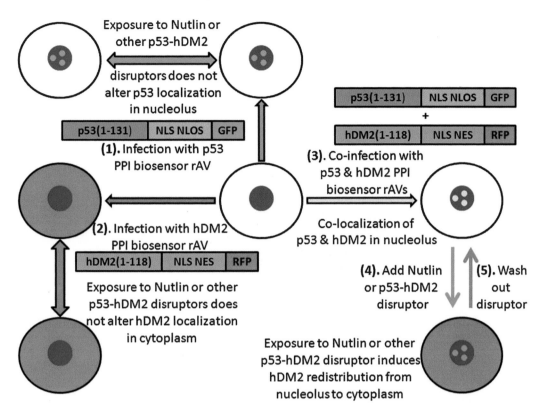

Fig. 1 p53–hDM2 protein-protein interaction biosensor design and distribution phenotypes. Recombinant adenovirus expression constructs were generated bearing the individual p53–GFP and hDM2–RFP PPI partners. The N-terminal residues 1–131 of p53 include the transactivating domain that contains the binding site for hDM2, and this protein fragment is expressed as a GFP fusion protein that is targeted and anchored in the nucleolus of infected cells by the inclusion of nuclear localization (NLS) and nucleolus localization (NLOS) sequences. The p53–GFP component is directed to the nucleolus to enhance and facilitate the image acquisition and analysis of the PPIs. *Route 1*: In cells infected with the p53–GFP "bait" biosensor adenovirus alone, the p53–GFP remains localized to bright fluorescent puncta in the nucleolus and its distribution does not alter upon exposure to Nutlin-3 or other p53–hDM2 disruptors. The N-terminal residues 1–118 of hDM2 encode the domain for binding to the p53 transactivating domain, and this fragment is expressed as a RFP fusion protein that includes both an NLS and a nuclear export sequence (NES). *Route 2*: In cells infected with the hDM2–RFP adenovirus "prey" biosensor alone, hDM2–RFP expression is localized only in the cytoplasm of cells and does not change upon exposure to Nutlin-3 or other p53–hDM2 disruptors. *Route 3*: In U-2 OS cells that are co-infected with both adenovirus constructs, however, the binding interactions between the hDM2 and p53 components of the biosensors result in both proteins becoming localized to the nucleolus. *Route 4*: Upon disruption of the p53–hDM2 protein-protein interaction with a compound like Nutlin-3, the p53–GFP interaction partner remains nucleolar, while the shuttling hDM2–RFP interaction partner redistributes into the cytoplasm. *Route 5*: Upon removal of a disrupting agent like Nutlin 3, the shuttling hDM2–RFP interaction partner redistributes back into the p53–GFP containing nucleolus

localization (NLOS) sequences (Fig. 1) [1, 2], which enhances and facilitates image acquisition and analysis of the PPIs. In cells infected with the p53–GFP "bait" biosensor adenovirus alone, the p53–GFP remains localized to bright fluorescent puncta in

558   Yun Hua et al.

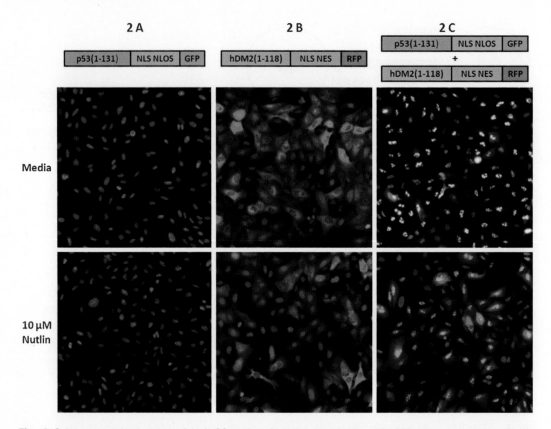

Fig. 2 Color composite images of U-2 OS cells infected with (**a**) the p53–GFP biosensor alone, (**b**) the hDM2–RFP biosensor alone, or (**c**) co-infected with both biosensors and then treated ± 10 μM Nutlin-3 for 90 min. U-2 OS cells were infected with (**a**) the p53–GFP biosensor alone, (**b**) the hDM2 biosensor alone, or (**c**) co-infected with both biosensors and 2,500 cells were seeded into the wells of 384-well assay plates, cultured overnight at 37 °C, 5 % CO$_2$, and 95 % humidity, and then treated ± 10 μM Nutlin-3 for 90 min. Cells were then fixed and stained with Hoechst 33342 and 20× 0.4 NA images of three fluorescent channels (Ch1—Hoechst *blue*, Ch2—p53–GFP *green*, and Ch3—hDM2–RFP *red*) were acquired on the ArrayScan V^{TI} automated imaging platform as described above. Representative color composite images of control and Nutlin-3 treated U-2 OS cells individually infected with or co-infected with both the p53–GFP and hDM2–RFP biosensors are presented. In cells infected with each of the biosensors alone, the p53–GFP biosensor remains localized to bright fluorescent puncta within the nucleoli of Hoechst stained nuclei and its distribution does not alter upon exposure to Nutlin-3, while the hDM2–RFP biosensor remains localized in the cytoplasm of cells and does not change upon exposure to Nutlin-3. In U-2 OS cells co-infected with both biosensors, however, the two biosensors become co-localized to the nucleolus, and upon exposure to Nutlin-3 the hDM2–RFP interaction partner redistributes into the cytoplasm while the p53–GFP interaction partner remains nucleolar

the nucleolus and its distribution does not alter upon exposure to Nutlin-3 or when co-infected with the hDM2–RFP biosensor (Figs. 1, route 1 and 2a). The N-terminal residues 1–118 of hDM2 encode the domain for binding to the N-terminal transactivating domain of p53, and this fragment is expressed as a RFP fusion protein that includes both an NLS and a nuclear export sequence (NES) (Fig. 1) [1, 2]. In cells infected with the hDM2–RFP adenovirus "prey" biosensor alone, hDM2–RFP expression is localized only in the cytoplasm of cells and does not change upon exposure to Nutlin-3 (Figs. 1, route 2 and 2b). In cells that are co-infected with both adenovirus constructs, however, the interaction of the hDM2 and p53 components of the biosensors results in both proteins becoming localized to the nucleolus (Figs. 1, route 3 and 2c) [1, 2]. Upon disruption of the p53–hDM2 protein-protein interaction with a compound like Nutlin-3, the p53–GFP interaction partner remains nucleolar, while the shuttling hDM2–RFP interaction partner redistributes into the cytoplasm (Figs. 1, route 4 and 2c) [1, 2].

The disruption of the p53–HDM2 interaction biosensor is measured by acquiring images on an automated HCS imaging platform, and an image analysis algorithm is used to quantify the relative distribution of the hDM2–RFP shuttling component of the biosensor between the cytoplasm and nuclear regions of compound treated cells that were co-infected with both adenovirus biosensors (Figs. 2 and 3) [1, 2]. The images of the three fluorescent channels (Hoechst, GFP, and RFP) of the p52–hDM2 PPIB were analyzed using the molecular translocation image analysis algorithm (Fig. 3a). The nucleic acid dye Hoechst 33342 was used to stain and identify the nucleus, and this fluorescent signal was used to focus the instrument and define a nuclear mask in channel 1 (Ch1). Objects in Ch1 that exhibited the appropriate fluorescent intensities above background and size (width, length, and area) characteristics were identified and classified by the image segmentation as nuclei (dark blue ring, Ch1). The nuclear mask was eroded 1 pixel to reduce cytoplasmic contamination within the nuclear area, and the reduced mask was used to quantify the amount of target channel, p53–GFP (Ch2) or hDM2–RFP (Ch3), fluorescence within the nucleus (yellow Circ masks in Ch2 and Ch3). The nuclear mask was then dilated to cover as much of the cytoplasmic region as possible without going outside the cell boundary. Removal of the original nuclear region from this dilated mask creates a Ring mask that covers the cytoplasmic region outside the nuclear envelope (magenta Ring mask in Ch2 and cyan Ring mask in Ch3). The image analysis algorithm outputs quantitative data such as: the total and average fluorescent intensities of the Hoechst stained objects (Ch1) the selected object or cell count from Ch1, the total and average fluorescent intensities of the p53–GFP (Ch2) and the hDM2–RFP (Ch3) signals in the nucleus (Circ) or cytoplasm (Ring) regions as an overall well average value, or on an individual cell basis. To quantify the translocation of the hDM2–RFP from the nucleus to the cytoplasm induced by disruptors of the p53–hDM2 protein-protein interaction the image analysis algorithm calculates a mean average intensity difference by subtracting the average hDM2–RFP intensity in the Ring (Cytoplasm) region from the average hDM2–RFP intensity in the Circ (Nuclear) region of Ch3; Mean Circ-Ring Average Intensity Difference Channel 3

a   **Molecular Translocation Image Analysis Algorithm**

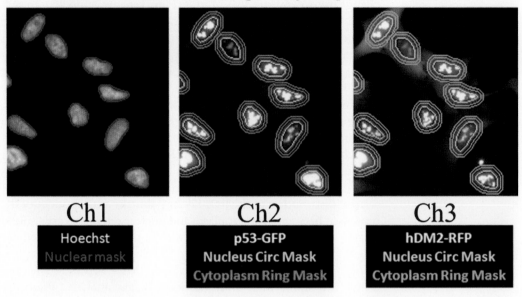

Ch1	Ch2	Ch3
Hoechst Nuclear mask	p53-GFP Nucleus Circ Mask Cytoplasm Ring Mask	hDM2-RFP Nucleus Circ Mask Cytoplasm Ring Mask

b   **p53-hDM2 Protein-protein Interaction Parameter:**
Mean Circ-Ring Average intensity Difference Ch3

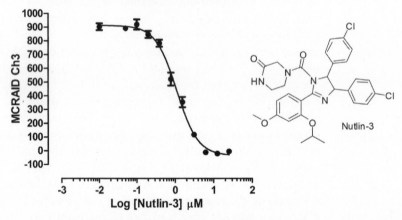

Fig. 3 Molecular translocation image analysis algorithm. U-2 OS cells co-infected with the p53–hDM2 PPIB adenoviruses were seeded at 2,500 cells per well in 384-well Greiner collagen-coated assay plates, cultured overnight at 37 °C, 5 % CO_2, and 95 % humidity, and were then exposed to the indicated concentrations of Nutlin-3 in 0.5 % DMSO for 90 min prior to fixation with 3.7 % formaldehyde containing 2 μg/mL Hoechst 33342. Images in three fluorescent channels were sequentially acquired on the ArrayScan VTI platform using a 20× 0.4 NA objective with the XF93 excitation and emission filter set and were analyzed with the molecular translocation (MT) image analysis algorithm. (**a**) Image segmentation. In Ch1, Hoechst stained objects that exhibited the appropriate morphological characteristics (width, length, and area) with fluorescent intensities sufficiently above a background threshold were identified and classified by the image segmentation as nuclei. The nuclear mask derived from Ch1 (*dark blue ring*) was then used to segment the images from Ch2 and Ch3

(MCRAID-Ch3) (Fig. 3b). On average, Nutlin-3 exhibited an IC_{50} of 0.608 ± 0.382 μM ($n = 5$) in the p53–hDM2 PPIB assay (Fig. 3b). To implement the p53–hDM2 PPIB assay in screening, the mean MCRAID-Ch3 value of the DMSO maximum plate control wells ($n = 32$) and the mean MCRAID-Ch3 value of the 10 μM Nutlin-3 minimum plate control wells ($n = 24$) were utilized to normalize the MCRAID-Ch3 compound data and to represent 0 % and 100 % disruption/inhibition of the p53–hDM2 interactions, respectively [1, 2].

2 Materials

1. Nutlin-3.

2. Formaldehyde.

3. Dimethyl sulfoxide (DMSO) (99.9 % high performance liquid chromatography-grade, under argon).

4. Hoechst 33342.

5. U-2 OS osteosarcoma cell line.

6. Culture medium: McCoy's 5A medium, 2 mM L-glutamine, 10 % fetal bovine serum, and 100 U/mL penicillin and streptomycin.

7. Humidified incubator at 37 °C, 5 % CO_2, and 95 % humidity.

8. Dulbecco's Mg^{2+} and Ca^{2+} free phosphate buffered saline (PBS).

Fig. 3 (continued) into nuclear (*Circ*) and cytoplasmic (*Ring*) regions. To reduce cytoplasm contamination within the nuclear area the nuclear mask was eroded, and the reduced mask (*yellow ring*) was used to quantify the amount of target channel, p53–GFP in Ch2 and hDM2–RFP in Ch3, fluorescence within the nuclear region. To measure fluorescence in a region of the cytoplasm in Ch2 and Ch3, the nuclear mask derived from Ch1 was dilated to cover as much of the cytoplasm region as possible without going outside the cell boundary. Removal of the original nuclear region from this dilated mask creates a ring mask that covers the cytoplasm region outside the nuclear envelope. The number of pixels away from the nuclear mask and the number of pixels (width) between the inner and outer ring masks were selectable within the MT bio-application software. The ring masks were then used to quantify the amount of target channel, p53–GFP (Ch2, *mauve rings*) or hDM2–RFP (Ch3, *light blue rings*), fluorescence within the cytoplasm region. (**b**) Chemical structure of Nutlin-3 and concentration dependent disruption of p53–hDM2 PPIs. The molecular translocation image analysis algorithm produces a mean average intensity difference in Ch3 (MCRAID-Ch3) parameter calculated by subtracting the average hDM2–RFP intensity in the *Ring* (Cytoplasm) region from the average hDM2–RFP intensity in the *Circ* (Nuclear) region of Ch3. High MCRAID-Ch3 values indicate that the hDM2–RFP biosensor is predominantly localized within the nuclear region, while low MCRAID-Ch3 values indicate a more prominent localization within the cytoplasm. Nutlin-3 treatment induced a concentration dependent decrease in the hDM2–RFP MCRAID-Ch3 signal that was consistent with the redistribution of the hDM2–RFP from the nucleus to the cytoplasm. Nutlin-3 consistently exhibited an IC_{50} of 0.608 ± 0.382 μM ($n = 5$) for the disruption of the p53–hDM2 PPIs

9. Trypsin 0.25 %, 1 g/L EDTA solution.

10. Recombinant adenovirus p53–GFP and hDM2–RFP Biosensors: Recombinant adenovirus expression constructs bearing the individual p53–GFP (TagGFP, Evrogen, Inc.) and hDM2–RFP (Tag RFP, Evrogen, Inc.) protein-protein interaction partners were obtained from Cyprotex [1, 2].

11. 384-well collagen-coated barcoded assay microplate.

3   Methods

1. Aspirate spent tissue culture medium from U-2 OS cells in tissue culture flasks that are <70 % confluent (*see* **Note 1**), wash cell monolayers 1× with PBS, and expose cells to trypsin-EDTA until they detach from the surface of the tissue culture flasks. Add serum containing tissue culture medium to neutralize the trypsin. Transfer the cell suspension to a 50 mL capped sterile centrifuge tube and centrifuge at $500 \times g$ for 5 min to pellet the cells. Resuspend cells in serum containing tissue culture medium and count the number of trypan blue excluding viable cells using a hemocytometer.

2. Co-infect 1×10^7 U-2 OS cells with p53–GFP and hDM2–RFP adenovirus by incubating cells with the manufacturer's recommended volume of virus (*see* **Note 2**), typically $5 \mu L/10^7$ cells, in 1.5 mL of culture medium for 1 h at 37 °C, 5 % CO_2 in a humidified incubator with periodic inversion (every 10 min) to maintain cells in suspension.

3. Dilute co-infected cells to 5.6×10^4 cells/mL and dispense 45 μL (2,500 cells) in each well of a 384-well collagen-coated barcoded assay microplate using a liquid handler (*see* **Note 3**).

4. Incubate assay plates overnight at 37 °C, 5 % CO_2 in a humidified incubator (*see* **Note 4**).

5. Use an automated liquid handling device to add diluted compounds or plate controls (Nutlin3, 10 μM final, or 5 μL of DMSO, 0.25–0.5 % final) to appropriate wells for a final screening concentration of 25 μM (*see* **Note 5**).

6. Incubate plates at 37 °C, 5 % CO_2 in a humidified incubator for 90 min (*see* **Note 6**).

7. Fix samples by the adding 50 μL of pre-warmed (37 °C) 7.4 % formaldehyde and 2 μg/mL Hoechst 33342 in PBS using a liquid handler and incubate at room temperature for 30 min (*see* **Note 7**).

8. Aspirate the liquid and wash the plates twice with 85 μL of PBS using a liquid handler (*see* **Note 8**). Seal with adhesive aluminum plate seals (Abgene) with the last 85 μL wash of PBS in place.

9. Acquire fluorescent images in three channels on an automated imaging platform (*see* **Note 9**) (e.g. ArrayScan VTI, Thermo-fisher Scientific) (Fig. 2). Images of three fluorescent channels (Hoechst, GFP, and RFP) are sequentially acquired on an ArrayScan VTI platform using either a 10× 0.3 NA or a 20× 0.4 NA objective with the XF93 excitation and emission filter set (Hoechst, FITC, and TRITC) (Fig. 2). In our system, excitation is provided by an X-CITE™ 120 W high pressure metal halide arc lamp with Intelli-Lamp™ technology (Photonic Solutions Inc. Mississauga, Canada), and images are captured on a high sensitivity cooled Orca CCD Camera (Photometrics Quantix). Typically with the 10× 0.3 NA objective, the ArrayScan VTI platform is set up to acquire 250 selected objects (nuclei) or two fields of view, whichever comes first. With the 20× 0.4 NA objective, the ArrayScan VTI platform is set up to acquire four fields of view.

10. Analyze the images of the three fluorescent channels (Hoechst, GFP, and RFP) of the p53–hDM2 PPIB using the molecular translocation image analysis algorithm as described above (Fig. 3) (*see* **Note 10**). The MCRAID-Ch3 output is utilized to quantify the disruption of p53–hDM2 PPIs (Fig. 3b).

4   Notes

1. Typically better responses are obtained when the p53–hDM2 adenovirus biosensors are used to co-infect U-2 OS cells harvested from tissue culture flasks that are <70 % confluent.

2. The optimal volume of each batch of recombinant adenovirus biosensor per 10^6 U-2 OS cells is determined empirically in virus titration experiments. Increasing amounts of virus are incubated with the same number of cells and then the levels of biosensor expression and % of cells infected are determined on the HCS platform after 24 h in culture. Performing infections in cells suspended in a low volume of media combined with periodic inversion (every 10 min) to maintain cells in suspension enhances both the rate of infection and expression levels.

3. Determining the optimal cell seeding density is a critical assay development parameter for any cell-based assay, and especially for HCS assays. The objective is to minimize the cell culture burden while ensuring that sufficient cells are captured per image to give statistical significance to the image analysis parameters of interest. Typically variability increases as the number of cells captured and analyzed decreases.

4. The optimal length of time in culture post viral infection should be determined empirically for each adenovirus biosensor. In general, we have found 24 h post infection to be optimal for most recombinant adenovirus biosensor constructs.

5. Determining the DMSO tolerance is a critical assay development parameter for any cell-based assay, and especially for HCS assays. DMSO has two major effects on HCS assays [1–3, 8, 9, 12, 13]. At DMSO concentrations ≥5 % there is a significant cell loss due to cytotoxicity and/or reduced cell adherence. At DMSO concentrations >1 % but <5 %, cells change from a well-spread and well-attached morphology to a more rounded loosely attached morphology that interferes with the ability of the image analysis algorithm to segment images into distinct cytoplasm and nuclear regions. The maximum DMSO tolerance of the HCS assay and the stock concentration of the compound library are the major determining factors of the compound concentration selected for primary screening, confirmation, and follow-up studies.

6. Determining the appropriate compound exposure period is a critical assay development parameter for any cell-based assay, including HCS assays, and should be determined empirically for each new assay. Longer compound exposure periods can result in increased levels of cytotoxicity, which may significantly hinder the ability to make reliable measurements.

7. Determining appropriate cell fixation and nuclear staining conditions is a critical assay development parameter for all end point HCS assays. Combining cell fixation with Hoechst nuclear staining in a single procedure saves time and reduces the number of steps in the protocol. Cell fixation is also important because the scanning and acquisition of a full 96-well or 384-well assay plate on an automated imaging platform depends upon the number of fluorescent channels and images captured per well and the focusing options selected. Scanning times may range anywhere from 10–15 min to 2–3 h per plate depending upon the complexity of the image acquisition procedure.

8. To control and reduce environmental exposure levels to formaldehyde and for long term storage of fixed assay plates, we recommend the aspiration and washing of plates on an automated plate washing platform.

9. Although we have described the image acquisition process on the ArrayScan VTI platform, most automated imaging platforms designed for HCS would be capable of capturing these images.

10. Although we have described the image analysis of the molecular translocation algorithm provided by the ArrayScan VTI platform, most automated imaging platforms designed for HCS provide similar image analysis programs that would be capable of analyzing these images.

References

1. Dudgeon D, Shinde SN, Shun TY, Lazo JS, Strock CJ, Giuliano KA, Taylor DL, Johnston PA, Johnston PA (2010) Characterization and optimization of a novel protein-protein interaction biosensor HCS assay to identify disruptors of the interactions between p53 and hDM2. Assay Drug Dev Technol 8:437–458

2. Dudgeon D, Shinde SN, Hua Y, Shun TY, Lazo JS, Strock CJ, Giuliano KA, Taylor DL, Johnston PA, Johnston PA (2010) Implementation of a 220,000 compound HCS campaign to identify disruptors of the interaction between p53 and hDM2, and characterization of the confirmed hits. J Biomol Screen 15:152–174

3. Johnston PA, Shinde SN, Hua Y, Shun TY, Lazo JS, Day BW (2012) Development and validation of a high-content screening assay to identify inhibitors of cytoplasmic dynein-mediated transport of glucocorticoid receptor to the nucleus. Assay Drug Dev Technol 10:432–456

4. Knauer S, Moodt S, Berg T, Liebel U, Pepperkok R, Stauber RH (2005) Translocation biosensors to study signal-specific nucleocytoplasmic transport, protease activity and protein-protein interactions. Traffic 6:594–606

5. Knauer S, Stauber RH (2005) Development of an autofluorescent translocation biosensor system to investigate protein-protein interactions in living cells. Anal Chem 77:4815–4820

6. Lundholt B, Heydorn A, Bjorn SP, Praestegaard M (2006) A simple cell-based HTS assay system to screen for inhibitors of p53-Hdm2 protein-protein interactions. Assay Drug Dev Technol 4:679–688

7. Stauber R, Afonina E, Gulnik S, Erickson J, Pavlakis GN (1998) Analysis of intracellular trafficking and interactions of cytoplasmic HIV-1 rev mutants in living cells. Virology 251:38–48

8. Trask O, Nickischer D, Burton A, Williams RG, Kandasamy RA, Johnston PA, Johnston PA (2009) High-throughput automated confocal microscopy imaging screen of a kinase-focused library to identify p38 mitogen-activated protein kinase inhibitors using the GE InCell 3000 analyzer. Methods Mol Biol 565:159–186

9. Trask OJ Jr, Baker A, Williams RG et al (2006) Assay development and case history of a 32 K-biased library high-content MK2-EGFP translocation screen to identify p38 mitogen-activated protein kinase inhibitors on the ArrayScan 3.1 imaging platform. Methods Enzymol 414:419–439

10. Kumar S, Saradhi M, Chaturvedi NK, Tyagi RK (2006) Intracellular localization and nucleocytoplasmic trafficking of steroid receptors: an overview. Mol Cell Endocrinol 246:147–156

11. Stauber R (2002) Analysis of nucleocytoplasmic transport using green fluorescent protein. Methods Mol Biol 183:181–198

12. Nickischer D, Laethem C, Trask OJ Jr et al (2006) Development and implementation of three mitogen-activated protein kinase (MAPK) signaling pathway imaging assays to provide MAPK module selectivity profiling for kinase inhibitors: MK2-EGFP translocation, c-Jun, and ERK activation. Methods Enzymol 414:389–418

13. Williams RG, Kandasamy R, Nickischer D et al (2006) Generation and characterization of a stable MK2-EGFP cell line and subsequent development of a high-content imaging assay on the Cellomics ArrayScan platform to screen for p38 mitogen-activated protein kinase inhibitors. Methods Enzymol 414:364–389

Chapter 38

Case Study: Discovery of Inhibitors of the MDM2–p53 Protein-Protein Interaction

Liu Liu, Denzil Bernard, and Shaomeng Wang

Abstract

The p53 protein, a tumor suppressor, is inactivated in many human cancers through mutations or by its interaction with an oncoprotein, MDM2. Blocking the MDM2–p53 protein-protein interaction has the effect of activating wild-type p53 and has been pursued as a novel anticancer strategy. Small-molecule inhibitors of the MDM2–p53 interaction have been discovered through various approaches, and a number of them have progressed into clinical trials for cancer treatment. Here, we describe the methods and techniques used in the discovery of small-molecule inhibitors of the MDM2–p53 interaction.

Key words p53, MDM2, Protein-protein interaction, Cancer therapy, High-throughput screening, Virtual screening

1 Introduction

Protein-protein interactions (PPI) regulate many biological processes, and targeting those interactions that have been implicated in human diseases has become an important strategy in drug discovery [1, 2]. One such PPI target is the interaction between murine double minute 2 (MDM2) and the tumor suppressor protein p53. p53 is involved in processes such as regulation of the cell cycle, apoptosis, DNA repair, and senescence [3–6]. The critical role of p53 as a tumor suppressor protein is evident from the fact that it is among the most frequently mutated proteins in human tumors and alterations of its precursor *p53* gene are seen in about 50 % of human cancers and lead to either inactivation or loss of p53 protein [7, 4]. In many cancers, wild-type p53 is effectively inhibited by the MDM2 (HDM2 in humans) protein, which was originally discovered as an overexpressed protein in a mouse tumor cell line [4, 7–10]. The interaction of MDM2 with p53 [11–13] was shown to inhibit transactivation mediated by p53 [8] enhancing the tumorigenic potential of cells overexpressing MDM2 [14–20]. Thus, overexpression of MDM2 in multiple types of human cancers

Cheryl L. Meyerkord and Haian Fu (eds.), *Protein-Protein Interactions: Methods and Applications*, Methods in Molecular Biology, vol. 1278, DOI 10.1007/978-1-4939-2425-7_38, © Springer Science+Business Media New York 2015

is related to their poor prognosis and response to therapy [14–19]. It was also demonstrated by the rescue of MDM2-null mice with simultaneous deletion of the *p53* gene [21, 22] that the MDM2–p53 interaction is the crucial determinant in death associated with the absence of MDM2. In addition, a study of over 3,000 tumors [18] revealed a negative correlation between MDM2 amplification and p53 mutations. Thus in cancers retaining wild-type p53, the MDM2–p53 protein-protein interaction is an important therapeutic target [23–25].

1.1 Regulation of p53 and MDM2

The interaction between MDM2 and p53 regulates the cellular basal levels and activity of p53 through an autoregulatory feedback loop (Fig. 1). The expression of MDM2 protein is induced by p53 which binds to the P2 promoter of the *MDM2* gene leading to its transcriptional activation. MDM2 binds to p53 protein and inhibits its activity by various mechanisms such as inhibition of the transactivation function of p53, export of p53 out of the nucleus, or direct ubiquitination of p53 leading to its proteasomal degradation [26–28]. Thus inhibition of the MDM2–p53 interaction can lead to activation of p53 and control of its tumor suppressing capacity.

1.2 Structural Basis of the MDM2–p53 Interaction

In 1996, the crystal structure of HDM2 in a complex with an N-terminal fragment of the p53 protein was published [29], revealing the structural basis for the MDM2–p53 protein-protein interaction. On the MDM2 protein surface, there was observed a well-defined binding pocket involving mostly hydrophobic residues, including Met[50], Leu[54], Leu[57], Gly[58], Ile[61], Met[62], Tyr[67], His[73], Val[75], Phe[91], Val[93], His[96], Ile[99], and Tyr[100], that complemented the hydrophobic surface of the α-helical p53 peptide. Interestingly, three residues, Phe[19], Trp[23], and Leu[26], from the p53 peptide were found to occupy this hydrophobic pocket, with the formation of a hydrogen bond between the backbone carbonyl of Leu[54] in MDM2 and the indole nitrogen of Trp[23] in p53 (Fig. 2). The compact binding pocket on the MDM2 protein surface is suitable for the development of small-molecule compounds blocking the MDM2 interaction with p53, and a variety of such inhibitors have been reported in the literature and in patents [30–32].

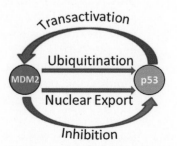

Fig. 1 MDM2–p53 autoregulatory feedback loop

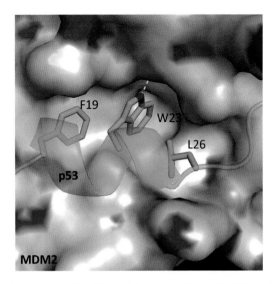

Fig. 2 Crystal structure of MDM2 protein in complex with p53 peptide (PDB ID: 1YCR). The MDM2 protein surface is shown with the p53 peptide in a *green cartoon* representation. The p53 residues Phe[19], Trp[23], and Leu[26] are shown in *stick* form. The hydrogen bond between Leu[54] of MDM2 and Trp[23] of p53 is indicated by a *dashed line*

2   Discovery of Small-Molecule Inhibitors Targeting the MDM2–p53 Interaction

Since publication of the crystallographic structure indicating the small, well-defined p53 binding pocket on MDM2, several classes of small-molecule inhibitors have been reported. The most studied representatives are *cis*-imidazolines [33], spiro-oxindoles [34, 35], benzodiazepinediones [36–38], chromenotriazolopyrimidines [39], *m*-chloropiperidinones [40], indolyl hydantoins [41], imidazothiazoles [42], terphenyl compounds [43, 44], quinolinols [45], and sulfonamides [46]. In most cases the initial lead compounds were discovered by application of protein-protein interaction exploration methods which are discussed below in detail. Generally, the methods applied may be classified as experimental (both high and regular throughput) and computational (virtual) screenings.

2.1   Experimental Screening of Both Large Diverse and Small Focused Chemical Libraries

Screening of large diverse chemical libraries using high-throughput screening (HTS) methods to identify hit compounds is an approach commonly employed in protein-protein interaction studies and drug discovery. HTS methods are able to test millions of compounds with largely diverse structural scaffolds, from which hit compounds may be identified as starting points for further design and optimization of chemical compounds. HTS is particularly useful for those cases in which no detailed structural information of the interacting interfaces is available. Screening small libraries with certain preferred properties is also quite useful and may be more valuable than HTS of large libraries. Structural insights regarding the target or the nature of ligands binding to it can often permit screening of a discreet segment of the chemical space, thereby focusing the search on likely hits. In addition to methods utilized for HTS, methods which cannot be adapted to an HTS format may be developed and used here. These methods, unsuited to HTS, can also be utilized as secondary assays to confirm hit compounds identified by HTS. Both approaches have been utilized successfully for the screening of nonpeptidic small molecules capable of disrupting the MDM2–p53 interaction and have been responsible for the discoveries of various lead compounds, including those mentioned above.

2.1.1   Surface Plasmon Resonance (SPR)

Surface plasmon resonance (SPR) currently is one of the most widely used label-free methods to evaluate protein–protein and protein–small molecule interactions. In addition to affinities, it can determine kinetic parameters, including k_{on} and k_{off} rates, which provide additional useful information for protein-protein interaction studies and drug discovery.

1. *SPR as a primary assay*
At Hoffmann-La Roche, lead compounds of the first class of potent and selective small-molecule inhibitors of the MDM2–p53 interaction, named Nutlins, were discovered by HTS of a diverse library of synthetic compounds using SPR [47]. Although the throughput of SPR is typically not high enough to be compatible with HTS, researchers of Roche successfully integrated a competitive interaction strategy into the SPR assay to increase its throughput. Instead of directly evaluating the interaction between compounds and the MDM2 protein, the interaction between MDM2 and p53 in the presence of test compounds was evaluated to determine the inhibitory abilities of those compounds. No details of this work were provided however.

Materials

The popular Biacore series of SPR instruments was initially introduced by Sweden's Biacore AB Corporation, which was purchased by GE Healthcare in 2006. Currently, SPR instruments are available from a variety of manufacturers, such as Biosensing Instrument, Sensia, Thermo, and Bio-Rad. Most of these are much less expensive than those from GE/Biacore [48]. In the Roche study, competitive assays were performed on a Biacore S51 SPR instrument. CM5 sensor chips were used for protein immobilization. Reagents included an EDC/NHS activation kit from GE Healthcare, PentaHis antibody from Qiagen, His-tagged p53 protein, and MDM2 protein. The running buffer was 10 mM HEPES with 0.15 % NaCl and 2 % DMSO. The Evaluation software provided with the instrument was used to fit the kinetic sensorgrams and Microsoft Excel was used to calculate the IC_{50} values of the compounds tested.

Methods

In this study, the capture approach rather than the commonly used direct immobilization approach was used to load p53 protein onto the sensor chip. PentaHis antibody (Qiagen) was pre-immobilized on the Biacore CM5 sensor chip by amine coupling chemistry, which subsequently was able to capture the His-tagged p53 protein. Since the location of the HisTag can be precisely controlled on either the N- or C-terminus of the p53 protein, the protein orientation in the immobilization/capture process can be controlled and the influence on the p53–MDM2 interacting model is minimized. To assure data consistency, the level of p53 captured was controlled at around 200 response units (RU; one RU corresponds to 1 pg of protein per mm^2). In all the assays, 300 nM of MDM2 protein was included and 10 mM DMSO stock solutions of test compounds were further serially diluted. The binding of immobilized p53 protein and 300 nM of MDM2 protein in the presence of test compounds was determined as a percentage of binding in the absence of inhibitor. IC_{50} values were calculated using Microsoft Excel.

Notes

This class of Nutlin compounds, developed from hits identified through competitive SPR, has a *cis*-imidazoline core structure. Structure-based chemical modifications of the lead compound, monitored by NMR [49], led to compounds with high affinity. Among two possible enantiomers that were formed, only one was active, **Nutlin-3a** (**1**, Fig. 3), which was isolated from racemic Nutlin-3 and found to disrupt the MDM2–p53 interaction with an IC_{50} value of 90 nM. Subsequent crystallization of MDM2 in a complex with a Nutlin compound (PDB ID: 1RV1) showed that it binds to the p53 binding site in MDM2. The interaction of the p53 peptide with MDM2 is mimicked by **Nutlin-3a** placing one of its bromophenyl moieties in the Trp pocket, the other bromophenyl group in the Leu pocket, and the ethyl ether side chain in the Phe pocket. **Nutlin-3** has been extensively evaluated for its therapeutic potential and mechanism of action in human cancer. A derived compound, **RG7112** (**2**, Fig. 3), developed by Roche scientists, binds to MDM2 with an IC_{50} value of 18 nM from a Homogeneous Time Resolved Fluorescence (HTFR) assay and was the first MDM2–p53 inhibitor to enter clinical trials [50].

However, this competitive SPR method cannot determine the actual K_i values of compounds tested and the IC_{50} values determined rely significantly on the experimental conditions. Thus, this method is mainly suitable for parallel comparison, with assay conditions carefully controlled by experienced operators; extreme caution needs to be taken when comparing results from different labs or scientists.

2. *SPR as a secondary assay*

SPR is however an excellent method when used as a secondary assay to confirm hit compounds identified from HTS. Researchers from Amgen developed SPR direct binding methods which were able to

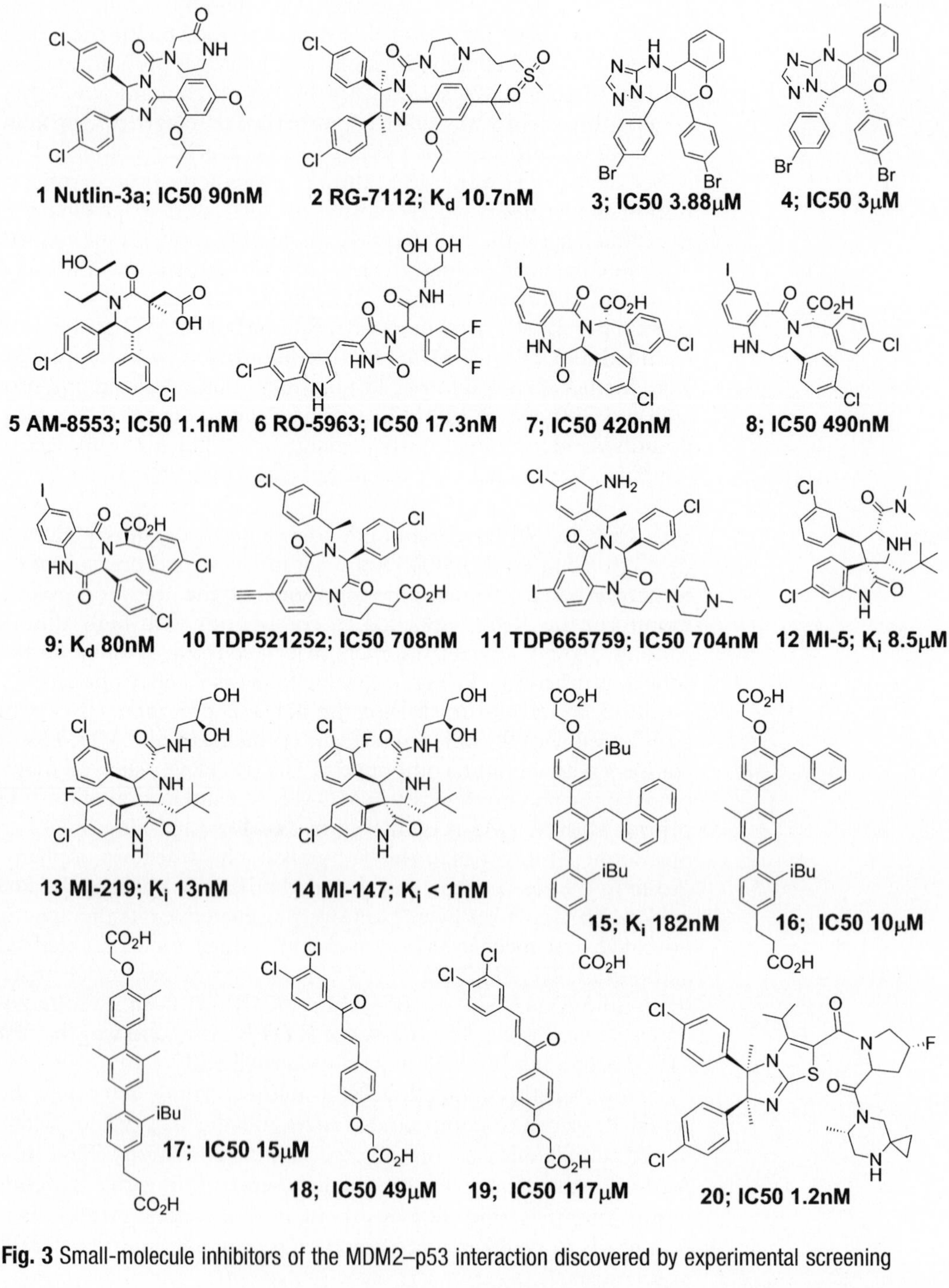

Fig. 3 Small-molecule inhibitors of the MDM2–p53 interaction discovered by experimental screening

accurately determine the K_d values of compounds. This led to the development of two classes of hit compounds, chromenotriazolo-pyrimidines [39] and *m*-chloropiperidinones [40], identified from Homogeneous Time-Resolved Fluorescence (HTRF), that potently target the p53–MDM2 interaction. Details of this HTRF method and these two classes of compounds are discussed below. For the SPR direct binding method, the p53 protein used in the competitive SPR method discussed above was no longer required. The binding target, MDM2 protein, was loaded on the sensor chip and the interaction between the tested compounds and MDM2 protein was monitored directly.

Materials

Assays were performed on the Biacore S51 SPR instrument and CM5 sensor chips used just as in the Roche competitive assay. As mentioned above, the major difference of materials used here was that p53 protein was no longer needed. Both GST and AviTAG labeled MDM2 proteins, goat anti-GST antibody, and Streptavidin (SA) were used to prepare MDM2 surface on the sensor chips through the capture approach. Other common reagents for SPR included an EDC/NHS amine activation kit, acetate buffer for Anti-GST antibody or SA immobilization, MDM2 protein capture, and assay running buffer (25 mM Tris–HCl, pH 7.5, 150 mM NaCl, 0.2 mM TCEP, and 0.005 % Tween 20). Instead of the Evaluation software, Scrubber2 software (BioLogic Software Pty Ltd.) was used to fit the kinetic data.

Methods

CM5 chips were activated with the EDC/NHS activation kit following protocols provided by Biacore. Immediately after activation, Anti-GST antibody or SA was immobilized on the CM5 chips which would capture GST or AviTAG labeled MDM2 proteins to prepare a stable MDM2 surface where interactions with small molecules would take place. Serial dilutions of p53 peptide, as the positive control, and compound stock solutions in DMSO were further diluted to the desired concentration range in the assay buffer. Compound working solutions were then injected over captured MDM2 surfaces. Two minutes of association and 3–4 min of dissociation were monitored and raw data collected were fit, using Scrubber2, to a typical 1:1 binding model from which K_d, k_{on}, and k_{off} were obtained.

Notes

Recently, kinetic parameters, particularly the dissociation rate constants (k_{off}) have captured the attention of quite a few researchers. k_{off} is the key component with the most potential to enhance compound affinity against the target protein. Furthermore, besides potent affinity, relatively slow off-rate of the compound is desired for drug action as well. Significant progress has been made to design and develop screening assays based on k_{off} values.

Experimentally, direct binding assay is more straightforward and relatively easier to be developed and optimized than the competitive assay discussed above. However, sensorgram fitting may be confusing and even misleading since quite a few matters, such as compound aggregation, DMSO drift, especially non-specific interactions while compound concentrations are relatively high, can produce artificial SPR responses. Raw SPR sensorgrams need to be carefully checked before performing kinetic fitting in order to obtain accurate k_{on}, and k_{off} values.

2.1.2 Homogeneous Time-Resolved Fluorescence (HTRF)

Homogeneous time-resolved fluorescence, also known as Time-Resolved Fluorescence Resonance Energy Transfer (TR-FRET), is one of the most widely used homogenous methods for primary HTS and also for secondary confirmation. It is a quite simple and versatile method, and due to the bounty of various robust fluorescent donor–acceptor combinations and protein labeling protocols, it is suitable for most protein-protein interaction studies.

A class of chromenotriazolopyrimidine compounds was discovered at Amgen using HTRF-based HTS of about 1.4 million compounds. This screen detected the quenching by the compounds tested of the FRET signal from fluorescent donor-labeled MDM2 to acceptor labeled p53 protein. The initial hit compound, chromenotriazolopyrimidine 3 (Fig. 3), a racemic mixture of the syn- and anti-diastereoisomers was identified, with an IC_{50} of 3.88 μM [39]. The binding of this compound to MDM2 was further confirmed by the SPR direct binding method as described above. Further modifications and separation of isomers led to compound 4 (Fig. 3), which has an IC_{50} of 3.0 μM in a luciferase reporter assay.

Materials

The essential materials for an HTRF assay are interacting partners labeled with fluorescent donor and acceptor, respectively. In this case, GST-MDM2 protein which would be recognized by anti-GST antibody labeled with europium cryptate and Avi-p53 protein tethered to SA-Xlent were designated as the donor–acceptor pair. Besides these carefully designed fluorescently labeled proteins, other materials included phosphate assay buffer, an Envision plate reader, Perkin Elmer white 384 Opti plates, and some typical data processing software. All materials are commonly needed for typical in vitro biochemical assays and excellent evidence of the simplicity of HTRF method.

Methods

Generally, HTRF assays are performed through a competitive approach in which compounds are tested for their ability to inhibit the interaction between protein partners of interest. In the Amgen screen, serial dilutions of compounds in DMSO were added to the reaction plate followed by assay buffer, GST-MDM2 and Avi-p53 proteins. Reaction mixtures were incubated at room temperature for 60 min during which equilibrium amongst the three components, compound-MDM2–p53, was established. A detection mixture composed of both SA-Xlent and Eu-anti-GST was added to the reaction mixture. After incubation at room temperature for 18 h, the reaction mixture was transferred to assay plates, which were read on an Envision plate reader with an excitation wavelength of 320 nm that is specific for Eu. Emissions were measured at 665 nm and 615 nm, corresponding to Xlent and Eu, respectively, and the ratio of Em665/Em615 represents the binding between MDM2 and p53. IC_{50} values were calculated from the inhibition curves obtained.

Notes

Further co-crystallization studies on the anti-diastereoisomer of compound **3** and the MDM2 protein illustrated that the three hydrophobic pockets of MDM2, where key p53 residues bind, were successfully occupied by the hydrophobic groups of compound **3**, indicating that this compound inhibits p53 competitively. The discovery and validation of this new class of inhibitors was an important step towards testing the hypothesis of the utility of p53–MDM2 inhibition in the treatment of cancer. However, some fundamental disadvantages of this class of compounds include their chemical instability, poor physical properties, and moderate affinities, which compromised their potential for development as clinical lead compounds. Based on this class of compound, another class of *m*-chloropiperidinone compounds was developed recently by Amgen. The most potent **AM-8553** (**5**, Fig. 3) had an IC_{50} value of 1.1 nM in the same HTRF assay and excellent pharmacokinetic properties and in vivo efficacy [40].

Recently, researchers from Hoffmann-La Roche, using an HTRF (TR-FRET) method, discovered a new class of small-molecule inhibitors targeting the p53–MDM2 interaction [41]. Initial HTS was performed for the p53–MDMX interaction. An indolyl hydantoin class was identified as a hit, and compounds in this class were further evaluated in a p53–MDM2 HTRF assay. A dual inhibitor, **RO-5963** (**6**, Fig. 3), was discovered which targets both MDM2 and MDMX with IC_{50} values to MDM2 and MDMX of 17.3 nM and 24.7 nM, respectively. These cases demonstrated that HTRF is versatile and efficient, quite straightforward to perform, and suitable for both HTS and secondarily for hit confirmation. However, as in the competitive SPR method discussed above, the most important questions concerning the HTRF method include its inability to determine K_i values of compounds accurately and the significant influence of assay conditions on the IC_{50} values obtained. Again, data comparison between different laboratories or different methods should be carried out with extra caution.

2.1.3 ThermoFluor Microcalorimetry

ThermoFluor microcalorimetry is another widely used method for protein–protein and protein–small molecule interaction studies. Different from most other methods, which typically monitor the change of a certain physical or chemical property of the protein or complex before and after the introduction of compounds, Thermo-Fluor monitors protein unfolding as a function of temperature. Protein unfolding proceeds differently in the presence of small molecules that interact with the protein. A fluorescent dye, which exhibits different fluorescence intensities depending on the folding status of the protein binding to it, is used in the assay. Compounds binding to the target protein are detected by a consequent increase in thermal stability, which is determined as the change in midpoint transition temperature (ΔT_m) upon the addition of the compound at a specific concentration. The larger the shift in T_m the higher the affinity of the binder. Due to its simplicity, this method works perfectly as an HTS method enabling very high throughput.

A series of benzodiazepinedione-based MDM2 inhibitors were discovered by researchers at Johnson and Johnson using ThermoFluor-based HTS. This involved the use of ThermoFluor microcalorimetry technology to screen a library of 22,000 benzo-diazepinediones, designed using Directed Diversity software [37], and also a library of 338,000 compounds developed by combinatorial chemistry [36].

Materials

Based on the principles of this assay, an instrument which is able to precisely control reaction temperatures and simultaneously monitor fluorescence is required. Although initially a specially designed ThermoFluor instrument was required, which was used in these studies, it was subsequently demonstrated that a typical real-time quantitative PCR (qPCR) worked well for the ThermoFluor method [51, 52]. This discovery made ThermoFluor accessible in most biochemical and biological laboratories and has significantly enhanced the application of ThermoFluor.

Since the method directly monitors the interaction between the target protein and a small molecule, MDM2 is the only protein needed (p53 protein is not needed). Besides the required compounds, the only reagent needed is an appropriate fluorescent dye, dapoxyl sulfonic acid, which was used in studies from the same group. Two other commonly used dyes are 1-anilinonaphthalene-8-sulfonic acid (ANS) and SYPRO Orange, the latter being used typically for assays performed with qPCR.

Methods

The procedure of ThermoFluor is quite straightforward. In the work done by Johnson and Johnson, purified MDM2 protein was added to each well of a 384-well polypropylene plate followed by the fluorescent probe, dapoxyl sulfonic acid, and the compound to be assayed. After mixing, the plates were heated in the Thermo-Fluor instrument to a maximum temperature of 85 °C and the fluorescence signal monitored at 500–530 nm with 8 and 12 bit CCD cameras. In order to minimize evaporation, particularly at higher temperatures, 5–10 μL of mineral oil may be added to each well to overlay the reaction mixture. The midpoint transition temperature (ΔT_m) was obtained by fitting the transition curve of fluorescence intensities versus reaction temperatures.

Notes

Although ThermoFluor is a quick and easy method and feasible for most laboratories, one must understand that the correlation between melting temperature change obtained from ThermoFluor and affinity is qualitative, and an increase in melting temperature does not necessarily correspond to a specific interaction of interest. A secondary specific and quantitative assay is required to obtain accurate affinity information. In these studies, MDM2 binders were tested with an FP-based peptide displacement binding assay (see below) to identify *bona fide* inhibitors of the MDM2–p53 interaction. This led to the identification of some benzodiazepinedione-based MDM2 inhibitors: compounds **7** ($IC_{50} = 420$ nM) and **8** ($IC_{50} = 490$ nM) which disrupt the MDM2–p53 interaction, and the optimized compound **9** whose K_d value is 80 nM (Fig. 3). Crystal structures of **9** in a complex with MDM2 revealed that **9** fits into the same pockets in the MDM2 binding cleft used by the p53 peptide side chains Phe[19], Trp[23], and Leu[26]. As with the Nutlins, the interactions of these inhibitors with the MDM2 protein are primarily through van der Waals contacts. Another series of benzo-diazepinedione compounds identified by the same ThermoFluor method followed by optimization led to **TDP521252** (**10**) and **TDP665759** (**11**) (Fig. 3) which bind to MDM2 with IC_{50} values of 708 nM and 704 nM, respectively [38].

2.1.4 Fluorescence Polarization (FP)

The FP assay, which can involve a direct binding or a competitive approach, is probably the most widely used method for screening small molecules targeting protein-protein interactions. Through the direct binding approach, K_d values can be easily determined accurately by monitoring the change in FP of the fluorescently labeled small molecule or peptide while binding to different concentrations of protein. This approach is not suitable for HTS since it is not practical to fluorescently label each compound tested. However, it can be easily adapted to HTS by employing a competitive approach in which compounds compete with a fluorescently labeled small-molecule compound or peptide, typically termed a tracer, for binding to the target protein. In most cases, the K_i values of compounds based on the experimental IC_{50} values obtained can be accurately determined, as can the K_d value of the tracer to the target protein, the latter being determined with the direct binding approach.

There are many successful cases reported of discovery and development of MDM2–p53 interaction inhibitors using the FP method. Among them, the most successful one probably is the spiro-oxindole class of compounds developed at the University of Michigan [34, 35, 53, 54]. Starting from the crystal structure of the MDM2 protein complexed with the p53 peptide, a structure-based de novo rational strategy was employed to design new classes of small-molecule inhibitors targeting the MDM2–p53 interaction. With the assistance of computational screening, a spiro-oxindole core structure was discovered that mimics the interaction between the p53 peptide and the MDM2 protein, which subsequently was used as the starting point for rational design. A sensitive and quantitative FP-based binding assay was designed and developed to determine the accurate binding affinities of newly synthesized compounds. Accurate SAR data guided the de novo design precisely. Starting from compound **MI-5** (**12**, Fig. 3) with a K_i value of 8.5 µM, a new class of highly potent, small-molecule inhibitors, such as **MI-219** (**13**, Fig. 3) and **MI-147** (**14**, Fig. 3) with K_i values of 13 nM and 0.6 nM, respectively, were obtained. A fully optimized compound, **SAR405838**, in this class entered clinical trials in 2012.

Materials

Since the FP assay was performed as a competitive approach, in addition to the target MDM2 protein, a fluorescently labeled modified p53 peptide serving as the tracer was the most important reagent in the competitive assay [55]. Other materials included phosphate assay buffer, typical Microtiter assay plates, a plate reader capable of reading FP, and typical data processing software, such as Prism, all of which are common in most biochemical laboratories. Compared to the similar HTRF method discussed above, the FP assay is even easier to perform since only one fluorescently labeled component is required and this molecule is typically a small peptide or organic molecule which is much more easily labeled than proteins used in the HTRF method.

Methods

One of the major advantages of the FP method is that K_i values of compounds tested can generally be determined accurately. In order to calculate K_i, the K_d value of the tracer to the target protein must first be determined. This can be achieved by monitoring the total FP of mixtures containing the tracer at a fixed concentration and the target protein at increasing concentrations up to full saturation. Thus, serial dilutions of the MDM2 protein in the assay buffer (100 mM potassium phosphate, pH 7.5, 100 µg/ml bovine γ-globulin, 0.02 % sodium azide, with 0.01 % Triton X-100 and 4 % DMSO) were prepared in a 96-well black assay plate followed by addition of the tracer solution with a fixed concentration in each well. The wells were mixed and incubated at room temperature for 60 min with gentle shaking to ensure equilibrium. The FP values in millipolarization units (mP) were measured at an excitation wavelength of 485 nm and an emission wavelength of 530 nm, with Fluorescein amidite (FAM) as the fluorophore. Equilibrium dissociation constants (K_d) were calculated with Graphpad Prism 5.0 software (Graphpad Software, San Diego, CA) by measuring the sigmoidal dose-dependent FP increases as a function of protein concentrations.

The K_i values of compounds tested were determined in a dose-dependent competitive binding experiment. Mixtures of serial dilutions of compounds dissolved in DMSO with preincubated MDM2/tracer complex at fixed concentrations in the assay buffer were added into assay plates and incubated at room temperature with gentle shaking. Negative controls containing MDM2/tracer complex only (equivalent to zero inhibition), and positive controls containing only free tracer (equivalent to 100 % inhibition), were included in each assay plate. Using Prism, IC_{50} values were determined by nonlinear regressive fitting of the sigmoidal dose-dependent FP decreases as a function of total compound concentrations. The K_i values of tested compounds were calculated using the measured IC_{50} values, the K_d value of the tracer, and the concentrations of the MDM2 protein and the tracer in the competitive assay [56].

Notes

As previously mentioned, K_i values can be calculated in most cases. However, when the compound is more potent than the tracer used, the K_i value cannot be determined accurately since under this circumstance, the experimental IC_{50} value is not correlated linearly with K_i [57]. Mathematical solutions have been proposed to deal with this issue [57–60] but none of them performs perfectly with extremely potent compounds. Among all these interpretations, that proposed by Zhang [60] utilized a unique approach in which K_i values were obtained by directly fitting the inhibition curves without involving IC_{50} values. This approach was beneficial in some cases in which calculations based on the IC_{50} produced meaningless negative values.

FP was also used by other laboratories in which different classes of small-molecule inhibitors targeting MDM2–p53 interaction were discovered. The p53 peptide binds to MDM2 in an amphiphatic manner via an alpha helix that locates to the hydrophobic residues in the binding pocket. Using this knowledge, a small library of terphenyl compounds which can mimic the alpha helical peptide [61] was developed and screened using the FP method. By substituting alkyl or aryl groups at the *o*-positions of the terphenyl scaffold, a small library of 21 terphenyl derivatives was synthesized [43]. The most potent of these, compound **15** (Fig. 3) had a K_i value with MDM2 of 182 nM. In addition to the displacement of p53 peptide in the FP assay, ^{15}N HSQC NMR spectroscopy showed that the terphenyl **15** interacts with the p53 binding region of MDM2.

In a patent, Novartis disclosed compounds from an imidazole-indole class as inhibitors of MDM2 and MDMX. The most potent compound has an IC_{50} of 15 nM for MDM2 in an FP binding assay and interestingly also binds to MDMX with $IC_{50} = 1.32$ μM in a TR-FRET binding assay [62].

Other methods to study protein–protein and protein–small molecule interactions were used successfully on the MDM2–p53 interaction. These included the ELISA, ITC, and NMR methods. Compounds **16** and **17** (Fig. 3), which inhibit the MDM2–p53 interaction with IC_{50} values of 10 µM and 15 µM, respectively, are terphenyl compounds which were identified using an in vitro, quantitative, ELISA-based assay as a screen [44]. The screening of a small library of 16 chalcones, compounds known to have anticancer properties [63], with a similar ELISA-based assay monitoring disruption of the MDM2–p53 interaction resulted in the identification of two compounds **18** and **19** (Fig. 3) [64] which have IC_{50} values of 49 µM and 117 µM, respectively. NMR titration experiments were used to confirm the binding, and the compounds were also found to inhibit the MDM2–p53 interaction in an in vitro DNA binding electrophoretic mobility super-shift assay (EMSA). Recently, researchers from Daiichi Sankyo developed an imidazothiazole class of potent MDM2 inhibitors starting from the core structure of the Nutlins. Based on extensive SAR information obtained from ELISA, the most potent compound **20** (Fig. 3) had an IC_{50} value of 1.2 nM in their ELISA assay [65].

ITC, another quite popular and unique method for protein–protein and protein–small molecule interaction studies, is the only 100 % real homogeneous and label-free method available. In addition to the binding affinities, it can also produce very useful information on the stoichiometry and ΔH of the interaction of interest, data which are obtained only with great difficulty by any other method. However, it also has huge disadvantages over other methods. The most important of these are the significantly large amount of protein that is consumed and the very low throughput that is obtained. Thus the most suitable application for ITC is its use as a confirmatory assay for fully optimized compounds that are ready to enter the next research stage. Researchers from Amgen designed and developed a robust ITC assay and confirmed their new class of MDM2 inhibitors that were identified by HTRF, as discussed above. In this way, novel structural features were discovered of the MDM2–small molecule interaction model involving the MDM2 N-terminal residues that had previously been regarded as "structureless" [66].

2.2   Virtual Database Screening

In contrast to direct experimental screening of chemical libraries, when the structural basis of the interaction of the target proteins is known, computational database screening of millions of compounds can be performed to produce a select set of compounds for further experimental study.

2.2.1   Pharmacophore Screen

Pharmacophores represent the groups in a molecule and their spatial juxtaposition that are important for its interaction with a target protein. Using different sets of available data, namely (a) p53 mutagenesis data affecting MDM2 binding [67], (b) truncation studies of peptidic inhibitors [68], and (c) peptidic inhibitors with unnatural amino acids [55], multiple pharmacophore models can be generated and used to screen for hits [46]. In a typical study of this type, the compounds selected were then tested for inhibition of MDM2–p53 interaction with an ELISA-based assay. A sulfonamide, **NSC 279287** (**21**, Fig. 4), was identified and found to inhibit the interaction between full-length MDM2 and p53 proteins with an IC_{50} value of 31.8 μM. However, in a p53 reporter gene assay only a 20 % increase in p53 transcriptional function was seen after treatment with 100 μM of **NSC 279287**, indicating its low potency [46].

A set of MDM2–p53 inhibitors based on a pharmacophore model developed from the crystal structure of MDM2 was obtained by molecular dynamics simulation, which considered protein flexibility [69]. Hits obtained from pharmacophore screening of ~50,000 synthetic compounds were tested in a fluorescence-polarization MDM2 binding assay. This led to the discovery of five nonpeptidic, small-molecule inhibitors, the most potent of which, **22** (Fig. 4), had a K_i of 110 nM.

2.2.2 Combined Pharmacophore and Structure-Based Screening

A group of 2,599 compounds was obtained by application of a pharmacophore screen to a database of over 100,000 compounds, prefiltered for drug-like properties and including both synthetic and natural products [45]. In this case, the pharmacophore was derived from the MDM2–p53 crystal complex and known small-molecule inhibitors such as the Nutlins and benzodiazepinediones. The hits from the first screen were then subjected to a second screen using structure-based docking with the GOLD program [70, 71] to rank them by their predicted binding affinity to the p53 binding site. After rescoring with Chemscore [72] and X-Score [73], the top 200 compounds were then combined to obtain a set of 354 non-redundant compounds, each of which was complementary to the MDM2 target site. From a final set of 67 of these compounds, tested with a competitive FP MDM2 binding assay, ten hits were identified as compounds with a K_i value <10 μM. The most potent of these, **NSC 66811** (**23**, Fig. 4), had a K_i value of 120 nM and dose-dependently activated p53 in LNCaP prostate cancer cells.

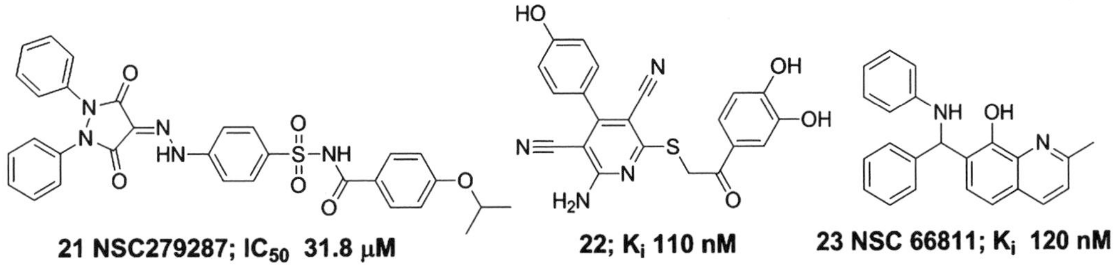

21 NSC279287; IC$_{50}$ 31.8 μM   **22; K$_i$ 110 nM**   **23 NSC 66811; K$_i$ 120 nM**

Fig. 4 Small-molecule inhibitors of the MDM2–p53 interaction discovered by virtual database screening

3   Concluding Remarks

Several small-molecule inhibitors targeting the MDM2–p53 protein-protein interaction have been successfully developed in the last decade with the assistance of various biochemical and biophysical methods including both experimental and virtual database screenings. Among them, **RG7112** and **RG7388**, two Nutlin derivatives from Roche, AMG 232 from Amgen, and **MI-77301/ SAR405838**, a spiro-oxindole discovered from our laboratory at the University of Michigan, are examples that have advanced to clinical trials. Targeting the MDM2–p53 protein-protein interaction using small molecules which reactivate the p53 is a potentially attractive cancer therapeutic strategy for the treatment of human cancers retaining wild-type p53. Clinical testing of these new agents will provide the ultimate test of the usefulness of this therapeutic strategy for the treatment of cancer and the critical role of PPI in key human disease-related biological processes.

These successful cases are excellent examples of the applications of some most widely used methods, which are valuable for those interested in other protein–protein and protein–small molecule interactions. To develop methods successfully, typical criteria that must be considered include assay throughput, the information that each method can provide, the suitability of each technology to the protein–protein and protein–small molecule interactions of interest, and cost. Some other essential issues that need also to be considered include accuracy, reproducibility, and the relative ease of performance.

Acknowledgements

Funding from the National Cancer Institute of the National Institutes of Health is greatly appreciated.

Disclosure statement

S.W. and the University of Michigan own equity in Ascenta Therapeutics, Inc., which has licensed the technologies related to the MDM2 inhibitors of the spiro-oxindole class. S.W. serves as a consultant for Ascenta.

References

1. Fry DC, Vassilev LT (2005) Targeting protein-protein interactions for cancer therapy. J Mol Med 83:955–963

2. Murray JK, Gellman SH (2007) Targeting protein-protein interactions: lessons from p53/MDM2. Biopolymers 88:657–686

3. Fridman JS, Lowe SW (2003) Control of apoptosis by p53. Oncogene 22:9030–9040

4. Hainaut P, Hollstein M (2000) p53 and human cancer: the first ten thousand mutations. Adv Cancer Res 77:81–137

5. Vogelstein B, Lane D, Levine AJ (2000) Surfing the p53 network. Nature 408:307–310

6. Vousden KH, Lu X (2002) Live or let die: the cell's response to p53. Nat Rev Cancer 2:594–604

7. Feki A, Irminger-Finger I (2004) Mutational spectrum of p53 mutations in primary breast and ovarian tumors. Crit Rev Oncol Hematol 52:103–116

8. Momand J, Zambetti GP, Olson DC et al (1992) The mdm-2 oncogene product forms a complex with the p53 protein and inhibits p53-mediated transactivation. Cell 69:1237–1245

9. Fakharzadeh SS, Trusko SP, George DL (1991) Tumorigenic potential associated with enhanced expression of a gene that is amplified in a mouse tumor cell line. EMBO J 10:1565–1569

10. Fakharzadeh SS, Rosenblum-Vos L, Murphy M et al (1993) Structure and organization of amplified DNA on double minutes containing the mdm2 oncogene. Genomics 15:283–290

11. Lane DP, Crawford LV (1979) T antigen is bound to a host protein in SV40-transformed cells. Nature 278:261–263

12. Linzer DI, Levine AJ (1979) Characterization of a 54 K dalton cellular SV40 tumor antigen present in SV40-transformed cells and uninfected embryonal carcinoma cells. Cell 17:43–52

13. DeLeo AB, Jay G, Appella E et al (1979) Detection of a transformation-related antigen in chemically induced sarcomas and other transformed cells of the mouse. Proc Natl Acad Sci U S A 76:2420–2424

14. Bond GL, Hu W, Bond EE et al (2004) A single nucleotide polymorphism in the MDM2 promoter attenuates the p53 tumor suppressor pathway and accelerates tumor formation in humans. Cell 119:591–602

15. Oliner JD, Kinzler KW, Meltzer PS et al (1992) Amplification of a gene encoding a p53-associated protein in human sarcomas. Nature 358:80–83

16. Zhou M, Gu L, Abshire TC et al (2000) Incidence and prognostic significance of MDM2 oncoprotein overexpression in relapsed childhood acute lymphoblastic leukemia. Leukemia 14:61–67

17. Rayburn E, Zhang R, He J et al (2005) MDM2 and human malignancies: expression, clinical pathology, prognostic markers, and implications for chemotherapy. Curr Cancer Drug Targets 5:27–41

18. Momand J, Jung D, Wilczynski S et al (1998) The MDM2 gene amplification database. Nucleic Acids Res 26:3453–3459

19. Gunther T, Schneider-Stock R, Hackel C et al (2000) Mdm2 gene amplification in gastric cancer correlation with expression of Mdm2 protein and p53 alterations. Mod Pathol 13:621–626

20. Bond GL, Hu W, Levine AJ (2005) MDM2 is a central node in the p53 pathway: 12 years and counting. Curr Cancer Drug Targets 5:3–8

21. Jones SN, Roe AE, Donehower LA et al (1995) Rescue of embryonic lethality in Mdm2-deficient mice by absence of p53. Nature 378:206–208

22. de Oca M, Luna R, Wagner DS, Lozano G (1995) Rescue of early embryonic lethality in mdm2-deficient mice by deletion of p53. Nature 378:203–206

23. Vassilev LT (2007) MDM2 inhibitors for cancer therapy. Trends Mol Med 13:23–31

24. Chene P (2003) Inhibiting the p53-MDM2 interaction: an important target for cancer therapy. Nat Rev Cancer 3:102–109

25. Shangary S, Wang S (2008) Targeting the MDM2-p53 interaction for cancer therapy. Clin Cancer Res 14:5318–5324

26. Freedman DA, Wu L, Levine AJ (1999) Functions of the MDM2 oncoprotein. Cell Mol Life Sci 55:96–107

27. Juven-Gershon T, Oren M (1999) Mdm2: the ups and downs. Mol Med 5:71–83

28. Wu X, Bayle JH, Olson D et al (1993) The p53-mdm-2 autoregulatory feedback loop. Genes Dev 7:1126–1132

29. Kussie PH, Gorina S, Marechal V et al (1996) Structure of the MDM2 oncoprotein bound to the p53 tumor suppressor transactivation domain. Science 274:948–953

30. Millard M, Pathania D, Grande F et al (2011) Small-molecule inhibitors of p53-MDM2 interaction: the 2006-2010 update. Curr Pharm Des 17:536–559

31. Zhan C, Lu W (2011) Peptide activators of the p53 tumor suppressor. Curr Pharm Des 17:603–609

32. Zak K, Pecak A, Rys B et al (2013) Mdm2 and MdmX inhibitors for the treatment of cancer: a patent review (2011–present). Expert Opin Ther Pat 23:425–448

33. Vassilev LT (2004) Small-molecule antagonists of p53-MDM2 binding: research tools and potential therapeutics. Cell Cycle 3:419–421

34. Ding K, Lu Y, Nikolovska-Coleska Z et al (2006) Structure-based design of spiro-oxindoles as potent, specific small-molecule inhibitors of the MDM2-p53 interaction. J Med Chem 49:3432–3435

35. Ding K, Lu Y, Nikolovska-Coleska Z et al (2005) Structure-based design of potent non-peptide MDM2 inhibitors. J Am Chem Soc 127:10130–10131

36. Grasberger BL, Lu T, Schubert C et al (2005) Discovery and cocrystal structure of benzodiazepinedione HDM2 antagonists that activate p53 in cells. J Med Chem 48:909–912

37. Parks DJ, Lafrance LV, Calvo RR et al (2005) 1,4-Benzodiazepine-2,5-diones as small molecule antagonists of the HDM2-p53 interaction: discovery and SAR. Bioorg Med Chem Lett 15:765–770

38. Koblish HK, Zhao S, Franks CF et al (2006) Benzodiazepinedione inhibitors of the Hdm2: p53 complex suppress human tumor cell proliferation in vitro and sensitize tumors to doxorubicin in vivo. Mol Cancer Ther 5:160–169

39. Allen JG, Bourbeau MP, Wohlhieter GE et al (2009) Discovery and optimization of chromenotriazolopyrimidines as potent inhibitors of the mouse double minute 2-tumor protein 53 protein–protein interaction. J Med Chem 52:7044–7053

40. Rew Y, Sun D, Gonzalez-Lopez De Turiso F et al (2012) Structure-based design of novel inhibitors of the MDM2-p53 interaction. J Med Chem 55:4936–4954

41. Graves B, Thompson T, Xia M et al (2012) Activation of the p53 pathway by small-molecule-induced MDM2 and MDMX dimerization. Proc Natl Acad Sci U S A 109:11788–11793

42. Miyazaki M, Kawato H, Naito H et al (2012) Discovery of novel dihydroimidazothiazole derivatives as p53-MDM2 protein-protein interaction inhibitors: synthesis, biological evaluation and structure-activity relationships. Bioorg Med Chem Lett 22:6338–6342

43. Yin H, Lee GI, Park HS et al (2005) Terphenyl-based helical mimetics that disrupt the p53/HDM2 interaction. Angew Chem Int Ed Engl 44:2704–2707

44. Chen L, Yin H, Farooqi B et al (2005) p53 alpha-helix mimetics antagonize p53/MDM2 interaction and activate p53. Mol Cancer Ther 4:1019–1025

45. Lu Y, Nikolovska-Coleska Z, Fang X et al (2006) Discovery of a nanomolar inhibitor of the human murine double minute 2 (MDM2)-p53 interaction through an integrated, virtual database screening strategy. J Med Chem 49:3759–3762

46. Galatin PS, Abraham DJ (2004) A nonpeptidic sulfonamide inhibits the p53-mdm2 interaction and activates p53-dependent transcription in mdm2-overexpressing cells. J Med Chem 47:4163–4165

47. Vassilev LT, Vu BT, Graves B et al (2004) In vivo activation of the p53 pathway by small-molecule antagonists of MDM2. Science 303:844–848

48. Schasfoort RBM, Tudos AJ (2008) Handbook of surface plasmon resonance. Royal Society of Chemistry Publishing, Cambridge, UK

49. Fry DC, Graves B, Vassilev LT (2005) Development of E3-substrate (MDM2-p53)-binding inhibitors: structural aspects. Methods Enzymol 399:622–633

50. Vu B, Wovkulich P, Pizzolato G et al (2013) Discovery of RG7112: A small-molecule MDM2 inhibitor in clinical development. ACS Med Chem Lett 4:466–469

51. Ericsson UB, Hallberg BM, Detitta GT et al (2006) Thermofluor-based high-throughput stability optimization of proteins for structural studies. Anal Biochem 357:289–298

52. Lo MC, Aulabaugh A, Jin G et al (2004) Evaluation of fluorescence-based thermal shift assays for hit identification in drug discovery. Anal Biochem 332:153–159

53. Shangary S, Qin D, McEachern D et al (2008) Temporal activation of p53 by a specific MDM2 inhibitor is selectively toxic to tumors and leads to complete tumor growth inhibition. Proc Natl Acad Sci U S A 105:3933–3938

54. Yu S, Qin D, Shangary S et al (2009) Potent and orally active small-molecule inhibitors of the MDM2-p53 interaction. J Med Chem 52:7970–7973

55. Garcia-Echeverria C, Chene P, Blommers MJ et al (2000) Discovery of potent antagonists of the interaction between human double minute 2 and tumor suppressor p53. J Med Chem 43:3205–3208

56. Nikolovska-Coleska Z, Wang R, Fang X et al (2004) Development and optimization of a binding assay for the XIAP BIR3 domain using fluorescence polarization. Anal Biochem 332:261–273

57. Huang X (2003) Fluorescence polarization competition assay: the range of resolvable inhibitor potency is limited by the affinity of the fluorescent ligand. J Biomol Screen 8:34–38

58. Kenakin TP (1993) Pharmacologic analysis of drug-receptor interaction, 2nd edn. Raven, New York

59. Munson PJ, Rodbard D (1988) An exact correction to the "Cheng-Prusoff" correction. J Recept Res 8:533–546

60. Zhang R, Mayhood T, Lipari P et al (2004) Fluorescence polarization assay and inhibitor design for MDM2/p53 interaction. Anal Biochem 331:138–146

61. Kutzki O, Park HS, Ernst JT et al (2002) Development of a potent Bcl-x(L) antagonist based on alpha-helix mimicry. J Am Chem Soc 124:11838–11839

62. Boettcher A, Buschmann N, Furet P, et al. (2008) 3-imidazolyl-indoles for the treatment of proliferative diseases. US Patent WO 2008119741

63. Go ML, Wu X, Liu XL (2005) Chalcones: an update on cytotoxic and chemoprotective properties. Curr Med Chem 12:481–499

64. Stoll R, Renner C, Hansen S et al (2001) Chalcone derivatives antagonize interactions between the human oncoprotein MDM2 and p53. Biochemistry 40:336–344

65. Uoto K, Kawato H, Sugimoto Y, et al. (2009) Imidazothiazole derivative having 4,7-diazaspiro[2.5]octane ring structure. US Patent WO/2009/151069

66. Michelsen K, Jordan JB, Lewis J et al (2012) Ordering of the N-terminus of human MDM2 by small molecule inhibitors. J Am Chem Soc 134:17059–17067

67. Lin J, Chen J, Elenbaas B et al (1994) Several hydrophobic amino acids in the p53 amino-terminal domain are required for transcriptional activation, binding to mdm-2 and the adenovirus 5 E1B 55-kD protein. Genes Dev 8:1235–1246

68. Bottger V, Bottger A, Howard SF et al (1996) Identification of novel mdm2 binding peptides by phage display. Oncogene 13:2141–2147

69. Bowman AL, Nikolovska-Coleska Z, Zhong H et al (2007) Small molecule inhibitors of the MDM2-p53 interaction discovered by ensemble-based receptor models. J Am Chem Soc 129:12809–12814

70. Jones G, Willett P, Glen RC et al (1997) Development and validation of a genetic algorithm for flexible docking. J Mol Biol 267:727–748

71. Verdonk ML, Cole JC, Hartshorn MJ et al (2003) Improved protein-ligand docking using GOLD. Proteins 52:609–623

72. Eldridge MD, Murray CW, Auton TR et al (1997) Empirical scoring functions: I. The development of a fast empirical scoring function to estimate the binding affinity of ligands in receptor complexes. J Comput Aided Mol Des 11:425–445

73. Wang R, Lai L, Wang W (2002) Further development and validation of empirical scoring functions for structure-based binding affinity prediction. J Comput Aided Mol Des 16:11–26

Chapter 39

Biophysical Methods for Identifying Fragment-Based Inhibitors of Protein-Protein Interactions

Samuel J. Pfaff, Michael S. Chimenti, Mark J.S. Kelly, and Michelle R. Arkin

Abstract

Fragment-based lead discovery complements high-throughput screening and computer-aided drug design for the discovery of small-molecule inhibitors of protein-protein interactions. Fragments are molecules with molecular masses ca 280 Da or smaller, and are generally screened using structural or biophysical approaches. Several methods of fragment-based screening are feasible for any soluble protein that can be expressed and purified; specific techniques also have size limitations and/or require multiple milligrams of protein. This chapter describes some of the most common fragment-discovery methods, including surface plasmon resonance, nuclear magnetic resonance, differential scanning fluorimetry, and X-ray crystallography.

Key words Fragment, Ligand, Discovery, Biophysics, SPR, NMR, DSF, Crystallography, Protein-protein, Interaction

1 Introduction

Fragment-based lead discovery (FBLD) is an increasingly popular screening modality for discovering small-molecule inhibitors of protein-protein interactions (PPI) [1–6]. "Fragments" are organic compounds roughly ½ the size of a drug-like molecule, with molecular masses roughly 120–280 Da; fragment-screening libraries range in size from ca 500 to 15,000 members. Fragments are typically screened using a structural or physical method that identifies binders to the protein of interest. These binders are then optimized using structure-guided design; in favorable cases, fragments that bind to adjacent sites are linked together to create a tight-binding lead compound.

There are several potential advantages to FBLD for PPI. First, small-molecule inhibitors of PPI are likely rare in a standard high-throughput screening (HTS) library because chemical space is vastly larger than an HTS library (10^{60} vs. 10^{6} members) and

Cheryl L. Meyerkord and Haian Fu (eds.), *Protein-Protein Interactions: Methods and Applications*, Methods in Molecular Biology, vol. 1278, DOI 10.1007/978-1-4939-2425-7_39, © Springer Science+Business Media New York 2015

selection criteria could be biased towards previously "drugged" targets. Since chemical space is smaller (ca 10^6) for fragment-sized compounds, fragment libraries can sample chemical space more broadly. Second, the binding surfaces of PPI tend to comprise 3–5 subsites, rather than one large, deep cavity, as is observed for many enzymes [7]. Because the binding surfaces are complex, it is difficult to satisfy all the necessary binding interactions through random screening. By contrast, a fragment just needs to fill one subsite. Third, PPI tend to contain regions of structural flexibility, and fragments have been particularly successful at selecting such regions. This chapter provides procedures for some of the most common technologies for fragment screening. For more information on the design of fragment libraries and chemical optimization of fragments, the reader is referred to published reviews [8–10].

FBLD screens usually employ binding-based methods such as surface plasmon resonance (SPR; biosensor) and nuclear magnetic resonance (NMR), protein-stabilization methods such as differential scanning fluorimetry (DSF), or structure-based methods for which X-ray crystallography is the gold standard. These biophysical approaches circumvent many of the artifacts common in high-throughput screening; in fact, the procedures outlined here are highly recommended for secondary assays following an inhibition- or activity-based screen. In contrast to activity-based screens, binding methods provide direct evidence of target engagement, and usually also provide important details about binding stoichiometry (generally, one fragment/protein is desired), binding kinetics, and binding affinity. In the case of protein-detected NMR, Tethering [11, 12], and X-ray crystallography, the binding site on the protein is also known; X-ray provides the most detailed characterization of the protein/fragment interaction.

The ideal methodology depends on the size of the library, the size and availability of protein, and the existence of NMR and/or X-ray structures. At least one of the methods here should be suitable for a given protein-of-interest, provided that the protein can be expressed and purified on the milligram scale.

1.1 Surface Plasmon Resonance

Surface plasmon resonance (SPR) is a sensitive biophysical technique that provides binding affinities, kinetics, and stoichiometry for PPI and protein/small-molecule interactions. SPR and biolayer interferometry belong to a broader class of technologies known as biosensors. These techniques aim to measure the kinetics and stoichiometry of a bimolecular interaction through physical changes at a measurement surface. An SPR assay commonly consists of a protein tethered to a gold surface by a flexible chemical layer, such as dextran. The tethered protein remains under a steady flow of buffer, and is interrogated by injection of small-molecule or protein "analyte." As analyte binds to the protein surface, the refractive index (RI) of the buffer above the gold layer changes, and this change in RI is measured. After an analyte injection is complete, buffer is flowed over the surface, allowing monitoring of the analyte's dissociation.

Binding analysis by SPR makes use of kinetic and equilibrium measurements. Most SPR instruments measure the signal at a rate of 1–10 Hz, allowing real-time measurement of complex formation (association) and dissociation. Importantly, the change in RI is directly proportional to the mass of analyte that accumulates on the surface; thus, by knowing the amount of material immobilized on the surface and the molecular weights of the protein and analyte, one can calculate the stoichiometry of the binding interaction. The binding affinity of the complex is then determined by measuring the binding stoichiometry at multiple doses or from fitting binding/dissociation kinetics. Fragments tend to bind to proteins with K_D values in the 0.1–10 mM range. For these low-affinity interactions, there is limited kinetic information in the data, and equilibrium is reached within ~1 s. Fragments are therefore typically measured at multiple doses, and K_D is determined by binding isotherm (Fig. 1).

The power of SPR for fragment discovery lies in its very high sensitivity and the ability to derive quantitative binding information. In addition, SPR requires only a few milligrams of protein, and modern instruments have sufficient throughput to monitor ~10^3–10^4 binding interactions in 1 week. GE Healthcare's Biacore 4000 instrument, for instance, can screen four independent targets against four compounds in a single 3–5 min injection cycle, and holds ten 384-well plates plus control solutions. Other instruments, such as Biacore's T200, ForteBio's OctetRed, Pioneer's SensiQ, and Bio-Rad's Proteon offer somewhat different capabilities in terms of throughput, automation, sensitivity, cost, and innovative features.

SPR has also found widespread use as a secondary assay for prioritizing compounds from HTS campaigns. Aggregation, covalent binding, and nonspecific binding are easy to identify by SPR, since such mechanisms show high stoichiometry, low selectivity, and/or slow kinetics. Thus, SPR is very sensitive to misbehavior by small molecule analytes [13]. These non-desirable binding modes commonly generate false positives in high-throughput biochemical or cell-based screens, and can lead to a loss of time and resources if incorrectly prioritized [14].

1.2 Differential Scanning Fluorimetry

Differential scanning fluorimetry (DSF; alternatively called thermal shift or ThermoFluor) is an inexpensive and flexible method that has recently become commonplace in biophysical drug screening. DSF uses a thermocycler and fluorescent dyes to provide a measure of the thermal stability of a target protein. In the context of drug screening, DSF is used to identify compounds that increase the thermal stability of a protein (Fig. 2). This thermal shift between apo and bound protein forms is proportional to the change in

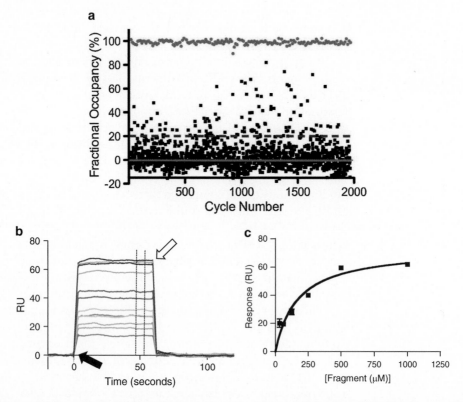

Fig. 1 Fragment screening by SPR. (**a**) Primary screening data (courtesy of Stacie Bulfer). *Green circles* are positive controls and *black squares* are fragments screened at 250 µM. The *red line* represents zero response and the *blue dashed line* represents the hit cutoff of three standard deviations. (**b**) Fragment hit follow-up: duplicate dose–response data for a fragment hit. This compound shows canonical fragment binding kinetics with rapid approaches to equilibrium at the beginning and end of each injection. The injection start is indicated by a *closed arrow* and the injection end (dissociation phase) is indicated by an *open arrow*. The data used to generate equilibrium binding values are the averages of the values between the *dashed lines*. (**c**) Equilibrium binding analysis of the compound used in (**b**). Data were fit to a simple 1:1 hyperbolic binding isotherm

Gibbs free energy upon binding of the compound [15]. The reagents that enable DSF are a class of environmentally sensitive dyes, including ANS, DAPOXYL derivatives, and SYPRO orange. Most critically, these dyes are strongly quenched in aqueous environments and bind preferentially and nonspecifically to hydrophobic surfaces. When bound to a hydrophobic surface, quenching is relieved and a fluorescent signal can be measured. As a protein denatures, hydrophobic core residues become exposed to solvent and are available for dye binding, allowing enhanced fluorescence. These characteristics allow protein unfolding to be monitored and quantified in a straightforward experiment.

DSF can be employed at many stages in a fragment screening campaign. Primary library screening by DSF has significant advantages over other biophysical techniques. Unlike SPR, X-ray, or NMR, DSF requires very little optimization of the initial conditions

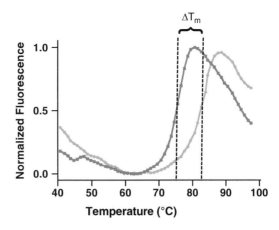

Fig. 2 Protein stabilization measured by DSF. Thermal denaturation of a cation-sensitive protein was measured by DSF in the presence of either 0.5 M K$^+$ (*blue*) or 0.5 M Na$^+$ (*orange*) (Data courtesy of Samuel Pfaff). Melting temperatures were calculated by fitting to the Boltzmann equation and are represented by *dashed lines*. The rightward shift in T_m seen for the Na$^+$ condition is analogous to that of a potent small molecule hit

[16]; the initial concentrations of protein and dye are the only parameters that typically require optimization. This allows divergent targets to be screened under very similar conditions. Buffer and salt conditions can have significant effects on protein melting curves, and can be optimized as well. However, it is usually desirable to employ a simple buffer system that best reflects the physiological environment of the target protein. Assay throughput is high, since 384 wells can be read simultaneously in 45–90 min, depending on assay parameters. Thus, an average-sized fragment library can be screened in a day with plate-handling automation. Protein consumption during screening is also quite low (typically 5 mg), as very small volumes (<5 µL/well) can be used during the assay. For targets that are challenging to isolate, this can be critical. Finally, the reagents and instrumentation are relatively inexpensive. RT-PCR instruments have become commonplace at research centers in the last decade, and they are far less complex to maintain than SPR, NMR, or in-house X-ray equipment. The dye necessary for a screen is very inexpensive, and 384 well PCR plates are comparable in price to most assay plates. On the other hand, For fragments, thermal shifts can be within 1 °C, and binding stoichiometry is not directly measured. Taking advantage of its speed, DSF screening is often used as a primary screen to identify binders, followed by higher resolution methods.

DSF is also valuable as a first step in structural biology experiments. Proteins can show wide ranges of thermal stabilities as a function of buffer pH, salt identity and concentration, reducing agents, and other additives. Using DSF, one can rapidly screen a matrix of conditions that can help to identify optimal storage buffers and stabilizing classes of reagents that suggest possible crystallization conditions. This type of prescreen is widely employed in structural genomics and drug discovery environments, and can decrease the time required to solve structures.

1.3 Nuclear Magnetic Resonance

Nuclear Magnetic Resonance (NMR) spectroscopy is a powerful and flexible biophysical technique that has assumed a key role in examining the structural biology of ligand–protein and protein-protein interactions. A sample containing a solution of protein is inserted into a strong electromagnet (9–21 T) and sensitive electronics record the resonances of NMR-active nuclei in response to carefully designed perturbations. A number of different properties of the NMR-active nuclei are recordable in this manner, yielding information about the *chemical environment* and *relaxation* of the protein or ligand nuclei under observation. Because the binding of a ligand to a protein will alter the environment of both partners, NMR-based techniques can report on the nature and extent of binding events. Methods have been developed to screen for either the disruption or stabilization of protein-protein interactions directly; however, that is outside the scope of this chapter [17]. Of the diverse palette of NMR experiments that aid small-molecule discovery efforts [18], we focus on ligand- and protein-observed experiments that detect the typically weak binding (μM–mM) of initial fragment hits (Fig. 3). Ligand-observed methods make use of the alterations in chemical shift and/or relaxation of the compound upon binding to the larger protein. Protein-observed methods generally focus on perturbations in chemical shift in the protein resonances.

NMR-based screening offers several potential benefits. First, the chemical shift is exquisitely sensitive to the chemical environment of each nucleus, making it possible to describe precisely which moieties of a ligand are binding to the protein, and vice versa, providing a detailed structural understanding of the interaction. Binding affinities can be measured and competition between fragments can also be determined. Second, both ligand and protein can be detected; hence, NMR allows visualization of the whole system, and is particularly sensitive to common artifacts such as precipitation or aggregation of the ligands or the target. Third, the NMR experiment is measured in solution, so interactions between proteins and ligands can be measured in vitro under near physiological conditions (pH, salt, and protein concentration). Furthermore, because proteins are not restricted by a support matrix (as in SPR or crystallography), they are free to adopt different conformations, and potential fragment-binding sites are not occluded. Recent developments in NMR, such as In-Cell-NMR and experiments using cell extracts, allow monitoring of the interactions between individual proteins and/or their ligands in complex milieu [19].

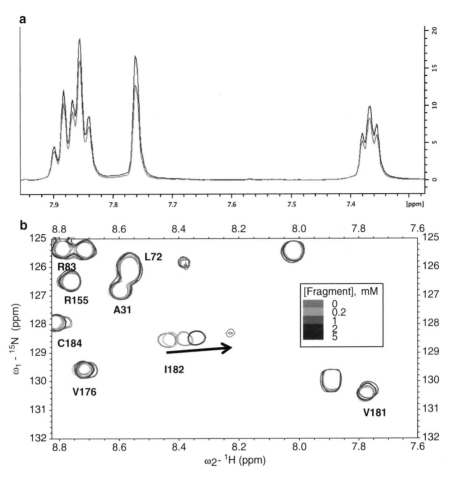

Fig. 3 NMR experiments for fragment screening. (**a**) Saturation transfer difference. 1H NMR spectrum of a fragment (*blue*) is reduced when it binds to protein (*red*). Measuring the difference spectrum then gives a negative signal with the same chemical shifts as the fragment (not shown). (**b**) 1H-15N HSQC. Each backbone NH has a characteristic chemical shift. When fragments bind, they induce changes to these shifts in dose-dependent manner (see legend) (Data courtesy of Mark Kelly, Michael Chimenti, and Rich Tjhen)

Several technological developments have given NMR the throughput and sensitivity to serve as a tool for primary fragment screening. High-field magnets and cryogenically cooled electronics provide increased sensitivity and resolution; robotic sample preparation, automated data acquisition, and supporting software have increased throughput. Practically speaking, the throughput of NMR screens varies dramatically, depending on the instrumentation and assay format. For protein-detected and ^{19}F ligand-detected formats, primary screens of several thousand fragments have been described using mixtures of ~100 compounds [20–23]. Screening in mixtures biases towards tighter binders due to competition and often lower concentrations of each compound. Other common ligand-detected formats, such as saturation transfer difference

(STD), are best performed with smaller pools of ~10 compounds, reducing the overall throughput. Although a number of groups currently perform NMR primary screens with libraries on the order of 2,000–10,000 fragments, others prefer to apply NMR to smaller libraries and as a secondary assay to obtain confirmatory or higher resolution binding information.

Protein targets are suitable for NMR fragment screening if they have good solubility, are mono-disperse and are stable in solution for the duration of the screening run. Stability and consistency can be determined by including known ligands as positive controls throughout automated screening runs. For protein-observed experiments, one can also compare 2D inverse-correlation spectra (e.g., 2D [^{15}N, ^{1}H]-HSQC) for signs of protein denaturation and degradation. The size of the protein is not critical for ligand-detected experiments; for protein-observed NMR, the size dictates what methods (pulse sequences) are most suitable. NMR typically requires relatively large amounts of pure protein, so the expression protocol must allow access to 10s of mg of protein. Protein concentrations average 10–20 μM protein for ligand-observed methods and ~50–200 μM protein for protein-observed techniques (depending on target size). Below, we describe experiments performed with the most common formats, using 5 mm outside diameter standard or Shigemi NMR tubes (500 μl and 250 μl sample volumes, respectively). NMR accessories exist for smaller tubes (3 mm), capillaries, and LC systems that significantly reduce protein requirements; however, with the capillary systems, precipitation of fragments and/or protein can lead to blockage.

NMR has the potential to define the binding sites of fragments on the protein target. In the case of protein-observed methods, site-specific assignments of resonances are usually required. In some cases, these have been performed previously and are available in the literature. If not previously determined, assignments for key residues, i.e., "probe-residues," can be obtained by mapping perturbations from known ligands, residue type-specific labeling or mutagenesis (e.g., to identify specific methyl-groups). In the case of ligand-observed experiments, competition with natural ligands or substrates can be used to aid identification of binding sites.

Many considerations that apply generally to the design of fragment libraries are also important for NMR; however, a few special criteria are important to mention here. First, high solubility in aqueous buffer at acceptable d$_6$-DMSO concentrations (1–5 %) is essential, given NMR's unique ability to detect very weak binders in the 10 mM range. It is also critical to have adequate buffering capacity in samples and to monitor the pH of samples following addition of ligands. Compounds are commonly screened as mixtures of 5–10 fragments to reduce instrument time and protein requirements. Pool size is selected based on total compound solubility and concerns about possible interactions between fragments. For ligand-observed methods, the use of mixtures is very attractive because individual hits can be identified in NMR spectra from resonance changes associated with binding. For such experiments, fragments should be combined in mixtures such that as few resonances as possible overlap in the 1D ^{1}H NMR spectrum.

1D proton reference spectra NMR of fragments are acquired prior to screening for quality control (QC) to eliminate compounds that have low solubility and to enable fragments with minimal spectral overlap to be combined in mixtures. The solution concentration is estimated by comparing the fragment's signal intensities to either the residual ^{1}H signal from the 99.9 % d_6-DMSO or to the DSS chemical shift reference. For ligand observed experiments, at least one ligand signal must not overlap with the water resonance.

In principle, libraries of several thousand fragments can be screened in a reasonable time using the methods outlined here. However, when resources and/or protein are limiting, fragments can be preselected based on predicted solubility (logP), diversity (Tanimoto coefficient), and/or virtual screening (docking scores).

1.4 X-Ray Crystallography

The ideal situation for most fragment screening efforts is to obtain atomic-resolution structures of fragment hits bound to the target (Fig. 4). High-resolution structures allow medicinal chemists to improve compound potency by rationally altering chemical structure to better complement the target site. For small fragments, whose binding modes are difficult to model precisely or predict a

Fig. 4 High-resolution X-ray structure of a PPI inhibitor fragment hit. RadA bound to fragment hit indazole at 1.3 Å resolution (PDB: 4B2I) [3]. 2Fo-Fc electron density map is shown for the compound contoured at 1.5σ. The electrostatic surface and figure were generated using PyMOL [42]

priori, experimental structures by X-ray or NMR are almost essential. While X-ray crystallography is commonly a follow-up method attempted with a small set of preferred fragment hits, many groups use it very successfully for initial library screening [24, 25]. Placing structural efforts first is very attractive in that every screening hit provides fodder for chemical optimization.

In contrast to the other techniques discussed, knowledge of a hit compound's potency is completely devalued in this context. Compounds that produce interpretable electron density have a known binding mode and are known to be soluble enough to at least partially saturate the binding sites in a crystal. This is a wealth of information, even in the absence of initial potencies. It is not possible to rank order the binding affinity of hits by X-ray, but affinity can be obtained using a functional assay or one of the other binding methods described in this chapter.

The key prerequisites for screening by crystallography are large quantities of pure target protein and reproducible conditions for growing high-resolution crystals. Since most fragment discovery campaigns move to structure determination at some point, the availability of crystal structures is generally important for FBLD. When used as a screening methodology, however, crystallization methods must be solved up-front. Furthermore, diffraction-quality crystals must be available in a high-throughput and reproducible manner.

There are two primary means of achieving fragment occupancy in a crystal: crystal soaking and co-crystallization. Crystal soaking is far more common in fragment screening [26]. In this experiment, pre-grown crystals are transferred from their growth wells into wells containing the same buffer plus high concentrations of one or several fragments. Each test well requires one large, high-quality crystal, making screening of an average-sized library of singletons daunting. Five to ten fragments are commonly included in each well, both reducing the number of crystals needed and increasing the likelihood of finding a hit in any well.

1.5 Summary

The techniques overviewed above are the most commonly used techniques in FBLD and each technique has a proven record of success in identifying promising chemical matter against many targets. Given below are detailed experimental protocols for developing these assays and employing them in an FBLD campaign aimed at a protein or PPI target.

2 Materials

2.1 Surface Plasmon Resonance

1. 0.25 mg/mL neutravidin in 10 mM Na-Acetate, pH 4.5.
2. 0.2 μm bottle-top filter.

3. 1:1 mixture of 0.1 M *N*-hydroxysuccinimide (NHS) and 0.4 M 1-ethyl-3-[3-dimethylaminopropyl]carbodiimide (EDC).

4. 1 M ethanolamine, pH 8.5.

5. Biotinylated target protein.

6. Immobilization buffer: 10 mM HEPES, pH 7.4, 150 mM NaCl, 0.25 mM TCEP, 0.05 % (v/v) Tween 20.

7. 0.22 μm, 0.5 mL spin filter.

8. 0.2 mg/mL Amine-PEG-Biotin.

9. Running buffer: 10 mM Tris–HCl, pH 8.0, 150 mM NaCl, 0.25 mM TCEP, 0.05 % (v/v) Tween 20, 5 % (v/v) DMSO.

10. Positive control molecule, peptide, or protein.

11. Dimethyl sulfoxide (DMSO).

12. Assay plates.

13. Fragment library.

2.2 Differential Scanning Fluorimetry

1. PCR plates suitable for the instrument being used.

2. Fluorescent dye.

3. Protein of interest.

4. Fragment library.

5. Assay Buffer: 20 mM HEPES, pH 7.4, 150 mM NaCl, 0.25 mM TCEP.

6. Optically clear sealing tape.

2.3 Nuclear Magnetic Resonance

1. Ligand(s) of interest or fragment library.

2. Protein of interest.

3. d_6-DMSO.

4. D_2O.

5. NMR buffer: 100 % D_2O PBS, pH 7.5, 1 mM NaN_3, and 1 mM DTT, 0.5 % DMSO with 10 μM DDS.

6. 10 % (w/v) solution of NaN_3 for dilution into assay buffer.

3 Methods

3.1 Surface Plasmon Resonance

These methods assume that the user is familiar with the operation of the SPR instrumentation in his/her laboratory.

3.1.1 Select SPR Chip Surface

The first step in performing any SPR experiment is selecting a strategy for immobilizing the target protein to the sensor surface. The two most common immobilization strategies are (a) covalent coupling through lysine side chains and N-termini using EDC/NHS-activated carboxyl groups on the sensor surface (random-amine coupling) and

(b) capture of biotinylated proteins by avidin surfaces. Other methods for immobilization include thiol coupling, NiNTA coupling of hexa-histidine-tagged proteins, and capture by specific antibodies (*see* Chapter 7 for more details).

For fragment screening, it is essential to couple a high density of target protein on the sensor surface. The maximum SPR binding signal for an analyte is given by:

(Immobilization Density) × (MW analyte)/(MW protein target)

The mass ratio above becomes very small for fragments against average-sized protein targets (0.01 or less), necessitating a high immobilization density to compensate. This is one negative aspect of screening by SPR. High-density surfaces have often been associated with artifactual protein behavior, due primarily to the high concentration of protein at the surface. Additionally, some immobilized protein might be inactive towards small-molecule binding due to denaturation or orientation/immobilization to the SPR surface. For these reasons, it is critical to optimize the protein-immobilization procedure using a well-behaved control compound, peptide, or protein binding partner as a control. This positive control will also be used to monitor the stability of the protein surface during screening.

Our preferred method of immobilization is biotin–avidin capture using site-specifically biotinylated proteins [27]. The AviTag sequence is a short peptide that is a substrate for *E. coli* biotin ligase BirA. If protein expression is performed in *E. coli*, the AviTagged target protein can be coexpressed with BirA to yield protein that is singly biotinylated at a specific lysine in the AviTag sequence. Vectors for appending an AviTag to the N or C termini of targets are commercially available. We prefer this method to random biotinylation (coupling to lysine or cysteine residues) or covalent coupling, as it is a consistent, predictable method that requires no extra processing steps, and it does not compromise potentially relevant protein amines.

A generic protocol for immobilization using biotinylated protein follows. Streptavidin and neutravidin-coated SPR surfaces (chips) can be purchased from many instrument vendors or can be prepared from carboxymethyl chips, such as Biacore CM5 or CM7 (these two chips differ in density of carboxyl groups, *see* Chapter 7 for more details). We prefer neutravidin, a mutant form of streptavidin, as it contains fewer potential small-molecule binding sites, and is more amenable to covalent coupling due to its increased pI. **Steps 2–5** describe the protocol for immobilizing neutravidin; this protocol is similar for covalent coupling of any target to a carboxymethyl chip surface by EDC/NHS chemistry.

3.1.2   Prepare SPR Chip Surface

1. Install SPR chip into instrument per vendor recommendation.

Covalently modify SPR chip with neutravidin:

2. Dilute neutravidin to 0.25 mg/mL into 10 mM Na-Acetate, pH 4.5. Buffer should first be filtered through 0.2 μm bottle-top filter (*see* **Note 1**).

3. Inject a 1:1 mixture of 0.1 M NHS and 0.4 M EDC for 5 min to activate the chip surfaces.

4. Inject 0.25 mg/mL neutravidin for 4–7 min across all surfaces. Using a Biacore CM5 chip this should yield 8,000–12,000 RU of neutravidin density on all surfaces. Leave one surface as a neutravidin-only reference.

5. Inject 1 M ethanolamine, pH 8.5, for 5 min across all surfaces to block all remaining reactive sites.

Immobilize biotinylated protein

6. Dilute the biotinylated target protein to low concentration (5–50 μg/mL) in immobilization buffer.

7. Filter the target protein solution using a 0.22 μm, 0.5 mL spin filter.

8. Inject the target protein on an avidin-labeled flow cell for 1–5 min, or until the desired amount has been captured. A surface of ~10,000 RUs for a protein of average size (20–50 kDa) is a reasonable goal (*see* **Note 2**).

9. Inject 0.2 mg/mL Amine-PEG-Biotin (or a similar, soluble biotin analog) for 4 min across all surfaces to block any remaining biotin binding sites.

3.1.3   Establish the Activity of the SPR Surfaces

It is essential to determine the quality of the immobilized protein before proceeding with a fragment screening effort.

1. Select a running buffer. A typical buffer could contain 10 mM Tris–HCl, pH 8.0, 150 mM NaCl, 0.25 mM TCEP, 0.05 % (v/v) Tween 20, 5 % (v/v) DMSO (*see* **Note 3**).

2. Select a small molecule, peptide, or protein that binds to the site of interest to be used as a positive control.

3. Serially dilute the control in DMSO and transfer to a running buffer without DMSO, so that the % DMSO is identical in the sample well and running buffer. For example, for a final DMSO concentration of 5 % in 100 μL sample volume, dilute 5 μL of compound into 95 μL of running buffer without DMSO. Select a wide concentration range to span the expected binding constant. Inject samples over the chip surface and measure the SPR signal.

4. Calculate the K_D using kinetic or equilibrium measurements, and compare to the expected value. Determine the fraction of active protein on the surface using the following equation:

$$\text{surface activity} = (\text{RU analyte})/(\text{RU immobilized protein}) \\ \times (\text{MW protein})/(\text{MW analyte}) \\ \times 1/\text{expected binding stoichiometry}$$

Surface activities of >0.8 can often be achieved, but lower (ca 0.5) are also common.

3.1.4 Run Fragment Screen

1. Set up assay plates using a standardized format based on the capabilities of your instrument. As with the control compounds, the DMSO concentration and buffer should exactly match the running buffer. Typical concentrations for fragments are 100–500 µM in 2–5 % DMSO.

2. Run the screen using injection times between 10 and 30 s, followed by dissociation monitoring for 30 s. Because fragments should bind weakly and therefore reach equilibrium quickly, these short injections provide a significant amount of data with which to classify compounds.

3. Inject buffer blanks (containing DMSO) and positive control molecules regularly throughout the screen (e.g., at the end of a column or each plate), to evaluate assay stability. Blank injections should also be used during data analysis to "double-reference," which is especially critical in fragment screening where the signal is expected to be low [28]. The controls should also be used to calculate z' or other statistical factors to assess the quality of the screen [29].

3.1.5 Identify Screening Hits

SPR data allows for hit selection based on both the % fractional occupancy the compound achieves, and also the kinetic profile it produces.

1. Reference SPR data ("sensorgrams") to the unmodified flow cell and double-reference to blank injections using the instrument's software or third-party software such as Scrubber (Biologic Software, http://www.biologic.com.au/). Additional data smoothing and scaling can reduce complexity arising from variation in positive control binding and experiments run on separate surfaces [28].

2. Prepare a scatterplot (Fig. 1a) of the control and test compounds. Select a time-range near the end of the sample injection, where the positive control has achieved equilibrium. Use this time range to determine the RU for each control and test compound, and plot on the y-axis. Scatterplots can also be scaled to binding stoichiometry:

$$stoichiometry = (RU\ analyte)/(RU\ immobilized\ protein)$$
$$\times (MW\ protein)/(MW\ analyte)$$
$$\times (surface\ activity)$$

3. Select compounds for follow-up based on approximate binding stoichiometry and kinetics. The standard threshold for hit selection is three times the standard deviation of the non-interacting compounds (Fig. 1a) (*see* **Note 4**).

4. Cherry-pick or prepare fresh DMSO stocks of selected fragments and retest at multiple doses (Fig. 1c). As with positive control compounds, hits should be serially diluted in DMSO and then transferred to running buffer that does not contain DMSO. If possible, dose responses should start severalfold above the concentration for primary screening; eight 3-fold dilutions or ten 2-fold dilutions are common. The Kd is estimated by converting RU to stoichiometry at each concentration point and fitting the concentration vs. stoichiometry binding isotherm (or log[analyte] vs. stoichiometry). This experiment yields dissociation constants and stoichiometry for well-behaved compounds and serves to identify bad-acting compounds that looked passable at a single concentration.

3.1.6 Initiate Secondary Assays

At this stage, repurchase or synthesize the most potent compounds with low stoichiometry and move forward to secondary screening (such as functional activity), chemical optimization, and structure determination by NMR or X-ray crystallography (see below).

3.2 Differential Scanning Fluorimetry

3.2.1 Develop DSF Assay

1. Select reagents. The choice of plates will depend on the model of RT-PCR instrument to be used; in general, use 384-well PCR plates with optically clear sealing tape. There are also many potential dyes. No published study has systematically compared the appropriate dyes for overall performance, and target-specific effects of particular dyes may be possible. SYPRO Orange is cost-effective and widely used. It is provided as a 5,000× stock in DMSO, and our experience has shown it to work well at 1× final concentration across a range of targets and buffers.

2. Prepare a matrix containing several concentrations of protein and several concentrations of dye, (e.g., SYPRO Orange); each combination should be tested in triplicate. Typical starting parameters are 2 μM protein with 5,000× diluted SYPRO Orange [30]. Assay volumes can be as low as 5 μL with appropriate 384-well plates.

3. Measure the melting temperature of each dye–protein combination using a qRT-PCR instrument. Record fluorescence data at the appropriate dye wavelengths, using a temperature range

of 25–95 °C in 1° steps with 30 s per step. Load the plate into the thermocycler and initiate the experiment. Analyze the data by plotting the fluorescence intensity vs. temperature; the melting temperature (T_m) is the temperature at the inflection point (*see* Subheading 3.2.2 below for details).

4. Select dye/protein combinations. Thinking ahead to the screen, it is useful to minimize the amount of protein and dye required per well. Thus, select the minimal reagent concentrations and well volumes that give reproducible T_m values.

5. Establish stability of the protein in DMSO/buffer solutions (*see* **Note 5**).

3.2.2 Run Fragment Screen

The following procedure is a general method for fragment screening by DSF in 384-well plates with a final well volume of 20 μL and a final compound concentration of 1 mM.

1. Prepare reagents for screening as follows: dilute protein from a concentrated stock to 1.33× the predetermined optimal assay concentration (usually between 0.5 and 5 μM) into assay buffer without DMSO. Dilute dye to 5× the predetermined optimal concentration into assay buffer without DMSO. Prepare or store compounds at 20× final concentration in 100 % DMSO (20 mM for a final assay volume of 1 mM).

2. Add 15 μL diluted protein to each well of a 384-well PCR plate.

3. Add 1 μL DMSO to one column of the plate. These wells serve as the negative controls.

4. If a positive control compound is available, add 1 μL of 20× positive control in DMSO to one column of the plate.

5. Add 1 μL of test compound to each non-control well using a multichannel pipette or an automated liquid handler.

6. Add 4 μL of diluted dye to each well using a multichannel pipette or an automated liquid handler. Seal the plate with optically clear sealing tape, cover with foil, and mix by gentle shaking at room temperature for 10–20 min.

7. Remove foil and centrifuge plate for 2 min at 1,000–2,000 × *g* to remove bubbles. The plate is now ready to be assayed.

8. Set up the RT-PCR instrument to record fluorescence data at the appropriate dye wavelengths, for a temperature range of 25–95 °C in 1° steps with 30 s per step. Load the plate into the thermocycler and run the experiment. With these parameters, each plate should take between 35 and 45 min.

9. Determine the T_m for fragments. Plot the fluorescence vs. temperature data and fit to the Boltzmann equation:

$$F = \min + (\max - \min)/(1 + \exp((T_m - x)/\text{slope}))$$

Where F is the measured fluorescence, min is the minimum observed fluorescence, max is the maximum observed fluorescence, x is the temperature, T_m is the melting temperature, and slope is the slope of the curve at the T_m. T_m and slope are to be fit, while the other parameters are given by the data (*see* **Note 6**).

3.2.3 Identify Screening Hits

To select screening hits, prepare a scatter plot of T_m value vs. compound and control wells. The control wells should produce a precise T_m with low standard deviation ($<1°$). Fragments that produce a T_m increase greater than two times the standard deviation of control wells are typically considered hits. This threshold can be modified if the number of hits above 2 standard deviations is thought to be too high or too low. Setting too low of a hit threshold is, however, likely to generate many false positive compounds that cannot be confirmed. In this case it might be advantageous to rescreen at a higher compound concentration.

3.2.4 Initiate Secondary Assays

If desired, affinities can be directly determined by DSF using serial dilutions of fragments. Thermal shift dose–response data is not fitted to a standard hyperbolic curve, but rather to a more complicated expression that represents the free energy change of folding as well as that of compound binding. Matulis et al. have derived a model to fit DSF titration curves [31]. Accurate binding affinities may be more readily obtained from other biophysical methods, such as SPR, NMR, or ITC, as there is no reliance upon an independently determined or estimated parameter, as is the case with DSF. Since every assay format has advantages and limitations, it is generally advisable to follow up hits from any fragment screen using an orthogonal method.

3.3 Nuclear Magnetic Resonance

The detailed operation of NMR instruments is beyond the scope of this chapter. Omitted are basic preparative setup steps for acquiring NMR spectra, such as shimming and tuning, as it is assumed the user is familiar with these or will have support from an NMR spectroscopist. For further reading, see Till Maurer's excellent introduction to NMR methods in fragment-based ligand screening and Bertini for more detailed protocols and troubleshooting [32, 33].

3.3.1 Develop a Saturation Transfer Difference (STD) Assay

The Saturation Transfer Difference (STD) experiment relies on the nuclear Overhauser effect (nOe) between nuclei in the protein and in a small-molecule ligand that binds transiently. The compounds that interact with the protein will show reduced intensity in their NMR signals relative to non-binders, whose signals will be unaffected by the presence of protein. STD is performed as a subtractive measurement, where the 1-D ligand spectra are collected in the presence and absence of protein irradiation ("on-resonance" and "off-resonance," respectively). These spectra are subtracted, so that bound ligands yield an NMR signal, while non-binders are not observed.

Ligand–protein ratios, NMR irradiation time, and temperature are best optimized empirically with a known ligand; however, standard conditions may be applied initially. If no control is available, screening may proceed initially until a "hit" is found.

1. To avoid spectral overlap with fragments, buffers should contain deuterated organic salts and other additives (*see* **Note 7**).

2. For STD, it is critical to maintain a large excess of free ligand, ideally at a ratio of 10–50:1 ligand to protein. Typically, we screen with 1–5 % d_6-DMSO, 500–1,000 μM ligand, and 10–50 μM protein.

3. Generally, low temperatures (10 °C) are used for smaller proteins (6–20 kDa), since the strength of the nOe is dependent on the protein's correlation time, which increases with lower temperatures. Additionally, modifications of the standard pulse sequences are available which include filters to suppress protein resonances that may overlap with ligand signals.

4. Select a portion of the protein spectrum away from ligand resonances (typically near the methyl region, −0.5 to −1.5 ppm). Selectively irradiate protein using a shaped 270° pulse train (e.g., an 80 ms Gaussian pulse on a 500 MHz instrument) for several seconds. The NMR irradiation time is an optimizable parameter. It is good practice to record an on-resonance STD spectrum of the fragment solution (in the absence of protein) to serve as a reference.

5. In a second reference experiment, move the saturation pulse "off-resonance," so as not to excite any protein nuclei, usually by 20–50 ppm (we irradiate at 20 ppm). Care needs to be taken that the selective "saturation" pulse length, shape, and power are calibrated such that ligand resonances outside of the methyl region are not affected by the saturation pulse (to check this, the reference spectrum from **step 4** can be compared to a simple 1D ^1H spectrum of the ligand).

6. Collect the STD experiment as a "pseudo-2D" where both the on- and off-resonance free induction decays (FIDs) are collected in an interleaved manner and stored separately on disk. We recommend this approach because it allows both experiments to be handled separately. However, many commonly used pulses sequences subtract the on-resonance FID from the off-resonance FID (reference) during acquisition. For both experiment types, the data are apodized (1–3 Hz line-broadening), zero-filled once, Fourier transformed, and phased.

In the case of overlaid pseudo 2D data, peaks that show an STD effect will be reduced in intensity in the on-resonance experiment. By scaling the off-resonance spectrum to the on-resonance spectrum, the "percent STD" reduction for each peak can be measured. The two STD spectra can then be subtracted to generate a difference spectrum if desired.

3.3.2 Run an STD Fragment Screen

The following is a general outline for a protocol when performing STD experiments in standard 5 mm diameter NMR tubes. Typically ~100 samples (5–10 fragments each) or more can be screened in a 24 h period with automated hardware.

1. Prepare the sample(s) without protein in NMR buffer (final volume ~450 μL). Samples will include fragment(s), ~10 μM DSS (for referencing), a deuterated buffer and D_2O (99.96 %) as solvent (*see* **Note 8**).

2. Acquire a simple 1D spectrum at the temperature to be used for screening for each fragment to be used as a reference spectrum during data analysis.

3. Add protein to each sample. Place samples in a rack. Make a "master mix" containing protein (plus fresh DTT if indicated), and NaN_3 (to control bacterial growth) along with 10–20 μL of NMR buffer per sample. Dispense the correct volume of master mix (usually ~50 μL) into each tube with a pipette (by placing NMR tubes into a pipette tip box, an 8-channel pipette can be used for this). Carefully swing the rack up and down to force the solution to the bottom of each tube (cover the top of the rack with a piece of cardboard and grip it firmly). Allow the tubes to sit for 1–2 h or overnight at 4 °C to mix.

4. After tuning, shimming, etc., load parameters for the STD NMR experiment. Be sure to optimize the ^{1}H transmitter frequency to obtain good suppression of any residual water. The number of scans is typically set to 16 or 32, saturation time to 1.3 s, recycle delay to 3.7 s with a spectral width of 16 ppm and collecting 4,096 complex data points. Experiments can be acquired in 10–15 min per sample.

5. The data are then apodized (1–3 Hz line-broadening), zero-filled once, Fourier transformed, and phased. 1D on- and off-resonance STD are extracted from the pseudo 2D spectrum and overlaid for comparison. The "off-resonance" experiment is scaled down to the "on-resonance" experiment and the "% STD" reduction for each peak measured.

6. When working with mixtures of ligands, repeating the experiment with the individual fragments from a pool that showed a "hit" can confirm the identity of the binding fragment(s).

<table>
<tr><td>

3.3.3 Identifying Fragment Hits

</td><td>

In the STD and ligand-detected experiments, hit identification is determined by examining the on-resonance and off-resonance (reference) experiments, either by overlaying the two for comparison or by computing the difference spectrum (Fig. 3a). Ligands that have no interaction with the protein target should not exhibit changes between the on- and off-resonance experiments, and give no signals in the difference spectrum. Ligands with strong interactions will give large changes in intensity, and unambiguous negative signals in the difference spectra. Ligands that bind weakly (like most fragments) will give small, but interpretable changes in the spectrum. Unless a ligand with known affinity is available as a reference, a cutoff value is usually chosen to determine which fragments are candidates for "hit" follow-up. Typically fragments with STD difference values of >5 % are classified as a "hit".

Secondary assays for STD include measuring dose–response behavior of the signal to rank-order fragments. Binding selectivity can be tested by running the STD experiment against a control protein or homolog of the target. STD gives limited information about the binding site or stoichiometry, and thus this method is complemented by the orthogonal techniques (ITC, SPR, X-ray, protein-detected NMR).

</td></tr>
</table>

3.3.4 WaterLOGSY

Similar to STD NMR, WaterLOGSY NMR is a ligand-detected method. Details are described in [34]. WaterLOGSY relies on transfer of magnetization from bulk water to a ligand through waters bound in or adjacent to the small-molecule binding site, and from magnetization from bulk water that is transferred through protons in the target by spin-diffusion. For some targets, detection of small-molecule binding can be stronger by Water-LOGSY, and new PO-WaterLOGSY experiments offer higher sensitivity [35]. Implementation of the PO-WaterLOGSY can be more involved, however. Unlike STD-NMR, WaterLOGSY cannot be performed in 100 % D_2O; 10 % H_2O is usually sufficient to obtain a strong signal, and also allows an STD experiment to be run on the same sample.

3.3.5 HSQC and HMQC NMR

One of the first fragment screening methods, called "SAR by NMR," was based on a protein-observed HSQC experiment [36]. A key advantage of protein-observed experiments is that the structure and state of the target protein are monitored in each experiment.

HSQC and related chemical-shift perturbation methods rely on the availability of labeled protein, either ^{15}N-labeling or ^{13}C-labeling for larger systems. The collection of a $^{13}C/^{15}N$ 2D inverse-correlation or HSQC/HMQC type experiment yields a 2D spectrum, where cross-peaks corresponding to the frequencies of two directly bound nuclei (e.g., Hn–^{15}N or ^1H–^{13}C in a methyl group) are dispersed on a 2D plane based on their chemical shifts (Fig. 3b). Protein-detected techniques are limited by the need to obtain a well-resolved 2D correlation spectrum, placing an upper limit on the domain size of around 50 kDa when using ^{15}N-labeling. Large proteins or complexes can be screened by resorting to selective-labeling strategies where the methyl groups of Isoleucine, Leucine, Valine, Alanine, and Methionine are labeled with ^{13}C combined with deuteration of other aliphatic groups.

The details of running HSQC experiments are dependent on the NMR instrument and the target. Some general considerations for fragment screening by HSQC/HMQC follow:

1. The spectrum is collected with and without ligand or a pool of ligands. If the ligand binds to the target, frequency changes (shifts in peak positions) or broadening of peaks will be observed for groups close to the binding site.

2. Due to the sensitivity of chemical shifts of groups (in particular amide protons) and protein structure to changes in solvent pH, it is essential to monitor the pH of samples following the addition of fragments.

3. Deuterated DMSO (1–5 %) is usually used as a co-solvent as in ligand-observed experiments, but it is important to record a spectrum of the target in the presence of DMSO as a reference as it may induce shifts and also to test for changes in activity.

4. Where mixtures (5–10) of fragments are screened, it is necessary to deconvolute pools that contain a "hit" by screening individual fragments. The size of the pools of fragments is usually kept fairly small (5–10) to allow weak binding fragments to be screened at concentrations of 1–2 mM while reducing the chance of solubility limitations.

5. If chemical shift assignments are available or can be obtained, the location of the binding site can be identified at atomic resolution. If binding of a ligand causes allosteric changes in the target, peaks corresponding to residues remote from the binding site may also be affected. It is reasonable to predict however, that the strongest shifts will cluster to a region that forms the binding pocket.

6. The location of the binding pocket can be confirmed by identifying short-range interactions (<6 Å) between the ligand and target, such as by recording NOESY experiments.

Once a binding fragment has been identified, a second fragment that binds in close proximity can be sought by repeating the screen in the presence of a high concentration of the first ligand. This approach was originally described by Shuker et al. [36]. When the HSQC spectrum has been assigned for the target protein, HSQC-based screens offer a unique combination of throughput, simplicity of design, and high-resolution binding data.

3.4 X-Ray Crystallography

3.4.1 Validate Crystal Forms for Fragment Screening

1. Select the appropriate protein construct (*see* **Note 9**).

2. Select a storage buffer that confers good protein stability. Prescreening for protein thermal stability in a range of possible storage buffers can be easily accomplished using differential scanning fluorimetry (DSF; see above).

3. Screen crystallography conditions. Obtaining protein crystals that diffract to better than 2.5 Å resolution is almost always the result of brute force screening. In a typical case, 3–10 protein expression constructs are screened against >1,000 initial conditions from sparse-matrix, commercial crystal screens (*see* **Note 10**).

4. Assess the stability of crystals in cryogenic conditions [26]. Nearly all protein crystallography is carried out under cryogenic conditions, requiring the optimization of freezing conditions that prevent crystal damage. Typically, this involves temporary soaking of each crystal in a condition containing a high concentration (20–30 % (v/v)) of glycerol or high molecular weight PEG. This should be performed with several test solutions in a trial and error fashion.

5. Assess the stability of crystals in DMSO (~10–15 % (v/v)) to simulate the conditions that will be used for soaking in test compounds. Several crystals should be transferred to drops containing varying concentrations of DMSO and observed intermittently over the next 24 h. Some crystals cannot tolerate DMSO and will become visibly damaged. X-ray data sets should then be collected from crystals that remain intact in DMSO to test for more subtle damage caused by the addition of organic solvent.

6. Plate compounds for an X-ray based fragment screen. To increase throughput and reduce the number of crystals needed, compounds are typically pooled in sets of 5–10. Premixed and plated compounds can be stored in DMSO or as dried powder mixes. In either case, after the compounds are solubilized in DMSO, crystallization buffer is added to each well to a final volume of 2–5 μL. Crystals are then transferred into each test well and incubated for a variable length of time. The time necessary to soak compounds into a crystal is highly protein dependent.

3.4.2 Collect and Process X-Ray Diffraction Data

1. Prepare crystals for data collection. Data collection and crystal handling procedures are dictated by the X-ray generation site, be it a home source or an external light source. For screening, some measure of automation in crystal handling is critical, as it can greatly increase efficiency and decrease the risk of manual error and sample loss. Many beamlines are now outfitted with frozen crystal handling robots that mostly automate the

process of crystal mounting/dis-mounting and X-ray beam centering. The process of freezing crystals must still be carried out manually, but no further human contact with the crystals is required.

2. Determine a data-collection strategy and collect data. Data collection throughput is largely driven by the number of data frames that must be collected per crystal. This number primarily depends on the symmetry parameters of the crystal and can only be optimized to a certain degree (*see* **Note 11**).

3. Process data and generate a model. These steps have become much more automated since the advent of fragment screening. We recommend XDS [37], ELVES [38], or HKL2000 for indexing, scaling, and merging of diffraction data. ELVES is the most automated package for robust data processing, and is well suited for processing dozens of data sets quickly. For phasing and model refinement, we recommend Phenix [39], which is amenable to scripting through the command line interface, and has an advanced graphical interface as well. For model visualization and manual building, Coot [40] is the current standard.

3.4.3   Identify Hits from X-Ray Screens

1. Search electron density maps for bound fragments (Fig. 4). Fragment occupancy can be evaluated manually or by computational methods. Both Coot and Phenix contain modules for locating large unfilled regions of electron density.

2. When fragment density is observed, attempt to identify its structure from among the compounds in the screening pool of 5–10 fragments. This analysis is easiest when the fragment pools are designed to contain very dissimilar compounds, the fractional occupancy of the site is high, and the resolution of the crystal is good (~2 Å). When the identity is not obvious, one can build separate models for each possible compound and refine them; the refinement process should eliminate some compounds from consideration. More automated approaches to considering fragment electron density have been reported by industrial labs, but are not publicly available [25]. In many cases, it is best to recrystallize the fragment/protein complex from a pure solution of the compound.

3.4.4   Characterize Hits from X-Ray Screens

Developing secondary assays for an X-ray fragment screen depends on many factors, including target type, chemistry resources, and availability of secondary biophysical assays to measure binding affinities. For a PPI target, it is ideal to assess whether or not the fragments inhibit the interaction, using some of the techniques described elsewhere in this book. Biophysical competition assays by SPR, analytical ultracentrifugation, analytical gel filtration, or static light scattering can also give a measure of functional potency.

4   Conclusion

FBLD holds promise for the discovery of small-molecule inhibitors of PPI. Particular advantages include the ability to search chemical space efficiently and the use of binding-based assays to identify ligands. Each of the primary screening methods described in this chapter is widely utilized in FBLD, but each has its limitations. Judicious combination of orthogonal methods will ensure that fragments are selected based on lead-like criteria, including high ligand efficiency (binding affinity) and binding-site specificity.

5   Notes

1. For direct amine coupling of a test protein, prepare an immobilization buffer that is logical for your protein target, within the following guidelines: (a) select buffer pH below the pI for the protein. This allows the positively charged protein to "preconcentrate" at the carboxyl surface, thus increasing coupling efficiency, (b) avoid amine-based buffers, including Tris, (c) avoid dithiothreitol (DTT) because it reduces the reactivity of activated NHS-esters, (d) include a low amount of non-ionic detergent (Tween 20 at 0.05 % or less). All buffers should be kept at room temperature and filtered through 0.2 μm bottle-top filters prior to use.

2. Depending on the architecture of the chip and instrument, different configurations are possible. Using a Biacore CM5 chip and T200 instrument, there can be three active flow cells and one reference flow cell (where no test protein is added). Having a reference flow cell is mandatory, but the configuration of the three active flow cells is completely variable. It is common to add three different immobilization densities of the same protein. One can also screen three targets in parallel, or immobilize variants (such as point mutants) of the primary target [41].

3. This step is highly target-dependent, but the following guidelines should be considered: (a) the buffer should maximize the stability and activity of the immobilized protein, (b) addition of detergents, such as 0.01 % Tween or 0.1 % Triton is highly recommended to reduce nonspecific binding of test compounds to the protein and microfluidics, (c) carrier protein, such as Prionex or beta-gamma globulin, can further reduce nonspecific effects, (d) using the same buffer conditions in SPR as will be used for secondary assays can increase reproducibility across assays [28]. Finally, it is very important to include DMSO in the running buffer such that the %DMSO is identical

to that in the sample wells. Fragment screening is typically performed in a background of 2–5 % DMSO to aid in compound solubility.

4. Compounds that bind in excess of their maximum theoretical binding level are likely to be nonspecific; at the primary screening stage, we generally accept compounds that bind with apparent stoichiometries <3. All sensorgrams that pass these filters should be inspected by eye to take advantage of the high data content of an SPR-based screen. In almost all cases, fragments should bind and dissociate with rapid kinetics (Fig. 1b), which produces a sensorgram with a square pulse-like shape. Discernible kinetics have been reported for some low affinity fragments, but this is relatively uncommon.

5. As fragment concentrations in the screen will be high, a final concentration of 5 % (v/v) DMSO is typical. Therefore, monitor protein melting at several DMSO concentrations (e.g., 0, 1, 3, 5 %). If a significant reduction in T_m is seen at any DMSO concentration, it is advisable to stay below that concentration in the screen.

6. The T_m can also be estimated from the minimum of the negative first derivative of the raw data. This process has limited resolution, because the data is only collected in 1° steps, and the Tm is estimated by visual inspection. For this reason, we prefer to more robustly fit the T_m using the raw data. Several simple and free tools exist for fitting DSF data using this method. We use FIT, developed at the University of Washington (http://skuld.bmsc.washington.edu/FIT/). FIT can be used as a web server or the script can be downloaded and run on a local machine. Each well in a data set is individually fit to both single and double transition models, with figures and statistics automatically generated in useful formats.

7. Deuterated Good buffers are available for a range of pH values; deuterated acetate or phosphate buffer may also be used. Reductants such as DTT and TCEP are also available in deuterated form. Sufficient salt should be included to maintain protein activity and stability, however higher concentrations (>50 mM) reduce sensitivity, particularly for cryogenic probes. For a more detailed discussion of sample preparation, *see* Bertini Ch. 15 [33].

8. Tubes can be the inexpensive type, such as Norell XR series bulk NMR tubes. If using automation, it is essential to allow sufficient time (>3 min) for a sample to thermally equilibrate after it is inserted into the magnet. As noted, smaller targets (6–25 kDa) will yield stronger STD signals at lower temperatures, but may require longer equilibration times.

9. For targets that have previously been crystallized, it is often useful to create an expression construct that begins at the first modeled residue at the N-terminus and ends at the last modeled residue at the C-terminus of the published structure. By removing disordered termini, proteins can crystallize more readily and with higher resolution.

10. Miniature-scale crystallization robots such as the Mosquito and Honeybee systems have made the process of screening for crystal conditions much faster and less resource intensive. They allow for drop sizes lower than 100 nL, whereas traditional screening by manual pipetting is performed with at least 500 or 1,000 nL of concentrated protein stock per test condition.

11. X-ray test frames are analyzed computationally to give a preliminary guess at the space group and unit cell parameters. This information is then used to devise an efficient strategy for collecting a fully complete data set. As most crystals of the same target in the same condition will be isomorphous, the same collection strategy can be used throughout the screen. The type of X-ray detector also influences collection throughput [25].

References

1. Rees DC, Congreve M, Murray CW et al (2004) Fragment-based lead discovery. Nat Rev Drug Discov 3:660–672

2. Hajduk PJ, Greer J (2007) A decade of fragment-based drug design: Strategic advances and lessons learned. Nat Rev Drug Discov 6:211–219

3. Scott DE, Ehebauer MT, Pukala T et al (2013) Using a fragment-based approach to target protein-protein interactions. Chembiochem 14:332–342

4. Braisted AC, Oslob JD, Delano WL et al (2003) Discovery of a potent small molecule IL-2 inhibitor through fragment assembly. J Am Chem Soc 125:3714–3715

5. Arkin MR, Randal M, Delano WL et al (2003) Binding of small molecules to an adaptive protein-protein interface. Proc Natl Acad Sci U S A 100:1603–1608

6. Petros AM, Huth JR, Oost T et al (2010) Discovery of a potent and selective bcl-2-inhibitor using SAR by NMR. Bioorg Med Chem Lett 20:6587–6591

7. Fuller JC, Burgoyne NJ, Jackson RM (2009) Predicting druggable binding sites at the protein-protein interface. Drug Discov Today 14:155–161

8. Lau WF, Withka JM, Hepworth D et al (2011) Design of a multi-purpose fragment screening library using molecular complexity and orthogonal diversity metrics. J Comput Aided Mol Des 25:621–636

9. Na J, Hu Q (2011) Design of screening collections for successful fragment-based lead discovery. Methods Mol Biol 685:219–240

10. Chen IJ, Hubbard RE (2009) Lessons for fragment library design: analysis of output from multiple screening campaigns. J Comput Aided Mol Des 23:603–620

11. Erlanson DA, Wells JA, Braisted AC (2004) Tethering: fragment-based drug discovery. Annu Rev Biophys Biomol Struct 33:199–223

12. Wilson CG, Arkin MR (2013) Probing structural adaptivity at PPI interfaces with small molecules Drug Discovery Today: Technologies 10 (4):e501–e508

13. Giannetti AM, Koch BD, Browner MF (2008) Surface plasmon resonance based assay for the detection and characterization of promiscuous inhibitors. J Med Chem 51:574–580

14. Babaoglu K, Simeonov A, Irwin JJ et al (2008) Comprehensive mechanistic analysis of hits from high-throughput and docking screens against beta-lactamase. J Med Chem 51:2502–2511

15. Cimmperman P, Baranauskiene L, Jachimoviciute S et al (2008) A quantitative model of thermal stabilization and destabilization of proteins by ligands. Biophys J 95:3222–3231

16. Kranz JK, Schalk-Hihi C (2011) Protein thermal shifts to identify low molecular weight fragments. Methods Enzymol 493:277–298

17. Rizo J, Rosen MK, Gardner KH (2012) Enlightening molecular mechanisms through study of protein interactions. J Mol Cell Biol 4:270–283

18. Pellecchia M, Bertini I, Cowburn D et al (2008) Perspectives on NMR in drug discovery: a technique comes of age. Nat Rev Drug Discov 7:738–745

19. Ito Y, Selenko P (2010) Cellular structural biology. Curr Opin Struct Biol 20:640–648

20. Dalvit C, Fagerness PE, Hadden DT et al (2003) Fluorine-NMR experiments for high-throughput screening: theoretical aspects, practical considerations, and range of applicability. J Am Chem Soc 125:7696–7703

21. Dalvit C, Flocco M, Veronesi M et al (2002) Fluorine-NMR competition binding experiments for high-throughput screening of large compound mixtures. Comb Chem High Throughput Screen 5:605–611

22. Hajduk PJ, Meadows RP, Fesik SW (1999) NMR-based screening in drug discovery. Q Rev Biophys 32:211–240

23. Hajduk PJ, Gerfin T, Boehlen JM et al (1999) High-throughput nuclear magnetic resonance-based screening. J Med Chem 42:2315–2317

24. Murray CW, Blundell TL (2010) Structural biology in fragment-based drug design. Curr Opin Struct Biol 20:497–507

25. Spurlino JC (2011) Fragment screening purely with protein crystallography. Methods Enzymol 493:321–356

26. Bottcher J, Jestel A, Kiefersauer R et al (2011) Key factors for successful generation of protein-fragment structures requirement on protein, crystals, and technology. Methods Enzymol 493:61–89

27. Prakash O, Eisenberg MA (1979) Biotinyl 5'-adenylate: corepressor role in the regulation of the biotin genes of *Escherichia coli* k-12. Proc Natl Acad Sci U S A 76:5592–5595

28. Giannetti AM (2011) From experimental design to validated hits a comprehensive walk-through of fragment lead identification using surface plasmon resonance. Methods Enzymol 493:169–218

29. Zhang JH, Chung TD, Oldenburg KR (1999) A simple statistical parameter for use in evaluation and validation of high throughput screening assays. J Biomol Screen 4:67–73

30. Niesen FH, Berglund H, Vedadi M (2007) The use of differential scanning fluorimetry to detect ligand interactions that promote protein stability. Nat Protoc 2:2212–2221

31. Matulis D, Kranz JK, Salemme FR et al (2005) Thermodynamic stability of carbonic anhydrase: measurements of binding affinity and stoichiometry using thermofluor. Biochemistry 44:5258–5266

32. Maurer T (2011) Advancing fragment binders to lead-like compounds using ligand and protein-based NMR spectroscopy. Methods Enzymol 493:469–485

33. Bertini I, Molinari H, Niccolai N (1991) NMR and biomolecular structure, vol xvii. VCH, Weinheim, 209 p

34. Dalvit C, Pevarello P, Tato M et al (2000) Identification of compounds with binding affinity to proteins via magnetization transfer from bulk water. J Biomol NMR 18:65–68

35. Gossert AD, Henry C, Blommers MJ et al (2009) Time efficient detection of protein-ligand interactions with the polarization optimized PO-WaterLOGSY NMR experiment. J Biomol NMR 43:211–217

36. Shuker SB, Hajduk PJ, Meadows RP et al (1996) Discovering high-affinity ligands for proteins: SAR by NMR. Science 274:1531–1534

37. Kabsch W (2010) Xds. Acta Crystallogr D Biol Crystallogr 66:125–132

38. Holton J, Alber T (2004) Automated protein crystal structure determination using ELVES. Proc Natl Acad Sci U S A 101:1537–1542

39. Adams PD, Afonine PV, Bunkoczi G et al (2010) Phenix: a comprehensive python-based system for macromolecular structure solution. Acta Crystallogr D Biol Crystallogr 66:213–221

40. Emsley P, Cowtan K (2004) Coot: model-building tools for molecular graphics. Acta Crystallogr D Biol Crystallogr 60:2126–2132

41. Hamalainen MD, Zhukov A, Ivarsson M et al (2008) Label-free primary screening and affinity ranking of fragment libraries using parallel analysis of protein panels. J Biomol Screen 13:202–209

42. Schrodinger, Llc (2010) The PyMOL molecular graphics system, version 1.3r1

INDEX

Cheryl L. Meyerkord and Haian Fu (eds.), *Protein-Protein Interactions: Methods and Applications*, Methods in Molecular Biology, vol. 1278, DOI 10.1007/978-1-4939-2425-7, © Springer Science+Business Media New York 2015

Printed by Printforce, the Netherlands